中国蜻蜓大图鉴

DRAGONFLIES AND DAMSELFLIES OF CHINA

— 张浩淼 著 —
HAOMIAO ZHANG

上册 Vol.1

重庆大学出版社

内容提要

本书收录了中国蜻蜓3亚目23科175属820种，占中国已知蜻蜓物种总数的83.42%。其中束翅亚目13科65属293种，间翅亚目1科1属3种，差翅亚目8科106属524种。本书参照了最新的蜻蜓目分类系统，以中英双语的文字编排，分成上、下两册，包括“蜻蜓概述”和“中国蜻蜓图鉴”两大部分，内容覆盖了蜻蜓的形态学、生物学和分类学等多方面知识，每个物种配有简洁而精准的概括性描述，结合生态照片和必要的分类特征图像，为读者提供了简单快速的鉴定方法，另附有一份系统全面的“中国蜻蜓名录”。书中共选用彩色图片3500余幅，全面展示了蜻蜓迷人绚丽的生态影像，也是全球蜻蜓文献中收录蜻蜓种类最多的彩色图鉴。

本书可供昆虫学、动物学、生态学、生物学等相关专业的教师和学生，农林牧渔、环境保护、野生动物保护、艺术研究与鉴赏等专业人员与管理人员，以及蜻蜓爱好者、自然爱好者等人士在工作、学习和欣赏中借鉴与参考，是一本兼具科学性、实用性和艺术性的大型工具书和科普读物。

图书在版编目（CIP）数据

中国蜻蜓大图鉴：全2册 / 张浩淼著. -- 重庆：重庆大学出版社，2019.1

（好奇心书系. 图鉴系列）

ISBN 978-7-5689-1037-8

I. ①中… II. ①张… III. ①蜻蜓目一图解 IV. ①Q969.22-64

中国版本图书馆CIP数据核字(2018)第052628号

中国蜻蜓大图鉴

ZHONGGUO QINGTING DA TUJIAN

张浩淼 著

策 划：鹿角文化工作室

责任编辑：梁 涛 袁文华 版式设计：周 娟 钟 琛 刘 玲 何欢欢 代 艳 钟艳青

责任校对：张红梅 责任印刷：赵 晟

*

重庆大学出版社出版发行

出版人：易树平

社址：重庆市沙坪坝区大学城西路21号

邮编：401331

电话：(023) 88617190 88617185（中小学）

传真：(023) 88617186 88617166

网址：http://www.cqup.com.cn

邮箱：fxk@cqup.com.cn（营销中心）

全国新华书店经销

重庆新金雅迪艺术印刷有限公司印刷

*

开本：889mm×1194mm 1/16 印张：93.25 字数：3071千

2019年1月第1版 2019年1月第1次印刷

印数：1-4 000

ISBN 978-7-5689-1037-8 定价：1199.00元（上、下册）

Foreword 1

In China, people have long associated dragonflies with prosperity and harmony. They are also regarded as harbingers of good luck. Dragonflies were (and still are) also used in traditional medicine and thus several dragonfly groups were recognized and given Chinese names in an early pharmacopoeia published in China hundreds of years ago. However, from a scientific perspective, the beginning of the formal classification of Chinese dragonflies was the brief description of *Libellula chinensis* by Carolus Linnaeus in the 10th edition of his *Systema naturae* published in 1758, this name refers to the superb green-winged demoiselle damselfly, now known as *Neurobasis chinensis*. Actually Linnaeus himself never saw specimens of this insect himself, basing his description on a rather crude painting of a male specimen kept in a private collection in England. Not having examined the specimen Linnaeus failed to see that the species was related to his other demoiselle species, *Libellula virgo* (the present *Calopteryx virgo*), but considered it as what we would now regard as an anisopteran species.

The first scientific publications dedicated specifically to the Chinese dragonfly fauna were brief papers by Edmond de Selys Longchamps (1886), Robert McLachlan (1894, 1896) and W.F. Kirby (1900) on dragonflies collected in Beijing, Sichuan and Hainan, respectively.These were followed by the publication of a much more substantial work: *A manual of the dragonflies of China* (1930) by the American entomologist James G. Needham, who spent a year in China in 1927-1928. This handsomely printed book, with 20 plates of pen and ink drawings, was based on the author's studies of various collections both in China and elsewhere, on his own collections and on all previously published records. The book covered a total of 266 species in 89 genera, no fewer than 63 new species were described, of which 43 are presently considered valid.

The first Chinese entomologists to describe new dragonfly species were Hsiu-fu Chao and Chin-Wen Chen, both describing their first new species (in the genus *Megalestes*) in the same journal issue published in December 1947. Chen named only three other new species in 1950, but Chao continued his odonatological research and became a world renowned expert of the family Gomphidae. His series of papers on Chinese Gomphidae (1953-1955) and his later work on these insects, culminating in his famous book *The gomphid dragonflies of China* (1990), made this group the best known family among Chinese Odonata. With the help of Chao's contributions, many other Chinese odonatologists also studied this family and published new species. During the last 15-20 years, interest in dragonflies has greatly increased among Chinese entomologists. This has lead to a steady stream of research publications, including descriptions of numerous new species. Several books on local dragonfly faunas have been published, some of them illustrated with excellent photographs taken in the field. Several foreign scientists have also contributed significantly to the recent increase in knowledge of the Chinese Odonata fauna.

No doubt, the most productive local odonatologist in the continental China during the last few years has been the author of this book, Dr Hao-miao Zhang. Besides having undertaken very active field work in many provinces, most recently in the species-rich Yunnan, he has also authored and co-authored several important research papers on the taxonomy of the Chinese dragonflies. His studies on Aeshnidae, Chlorogomphidae and Calopterygoidea are especially noteworthy, but his work covers virtually all groups. In this book, the first of its kind for China, he presents an overview of the dragonfly diversity of the whole country with exquisite color photographs taken from living individuals either in the field or laboratory. This magnificent, colourful exposition will surely greatly increase the interest in dragonflies and their fascinating life history among both entomologists and insect hobbyists in China and elsewhere.

China, with its vast area, contains a great variety of landscapes and climates ranging from tropical rain forests in southern Yunnan to the deserts in the north-west. Mountains and hills cover over 40% of the landscape, this results in a great variety of freshwater habitats that support an enormously diverse dragonfly fauna. Over 900 species of Odonata have already been recognized, many of which are still undescribed, more species are expected to be found in remote, poorly studied areas, especially in Yunnan. Unfortunately due to the over-exploitation of natural resources by the ever increasing human population, forest streams and other fresh water habitats have suffered great degradation in most parts of China, many dragonfly species have lost their habitats and unfortunately this trend continues. Hopefully this work will raise social awareness of the beauty and wonder of these insects and the need to preserve them for future generations, especially within protected nature reserves.

Matti Hämäläinen
University of Helsinki, Finland

推荐序一
Foreword 1

中国人一直把蜻蜓作为繁荣、和谐和好运的象征。早在几百年前蜻蜓就被中医入药并记入药典，因此人们很早就开始认识蜻蜓并给予它们中文名字。但从科学的角度看，中国最早被认识的蜻蜓是一种翅上具有绿色金属光泽的豆娘，发表在林奈1758年出版的《自然系统》（第10版）中，当时被命名为*Libellula chinensis*，即我们今天所熟知的华艳色蟌。然而林奈本人并未见过这种蜻蜓标本，而是根据英国一个私人收藏的绘画来描述的。由于没有检查标本，林奈没有发现华艳色蟌与他所描述的另一种束翅亚目的色蟌十分相似，因此错误地把它当成一种差翅亚目的种类。

最早有关中国蜻蜓区系的研究报告包括Edmond de Selys Longchamps在1886年发表的北京蜻蜓、Robert McLachlan在1894年和1896年发表的四川蜻蜓以及W. F. Kirby在1900年发表的海南蜻蜓，然而这些报告非常简要。随后美国昆虫学家James G. Needham在1930年出版了《中国蜻蜓手册》，是他根据1927至1928年在中国的考察所获。这部精美的专著，包含了20个手绘图版，是他根据世界各地的蜻蜓收藏以及他在中国的采集和历史记录编写而成。这本书收录了中国蜻蜓89属共266种，包括63个以上新种，其中43种为有效种。

中国最早发表蜻蜓新种的昆虫学家是赵修复和陈锦文，他们在1947年12月31日同一期刊上描述了绿综蟌属新种，之后陈锦文仅在1950年又发表了3个蜻蜓新种，赵修复继续从事蜻蜓研究，并成为了春蜓科的世界顶级专家。从1953至1955年，赵修复发表的一系列春蜓科研究报告以及在1990年编写的著作——《中国春蜓分类》使中国的春蜓成为所有中国蜻蜓成员中最先被认知的类群。在赵修复的指导下，许多其他的中国蜻蜓学家开始关注并研究春蜓，更多的新种被发表。在过去的15～20年，中国投身于蜻蜓研究的昆虫学家数量迅速增加，大量的研究报告被发表，其中包括大量的蜻蜓新种。地区性的蜻蜓专著也陆续出版，一些彩色图鉴制作精美，附有野外拍摄的蜻蜓彩照。很多国外的科学家也投身于中国蜻蜓的研究队伍中，为中国蜻蜓学的发展作出了贡献。

毫无疑问，在当今中国研究成果最显著的蜻蜓学家是本书的作者——张浩淼博士。张博士具有非常丰富的野外工作经历，尤其是最近几年在云南省的考察意义非凡，已经发表了一系列非常重要的蜻蜓分类报告，在蜓科、裂唇蜓科和色蟌总科取得了非常重要的研究成果，他的研究领域涉及整个蜻蜓目。作为第一本非常全面的中国蜻蜓彩色图鉴，本书结合野外拍摄和实验室获得的彩色图片，细致地描述了整个中国异常丰富的蜻蜓多样性。这本华丽的彩色图鉴必定会大大提升中国乃至全世界的昆虫学家和爱好者对蜻蜓的兴趣。

中国幅员辽阔，从云南西部的热带雨林到西北地区的荒漠，涵盖了各种复杂多变的地形和气候，中国国土中有超过40%是山地，淡水栖息环境的多样化也缔造了丰富的蜻蜓区系。目前中国已经发现超过900种蜻蜓，但在研究匮乏的地区，尤其云南，仍将有大量的新种被发现。然而由于人口增长造成的自然资源的不合理开发，森林小溪和许多蜻蜓赖以生存的淡水生态环境正在逐渐消失，许多蜻蜓由于丧失了栖息环境正面临严峻的生存威胁。希望借以此书呼吁社会给予更多的关注，使更多的人看到蜻蜓的美丽和奇妙，保护蜻蜓和它们的生存环境。

Matti Hämäläinen
赫尔辛基大学，芬兰

推荐序二
Foreword 2

自幼，我就对昆虫有着浓厚的兴趣，蜻蜓自然是主要的关注点之一。从6到16岁，我的生活中心曾经位于北京城北某高校的校园内。夏日，硕大的足球场是蜻蜓集群飞舞的场所，而那两个并不很大的花园，更是我喜欢的去处。钻进花园，常常可见数十只黄蜻排队栖息在枝条上，三两只黑丽翅蜻犹如黑蝴蝶般在树梢翩翩飞舞，竖眉赤蜻在半人高的草丛中飞来飞去，很多都是多年之后再难见到的场景，至今令人难以忘怀。

1989年，我在全国集邮展览中展出了我的昆虫邮集《六足四翼，飞翔于天地之间》。这部邮集的名字出自《战国策·楚策》，庄辛说楚襄王："王独不见夫蜻蛉乎？六足四翼，飞翔乎天地之间，俛啄蚊虻而食之，仰承甘露而饮之。"这是我国早期对蜻蜓的记载之一，不仅对蜻蜓这种昆虫的形态有着准确的描述，而且对其生活习性也了如指掌。不过，翻看早期的文献史料，国人对蜻蜓的了解也多半局限于六足四翼、取食蚊蝇以及蜻蜓点水等最基本的特征和习性。

真正对中国蜻蜓的分类学研究始于外国学者，比利时动物学家Edmond de Selys Longchamps在1886年发表了关于北京蜻蜓的文章，美国人James G. Needham更是在1930年编写了中国蜻蜓的学术专著《中国蜻蜓手册》。

中国人最早的两本蜻蜓著作是1986年由隋敬之和孙洪国编写的《中国习见蜻蜓》，以及1990年赵修复教授编写的《中国春蜓分类》。这两本书我都是在其出版后第一时间购买的，也是我青年时代对蜻蜓最初的认识。

日本蜻蜓专家朝比奈正二郎曾高度评价赵修复教授的研究："研究亚洲的蜻蜓区系，没有中国的资料，等于胡闹。"但由于时代和年龄的限制，赵修复教授的研究仅仅止步于春蜓，对中国蜻蜓分类研究，不得不说是一大憾事！

在赵修复教授的当代继任者中，浩淼无疑是最为出色的！

我跟浩淼相识已有七八年的时间，一开始都是在网上联系。浩淼可以说是一个网络中的蜻蜓专家，对爱好者的问题，往往都能迅速准确地解答。后来有机会见面才发现他是一个非常阳光的大男孩！

浩淼的专业水准是毋庸置疑的，他与世界著名的蜻蜓专家都有着密切的往来。这些专家给了他极大的帮助，也使得他的研究更加国际化和令人信服。他不仅对世界各地的蜻蜓了如指掌（我曾多次拿欧美等地拍摄的蜻蜓照片来试探他），而且非常努力地去探索未知的蜻蜓世界。他每年在野外的时间长达半年以上，几乎可以这样形容他：不在野外，就是在去野外的路上！

浩淼的感染力也非同小可，在他关于中国蜻蜓的微信群里，有一大批蜻蜓爱好者，紧密地团结在他的周围，大家每天互通情报，交流蜻蜓观察的动向。我也曾受到他的感染，自告奋勇独自去云南山区，搜寻一种罕见的蜻蜓，最终无功而返，但虽败犹荣！

浩淼的优势还在于他的摄影技术，作为一个蜻蜓专家，他拍摄的蜻蜓生态照片，不仅构图精美、色彩逼真，而且鉴定特征清晰，容易辨识，充分体现了他在蜻蜓学和摄影两个方面的专业水准。

正因为这两方面的超常专业水准，以及非凡的组织协调能力，才使得这本世界规模最大的蜻蜓图鉴得以出版。

这本精美的《中国蜻蜓大图鉴》，不仅具有很强的科学性，而且具有很高的科普价值。对科研工作者来说，这是一本很好的研究专著；对爱好者以及青少年来说，这是一本不可多得的鉴定宝典。

目前，自然观察和博物教育在国内越来越受到重视，而昆虫无疑是最为贴近生活的动物类群，这其中蜻蜓必然是不可或缺的。

我坚信，随着本书的出版发行，“一起去观察、拍摄蜻蜓”，终将成为一种新的时尚！

2018年3月31日于重庆

作者序
与蜻蜓同行

我和蜻蜓的缘分或许与我的名字有关，“浩淼”既有水势浩大的含义，也预言了我的一生都将与水为伴。自幼年起，就常常溜到河边。家乡的牡丹江不算宽，却是一条生命之河。5岁时在野外的一次经历，开启了通往蜻蜓王国的旅程。那是我第一次见到蜻蜓在野外羽化，一次“丑小鸭变天鹅”的华丽变身，一只体型硕大、具有金属光泽的蜻蜓，深深触动了我。然而直到20岁时，我才知道了它的真正名字——圆大伪蜻。

2018年，我与蜻蜓结缘满30年，是个值得庆祝的时刻！然而这30年的蜻蜓之路走得十分曲折和艰辛。我从8岁起学习美术和音乐，在近10年的学习中，我似乎并没有特别深的造诣。随后放弃了艺术之路开始专攻学业。从小学到大学，蜻蜓几乎占据了我所有的时间。有时上学路上故意绕到江边，寻觅待羽化的蜻蜓幼崽，带到学校的书桌里羽化。时间久了，我对这些神秘昆虫的兴趣越来越浓。到大学以后，就经常利用空闲之余远足旅行去寻找和观察蜻蜓，也逐渐开始接触到了专业知识。本科和硕士阶段，我在大连工业大学接受的是化学工程和造纸工程方面的教育，直到有一天我的命运被完全改变。那是一个名为“蜻舞菲扬”的蜻蜓作品展，成为我人生的“变身”时刻。蜻蜓作品展最大的收获是让我认识了中国蜻蜓学家江尧桦先生。他多次寄给我重要的蜻蜓文献，教我认识蜻蜓，并推荐我到华南农业大学攻读昆虫学博士学位。在造纸工程专业硕士学习阶段的第二年，我开始为转入昆虫学专业备考。2008年3月，我第一次踏入华南农业大学的校园与我的恩师童晓立教授见面，他的出现对我的一生都有着深远的影响。童老师曾给我单独授课，讲昆虫分类，讲命名法规，让我充实各种基本理论。广东省无比优越的地理位置和气候条件，也为学习蜻蜓提供了充足的研究内容。2012年获得博士学位后，我继续到中国科学院水生生物研究所从事蜻蜓研究，师从蔡庆华研究员。我和蔡老师经常一起聊蜻蜓，他除了给予鼓励和支持，更重要的是给了我一个可以任性发展的空间。这期间，我开始学习生态学的相关知识，包括蜻蜓用于环境监测的研究方法，并开展在湖北省神农架林区和云南省的蜻蜓野外考察，尤其是在云南省的野外考察使我对大自然有了更深的感悟，也领略到了生物多样性热点区域的震撼力。

踏入专业领域10年，幸运地得到了多位国际蜻蜓学家的鼎力帮助。从我的第一份蜻蜓研究报告开始，就得到了芬兰著名蜻蜓学家Matti Hämäläinen博士的指导。我们曾多次见面，一起到野外考察，一起讨论蜻蜓，一起发表研究报告，他曾对我的博士学位论文提出了重要意见。澳大利亚的Albert Orr博士是国际蜻蜓学报和蜻蜓学杂志的主任编辑，在我的多篇重要研究报告中，他都给予了关键的修改意见，并经常协助新种的命名、英文的校正等工作。此外我们经常展开蜻蜓行为学方面的讨论。著名的美国蜻蜓学家Rosser W. Garrison博士，一直关注着我的工作，他多次为我寄来珍贵的蜻蜓文献和野外考察所需的工具。我和几位世界蜻蜓大师几乎每天都有书信来往，一起讨论蜻蜓学的各类疑难杂症。3位老友也都以审稿人的身份，投入到了这本图鉴的撰写工作中。

一个索尼的随身听，装满了王菲的歌曲，狂放在野外寻找蜻蜓，是我年少时最难忘的经历。时常追忆那些刺激的野外探索，无数次的一个人旅行，虽然无法挽留在田野里消逝的青春年华，但能与这些神秘的丛林精灵一次次地近距离接触，满足而骄傲。生命之河，蜿蜒流淌，有生之年，与蜻蜓结缘，是最大的荣幸。

这本彩色图鉴，从2009年开始筹备编写，耗时近10年。我把近30年在野外观察蜻蜓和10年里对蜻蜓分类学的研究成果全部汇入此书。从一个蜻蜓爱好者到专业研究者，角色变了，但不变的是对蜻蜓无比的热爱和执着。本图鉴收录了中国地区蜻蜓目昆虫共计820种，是目前全球同类图鉴中收录蜻蜓种类最多的一部。所包含的千幅彩照则是从超过20万幅生态照片中经过细致的考虑筛选出的最佳影像，从不同角度展示物种，结合简洁精准的概况性描述，有助于快速识别物种。分类特征图像，包括翅脉、肛附器、次生殖器等是根据个人收藏的标本进行实体拍摄和编制。从图片筛选到中英双语的文字编写，是一项繁重而复杂的工作。有时从早忙到晚，既要编写文字，又要检查标本，几乎是连续12小时的工作状态。然而书稿还有很多需要进一步完善和改进之处，而且随着野外考察工作的继续，更多的新种和新记录种会被陆续发现，中国也将成为世界首个突破千种的蜻蜓大国。期待这本彩色图鉴可以在世界范围提升中国蜻蜓区系的热度。

大数据时代，蜻蜓的研究手段和方法已大幅提升。借助先进的地理信息系统和全球定位系统可以搜寻到更多藏匿在深山的蜻蜓栖息环境。结合分子生物学的研究手段，可以更准确地把握它们的分类学地位及系统发育关系。中国蜻蜓学的发展急需更多的研究力量。愿本书可以号召更多的爱好者投入到研究队伍中，把对蜻蜓的热爱转化成科研力量，让这个小众的科研领域能够延续生命。

唤醒沉睡亿年的自然之美，把转瞬即逝的自然之魂化为永恒！谨以此书，献给所有支持蜻蜓研究工作的朋友们！

致读者

2018年3月18日于昆明

Preface
Living with dragonflies

It may be that by virtue of my name I was fated to study dragonflies. The Chinese word "Haomiao" means "plenty of water", perhaps predicting that I should have some affinity with aquatic ecosystems for the rest of my life. As a child, I often went to the river bank. The river called "Mudanjiang", the same name as my hometown, is not so wide but is "a river of life". At the age of five, I was enchanted and touched by the sight of the emergence of a large metallic insect. Not until I was 20 years old did I finally come to know its name: *Macromia amphigena*.

Now in 2018, I have been with my beloved dragonflies for 30 years; a moment to celebrate. But working with dragonflies was never an easy pursuit. I began to study fine arts and music when I was eight years old but after ten years of study, I realized this was not my vocation. Therefore I abandoned my career in the arts and turned to studying science and technology. Dragonflies continued to interest me from primary school to university and I devoted all my spare time to studying them. Sometimes on the way to school I detoured to the river bank where I found larvae which I took to my desk, observing their emergence. My interest in these mysterious insects grew as time passed. As a undergraduate student I often went to the field to search for dragonflies, and gradually began to acquire a professional knowledge of them. Although I studied chemical engineering for my Bachelor's and paper making for my Master's degrees, my interest in dragonflies remained as passionate as ever. Then, one day I hold a dragonfly exhibition called "Qingwufeiyang". This was an important event that was to change my life. By that exhibition, I became acquainted with the Chinese odonatologist, Mr Yaohua Jiang who provided me with many useful papers and taught me the essentials of odonatology. He then recommended me to study for a PhD in entomology at South China Agricultural University. At that time in my second year of postgraduate study towards a Master's degree in paper making, I began preparing for the entrance examination of entomology. In March of 2008, I went to the university and met my supervisor, Professor Xiaoli Tong, for the first time. He was to have a profound impact on my life. He instructed me in insect systematics, principles of Zoological nomenclature, as well as other aspects of entomology. I was also fortunate to be located in Guangdong Province, which, with its superior geographical location and climatic conditions provided me with ample material for study. I received my PhD in 2012 and then continued to work on dragonflies at the Institute of Hydrobiology, Chinese Academy of Sciences, under the guidance of my collaborative supervisor, Professor Qinghua Cai. We talked about dragonflies very often, he giving me much encouragement and support, as well as space to facilitate my eclectic development. At that time I began to study ecology of dragonflies, including the use of dragonflies as environmental indicators. Meanwhile I conducted extensive explorations in Shennongjia forest in Hubei and many parts of Yunnan. The fieldwork in Yunnan gave me a better understanding of natural history. I was amazed by its rich biodiversity.

I have been very lucky to receive much help from several eminent international odonatologists during the last ten years of professional Odonata study. Ever since my first dragonfly report, Dr Matti Hämäläinen (Finland) has freely given his time, help and special attention to me. We have met many times collecting in the field, discussed many aspects of dragonfly systematics and co-authored many papers. Dr Albert Orr (Australia), a subject editor of the journal *International Journal of Odonatology* and associate editor of *Odonatologica* provided many critical comments in my papers, and often helped me with the name of a new species, English expression, etc. Additionally we often discussed dragonfly behavior. The American odonatologist, Dr Rosser W. Garrison has helped me with literature and tools for ongoing fieldwork. I correspond with these well known authorities frequently and discuss problems in various dragonfly groups. These three old friends have all joined in the editorial work of the book as reviewers and English language editors.

Listening to my favorite songs by Faye Wong and collecting dragonflies were the abiding passions of my youth and remain so still. I have a rich fund of unforgettable and exciting memories of times in the field which I often recall-the many times I was alone with them. Although I can't keep my youth forever, nor go to the field so often, I will always feel enormous satisfaction and pride to had the intimate acquaintance of these beautiful and mysterious insects. From dragonfly enthusiast to professional odonatologist, my love and admiration for these insects has never waned.

I began to prepare this field guide in 2009, this work has taken almost ten years. It brings together field observations of dragonflies and the results of taxonomic studies. A total of 820 species from all over China are illustrated, more than in any comparable book worldwide. The thousands of beautiful photos have been carefully selected from over 200000 shots. They are chosen to show a species in the best view and help the readers to quickly identify species with the help of the generalized identification text. Diagnostic photos, including wing venation, anal appendages, secondary genitalia, etc., are taken from specimens in my collection. It was a difficult and complex task to select photos from my large photo library and edit both Chinese and English languages. However, this book is still not complete, for with continuing fieldwork, more new species and new records will be discovered, China will probably be the first country to have over a thousand dragonfly species. Hopefully this colourful field guide will stimulate further interest in the fascinating Chinese odonate fauna.

The Age of Big Data, methods for studying dragonflies have been greatly improved. We can now find more habitats with the aid of hand-held GPS, Google Earth and GIS applications. And the application of molecular techniques continues to enhance the study of these insects, to give a better understanding of their status and phylogenetic relationship. However, odonatologists in China are few, and more attention needs to be directed to the study of this ancient insect group. May this book bring more enthusiasts into the research team, transforming their love of dragonflies into scientific power, and keep this fascinating research area alive.

Awakening the beauty of nature sleeping for hundreds of millions of years, turning the fleeting spirit of nature into eternity! I dedicate this book to all my friends, family members as well as the dragonfly lovers all over the world.

To all my readers

Haomiao Zhang

18 March, 2018, Kunming

中国蜻蜓大图鉴

DRAGONFLIES AND DAMSELFLIES OF CHINA

目录
Contents

蜻蜓概述 Introduction

中国蜻蜓图鉴 Dragonflies and Damselflies of China

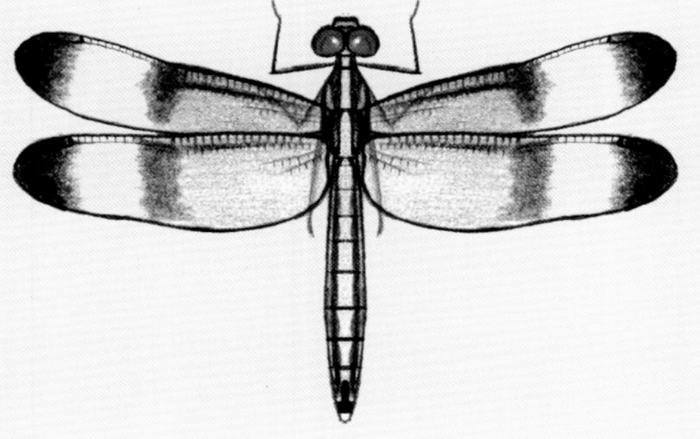

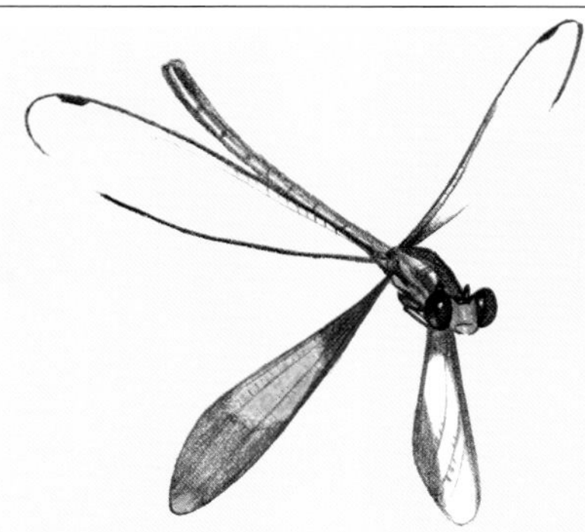

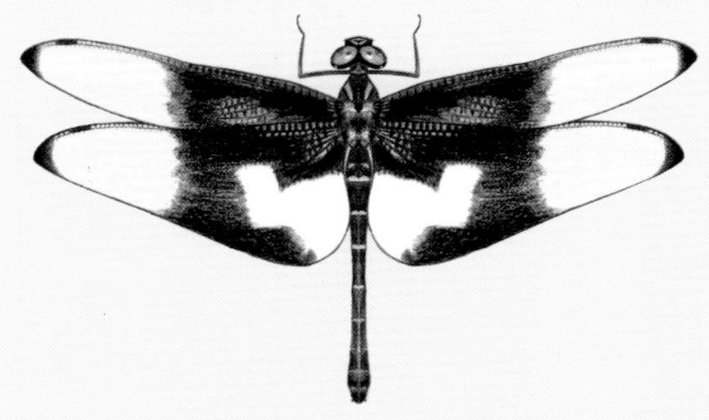

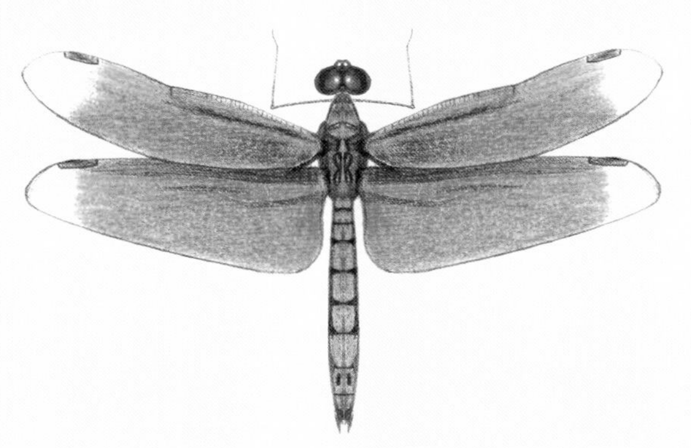

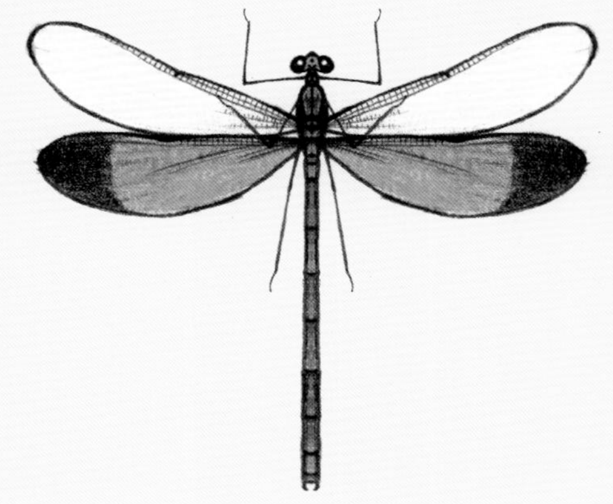

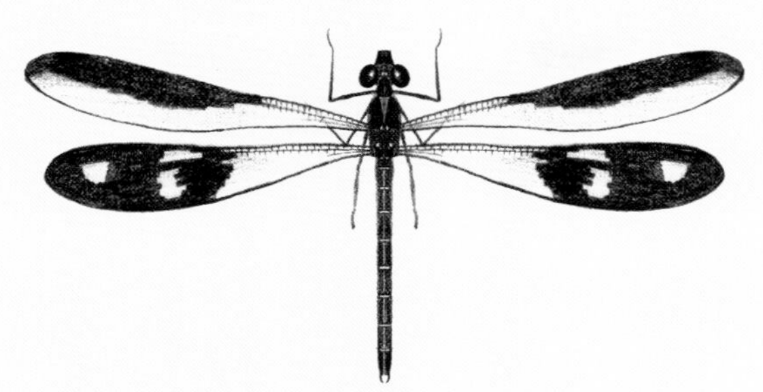

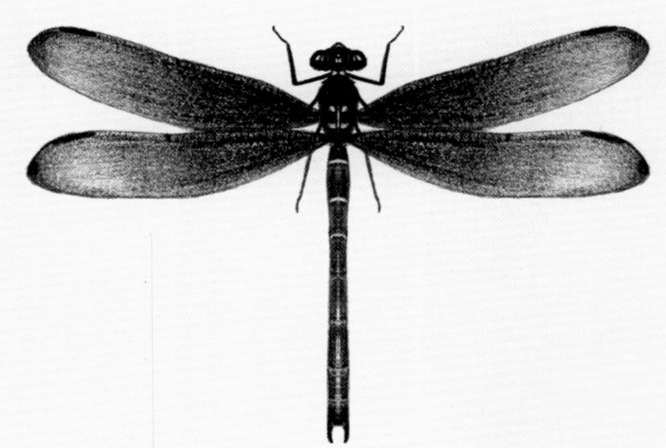

蜻蜓概述

INTRODUCTION

中国蜻蜓大图鉴

DRAGONFLIES AND DAMSELFLIES OF CHINA

蜻蜓目简介 Introduction to the order Odonata

蜻蜓是迄今最古老的飞行昆虫类群之一，最早的古蜻蜓发现于古生代石炭纪的化石中，距今至少已有3亿年的历史。蜻蜓在昆虫纲中容易辨识，头部、胸部和腹部分节明显。与昆虫纲其他昆虫区分的特征包括甚大的复眼、短小的触角、细长的腹部和狭长的翅，翅脉密集并在前缘脉中央具翅结。停歇时翅向体侧伸展或者合拢竖立于胸部背面。蜻蜓目可能会与脉翅目混淆，但后者具有更长的触角，翅上无翅结，停歇时翅合拢呈帐篷状。雄性蜻蜓是唯一在腹部第2节和第3节下方具次生殖器的昆虫。

蜻蜓的生活史包括卵、稚虫和成虫3个阶段，属于不完全变态发育。蜻蜓稚虫生活在水中，其一生在水下生活的时间最长，有些要经历数年才能发育成熟。稚虫的口器构造特殊，具有1个延长、折叠且可伸缩的下唇，下唇须叶的末端具尖刺。这个构造也称面罩，通常折叠于头部下方，但可突然伸出捕捉猎物。

Members of the order Odonata are among the most ancient of winged insects, the oldest odonate-like fossil dates back to the Carboniferous in the Paleozoic era, about three hundred million years ago. Dragonflies are easily identified within the Class Insecta with their body clearly divided into head, thorax and abdomen, the thorax bearing three pairs of legs. They may be distinguished from other insects by their large compound eyes, short antennae, long narrow abdomen and long narrow densely reticulated wings with a kink, or nodus along their leading edge. The wings are either held outspread or folded above the thorax when at rest. They can be confused only with certain lacewings (Order Neuroptera) which however have long antennae and fold their wings, which lack a nodus, tent like over the body when at rest. Male Odonata are the only insects to have secondary genitalia beneath abdominal segment 2 and 3.

Dragonflies have a life history consisting of three stages, the egg, the larva and the adult. Metamorphosis is regarded as incomplete because the wings develop gradually in the successive stages of the larvae. Their larvae live in water and spend the greatest part of their lifetime in this state. The larval stage may last several years in some species. The larvae have an unique, elongated, hinged labium armed with palps bearing long recurved spines at its apex. This structure, also termed the mask, is normally folded under the head but can be shot out extremely rapidly to catch prey.

除了极地地区，蜻蜓目广泛分布于世界各地，热带和亚热带地区种类最多。蜻蜓目是昆虫纲中较小的一个目，全世界已发现6000余种，分为3个亚目。

差翅亚目：俗称**蜻蜓**，包括一群体型较大且粗壮、停歇时翅向体侧展开的种类。头部侧面观半球形，正面观近似圆形或椭圆形；两复眼距离较近，多数类群在头顶交汇，少数类群复眼分离（如春蜓科、裂唇蜓科）；面部显著，额隆起。前翅和后翅的形状不同，后翅的臀区比前翅宽阔。雄性腹部末端具1对上肛附器和1个下肛附器。但有些种类下肛附器具显著的分枝，多是由下肛附器的中央缺刻造成，在春蜓科、裂唇蜓科等类群中常见。

蜻蜓的代表，碧翠蜓，雄
A representative of dragonfly, *Anaciaeschna jaspidea*, male

蜻蜓的代表，红腹异蜻，雄
A representative of dragonfly, *Aethriamanta brevipennis*, male

束翅亚目：俗称**豆娘**，包含一类体型相对较小且纤细的种类。很多豆娘休息时翅合拢竖立于胸部背面。除了少数类群，它们的头部正面观哑铃形，面部不明显，未显著隆起。前翅和后翅形状基本相同，有些种类翅向基方收窄形成翅柄。多数种类腹部较纤细。雄性腹部末端具1对上肛附器和1对下肛附器，雌性具产卵管。

间翅亚目：其体态集合了差翅亚目和束翅亚目的特征。前后翅形状相似，似豆娘，但身体粗壮，似蜻蜓。目前本亚目仅包含1科1属，全球已知4种，但其中2种的身份存疑。它们分布于日本、中国、印度和尼泊尔。间翅亚目被认为是古代蜻蜓现存的唯一后裔，在地球存活了超过1亿2000万年。

豆娘的代表，赤基色蟌，雄
A representative of damselfly, *Archineura incarnata*, male

豆娘的代表，蕾尾丝蟌，雄 | 宋睿斌 摄
A representative of damselfly, *Lestes nodalis*, male | Photo by Ruibin Song

间翅亚目的代表，日本昔蜓，雄 | 酒井正次 摄
A representative of Anisozygoptera, *Epiophlebia superstes*, male | Photo by Shoji Sakai

间翅亚目的代表，日本昔蜓，雌 | 酒井正次 摄
A representative of Anisozygoptera, *Epiophlebia superstes*, female | Photo by Shoji Sakai

一种蜻蜓的稚虫，老挝裂唇蜓
A dragonfly larva, *Chlorogomphus* (*Sinorogomphus*) *hiten*

一种豆娘的稚虫，黄肩华综蟌 | 宋睿斌 摄
A damselfly larva, *Sinolestes editus* | Photo by Ruibin Song

圆臀大蜓稚虫的面罩 | 宋睿斌 摄
The mask of *Anotogaster* larva | Photo by Ruibin Song

黄伟蜓稚虫捕食时伸出折叠的下唇 | 莫善濂 摄
Larva of *Anax immaculifrons* extends its folded labium | Photo by Shanlian Mo

The order Odonata has a worldwide distribution except for the polar regions but is most diverse in tropical and subtropical regions. Odonata represents a relatively small insect order comprised of a little over 6000 presently known species. The order is divided into three suborders.

Anisoptera: commonly called **dragonflies**, include large and robust species which generally perch with outstretched wings. The head is roughly hemispherical in lateral view, approximately circular or oblong when viewed frontally. The compound eyes are very large and broadly touching above in most families but somewhat smaller and well separated in the families Gomphidae and Chlorogomphidae. The face is prominent and protruding. The fore wings and hind wings are differently shaped, the hind wings being broader basally, commonly sharply angled in males. The males have one pair of superior appendages (cerci) and only one inferior appendage (epiproct). The inferior appendage is sometimes branched, especially in the families Gomphidae and Chlorogomphidae.

Zygoptera: commonly called **damselflies**, include relatively small and narrow species. Many species fold their wings above the thorax when perched. With few exceptions the head is dumb-bell shaped when viewed from the front, and the face does not obviously protrude. The fore wings and hind wings are of same shape, generally narrow and in some species the wings are weakly stalked at their bases. The abdomen is long and slender in most species. The males have a pair of superior (cerci) and inferior (paraprocts) appendages. Females have ovipositor.

Anisozygoptera: the morphology combines characters within the Anisoptera and Zygoptera. The wings are of similar shape as in Zygoptera but the body is robust as in Anisoptera. Currently only four species (two questionable) belonging to one genus and one family are known. Distributed in Japan, China, India and Nepal. Anisozygoptera is regarded as the only living descendants of ancient dragonflies, which lived on the Earth more than 120 million years ago.

蜻蜓头部
Heads of dragonflies

豆娘头部
Heads of damselflies

蜻蜓目学名“Odonata”一词的由来 The origin of the order name “Odonata”

蜻蜓目“Odonata”一词源自希腊语，其含义为“具齿的”，由Fabricius在1793年根据蜻蜓成虫具有尖锐的上颚这一特征提出，指1758年由林奈归入广义脉翅目Neuroptera中蜻属*Libellula*的一类昆虫。直到进入20世纪，“Odonata”一词与脉翅目的关系才逐渐有了变化，最终独立，被用来专门指代我们今天所认识的蜻蜓、豆娘和昔蜓。

The origin of the order name, “Odonata” is from the Greek, meaning “with teeth”. The term was coined by Fabricius in 1793 based on the sharp mandibles of adult dragonfly. “Odonata” referred to those species placed in the genus *Libellula* in the Order Neuroptera by Linnaeus in 1758. In the 20th century, the taxonomic interpretation of “Odonata” changed and was separated from Neuroptera to be used as the ordinal name for dragonflies, damselflies and *Epiophlebia* as we know today.

“蜻蜓”一词的含义 The word “dragonfly”

我们通常所说的“蜻蜓”一词，实际上是包括了3个亚目在内的所有成员，即包括俗称的豆娘在内。为避免混淆，本书中的差翅亚目部分，描述主要用“蜻蜓”一词；束翅亚目部分，则主要用“豆娘”一词。在前言部分，“蜻蜓”一词则指蜻蜓目的全体。

Commonly the word “dragonfly” is used to cover all members of the three suborders, including damselflies. To avoid confusion, in the Anisoptera species accounts of the book, the word “dragonfly” is used for description but in Zygoptera part replaced by the word “damselfly”. In the introduction part the word “dragonfly” is used for all members of the Order.

蜻蜓的重要性 Significance of odonates

宝贵的物种资源 Valuable species resources

作为宝贵的一类生物资源，蜻蜓的物种多样性在中国的优势显著，目前已经发现了900余种蜻蜓，堪称世界之最，相信经过进一步的科学考察，中国的蜻蜓总数最终会突破1000种，成为全球蜻蜓物种最丰富的国家。其中大量的特有和濒危物种，具有极高的保育价值。

As a biodiversity resource, the species richness of Odonata in China is remarkable. Over 900 species have so far been found in China, but it is expected this number will increase to over 1000 with further scientific investigation, potentially making China the most species rich country for Odonata in the world. A large number of endemic and endangered species are of great importance in terms of the need for their conservation.

旗舰物种 Flagship species

很多人说，蜻蜓拥有让人嫉妒的美，加上多姿优美的体态和卓越的飞行技能，使这些古老的昆虫位列旗舰物种。由于蜻蜓艳丽动人，且体型明显大于其他大多数昆虫，素有“鸟友之虫”的美誉。蜻蜓的美学价值使其与人类文化密切交织。作为一类重要的观赏性昆虫，蜻蜓也慢慢成了自然观察活动的主角，一些发达国家更是有一年一度的蜻蜓节。观鸟群体和其他自然观察者的加入，促进了蜻蜓美学价值的提升和野外观察活动的开展。蜻蜓的自然之美将继续被专业人员和业余爱好者所关注，并唤醒更多公众对自然百科的兴趣。

Odonates have a charismatic beauty. With their elegant appearance and superb flight skills, these ancient insects include some of the most important flagship species for conservation. Odonates are striking, diurnal and generally larger than most other insects. They thus deserve the appellation “bird watcher’s insects.” Their aesthetic appeal has seen them incorporated into human culture. As easily observed, visually attractive insects, odonates developed an appeal for many naturalists. Some developed countries have an Annual Dragonfly Festival. Bird watching groups and other nature observers promote their aesthetic value and the practice of observing them in nature. Their beauty continues to attract attention from both professionals and amateurs and increases public interest in natural history.

重要的环境指示生物 Important environmental indicators

蜻蜓是淡水生态系统的构成要素。由于蜻蜓幼年生活于各类淡水水域中，它们的生长受水质、水流速和水生植被等条件影响；成虫在羽化地点附近生活，很多种类终生不离开其生长的水域，其各项生命活动亦受周边植被状况影响。它们的生存不仅标志着水环境的健康状况，还可以反映森林的植被质量，因此蜻蜓目昆虫是重要的环境质量指示生物。许多敏感物种，对水体和植被的指标要求较高，可以准确评价各种干扰带来的环境问题。

Odonates are important components of aquatic ecosystems. Since their larvae live in all kinds of freshwater habitats, and their growth is affected by water quality, velocity, presence of aquatic plants, and many other factors. The adults live around water, many species never stray far from the vicinity of the habitat where they emerged. Their life activities may also be influenced by the presence and type of vegetation nearby. Therefore the presence of odonates at a site can serve as an important environmental indicator of the health of the habitat in terms of both water quality and the integrity of the surrounding vegetation. Many sensitive odonate species have very specific requirements for water conditions and nearby vegetation, hence may serve as indicators for accurately evaluating the environmental problems caused by various kinds of disturbance.

蜻蜓的身体结构 Dragonfly anatomy

蜻蜓的身体分成头、胸、腹3个明显的部分。头部具有2个发达的复眼和口器；胸部具有3对足和2对翅；腹部具有10个明显的体节，第10节末端具有肛附器。

The body of dragonfly is divided into three parts: head, thorax and abdomen. The head has two large compound eyes and the mouth parts, the thorax has three pairs of legs and two pairs of wings and the abdomen has ten segments with the last (10th) segment bearing the anal appendages.

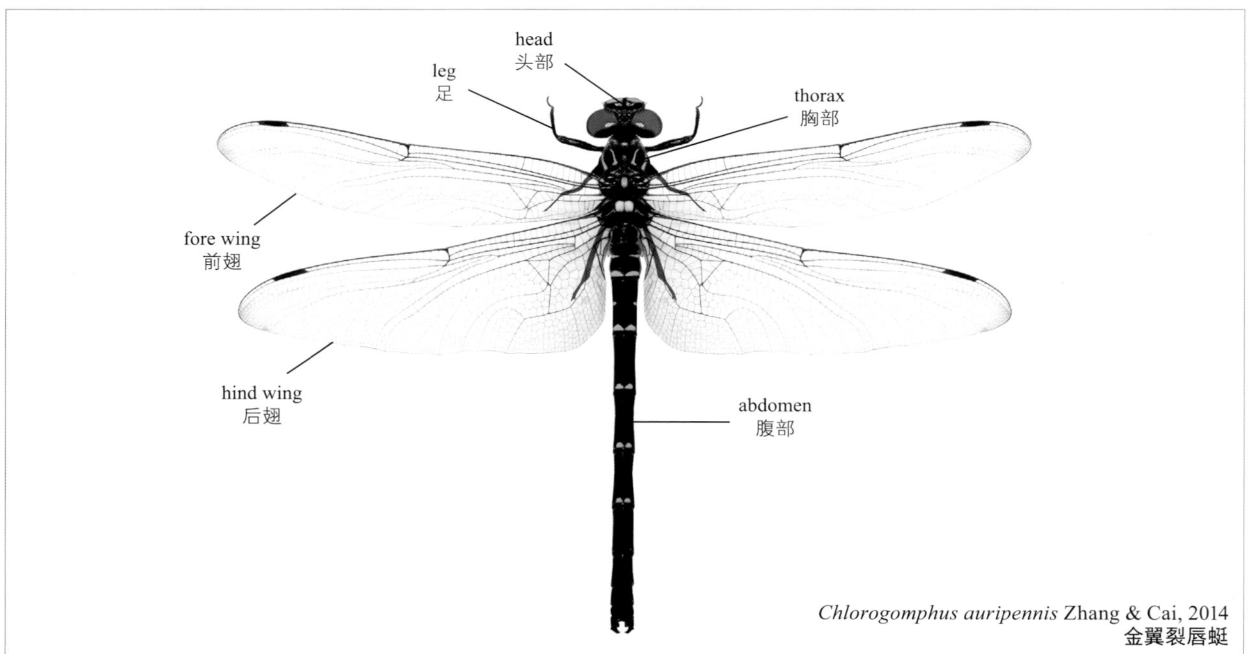

蜻蜓的整体结构
The general habitus of dragonfly

头部 Head

蜻蜓的头部主要包括**上唇**、**唇基**、**额**、**头顶**、**后头**和2个甚大的**复眼**。唇基又分为2个部分，下方的**前唇基**和上方的**后唇基**。额位于后唇基之上，下方倾斜的部分为**前额**，上方的平台为**上额**。在差翅亚目中，额向前突起至高于复眼的位置，但在束翅亚目中未见。差翅亚目的一些类群具有极为发达的额，如大蜓科和蜓科。蜓科头蜓属的一些种类具有极为宽阔而突起的额，是该属重要的识别特征。**头顶**相对较小，位于上额的上方，着生3个单眼和1对触角。**后头**是头部最后方的部分，位于头顶之上。在差翅亚目中的一些类群，后头被在头顶上方相交的复眼与头顶分离。

The head comprises **labrum**, **clypeus**, **frons**, **vertex**, **occiput** and two large **compound eyes**. The clypeus is divided into two parts, a lower **anteclypeus** and an upper **postclypeus**. The frons lies above the postclypeus, its lower declivity is called the **antefrons** and the dorsal surface is called the **top of frons**. In Anisoptera, the frons protrudes anteriorly beyond the level of the compound eyes, but not seen in Zygoptera. In some groups of Anisoptera, such as in some Cordulegastridae and Aeshnidae, the frons is very pronounced. Some species of *Cephalaeschna* (Aeshnidae) have broad and protruded frons, which is a reliable diagnostic character. The smaller **vertex** is located above the base of the top of frons. The paired antennae and three ocelli are located here. The **occiput** is the most posterior portion of the head and lies above the vertex and in some groups of Anisoptera is separated from the vertex by the a longitudinal eye seam.

头部附肢 Cephalic appendages

头部的附肢包括1对**触角**、1对**上颚**、1对**下颚**和**下唇**。

The cephalic appendages include a pair of **antennae**, a pair of **mandibles**, a pair of **maxillae** and **labium**.

胸部 Thorax

胸部分为明显的2个区域，前面较小的部分称为**前胸**，后面较大的盒形部分称为**合胸**，由**中胸**和**后胸**合并而成。前胸分为3个部分，背面的**前胸背板**、侧面的**前胸侧板**和腹面的**前胸腹板**。前胸着生1对前足。合胸具2对翅和2对足。**中胸**被**中胸侧缝**分为前方的**中胸前侧片**和后方的**中胸后侧片**，此外还包含1个**中胸前侧下片**。中胸前侧片在背面相连形成合胸**背脊**。后胸与中胸相似，由**后胸侧缝**分为**后胸前侧片**和**后胸后侧片**，还包含1个**后胸前侧下片**。

The thorax of the dragonfly is divided into two parts, the smaller **prothorax** and a larger box-like **synthorax**, the latter composed of the **mesothorax** and **metathorax**. The small prothorax, is composed of three parts, dorsally the **pronotum**, laterally the **propleuron** and ventrally the narrow **prosternum**. The prothorax bears a pair of fore legs. The synthorax bears two pairs of wings and two pairs of legs. The mesothorax is divided into an anterior **mesepisternum**, a posterior **mesepimeron** by the **mesopleural suture** and, in addition, a ventral **mesokatepisternum** (also called the **mesinfraepisternum**). The mesepisterna connect anteriorly forming the **dorsal carina**. Similarly, the **metathorax** is divided into a **metepisternum**, a posterior **metepimeron** by **metapleural suture** and, in addition, a ventral **metakatepisternum** (also called the **metinfraepisternum**).

胸部附肢 Thoracic appendages

胸部的附肢包括3对足。

The thoracic appendages include three pairs of legs.

足 Legs

足包括**前足**、**中足**和**后足**各1对。足从基部到端部依次为**基节**、**转节**、**腿节**、**胫节**、**跗节**和**前跗节**。前跗节是足的末端，主要是1对**爪**。足在停歇、羽化、捕食、交配等生命活动方面起重要作用。

Legs include paired **fore legs**, **mid legs** and **hind legs**. Main segments of the leg from the base to the end are: **coxa**, **trochanter**, **femur**, **tibia**, **tarsus** and **pretarsus**. Pretarsus mainly consists of a pair of **claws**. Legs play important role in life activities such as emergence, perching, prey capture and mating.

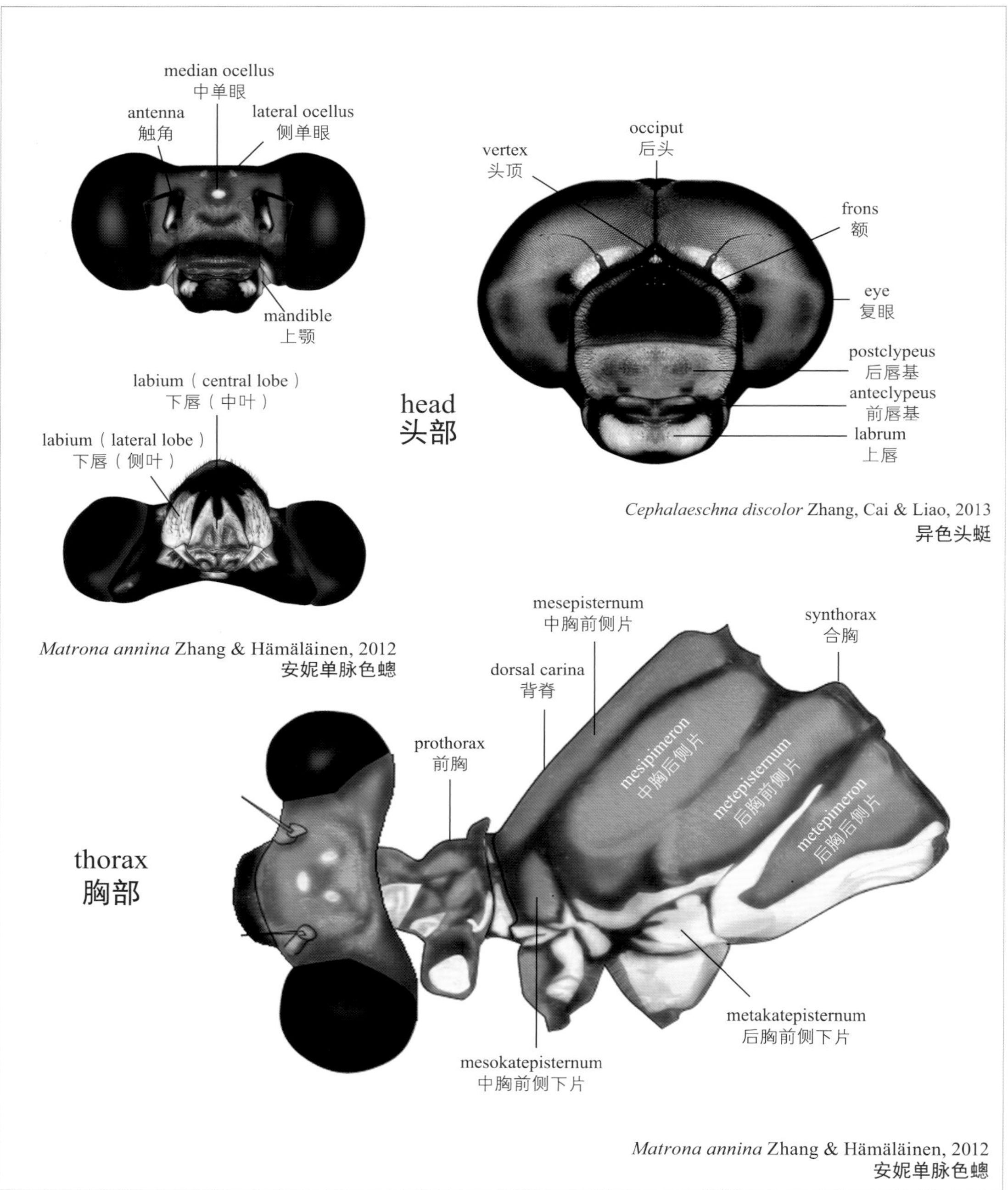

蜻蜓的头部和胸部构造
Head and thorax of dragonfly

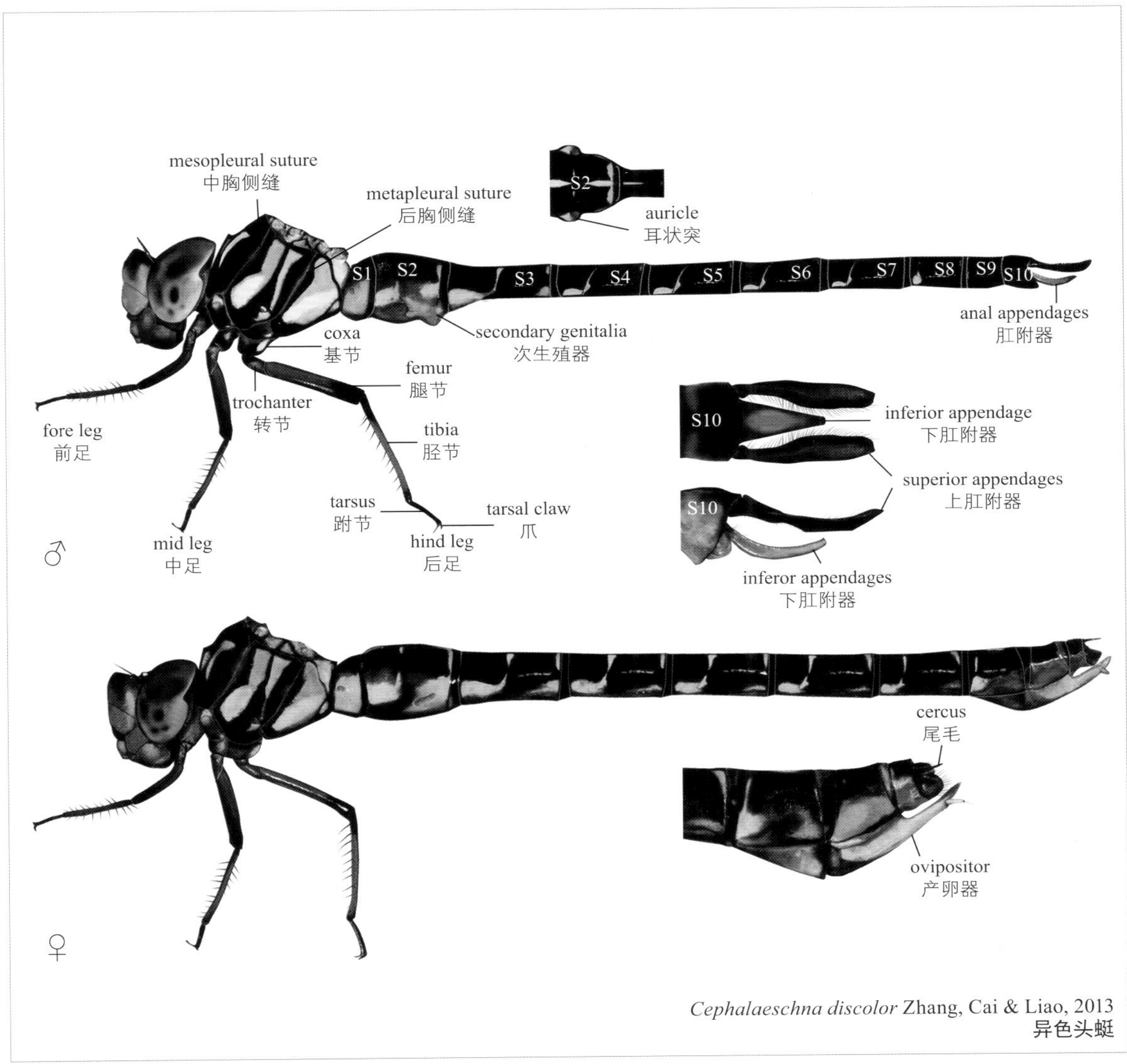

蜻蜓的整体结构
The general habitus of dragonfly

翅 Wings

翅是由复杂的网状翅脉交织而成，包括**前翅**和**后翅**各1对。前翅和后翅可以相对独立地运动，使它们可以多种不同的姿势和速度飞行及悬停。翅脉是蜻蜓目分类系统建立的重要依据。本书主要采用Tillyard & Fraser (1938—1940)的翅脉系统。

Wings have a complex reticulated venation supporting a thin, mostly clear membrane and comprise paired **fore wing**s and **hind wing**s. The two pairs of wings can flap independently, which allows the dragonfly to fly with great manou evreability and speed, including the ability to hover. Classification system is largely based on venation. Terminology used for venation in this book is mainly based on Tillyard & Fraser (1938-1940).

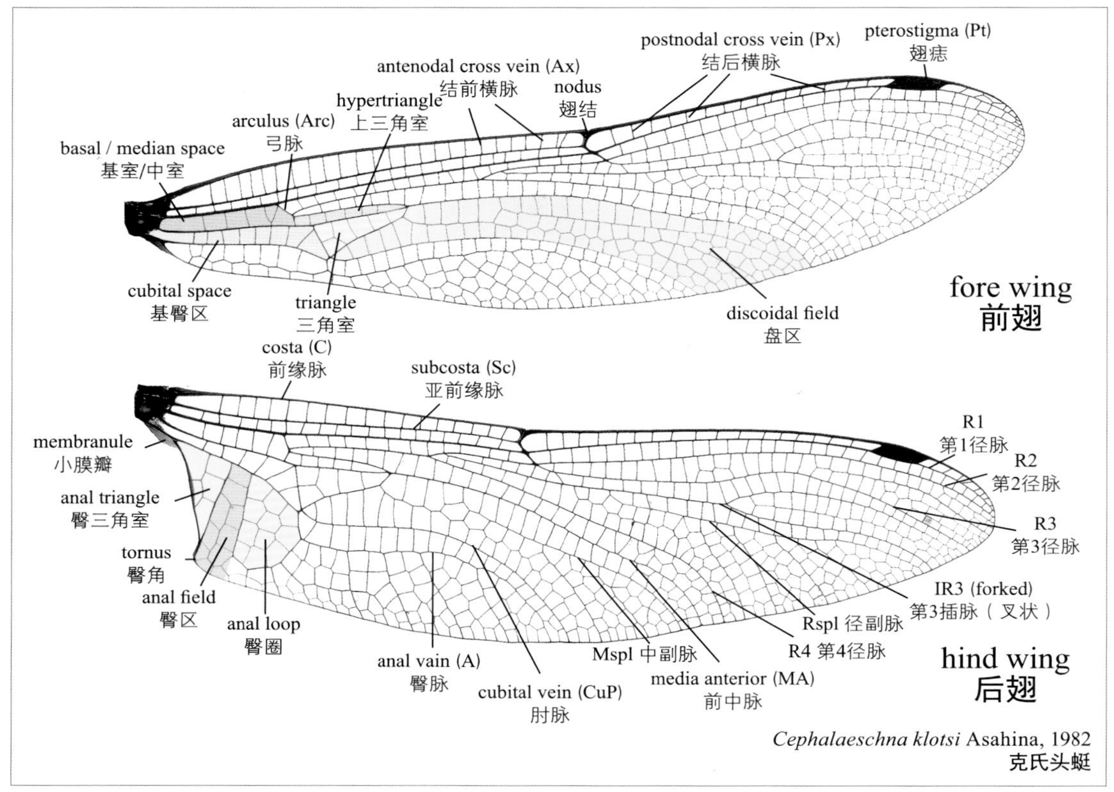

差翅亚目的翅脉

Wing venation of an anisopteran species

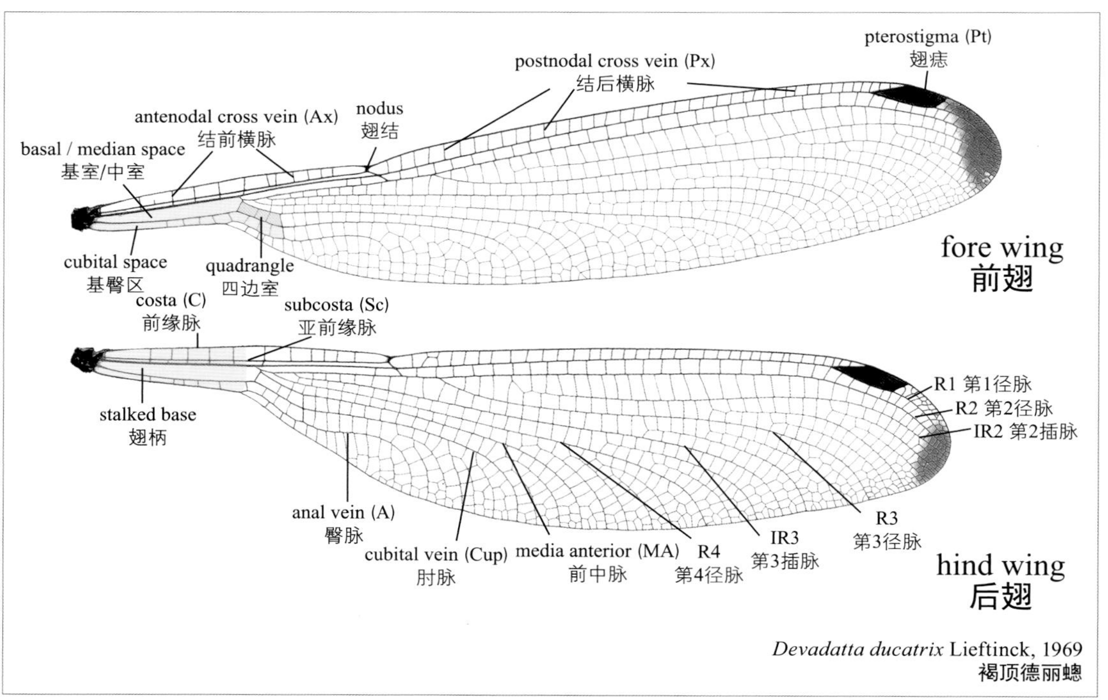

束翅亚目的翅脉

Wing venation of a zygopteran species

主要翅脉名称及其缩写 Main terms and abbreviations of venation

前缘脉(Costa, C)、亚前缘脉(Subcosta, Sc)、径脉(Radius, R)和中脉(Media, M)、弓脉(Arculus, arc)、前中脉(Media Anterior, MA)、肘脉(Cubital, CuP)、臀脉(Anal, A或1A)、翅结(Nodus, N)、亚翅结(Subnodus, sN)、翅痣(Pterostigma, Pt)、基室(Basal space, Bs)或称中室(Median space, Ms)、三角室(Triangle, T)、上三角室(Hypertriangle, hT)、下三角室(Subtriangle, sT)、四边室(Quadrilateral cell, q)、下四边室(Subquadrilateral cell, sq)、臀圈(Anal loop, Al)、臀角(Tornus)、盘区(Discoidal field)、基臀区(Cubital space)、臀三角室(Anal triangle)。

翅脉特征列举 Examples of wing features

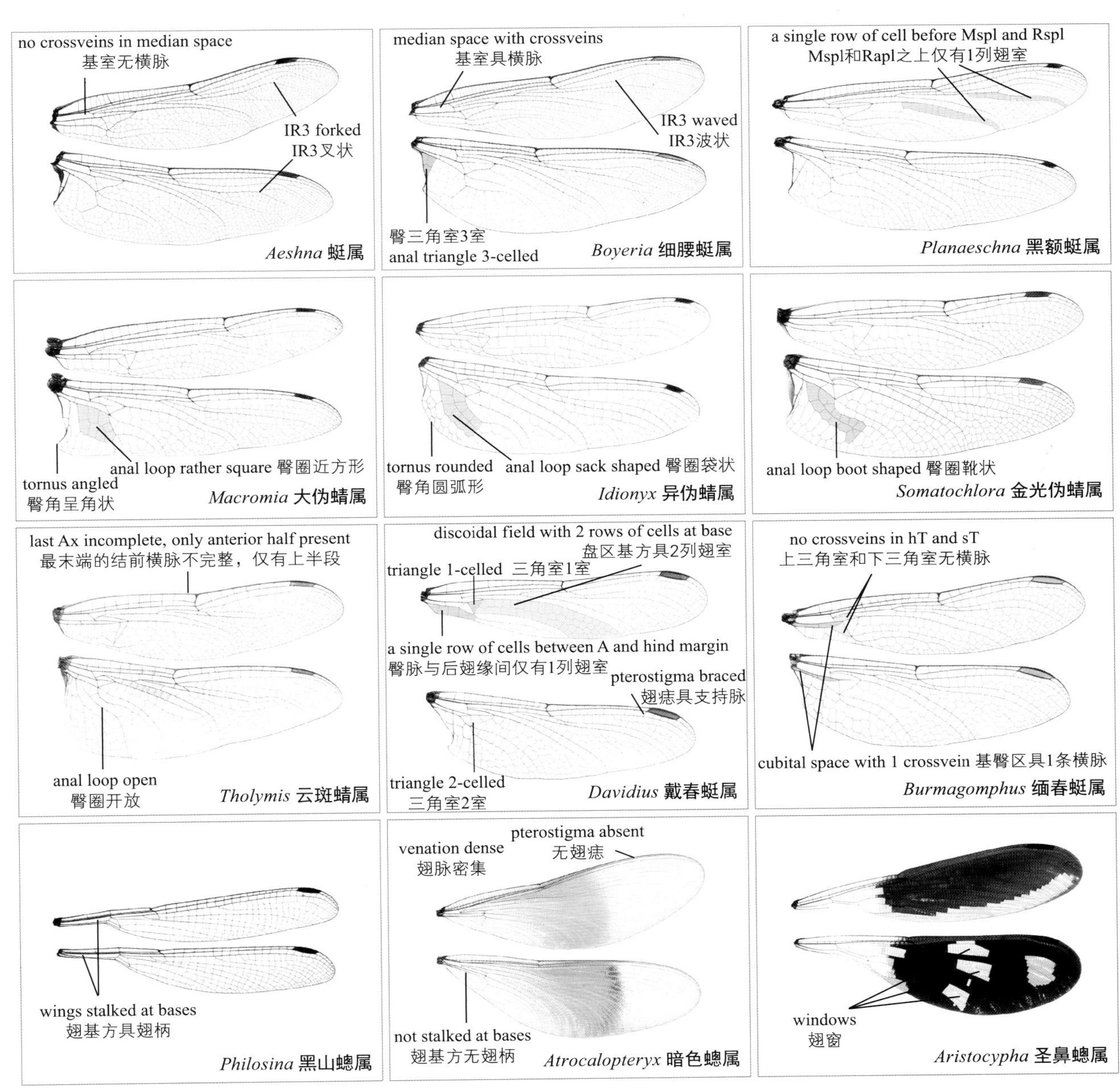

翅上的重要特征例举
Examples of useful wing features

腹部 Abdomen

腹部由10个体节组成，两性的构造不同。雄性第2节和第3节腹面具**次生殖器**；第10节末端具**肛附器**，通常是钳状，用于与雌性连结时抱握雌性的头部（差翅亚目）、前胸或中胸（束翅亚目）。雌性的附属器包括1个位于第8节和第9节腹面的**下生殖板**（部分差翅亚目）或锯形**产卵管**（所有的束翅亚目和蜓科）以及第10节末端的**尾毛**。雄性的肛附器、次生殖器及雌性的产卵器构造是重要的分类特征。腹部的色彩也是属级和种级较重要的分类特征。

The abdomen, consisting of ten segments and differs structurally according to sex. The male possesses **accessory genitalia** on segments 2 and 3 ventrally, and the terminal (10th) segment bears **anal appendages**, often forcipate, which are used to clasp the female by the head (Anisoptera) or portions of the pro- and mesothorax (Zygoptera) when achieving the tandem position. Appendages of the female include a **vulvar lamina** (several anisopteran families) or a saw-like **ovipositor** (all Zygoptera and Aeshnidae) on segments 8 and 9 ventrally and cerci on 10th segment. The structures of male anal appendages, accessory genitalia and female ovipositor are important taxonomic characters. The abdominal coloration is also useful for separating genera and species.

差翅亚目中雄性的肛附器包括1对**上肛附器**（也称**尾须**）和1个**下肛附器**（也称**肛上板**）。上肛附器的形状经常特化，有时为叉状并在腹面具齿状突起。束翅亚目的雄性拥有和差翅亚目相似的成对上肛附器，但和差翅亚目不同的是，它们具有1对下肛附器，也称**肛侧板**。雄性的次生殖器均位于腹部第2节和第3节，在2个亚目中相同。次生殖器包括**前钩片**、**后钩片**、夹在两者之间的**阳茎**，以及腹部第3节末端延伸出的**阴囊**。差翅亚目中，阳茎与阴囊相连，依次分成**茎节**、**中节**和**末节**，有时末节具成对的**鞭**。束翅亚目的阳茎由腹部第2节腹面生出。阳茎（或称**生殖舌**）的构造通常特化，是鉴定物种时依赖的重要特征。

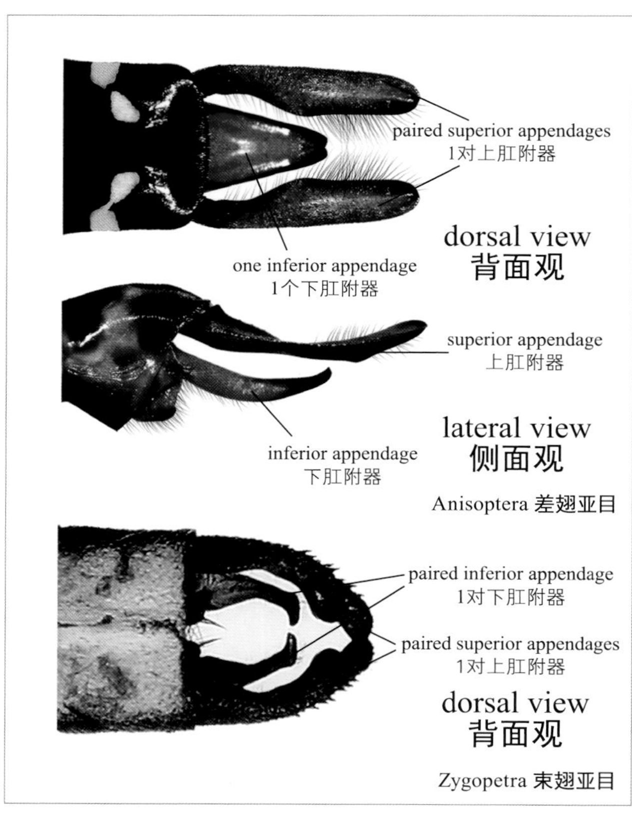

雄性蜻蜓的肛附器
Male anal appendages

The male anal appendages of an anisopteran contains a pair of **superior appendages** (also called **cerci**) (singular=**cercus**) and a single **inferior appendage** (also called the **epiproct**). The shape of the superior appendages often differs between closely related species; they may be forked and a series of small ventral teeth are often present. A typical male zygopteran has, as in the Anisoptera, paired superior appendages but unlike the Anisoptera, a ventral pair of inferior appendages called **paraprocts**. The male accessory genitalia for both suborders lie ventrally on S2 and S3. These structures consist of the **anteriorhamules**, **posterior hamules** and housed between them lies the **penis** and **vesica spermalis** which originates from the anterior margin of abdominal segment 3. The penis is connecting with the vesica spermalis and divided into the **stem**, the **median segment** and the **distal segment**. Sometimes the distal segment possesses paired **flagellum**. The damselfly penis arises ventrally from S2. The penis (or "**genital ligula**") is often species-specific and is sometimes important in differentiating between various species.

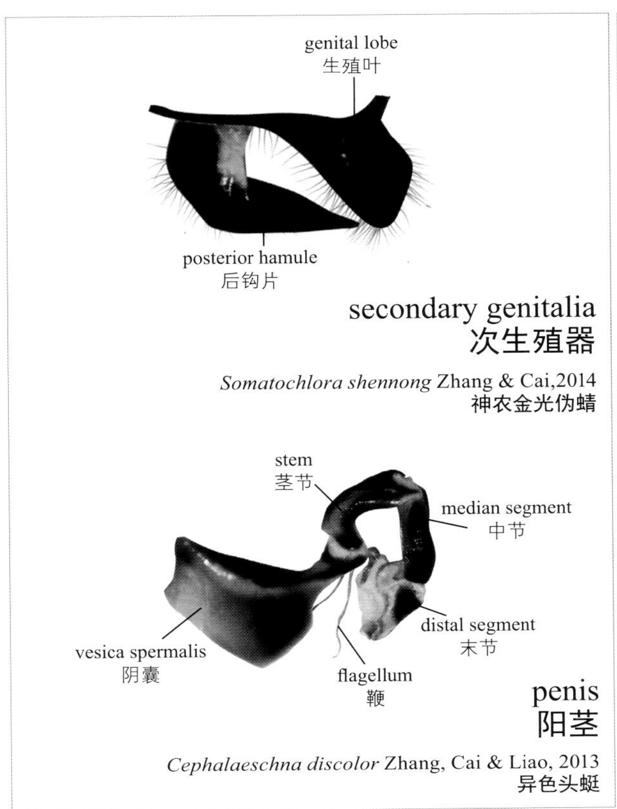

雄性蜻蜓的次生殖器
Male accessory genitalia

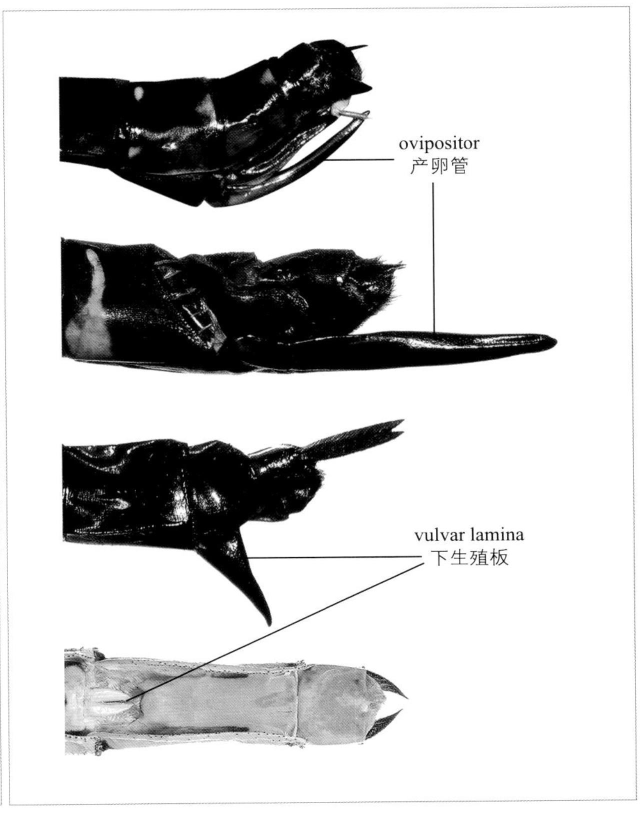

雌性蜻蜓的产卵器
Female ovipositor

胸部条纹 Thoracic markings

合胸色彩对于某些类群是重要的分类特征，尤其在春蜓科。合胸条纹的差异通常是种级分类特征。

The thoracic marking is a very useful diagnostic character within some groups, especially within Gomphidae. Difference of thoracic marking is often used for separating species.

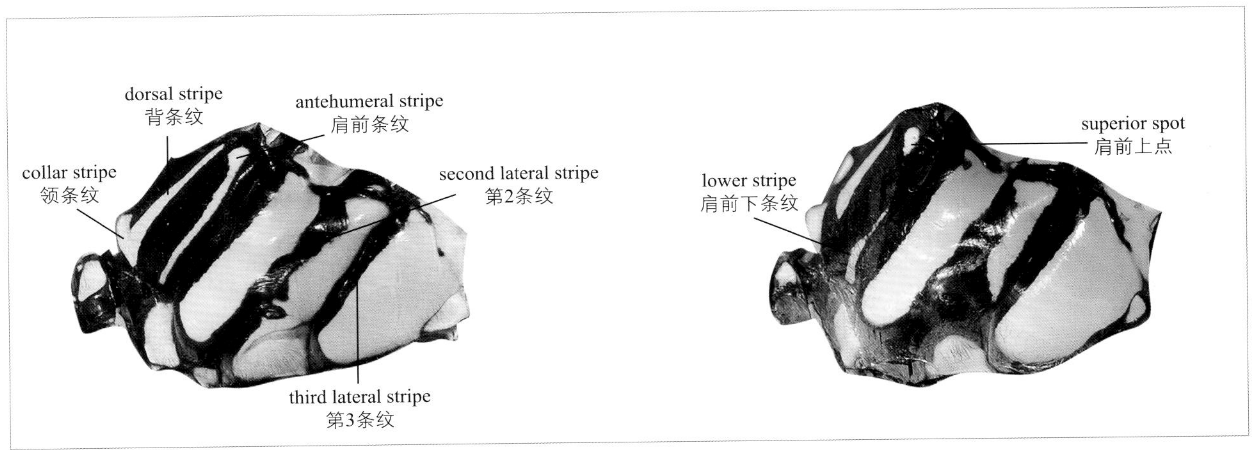

春蜓科胸部的重要条纹
Useful thoracic markings of Family Gomphidae

蜻蜓百科 Natural history of Odonata

蜻蜓素有“鸟友之虫”的美誉。这是由于其体型较大，体色艳丽，白天在固定的栖息环境活动，很多种类通过望远镜或者肉眼容易观察和识别，与鸟类的观察活动很相似。有趣的生活史和行为也给蜻蜓增添了几分魅力。与蜻蜓的互动，必须从识别蜻蜓开始。随着了解的深入，还可以关注它们的行为、生态学以及更多的生活常识。对蜻蜓稚虫的观察可以从饲养开始。

一个科学的自然观察活动，需要记录的信息包括：物种学名、性别、地点（地理坐标、海拔高度等）、日期、时间、天气等。数码相机可以抓拍并记录更多行为信息，有时需要采集标本以便深入研究。

Dragonflies are often called the bird watcher's insects. With their relatively large size, bright colors and frequent habit of perching in sunlight in well defined territories many species may readily be observed with binoculars or the naked eye, just as we observe birds. Their interesting life history and behavior also add to their fascination. In order to interact with dragonflies, one must start by learning to identify them. Then, as your knowledge increases you may focus more on studying aspects of their behavior, ecology and life histories. Larvae may be reared in captivity.

For scientifically useful information, one must record the following information concerning any sighting or observation: scientific name, sex, locality (coordinates, elevation, etc.), date, time of day, weather and any other observations of interest. A digital camera can help capture more detailed aspects of their behavior. Sometimes one needs to collect specimens for further study, even to establish their identity.

蝴蝶裂唇蜓稚虫潜伏在水底的细沙中捕食小鱼 | 宋睿斌 摄
Larva of *Chlorogomphus* (*Aurorachlorus*) *papilio* hides in the sandy substrate and preys on small fish | Photo by Ruibin Song

凶猛春蜓稚虫捕食蚯蚓 | 宋睿斌 摄
Larva of *Labrogomphus torvus* preying on fishworm | Photo by Ruibin Song

生活史 Life cycle

几乎所有蜻蜓的幼年都是在水中度过的，蜻蜓的幼虫期称为**稚虫**。交尾结束之后，雌性会在水面上或水附近选择合适的产卵地点。卵一般在几天至数周后孵化。刚孵化的稚虫足贴伏在体下并不能自由活动，称为**前稚虫**，但这阶段很短，仅持续几分钟的时间，这是稚虫的第1龄期。之后稚虫经历第1次蜕皮，进入第2龄期，此时具有活动自如的足。稚虫的生长期要经历数周或者数月，而一些生活在高山地区的种类则需要几年的时间，在这期间稚虫会多次蜕皮，通常为9~15次。蜻蜓稚虫是凶猛的水下捕食者。它们用其下唇特化出的面罩或者折叠钳钩，捕食各种小型无脊椎动物和脊椎动物，甚至同类相残。差翅亚目的稚虫用腹末的排泄孔，通过水的吸入和喷出获取氧气。束翅亚目的稚虫腹部末端具2根或3根尾鳃，有时具腹鳃，通过这些鳃片获取氧气。

Almost all the odonates spend their life underwater as **larvae**. After mating, females lay eggs in or nearby water surfaces or inserted into plant matter, including leaves stems and deadwood. Eggs hatch within several days to weeks. The first stage of larva, called the "**prolarva**", legs are appressed against the body and are not used, but this stage lasts for only a few minutes. The prolarva molts almost at once and the legs become functional during this second stage. The aquatic larval stages may last for weeks or months; some montane species may last several years. The larvae moult many times, usually ranges from 9-15. Odonate larvae are fierce underwater predators and feed on small invertebrates and vertebrates by the mask or folded forceps specialized from the labium. The larvae of Anisoptera obtain oxygen by movements of induction and expulsion from excretory pore. But larvae of Zygoptera possess 2-3 gills on tip of abdomen, some species also have ventral gills, the tools for obtaining oxygen.

沃氏短痣蜓稚虫捕捉住猎物之后将其托举出水面进食 | 宋睿斌 摄
Larva of *Tetracanthagyna waterhousei* holding up its prey out of water and eating it | Photo by Ruibin Song

稚虫漫长的生长历程在一个重要时刻宣告结束。**末龄稚虫**，即水下生活的最后龄期，在最后一次蜕皮之前会爬出水面，完成生命最重要的变态过程——**羽化**。整个过程会持续几个小时，这期间新生的成虫会从蜕中脱离。一些种类，比如春蜓科，会在水平位置羽化，它们喜欢水面上突出的大岩石表面或植物；另一些则会在垂直位置如树干上羽化。成功羽化的成虫会很快飞离羽化地点并隐藏进丛林中。

As their long underwater life is coming to an end, the **final stage larva** crawls out of the water in order to moult to the adult stage, the **emergence**, a process that can last for several hours. Some species, such as gomphids, emerge in a horizontal position preferring sand banks or rocks or some other horizontal substrate. Other species including most Zygoptera, aeshnids and libellulids emerge from a vertical position, using tree trunks or vertical stems. After their wings have fully expanded, the fragile adults fly away from the aquatic habits from which they emerged in order to mature.

赵氏显春蜓稚虫在岩石的水平位置羽化 | 宋睿斌 摄
Larva of *Phaenandrogomphus chaoi* emerging horizontally on a rock | Photo by Ruibin Song

汉森安春蜓稚虫在岩石的垂直位置羽化 | 宋睿斌 摄
Larva of *Amphigomphus hansoni* emerging vertically on a rock | Photo by Ruibin Song

金翼裂唇蜓羽化过程 | 宋睿斌 摄
Emergence process of *Chlorogomphus* (*Orogomphus*) *auripennis* | Photo by Ruibin Song

金翼裂唇蜓羽化过程 | 宋睿斌 摄
Emergence process of *Chlorogomphus* (*Orogomphus*) *auripennis* | Photo by Ruibin Song

长鼻裂唇蜓指名亚种初成虫体色较淡 | 宋睿斌 摄
Teneral *Chlorogomphus* (*Sinorogomphus*) *nasutus nasutus* with pale body color | Photo by Ruibin Song

刚刚从蜕脱离的成虫身体色彩较淡，身体柔软，称为“**初成虫**”；几个小时后，身体和翅硬化，而它们要经历数日或几星期才能发育出艳丽的体色，这个阶段称为“**未熟期**”。有些种类在发育阶段体色会发生剧烈的变化，例如雌性的杯斑小蟌未熟时身体为红色，成熟后为绿黄色。完全成熟的成虫体壁坚硬，通常色彩鲜艳，有些身体会覆盖上蓝色或白色粉霜，它们会花费大部分时间捕食和繁殖。从蜕脱离，成为具有飞行能力的成虫至死亡的阶段称为“**成虫期**”，通常历时数周。一种蜻蜓从第一只成虫出现至最后一只成虫死亡，这段时期称为“**飞行期**”。

Newly emerged adults, called **tenerals**, are fragile and lack fully developed colors. The teneral stage may last several hours to a few days and then entering the **immature stage**. During this transition, the wings and body cuticle gradually harden and the young adult gradually assumes its full adult body coloration. Body color can change through time for some species. For example immature female *Agriocnemis femina* are almost reddish but its body color changes to greenish yellow when mature. Fully mature reproductive adults become have a fully hardened exoskeleton and have assumed their full adult coloration. Upon maturing, in many species parts of the body and occasionally the wings develop a pale dull white or blue wax coating termed pruinosity. Fully mature adults return to their aquatic environments in order to forage and mate. **Adult lifespan** is measured from the moment of emergence to death and is normally a few weeks. **Flight season** is the time from the emergence of the first adults in a population to the death of the last individual.

雌性的杯斑小蟌未熟时身体红色
Iimmature female *Agriocnemis femina* body red

雌性的杯斑小蟌成熟后身体大面积黄色
Mature female *Agriocnemis femina* body largely yellow

雄性的赤褐灰蜻中印亚种未熟时身体黄色 | 宋睿斌 摄
Immature male *Orthetrum pruinosum neglectum* body yellow | Photo by Ruibin Song

雄性的赤褐灰蜻中印亚种成熟时胸部深褐色，腹部粉红色
Mature male *Orthetrum pruinosum neglectum* thorax dark brown, abdomen pink red

性成熟的雄性常会靠近水面，巩固自己掌控的领地。雌性相对少见但会多次回到水边交尾和产卵。交尾时以一种独特的心形环式连结来完成。

Mature males usually occur at bodies of water where they may defend a territory. Females are less in evidence but return to water for mating and oviposition, typically many times. Mating involves the pair forming of a heart-shaped ring with their bodies, a position unique to Odonata.

领地行为 Territorial behavior

雄性蜻蜓为争夺配偶，在繁殖地占据领地以获得交配机会。这种行为在蜻蜓目极为常见，但并非所有种类都具有领地行为，比如一些蜓科的雄性可以在森林中大范围游荡，而许多豆娘频繁移动停落位置并不会在一个特定地点逗留很久。领地行为根据其护卫领地方式的不同可分为停栖式领地占据和飞行式领地占据两种模式。

Males of many odonates defend a territory in order to guard prime oviposition sites for females. They defend their territory while perching or by constantly flying a certain beat. However not all adult males display territoriality. For example, some male aeshnids fly in forests searching for females and many male damselflies frequently change perches without remaining in a particular place for long.

停栖式领地占据 Defending a territory while perching

常见于束翅亚目、春蜓科和蜻科种类。成熟的雄性选择在水面或附近的岩石、植物或漂浮物上停落。它们也会偶尔巡飞，确保领域内无同种雄性个体或寻找领域内的雌性。

This behavior is common in Zygoptera, Gomphidae and Libellulidae. Mature males usually perch on rocks, plants or floating debris from which they may make occasional sorties checking to make sure no other males are present or in search of females entering their territory.

雄性的迈尔丽扇蟌停落在水面上的植物叶片占据领地
Male *Calicnemia miles* defending his territory while perching on a plant

雄性的三斑阳鼻蟌停落在漂浮的朽木上占据领地
Male *Heliocypha perforata* defending his territory while perching on floating deadwood

雄性的黑斑暗溪蟌停落在河面上悬挂的枝条上占据领地
Male *Dysphaea basitincta* defending his territory while perching on a hanging branch

雄性的劳伦斯尖尾春蜓停落在大岩石上占据领地
Male *Stylogomphus lawrenceae* defending his territory while perching on a large rock

飞行式领地占据 Defending a territory while flying

常见于飞行能力较强的差翅亚目。按照飞行方式的不同可以分为巡飞式和定点式两类。巡飞式的雄性通常在繁殖地反复来回飞行寻找雌性，它们的飞行往往具有固定的轨迹。定点式的雄性通常在水面上一个固定点低空悬停飞行。

This behavior is common in strong flying Anisoptera. They defend their territory by constantly patrolling or hovering. Patrolling males usually fly up and down following a definite route looking for a female. Males of other species may hover above water at a fixed point.

雄性的金黄显春蜓在溪流上方定点悬停占据领地
Male *Phaenandrogomphus aureus* hovering above the water in his territory

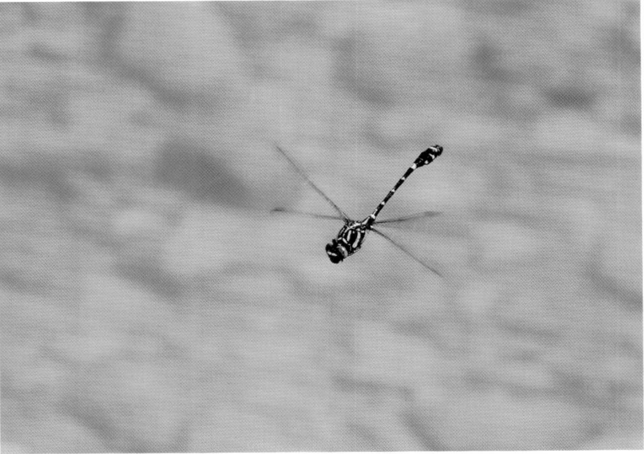

雄性的双髻环尾春蜓在溪流上方定点悬停占据领地
Male *Lamelligomphus tutulus* hovering above the water in his territory

雄性的长鼻裂唇蜓指名亚种在溪流的一段来回飞行占据领地
Male *Chlorogomphus nasutus nasutus* patrolling along a section of stream

争斗 Fighting

守卫领地的雄性一般具有攻击性，它们驱赶其他入侵者以获得交配权。通常一种雄性仅对同种个体或近似种个体具有攻击性。但有些凶猛的种类也会驱赶所有其他的种类。争斗的方式有许多种，有时两只对抗的雄性以头对头的方式在空中对抗，一进一退，最后强者将弱者赶出领地。雌性在产卵时，也会驱逐同种雌性。曾在贵州遇见红褐多棘蜓在林荫处的静水潭产卵，但有同种雌性在同一水潭产卵时，双方会展开争斗，弱方离开失去产卵权。

Males often compete for an established territory often driving away other conspecific males or even other species of similar size. Fighting may involve face to face encounters in flight with the weaker male eventually leaving. Some females also display territoriality when laying eggs evicting weaker conspecific females. An example observed in Guizhou, a female *Polycanthagyna erythromelas* was laying eggs in a shady pools in forest, a second female came to the same site, they competed and finally the weaker was evicted from the site.

多横细色蟌两雄互相追逐争斗
Two males of *Vestalis gracilis* chasing each other while competing for territory

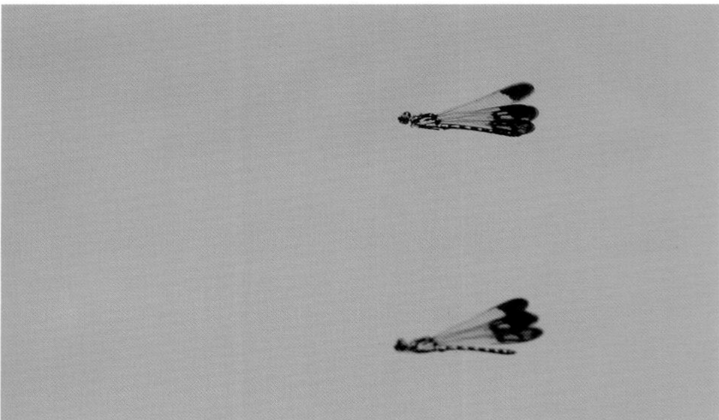

两只雄性的三斑阳鼻蟌平行飞行争斗
Two males of *Heliocypha perforata* in parallel flight during a territorial dispute

丽拟丝蟌雄性头对头争斗
Males of *Pseudolestes mirabilis* in a face to face encounter during a territorial dispute

求偶 Courtship

求偶行为仅在束翅亚目的少数类群中可见，主要是色蟌科和鼻蟌科的种类。

Courtship is only seen in a few groups in Zygoptera predominately Calopterygidae and Chlorocyphidae.

雄性的黄脊圣鼻蟌向雌性展示其白色的足
Male *Aristocypha fenestrella* displaying its white legs to the female

交尾 Mating

当成熟的雄性遇见同种雌性，它便企图交配。雄性通常从雌性后面靠近，先用足擒获雌性。然后雄性的肛附器抱握住雌性的后头或者前胸，雌雄连结。最终它们以蜻蜓目独有的心形或椭圆形环式连结来完成交尾。

许多种类在连结的开始，雄性要先进行授精。精子从腹部第9节精巢下方的开孔，传递到腹部第2节的次生殖器。此时雄性将腹部末端弯曲伸至第2节。授精结束之后才会以环式连结完成交尾。交尾可以在停歇时完成，也可以在空中进行。

云南异翅溪蟌交尾
A mating pair of *Anisopleura yunnanensis*

黄翅溪蟌交尾
A mating pair of *Euphaea ochracea*

When a mature male odonate encounters a female, he may attempt mating. The male approaches the female from behind seizing the female with his legs, then clasping the female by her head (Anisoptera) or prothorax (Zygoptera) using his anal appendages thus achieving the tandem position. Copulation is finalized when the female, if she is willing, moves the tip of her abdomen to join the male intermittent organ at the base of his abdomen (genital ligula in Zygoptera, vesica spermalis in Anisoptera) thus assuming the wheel position, a system unique to the order Odonata.

In many species, at the beginning of tandem linkage, males need several seconds to transfer sperm, from the genital opening under the ninth abdominal segment to the secondary genitalia beneath the second abdominal segment. Males bend the abdomen tip to the second segment. After this process, the wheel position is accomplished whereby the male transfers the sperm to the female. Mating can take place when perching or during flight.

金翅裂唇蜓交尾 | 莫善濂 摄
A mating pair of *Chlorogomphus* (*Neorogomphus*) *canhvang* | Photo by Shanlian Mo

连结 Pair in tandem

雄性授精 Male transferring sperm to secondary genitalia

环式连结 Pair in wheel

白尾野蟌交尾过程 | 莫善濂 摄
Copulatory process of *Agriomorpha fusca* | Photo by Shanlian Mo

产卵 Oviposition

点水式产卵 Exophytic oviposition: dipping abdomen into water

这种产卵行为可见于春蜓总科、大蜓总科和蜻总科。这些雌性蜻蜓中大多数没有发达的产卵管，通过腹部第8节下生殖板的开孔直接将卵排出。一些雌性通过腹部拍打水面产卵。另一些，比如大蜓科的雌性，拥有长而锋利的刀状产卵管，它们通过身体直立的插秧式产卵，将卵埋在浅溪流或渗流的底层。

This behavior occurs in superfamilies Gomphoidea, Cordulegastroidea and Libelluloidea. Most of these females lack a saw-like ovipositor and they extrude eggs from the genital pore in their eighth segment. Some females tap the water's surface discharging a clump of eggs. Others, for example females of Cordulegastridae possess a long pointed ovipositor and repeatedly dip their abdomen vertically into the water depositing their eggs into soft mud in shallow streams or seepages.

米尔蜻雌性点水产卵
Female *Libellula melli* laying eggs by dipping the tip of abdomen in the water

黑尾灰蜻雌性点水产卵 | 莫善濂 摄
Female *Orthetrum glaucum* laying eggs by dipping the tip of abdomen in the water | Photo by Shanlian Mo

缘斑毛伪蜻雌性先停落在水边的枝条上排卵，然后将卵块投入水中
Female *Epitheca marginata* perched preparing a long string of eggs, which will then be deposited into water

双斑圆臀大蜓雌性通过身体直立、反复地上下移动将卵插入底部的泥沙中
Female *Anotogaster kuchenbeiseri* repeatedly inserting her eggs into the muddy substrate

铃木裂唇蜓雌性通过身体直立将卵块插入水中
Female *Chlorogomphus suzukii* dipping her abdomen into water and releasing eggs while her body is held erect

空投式产卵 Exophytic oviposition: dropping eggs over water's surface while flying

这种产卵行为仅见于春蜓科和蜻科，包括环尾春蜓属、副春蜓属、日春蜓属和赤蜻属。这些雌性的春蜓在溪流上低空选择合适产卵点定点悬停，将卵空投入水中。

This behavior is only seen in Gomphidae and Libellulidae, including genera *Lamelligomphus*, *Paragomphus*, *Nihonogomphus* and *Sympetrum*. These females choose a suitable position above water, and drop their eggs above the water surface while hovering.

双髻环尾春蜓雌性在溪流上空中投蛋
Female *Lamelligomphus tutulus* dropping eggs above a stream while hovering

一对褐顶赤蜻在沼泽地空中投蛋 | 金洪光 摄
A pair of *Sympetrum infuscatum* dropping eggs in flight above a marsh | Photo by Hongguang Jin

插入式产卵 Endophytic oviposition: inserting eggs into plants or mud

这种产卵行为见于所有的束翅亚目、间翅亚目和差翅亚目中的蜓科种类。这些雌性都拥有锯形产卵管，可以将卵插入泥土和植物内。一些蜓科的雌性，如头蜓属、佩蜓属，拥有非常发达的产卵管，长而锋利，它们多会选择在溪流边缘具苔藓的陡峭石壁或湿润的泥土上产卵。

This behavior occurs in all Zygoptera, Anisozygoptera and family Aeshnidae in Anisoptera. These females have a saw-like ovipositor which can be inserted into plants or mud. Some females of family Aeshnidae, for example species of genera *Cephalaeschna* and *Periaeschna*, possess strongly developed, long and sharply pointed ovipositors, from which they usually lay their eggs into mossy banks or mud at the water's edge.

雌性的丽拟丝蟌在溪流边缘的朽木上产卵
Female *Pseudolestes mirabilis* inserting its eggs into deadwood near a stream

雌性的红褐多棘蜓在水潭边缘陡坡的苔藓上产卵
Female *Polycanthagyna erythromelas* laying eggs into the moss on a slope near a pool

雌性的崂山黑额蜓在溪流边缘的朽木上产卵
Females of *Planaeschna laoshanensis* inserting eggs into deadwood near a stream

一对方氏赤蜻在水面连结点水产卵
A pair of *Sympetrum fonscolombii* laying eggs in tandem onto the water surface

连结产卵 Exophytic and exophytic oviposition: ovipositing in tandem with contact guarding

交尾结束以后仍保持雌雄连结的状态，雄性携带雌性产卵。这种行为较常见，它们可以连结点水产卵，也可以连结插入产卵。产卵地点的选择则完全由雄性蜻蜓掌控。

After mating, the pair assumes the tandem position with the male holding the female while she oviposits at a suitable site. This type of oviposition, which is common in both suborders, allows the male to physically prevent other males from mating with his female. Tandem pairs can drop eggs into water or insert eggs into plants or mud.

一对黄狭扇螅停落在水面的水草上连结产卵
A pair of *Copera marginipes* laying eggs in tandem by perching on emergent plants above water

护卫产卵 Endophytic and exophytic oviposition: oviposition via non-contact guarding by males

雌性在产卵时，雄性虽未与雌性连结，但仍在其旁守护其产卵，并驱赶其他雄性个体。雄性护卫的方式可以是围绕雌性飞行，也可以是停落在雌性附近。

The pair dragonflies uncouple after mating but the male still guards the female while she is laying eggs, chasing off other males. Male can fly around the female or perch near her when she lays her eggs.

赤褐灰蜻中印亚种雄性围绕雌性飞行护卫其产卵
Male *Orthetrum pruinosum neglectum* flying around the female and guarding her during oviposition

三斑阳鼻螅雄性停落在雌性对面护卫其产卵
Male *Heliocypha perforata* perching in front of the female and guarding her during oviposition

集群产卵 Endophytic oviposition: group oviposition

这种产卵行为较常见于溪蟌科的尾溪蟌属。它们喜欢聚集在溪流边缘的灌木或具苔藓的土坡上产卵，有时可以多达几百对同时产卵。

This is commonly observed in genus *Bayadera* (Euphaeidae). Groups of these damselflies oviposit in bushes or the mossy banks in groups at the stream's margin where hundreds of pairs can sometimes be seen together.

巨齿尾溪蟌集群产卵
A group of *Bayadera melanopteryx* ovipositing

潜水产卵 Endophytic oviposition: underwater oviposition

这种产卵行为在束翅亚目和蜓科中可见，在色蟌科、溪蟌科和蟌科中较常见。这些雌性豆娘可以将半个身体甚至整个身体潜入水中产卵，通常可以持续数10分钟甚至1小时以上。少数雄性可以随雌性一同潜水。

This behavior, confined to Zygoptera and Aeshnidae, occurs commonly within Calopterygidae, Euphaeidae and Coenagrionidae. Females can oviposit half or entirely submerged underwater, a behavior that can last minutes to over an hour. A few males, may also become submerged under water, while in tandem following the females.

雌性的华艳色蟌全身潜水产卵
Female *Neurobasis chinensis* ovipositing entirely underwater

雌性的霜基色蟌半身潜水产卵
Female *Archineura hetaerinoides* ovipositing with body half submerged

一对褐斑蟌潜水产卵
A pair of *Pseudagrion spencei* ovipositing underwater

黏附式产卵 Exophytic oviposition: adhesive ovipostion

雌性将卵黏附在水面上悬挂的树枝或藤条，在中国仅见于蜻科的方蜻属。

Females attach the eggs to branches or rattans overhanging water, only seen in genus *Tetrathemis* (Libellulidae).

雌性的宽翅方蜻把卵黏附在水面的树枝上 | 莫善濂 摄
Female *Tetrathemis platyptera* attaching her eggs to branches above water | Photo by Shanlian Mo

捕食 Feeding

所有蜻蜓的成虫和稚虫都是捕食者。多数成虫取食小型猎物，例如一些小型飞虫、叶蝉和蚊。一些习性凶猛的种类取食大型猎物，也包括捕食其同类，一些大型蜻蜓可以捕食豆娘或者小型蜻蜓。

All odonates are predators in both the adult and larval stage. Most adults take small prey, including small flies, leafhoppers and mosquitoes. Some ferocious species feed on larger prey, including members of their own order, larger dragonflies prey on damselflies and smaller dragonflies.

雨林爪蜻捕食蝴蝶
Onychothemis testacea preying on a butterfly

锥腹蜻捕食蜘蛛 | 金洪光 摄
Acisoma panorpoides preying on a spider | Photo by Hongguang Jin

霸王叶春蜓捕食红蜻 | 袁屏 摄
Ictinogomphus pertinax preying on *Crocothemis servilia* | Photo by Ping Yuan

七条尾蟌捕食长叶异痣蟌 | 陈炜 摄
Paracercion plagiosum preying on *Ischnura elegans* | Photo by Wei Chen

蜻蜓的天敌 Predators

在脊椎动物中，鸟类是蜻蜓最主要的天敌。比如蜂虎偏爱捕食蜻蜓。两栖动物和爬行动物也经常捕食水面附近的蜻蜓。无脊椎动物中，蜘蛛和很多昆虫是蜻蜓主要的天敌。鱼可以捕食蜻蜓稚虫。

Birds are major predators. For example, nesting bee-eaters often prey heavily on odonates. Amphibians and reptiles also often prey on the odonates approaching water. Among invertebrates, spiders and many insects are main predators. Fish can also eat the larvae.

栗喉蜂虎捕食高翔漭蜻 | 黄海燕 摄
Merops philippinus preying on *Macrodiplax cora* | Photo by Haiyan Huang

中国林蛙捕食透顶单脉色蟌 | 陈炜 摄
Rana chensinensis preying on *Matrona basilaris* | Photo by Wei Chen

一群蚂蚁捕食羽化中的凶猛春蜓 | 宋睿斌 摄
Ants eating emerging *Labrogomphus torvus* | Photo by Ruibin Song

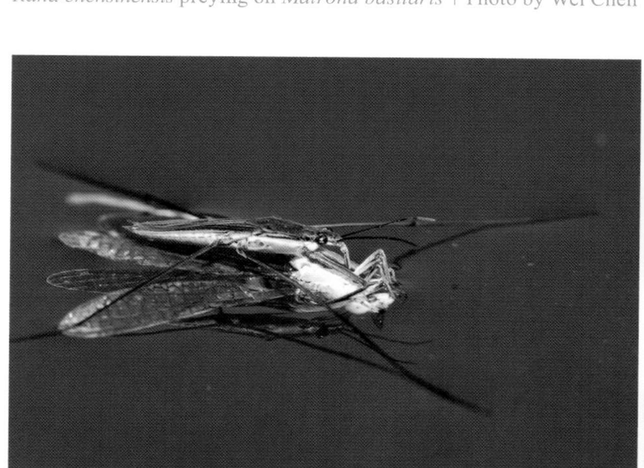

圆臀大黾蝽捕食蓝纹尾蟌 | 刘辉 摄
Aquarius paludum preying on *Paracercion calamorum* | Photo by Hui Liu

拟环纹豹蛛捕食蓝纹尾蟌 | 刘辉 摄
Pardosa pseudoannulata preying on *Paracercion calamorum* | Photo by Hui Liu

角类肥蛛捕食长叶异痣蟌 | 金洪光 摄
Larinioides cornuta preying on *Ischnura elegans* | Photo by Hongguang Jin

中华单羽食虫虻捕食长叶异痣蟌 | 陈炜 摄
Cophinopoda chinensis preying on *Ischnura elegans* | Photo by Wei Chen

雌雄二型和多型现象 Sexual dimorphism and polymorphism

雌雄二型 Sexual dimorphism

雌雄二型现象在蜻蜓目中十分普遍。两性在体型、体态和体色上有显著差异，雄性艳丽但体型较小，雌性暗淡但体型较粗壮。

Sexual dimorphism is common in Odonata. Males and females often differ in body color and pattern. Males usually possess brighter colors and smaller while females are often darker, paler and stronger.

雄性的晓褐蜻通体紫红色
Male *Trithemis aurora* body purple red throughout

雌性的晓褐蜻身体大面积褐黄色
Female *Trithemis aurora* body largely brownish yellow

雄性的网脉蜻身体大面积红色
Male *Neurothemis fulvia* body largely red

雌性的网脉蜻身体大面积黄褐色
Female *Neurothemis fulvia* body largely yellowish brown

雄性的叶足扇蟌中足和后足的胫节膨大呈片状
The mid and hind tibiae of male *Platycnemis phyllopoda* expanded and leaf-shape

雌性的叶足扇蟌足的胫节未膨大
The mid and hind tibiae of female *Platycnemis phyllopoda* not expanded

多型现象 Polymorphism or polychromatism

雄性的多型现象在蜻蜓目中少见，仅在几个属中发现。最常见的是色蟌科绿色蟌属。该属雄性个体都具二型，一种型翅具色彩，另一种型翅透明。雌性多型则较常见。其中与雄性体色差异较大的雌性称为异色型，而与雄性色彩极为相似的称为同色型。有些同色型的雌性能模仿雄性的行为，甚至追逐异色型雌性。

Male polymorphism is uncommon and only seen in a few genera. This is most common in genus *Mnais*, where males have two color forms, one morph with tinted wings and the other with clear wings. Female polymorphism is more common. Females with clearly different body color from male are the heterotypic morphs, whereas females with coloration similar to the male are termed homochromatic morphs. Some homochromatic females can imitate the behavior of males, or even chase the heterotypic females.

霜白疏脉蜻，雌，黄色型
Brachydiplax farinosa, female, the yellow morph

霜白疏脉蜻，雌，黑色型
Brachydiplax farinosa, female, the black morph

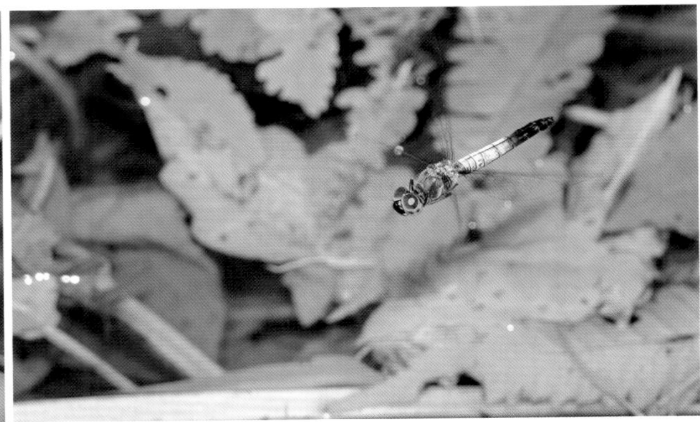

霜白疏脉蜻，雌，同色型
Brachydiplax farinosa, female, the homochromatic morph

安氏绿色蟌，雄，橙翅型
Mnais andersoni, male, the orange winged morph

安氏绿色蟌，雄，透翅型
Mnais andersoni, male, the hyaline winged morph

蜻蜓的栖息环境 Dragonfly habitats

蜻蜓生活在各类淡水环境中，稚虫水生。它们的栖息环境大致分为两大类。

Dragonflies are intimately associated with aquatic habitats since their larvae are aquatic. They breed in freshwater, including standing water and running water habitats.

静水环境：这类环境大至大型的湖泊和水库，小至树洞中的水潭，多是常见和广布型物种的繁殖场所。一些特殊的在高山上孤立的湿地环境也是最珍稀物种的出没地点。

Standing water (lentic) habitats: range from large lakes and reservoirs to small water-filled tree holes. Such habitats are usually occupied by widespread and common species, but some isolated wetland habitats in high mountainous regions are also home to some specialised species.

树洞中的水潭：这类环境可见于热带或亚热带森林中，包括从小型的孔形水潭到树干上支撑的大型水盆。这类生境容易被忽视。宽腹蜻属比较偏爱这类环境。

Tree holes in forest: these occur in tropical or subtropical forests and range from small rot holes to large buttress pans. They are easily overlooked. Species of *Lyriothemis* (Libellulidae) may utilize these habitats.

树洞中的水潭，海南（吊罗山）
Tree hole in forest, Hainan (Diaoluoshan Mountains)

森林中的小型孤立水潭：这类水潭通常四周有较陡峭的石壁和土坡，如果石壁上有苔藓就成了蜓科多棘蜓属和扇蟌科长腹扇蟌属的重要生境，这类蜻蜓仅在此类环境繁殖。

Isolated small pools in forest: such pools are usually located in steep precipices or earthen slopes, if accompanied by moss growing such habitats are attractive for females of *Polycanthagyna* (Aeshnidae) and *Coeliccia* (Platycnemididae).

林道的积水潭：热带雨林中的林荫小路常见一些小型的积水潭，有些是季节性的，但在雨季有水时经常可见线纹林蜻在四周徘徊。此外少数春蜓，如闽春蜓属蜻蜓也会在林道积水潭繁殖。

Small pools on paths or forest clearings: such small pools are seasonal in tropical forest habitats occurring primarily during the rainy season. *Cratilla lineata* (Libellulidae) and species of *Fukienogomphus* (Gomphidae) favor these habitats.

季节性浅水池塘：这些开阔地的低地浅水池塘是黄蜻、斜痣蜻、狭翅蜻和各类灰蜻非常偏爱的环境。

Seasonal shallow ponds: Lowland shallow ponds are favored by common, wide-ranging libellulid genera *Pantala*, *Tramea*, *Potomarcha* and *Orthetrum*.

森林中的孤立水潭，贵州
Isolated small forest pools, Guizhou

林道上的积水潭，云南（西双版纳国家级自然保护区）
Small pools on path in forest, Yunnan (Xishuangbanna National Nature Reserve)

浅水池塘，云南（西双版纳）
Shallow ponds, Yunnan (Xishuangbanna)

水稻田和沼泽地：许多蜻科种类选择水稻田繁殖，最常见的是红蜻、黄蜻、锥腹蜻、各种赤蜻，还有一些小型豆娘，如异痣蟌属的种类。

Paddy fields and marshes: such habitats are occupied by various libellulids and small damselflies, including common wide-spread species of *Crocothemis*, *Pantala*, *Acisoma*, *Sympetrum* (Libellulidae) and *Ischnura* (Coenagrionidae).

水稻田，贵州
Paddy field, Guizhou

高山沼泽地，贵州
Marshes in high mountain, Guizhou

水草茂盛的池塘和湿地：不同的水生植物种类和周边植被状况会影响物种多样性，通常水草茂盛的池塘会有较多的蟌科和蜻科种类，这是一类种类繁盛的蜻蜓生境，有时一个池塘可以容纳几十个种类，一些大型的蜓科种类如伟蜓属也偏爱这类环境。

Well vegetated ponds and wetlands: species richness may be augmented by the presence of aquatic plants and emergent vegetation. Various libellulids and small damselflies prefer such habitats as well as some large species, like *Anax* (Aeshnidae).

水葫芦滋生的池塘，云南（西双版纳）
Water hyacinth covered pond, Yunnan (Xishuangbanna)

水草茂盛的湿地，海南（五指山）
Well vegetated pond, Hainan (Wuzhishan Mountains)

杂草丛生的湿地，海南（吊罗山）
Pond with plenty of weeds, Hainan (Diaoluoshan Mountains)

茂盛森林中的池塘，云南（西双版纳）
Pond in dense forest, Yunnan (Xishuangbanna)

中至大型水草匮乏的池塘、水库和湖泊：少数蜻蜓会在周边具有水草的区域活动，通常这类生境下栖息的种类很少，春蜓科的叶春蜓属、新叶春蜓属以及大伪蜻科的丽大伪蜻属比较偏爱这种环境。

Medium to large ponds, reservoirs and lakes: a few species prefer these large water body habitats where species of *Ictinogomphus* (Gomphidae), *Sinictinogomphus* (Gomphidae) and *Epophthalmia* (Macromiidae) prefer to patrol the water's margin.

水库，海南（吊罗山）
Reservoirs, Hainan (Diaoluoshan Mountains)

人工水潭：这类环境会有许多小型豆娘栖息，也是多棘蜓属和方蜻属非常喜欢的环境。

Man made ponds: many small damselflies occur at these ponds, it is also home to various species of *Polycanthagyna* (Aeshnidae) and *Tetrathemis* (Libellulidae).

人工水潭，云南（西双版纳）
Man made pond, Yunnan (Xishuangbanna)

流水环境：主要包括河流、溪流、瀑布等环境，是一些较珍稀类群的重要栖息地，如大蜓科、裂唇蜓科和色蟌科等，同时中国最庞大的蜻蜓家族春蜓科的绝大多数种类都是选择此类生境繁殖。

Running water (lotic) habitats: includes rivers, streams, waterfalls, etc. Many rare species breed in such habitats, including most species within families Gomphidae, Cordulegastridae, Chlorogomphidae and Calopterygidae.

渗流地：仅有很小的水流或者仅是潮湿的环境而不见流水，却是许多珍稀物种偏爱的环境，比如稀有的亮翅色蟌属、圆臀大蜓属等。在茂盛的热带雨林中，这类环境通常被扇蟌科等森林豆娘占据。

Seepages: habitats with only a small trickle of water or marshy habitats supports some infrequently encountered species of *Echo* (Calopterygidae), and *Anotogaster* (Cordulegastridae). Many species of Platystictidae are denizens of these habitats.

渗流地，云南（德宏铜壁关国家级自然保护区）
Seepages, Yunnan (Tongbiguan National Nature Reserve in Dehong)

渗流地，云南（德宏）
Seepages, Yunnan (Dehong)

小型瀑布和具有滴流的石壁：这是扇蟌科丽扇蟌属豆娘的重要生境，偶尔也会有蜓科头蜓属种类选择这类环境，它们的稚虫生活在陡峭的石壁上，有时可以脱离水流到石壁上捕食。

Small waterfalls and trickles from steep precipices: they are prime habitats for *Calicnemia* (Platycnemididae) and *Cephalaeschna* (Aeshnidae) where their larvae breed in spray zones on nearly steep precipices surfaces.

小型瀑布，湖北（神农架）
Small waterfall, Hubei (Shennongjia)

具有滴流的石壁，云南（红河）
Steep precipices with trickle, Yunnan (Honghe)

狭窄溪流和沟渠：这种小型的流水环境是春蜓、大蜓、色蟌等类群所喜爱的生境。

Narrow streams and ditches: species of Gomphidae, Cordulegastridae and Calopterygidae prefer these habitats.

林荫溪流：热带雨林中的林荫小溪拥有最丰富的物种，珍稀的裂唇蜓科和蜓科种类可以在此环境找到，茂盛的林荫为这些大型蜻蜓提供庇护。

Shady streams: the shady streams in tropical forest are home to various uncommon and rare species, including species of Aeshnidae, Chlorogomphidae.

狭窄溪流，云南（德宏）
Narrow stream, Yunnan (Dehong)

森林中的沟渠，云南（大理）
Ditches in forest, Yunnan (Dali)

林荫溪流，海南（吊罗山）
Shady stream, Hainan (Diaoluoshan Mountains)

林荫溪流，云南（西双版纳）
Shady stream, Yunnan (Xishuangbanna)

林荫溪流，云南（德宏）
Shady stream, Yunnan (Dehong)

开阔溪流：一些具有细砂底质的开阔溪流是许多春蜓的理想生境，中国著名的蝴蝶裂唇蜓也在此环境出没。一些具有大型岩石的溪流，是溪蟌科、鼻蟌科和春蜓科种类偏爱的环境。

Exposed streams: streams with fine sandy substrate are ideal habitats for gomphids and the famous species *Chlorogomphus papilio* can be found here. Streams with big rocks are home for various Euphaeidae, Chlorocyphidae and Gomphidae.

开阔溪流，贵州
Exposed stream, Guizhou

开阔溪流，云南（普洱）
Exposed stream, Yunan (Pu'er)

中至大型河流：这类生境是大伪蜻属的理想繁殖地点，河岸植被茂盛的河段也是各类春蜓和多种豆娘的生境，如印鼻蟌属。

Medium to large rivers: prime habitats for *Macromia* (Macromiidae); many gomphids and damselflies, such as *Indocypha* species, occur along such rivers when banks are lined with emergent vegetation.

大型瀑布：很少种类选择此类生境，但虹蜻属种类较偏爱。

Large waterfalls: a few species such as *Zygonyx* (Libellulidae) will be found here.

罗梭江，云南（西双版纳）
Luosuojiang river, Yunnan (Xishuangbanna)

南溪河，云南（红河）
Nanxihe river, Yunnan (Honghe)

大型瀑布，云南（红河）
Large waterfall, Yunnan (Honghe)

中国蜻蜓研究简史 Brief history of Chinese Odonatology

中国第一种正式记录的蜻蜓是*Libellula chinensis*，收录在林奈1758年出版的《自然系统》（第10版）中，这种蜻蜓即是今天我们熟知的华艳色蟌。这本专著也被认为是现代分类学的起点。18世纪末，Dru Drury和Johan C. Fabricius描述了中国地区6个差翅亚目种类，包括广布的红蜻、网脉蜻和狭腹灰蜻。19世纪下半叶，一些欧洲蜻蜓学者开始描述中国蜻蜓，包括Edmond de Selys Longchamps、Robert McLachlan和Hermann A. Hagen。在Selys描述的27个有效种中，包括了常见而美丽的黑暗色蟌、透顶单脉色蟌和异色灰蜻。随后McLachlan描述了9种蜻蜓，包括产自中国台湾的艳丽豆娘褐顶色蟌。1880—1890年，Selys和McLachlan各自发表了其第一份区系调查报告，标本分别采自北京和四川。1900年William F. Kirby发表了海南蜻蜓报告，在其发表的为数不多的新种中包含了最奇特的丽拟丝蟌。至1900年底，根据采自中国的标本，一共有77种蜻蜓新种（含亚种）被描述，其中58种至今为有效种。

The first formal record of a dragonfly from China was *Libellula chinensis* by Carolus Linnaeus (Carl von Linné) in the 10th edition of his *Systema naturae* published in 1758, now known as *Neurobasis chinensis*. This publication is now recognized as the starting point for formal taxonomy. At the end of 18th century half 12 common anisopteran species from China were named and described by Dru Drury and Johan C. Fabricius. These include the widespread libellulids *Crocothemis servilia*, *Neurothemis fulvia* and *Orthetrum sabina*. In the latter half of the 19th century, a few European odonate specialists, such as Edmond de Selys Longchamps, Robert McLachlan and Hermann A. Hagen also began describing species from China. The 27 valid species that Selys described from China are common and striking species such as *Atrocalopteryx atrata*, *Matrona basilaris* and *Orthetrum melania*. McLachlan, in his turn, described nine valid species from China, including the splendid Taiwanese *Psolodesmus mandarinus*. In 1880-1890's, Selys and McLachlan also wrote the first faunistic reports on Chinese Odonata, based on specimens collected in Beijing and Sichuan, respectively. This was followed by William F. Kirby's (1900) paper on dragonflies from Hainan; among the few new species was the extraordinary *Pseudolestes mirabilis*. By the end of the year 1900, a total of 77 species and subspecies had been described from Chinese specimens; of these 58 are presently considered valid full species.

争斗中的雄性华艳色蟌，中国最早记录的蜻蜓
Males of *Neurobasis chinensis* fighting over territory, this is the first odonate described from China

进入20世纪初，更多的国外研究人员投身中国蜻蜓的研究工作。瑞士昆虫学家Friedrich Ris分别于1912年和1916年发表了中国南部和台湾地区的重要报告，文中包括若干新种。Ris共发表了25个新种（其中22个为有效种），这其中包括体型巨大而著名的蝴蝶裂唇蜓，发表于1927年。同一时期，美国康奈尔大学教授James G. Needham访问了许多高校和博物馆并组织了野外考察。1930年，他出版了《中国蜻蜓手册》，共记载中国蜻蜓89属266种，含63个新种。该书虽然存在较多的错误和遗漏，且其中1/3的新种是异名，但是首部涉及整个中国地区的蜻蜓专著，作者除了未获得云南和贵州的标本，其他各省的标本都有研究。这本专著也为中国蜻蜓的分类系统研究奠定了基础。1910—1930年描述中国新种的国外蜻蜓分类学者还包括Frederic C. Fraser（一位著名的印度蜻蜓研究专家）以及Kan Oguma。1937—1997年，日本学者朝比奈正二郎也为中国蜻蜓区系研究贡献巨大，他主要发表了有关中国北部、浙江、湖南、湖北、福建、四川、台湾和香港等地区的蜻蜓研究论文。朝比奈正二郎在整个亚洲地区有广泛的研究并发表了诸多新种。

From the beginning of twentieth century, an increasing number of western odonatologists worked on the dragonfly fauna of China. The eminent Swiss taxonomist Friedrich Ris published (1912 and 1916) two important papers on dragonflies of southern China and Taiwan of China with several new species. Altogether Ris described 25 new species (22 of them valid) from China, including the huge and famous *Chlorogomphus papilio*, described in 1927. At this time, in 1927-1928, the American entomologist James G. Needham from Cornell University spent one year in China visiting several universities and museums in different parts of the country and also conducting field work. In 1930 his handsomely printed book *A Manual of the Dragonflies of China* appeared, containing accounts of 266 species in 89 genera, of which 63 species were new to science. Although the book has many mistakes and omissions, and as many as one third of its new species have since proved to be synonyms, the volume was a very important contribution presenting the first treatment of the entire odonate fauna of our huge country. Needham had the benefit of studying material from all over China excepting Yunnan and Guizhou provinces. His book was successful in laying a good foundation for the systematic study of dragonflies in China. Other taxonomists publishing on Chinese Odonata and describing new species in 1910-1930 include Frederic C. Fraser, the prominent expert (on Indian odonates) and Kan Oguma. The well-known Japanese entomologist Syoziro Asahina published many papers on Chinese odonates between 1937-1997 concentrating mainly on the north of China, Zhejiang, Hunan, Hubei, Fujian, Sichuan, Taiwan and Hong Kong. He was a prolific author who published widely on the odonate fauna of many parts of Asia, describing numerous new species.

从18世纪中叶至20世纪中叶近200年的时间，中国蜻蜓的研究工作均由国外学者承担。国内最早涌现的蜻蜓学家是赵修复和陈锦文，在1947年底他们同时发表蜻蜓新种。截至1947年，国外学者根据中国蜻蜓标本已经发表257个新种（包含新亚种），但仅有174种至今仍然为有效种。1951年，赵修复（1917—2001）以《中国棍腹蜻蜓分类的研究》学位论文获得了美国马萨诸塞大学博士学位。他博士论文的部分内容（1953年发表）是有关双峰钩尾春蜓（现为双峰弯尾春蜓）的外部形态研究。随后赵修复回到福建农学院（现为福建农林大学）开始中国春蜓科的研究工作。1953—1955年，他陆续发表了题为《中国棍腹蜻蜓分类的研究》的系列论文。之后在1980—1990年他再次开展春蜓研究。1990年他出版了名为《中国春蜓分类》的专著，记述了中国春蜓145种，他所建立的春蜓科分类系统至今仍被广泛接受和采纳。赵修复在该专著中一共描述了65个新种，其中58种至今为有效种，主要集中在春蜓科。

From the mid-eighteenth to the mid-twentieth century, a period of nearly 200 years, systematic work on Chinese dragonflies was carried out solely by foreign scientists. The first Chinese entomologists to describe new species of Odonata were Hsiu-fu Chao and Chin-Wen Chen, both describing their first new species at the end of 1947. At that time a total of 257 new species (including a few subspecies) had been described by foreign authors from Chinese material, but only 174 of these taxa are presently ranked as valid full species. Hsiu-fu Chao (1917-2001) received his doctoral degree from the University of Massachusetts (United States) in 1951 after three years of study. Part of his thesis (published in 1953) was on the external morphology of the gomphid *Onychogomphus ardens* (presently known as *Melligomphus ardens*). Chao returned to Fujian Agricultural College (now Fujian Agriculture and Forestry University) and began to study the gomphid fauna of China, publishing a five part series of *Classification of the Chinese species of the family*

Gomphidae in 1953-1955. Later in 1980-1990 he returned to work on the Gomphidae. His seminal book '*The gomphid dragonflies of China*' was published in 1990. There he described and illustrated 145 species of Gomphidae from China. The book is an outstanding achievement and is still used by odonatologists and naturalists today. Altogether Chao described 65 new species or subspecies of Odonata, of which 58 are presently ranked as valid full species. Most of these are members of Gomphidae.

继赵修复之后，中国最有影响力的蜻蜓学家是周文豹。1979—2008年，他单独或合名发表43个新种，涉及华中和西南地区的多个类群，其中的35种至今有效。20世纪90年代初，投入蜻蜓研究的人数快速增长。至今天（2018年10月31日），国内共有65位昆虫学家参与了蜻蜓新种或新亚种的命名，其中25人为第一作者或者单独作者。在赵修复和周文豹之后，成果较多的作者包括张浩淼、朱慧倩、刘祖尧、于昕、徐奇涵和叶文琪等。于昕从2007年起开始分类学工作，研究类群主要集中在色蟌总科、蟌总科和丝蟌总科上，并发表了诸多新种。叶文琪主要从事蜓科的分类研究。徐奇涵专注于福建地区，主要研究内容包括蜓科和稚虫的分类。与此同时，国外专家继续发表中国蜻蜓报告，并经常与国内专家合作。最早的合作报告在1993年和1996年由杨兵和David A. L. Davies（英国）共同发表。英国蜻蜓学家Keith D. P. Wilson扎根香港，从1993年起，多次与Graham Reels、3次与许再福联名发表了多篇有关中国香港、海南、广东和广西的研究论文，描述了50多个新种。杨国辉主要从事云南蜻蜓区系和分类学研究，并曾和作者及其他作者一起合名发表云南蜻蜓新种。官昭瑛和比利时的蜻蜓专家Henri J. Dumont的研究主要集中在色蟌科的系统发育上。

After Chao, the second most prolific Chinese describer of new species is Wenbao Zhou, who in 1979-2008 authored or co-authored 43 species of various dragonfly and damselfly families from Central and Southwest China; of these 35 are presently ranked as valid full species. In the beginning of 1990's the flow of new species descriptions by Chinese authors started to increase and it has greatly accelerated during this century. Altogether, up to now (31 October 2018) a total of 65 Chinese entomologists have authored or co-authored species or subspecies; 25 of them as the first or sole author. After Hsiu-fu Chao and Wenbao Zhou the most prolific describers have been Haomiao Zhang, Huiqian Zhu, Zuyao Liu, Xin Yu, Qihan Xu and Wen-Chi Yeh. Xin Yu began his taxonomic work in 2007, and has published several papers concentrating on the superfamilies Calopterygoidea, Coenagrionoidea and Lestoidea. Wen-Chi Yeh has worked mainly on Aeshnidae. Qihan Xu has mainly focused on odonates from Fujian, and published papers mainly on Aeshnidae and larvae. Also foreign authors have continued to describe new species from China, often jointly with their Chinese colleagues, the first joint descriptions (1993, 1996) were by Bing Yang and David A. L. Davies (England). The English odonatologist Keith D. P. Wilson based in Hong Kong, published a series of papers on the Odonata of Hong Kong, Guangdong, Guangxi, Hainan (1993-2008), often with Graham T. Reels and three times with Zaifu Xu. In these papers nearly 50 new species were described. Guohui Yang has studied on the Odonata fauna of Yunnan, and published some new species from Yunnan with me and others. Zhaoying Guan, usually with Henri J. Dumont from Belgium, has published papers mainly on phylogeny of the family Calopterygidae.

继Needham的《中国蜻蜓手册》之后，第二本覆盖整个中国地区的蜻蜓专著是1984年隋敬之和孙洪国编著的《中国习见蜻蜓》。书中包含了分类特征绘图和26个质量较低的标本照黑白图版，共描述了中国常见蜻蜓208种。之后华南地区的蜻蜓通过彩色图鉴向世人展现。2003年Keith D. P. Wilson出版了《香港蜻蜓图鉴》，记载了香港蜻蜓111种。有关台湾地区蜻蜓的著作中，比较有代表性的是2000年汪仲良出版的《台湾蜻蜓》。两本有关中国广东地区的蜻蜓彩色画册分别是2012年吴宏道出版的《惠州蜻蜓》以及2014年崔晓东等人出版的《中国·南昆山蜻蜓》。2015年Graham Reels和张浩淼合著了《蜻蟌之地——海南蜻蜓图鉴》，是一本关于海南165种蜻蜓的野外识别手册，其中包括了许多海南特有种。

Since Needham's 1930 *Manual*, the next book to cover all odonate families in China was *Common species of dragonflies from China* (in Chinese) by Jingzhi Sui and Hongguo Sun, published in 1984. The book, including structural drawings and 26 low quality black & white plates of specimen photos, covers 208 species. The rich fauna of southern

China has later been documented in several books, often lavishly illustrated with color photos taken in the field. Keith D. P. Wilson wrote his books on Hong Kong species: *Field Guide to the dragonflies of Hong Kong* (2003), documenting 111 species. Several books have been published on the odonates in taiwan, most notably *Dragonflies of Taiwan* by Liang-Jong Wang (2000). Two books with plenty of color photos treat the Guangdong fauna: *Huizhou Dragonflies* by Hongdao Wu (2012) and *Dragonflies of Nankunshan, China* by Xiaodong Cui *et al.* (2014). *A field guide to the dragonflies of Hainan* by Graham Reels & Haomiao Zhang (2015) enables identification of most of the 165 species recorded from this tropical island with many endemic species.

作者于2008年开始投入蜻蜓分类学工作，至今已经发表蜻蜓研究报告30余篇，新种32种。这期间作者与国内外多名蜻蜓学家展开研究。早在2008年Graham Reels（英国）推荐作者与从事东南亚地区的研究专家Matti Hämäläinen（芬兰）展开合作，Matti的成果集中在色蟌总科。作者与Matti已合名发表了一些产自中国的色蟌、鼻蟌和溪蟌新种。Oleg E. Kosterin（俄罗斯）主要在柬埔寨开展研究，与作者主要在春蜓科和大伪蜻科一同合作。Rory Dow（英国）主要研究东南亚地区的小型豆娘，近日与作者一同建立了扁蟌科新属——云扁蟌属。

I began to study the taxonomy of dragonflies in 2008 and have so far published over 30 papers describing 32 new species. I have collaborated with many odonatologists both from abroad and from China. In 2008 Graham Reels (England) introduced me to Matti Hämäläinen (Finland), an expert of southeast Asian Odonata, especially of the superfamily Calopterygoidea, with whom I have described several new calopterygid, chlorocyphid and euphaeid species from China. With Oleg Kosterin (Russia), who has recently being studying the odonate fauna of Cambodia, I am working on some genera of Gomphidae and Macromiidae. With Rory Dow (England), an expert on Odonata of southeast Asia, I am working on several groups of small damselflies. We have recently erected a new genus in the family Platystictidae: *Yunnanosticta* Dow & Zhang, 2018.

随着分子生物学的兴起，全世界的蜻蜓研究正日益深入。2014年，Dijkstra等人通过对16S rRNA和28S rRNA提取分析所获得的结果对束翅亚目分类系统重新厘定。在2015年，Carle 等人对差翅亚目的分类系统也进行了修订。这些研究大幅度改进了之前的分类系统，比如Davies和Tobin（1984, 1985）的世界蜻蜓名录。21世纪以来，中国国内的蜻蜓研究报告日益增多，但多是基于一些简单的本底调查所总结的区系报告，缺乏系统性，而且无法验证物种鉴定的准确性。中国幅员辽阔，物种丰富，是全世界蜻蜓物种多样性最高的国家之一。然而目前中国蜻蜓的本底资源调查工作严重不足，许多地区从未开展过野外考察，多数地区的调查不够深入。因此急需更多的研究力量，开展对中国蜻蜓的系统性研究工作。

The development of molecular systematics has opened a new area in the systematics and classification of Odonata. In 2014, Dijkstra *et al.* provided a revised classification of Zygoptera based largely on the use of 16S rRNA and 28S rRNA genes. Then in 2015, Carle *et al.* gave a further revision of classification of Anisoptera. Those works providing a dramatic revision of previous classifications, such as that used by Davies & Tobin (1984, 1985) in their World Odonata checklists. Since the beginning of this century, an increasing number of reports on dragonflies from various parts of China have been published. Unfortunately most of these are faunal lists based on short surveys and lack a detailed systematic study; it is uncertain how correct the identifications are. China possibly has the highest diversity of dragonfly species in the world, but currently investigations of dragonflies are very inadequate. A large part of China has never been surveyed at all, and few areas have been comprehensively surveyed. Thus, more researchers are urgently needed for the systematic study of Chinese Odonata.

中国蜻蜓区系与生物地理 Odonata fauna of China and biogeography

中国地处太平洋西岸、亚洲东部，陆地面积约为960万平方千米，约占世界陆地总面积的1/15。全国共划分为23个省、5个自治区、4个直辖市和2个特别行政区。地势西高东低、复杂多样。山地、高原和丘陵约占陆地面积的67%，盆地和平原约占陆地面积的33%。气候复杂多样，从南到北跨热带、亚热带、暖温带、中温带、寒温带等气候带。淡水资源丰富，河流湖泊遍及全国，第一大河流长江，全长6300千米，为世界第三长河。

从世界动物地理区划上看，中国地跨古北和东洋两界，经度和纬度的跨度都很大。由于地形复杂，气候多变，水系发达，几乎涵盖了温带、亚热带和热带的各种生境类型，因此蜻蜓资源非常丰富。中国蜻蜓的分布特点按照中国昆虫地理区划有如下特征。

东北区：包括大小兴安岭、张广才岭、老爷岭、长白山以及辽河平原等地区。气候较寒冷。主要分布类群包括蜓科、春蜓科、伪蜻科、蜻科、色蟌科、丝蟌科、蟌科和扇蟌科等一些喜欢寒冷气候的古北种。

华北区：北界东起燕山山地、张北台地、吕梁山、六盘山北部，向西至祁连山脉东部，南抵秦岭、淮河，东临黄河、渤海，包括黄土高原、冀热山地及黄淮平原。气候夏热冬寒，四季显著。物种多样性不高，拥有一些中国特有种，例如著名的北京角臀大蜓、山西黑额蜓、长者头蜓和一些春蜓科种类。

青藏区：由帕米尔高原向东延伸，到其北缘的祁连山，南界为喜马拉雅山，东与东南则以四川西部及云贵高原西北部高山及康滇峡谷森林草地相隔。本区昆虫大多属喜马拉雅区系的东方种。目前对本区开展的蜻蜓调查非常匮乏。

蒙新区：包括内蒙古高原、河西走廊、塔里木、准噶尔盆地和天山山地，在大兴安岭以西，大青山以北，由呼伦贝尔草原直到新疆西陲。气候半干燥。主要分布着一些广布的古北种，包括蜓科、春蜓科、伪蜻科、蜻科、色蟌科、丝蟌科和蟌科。

华中区：包括四川盆地及长江流域各省，西部北起秦岭，东部为长江中、下游，包括东南沿海丘陵的半部，南面与华南区相邻。亚热带暖湿气候。为古北界与东洋界的交错地带，物种较丰富。中国特有种比例高，蜓科、春蜓科尤为突出。

西南区：包括四川西部、昌都东部、北起青海、甘肃南缘，南抵云南中北部，向西达藏东喜马拉雅南坡针叶林带以下山地。昆虫组成既非常复杂，又最丰富。中国蜻蜓总数的一半以上分布于此区域，仅云南省已发现超过400种蜻蜓。

华南区：包括广东、广西、海南和云南南部、福建东南沿海、台湾及南海各岛。气候温和，雨量丰富，生长着大量的热带雨林和亚热带森林。蜻蜓物种多样性仅次于西南区，也是中国诸多旗舰性物种的主要分布区域，如海南特有种丽拟丝蟌和体型巨大的蝴蝶裂唇蜓。

China is located in east Asia on the north-western side of the Pacific Ocean and has a land area of 9 600 thousand square kilometers, about 1/15 of the world land area. It is divided into 23 provinces, five autonomous regions, four municipalities directly under control by the Central Government, and two special administrative regions. The complex and diverse terrain ranges from very high in the west to low in the east. Mountains, plateaux and hills make up about 67% of the land area while the remaining 33% consists of lowland habitats. The climate is diverse and supports many biomes with diverse habitats. Climate zones, from south to the north, include tropical, subtropical, warm temperate, temperate and cold temperate. Fresh water resources are extremely rich, rivers and lakes are widely distributed all over the country, the largest river, the Yangtse River, is about 6300 kilometers long and is the third longest in the world.

China falls within both the Palearctic and Oriental Realms, and occupies a broad span of both longitude and latitude. Owing to its complex terrain, diverse climate and abundant freshwater resource, China has a diverse odonate fauna. Following the Chinese Insect Geographical Division the distribution of Odonata within China is as follows.

Northeast China Region: including the Daxinganling Mountains, Xiaoxinganling Mountains, Zhangguangcai Mountains, Laoye Mountains, Changbai Mountains and the Liaohe Plain. The climate is cold. Palaearctic species tolerant of cold weather, including species of Aeshnidae, Gomphidae, Corduliidae, Libellulidae, Calopterygidae, Lestidae, Coenagrionidae and Platycnemididae.

North China Region: the northern boundary east from the Yanshan Mountains, Zhangbei Platform, Lvliang

Mountains, and northern Liupan Mountains, west to the east of the Qilian Mountains, south to the Qinling Mountains and the Huaihe River, east to the Yellow River and the Bohai Sea, including the Loess Plateau, the Hebei Hot Mountains and the Huanghuai Plain. Seasons are pronounced here, summer being hot and winter cold. Species diversity is not high here, with a few endemic species including *Neallogaster pekinensis*, *Planaeschna shanxiensis*, *Cephalaeschna patrorum* and some gomphids.

Qinghai-Tibet Region: from the Pamir Plateau extending eastward, north to the Qilian Mountains, south to the Himalaya, east and southeast to high mountains in the west of Sichuan and the northwest of the Yunnan-Guizhou Plateau and forest grassland in the Kang Dian Canyon. Insects in this area belong to the Oriental component of the Himalayan fauna. To date dragonfly surveys of this area have been very inadequate.

Inner Mongolia-Xinjiang Region: including the Inner Mongolia Plateau, the Hexi Corridor, the Tarim, the Junggar Basin and the Tianshan Mountains, to the west of Da Hinggan Mountains, to the north of the Daqing Mountains, from the Hulunbuir Pasture Land to the west of Xinjiang. The climate is semi-arid. Some widespread Palaearctic species occur here, mainly including species of Aeshnidae, Gomphidae, Corduliidae, Libellulidae, Calopterygidae, Lestidae and Coenagrionidae.

Central China Region: including the Sichuan Basin and the provinces of the Yangtze River Basin, the west boundary north from the Qinling Mountains, the eastern part is the middle and lower reaches of the Yangtze River, including half of the hills of the southeast coast, south adjacent to South China Region. The climate is subtropical warm and wet. Located between the Palearctic and the Oriental Realms, this region has a diverse assemblage of species with a high proportion of endemics, especially in the families Aeshnidae and Gomphidae.

Southwest China Region: including the west of Sichuan, east of Changdu, north from Qinghai and the south of Gansu, south to north central Yunnan, west to the mountains below the coniferous forest belt of the Himalaya in the east of Tibet. The fauna is the richest in China but the affinities are very complex. More than half the odonate species of China occur here, and over 400 species have been recorded from Yunnan Province alone.

South China Region: including Guangdong, Guangxi, Hainan, the south of Yunnan, southeast coast of Fujian, Taiwan and islands in the South China Sea. This area has a mild climate and plentiful precipitation, with several areas of tropical rain forest and subtropical forest. Diversity of odonates is second only next to the Southwest China Region, but this is the main range for many flagship species, for example the unique Hainanese endemic species *Pseudolestes mirabilis* as well as the huge and charismatic *Chlorogomphus papilio*.

蜻蜓之地——中国云南 A paradise for dragonflies–Yunnan province

云南省位于中国西南边陲，云贵高原和青藏高原的结合部。西部与缅甸接壤，南部和老挝、越南毗邻。云南地形复杂，尤以山地为主，占总面积的94%。境内山脉河流纵横交错，主要山脉包括横断山、高黎贡山、怒山、云岭、哀牢山、无量山等，主要河流包括金沙江、澜沧江、怒江、元江等，境内有高原湖泊40余个，最大的湖泊是滇池和洱海。全省海拔最高点为梅里雪山，其最高峰海拔6740 m，而东部地区拥有海拔不足100 m的低地，差异显著。

云南省气候条件优越，北回归线横贯南部。气候在水平方向上，从北到南，相当于中国从黑龙江到海南岛的气候差异；在垂直方向上，由于众多山脉的海拔起伏较大，通常一座山脉可以感受几种不同的气候类型，即“一山分四季，十里不同天”。全省气候受到西南季风的显著影响，旱季和雨季分明，每年5—10月为雨季，11月至次年4月为旱季，西部尤为突出。在西部的德宏州、普洱市西部地区，受到西南季风的深度影响，而南部的西双版纳州和东部的红河州，同时受到西南季风和东南季风的影响。每一处的气候均有差异。

作者从2009年开始在云南进行蜻蜓考察，亲身感受到了云南蜻蜓区系的魅力并为之震撼。最初的科学考察都集中在西双版纳州，从2009—2012年考察期间，记录了西双版纳州近160种蜻蜓，每次开展新的考察都会有新发现。西双版纳州的蜻蜓区系和泰国北部非常接近，大量的中国新记录种被发现。随后在红河州的搜索也收获颇丰，许多描述于越南的蜻蜓都可以在金平县、河口县以及相邻的文山州发现。从2013年起，作者开始大规模考察滇西和滇西北地区，包括丽江市、大理州、德宏州、普洱市和临沧市，每年有半年的时间在云南工作。结果很明显，在西部边缘，渗入了大量印缅区系的物种。以德宏州的盈江县为例，从县城到边境的那邦镇大概有90 km的山路，这期间要穿过著名的铜壁关国家级自然保护区。那邦镇是一块低地河谷，海拔最低点不足200 m，是很多热带物种在中国唯一的可见区域。从那邦至铜壁关的山路很曲折，绕山而行，海拔高度迅速提升，在40 km后迅速升高至1300 m，在陡峭的山体和神秘的栖息环境中，隐藏着大量未知物种。仅在2013年7月的一次考察中，盈江县就记录到蜻蜓超过130种。普洱市西部、临沧市亦是如此。作者汇总了一份几年间在云南考察的蜻蜓名录（未发表），超过440种蜻蜓已经被发现，然而对于整个云南省来说，考察工作才仅仅是个开始。当然，全面和系统性地研究云南蜻蜓，可能还需要几十年甚至更长的时间。但毫无疑问的是，云南省是全世界最重要的蜻蜓栖息地之一，名副其实的“蜻蜓之地”。

Yunnan Province is located in the southwestern part of China, where the Yunnan-Guizhou Plateau merges with the Tibetan Plateau. The western border is adjacent to Myanmar, and the southern border with Laos and Vietnam. The terrain is complex with up to 94% of the land mountainous and many intersecting rivers. Prominent landmarks include Mt. Hengduan, Mt. Gaoligong, Mt. Nushan, Mt. Yunling, Mt. Ailao and Mt. Wuliang. Major rivers include the Jinsha, Lantsang, Nujiang and Yuanjiang Rivers. The highest peak, about 6740 m high, is the Meri Snow Mountain in the northwest and the lowest elevations of less than 100 m occur in the east. This is an enormous altitudinal range.

The climate of the province is generally mild, the southern part being traversed by the Tropic of Cancer. The climatic variation with altitude is equivalent to a similar latitudinal climate change from Heilongjiang to Hainan. Different climatic regimes occur on a single mountain depending on elevation and aspect. The entire province is affected by the Southwest monsoon, with the rainy season from May to October, and the dry season from November to April. At its western border, the impact of Southwest monsoon is more pronounced and the southern and eastern parts are also affected by Southeast monsoon. Climatic variation is clearly seen in different locations.

My fieldwork in Yunnan began in 2009, where I witnessed an extraordinary diversity of dragonflies. My preliminary fieldwork concentrated mainly on the Xishuangbanna Autonomous Prefecture. From 2009 to 2012, I found there 160 species and still more species were recorded during later surveys. The Odonata fauna of Xishuangbanna is very similar to that of northern Thailand, and many species are the new records for the country. Similarly species described from North Vietnam can be found in Jingping and Hekou counties of Honghe Autonomous Prefecture and adjacent Weshan Autonomous Prefecture. Since 2013 my fieldwork expanded to include the western and northwestern portions of Yunnan, including Lijiang, Dali, Dehong, Pu'er and Lincang, to which I allocated half my time spent in fieldwork every year. All of these surveys along the western border were productive, with species previously recorded only from Myanmar and India being recorded. One example involves the surveys in Yingjiang County of Dehong Autonomous Prefecture. The distance from the Yingjiang County to Nabang Town is about 90 km and across the Tongbiguan National Reserve. Nabang Town is low, an altitude of less than 200 m, it is the only location for many tropical species in China. High

西双版纳热带植物园
Xishuangbanna Tropical Botanic Garden

mountains, about 40 km from Nabang, rise to an altitude of about 1300 m and many previously unknown species with secretive habits were found along the altitudinal gradient. The survey in July 2013, a trip to Yingjiang County yielded 130 species. Western Pu'er and Lincang cities provided similar collecting results. My current list of species from Yunnan includes over 440 species but compared to the large size of the province, my fieldwork represent only a beginning. Certainly the complete and systematic study on dragonflies of Yunnan will take decades or even longer. But no doubt Yunnan is one of the most important dragonfly habitats in the world, a true paradise for odonates.

中缅边境上的绿洲——德宏铜壁关国家级自然保护区
The oasis on Sino-Burmese border—Tongbiguan National Nature Reserve of Dehong autonomous prefecture

北回归线上的绿洲——云南普洱，世界茶源
The oasis on the Tropic of Cancer—The world tea source, Pu'er in Yunnan

中国蜻蜓分类系统 The classification of Odonata from China

本书按照Dijkstra等（2013，2014）和Carle等（2015）对分类系统的修订，将中国蜻蜓归入3亚目23科中。由于原来的山蟌科已经被重新分类，但部分属的关系尚未明确，未被进行科级归类，它们被归入“分类位置待定”。这其中包括中国豆娘6属30余种。此处将它们放在“色蟌总科待定科”描述。

Based on the latest revision of classification by Dijkstra *et al.* (2013, 2014) and Carle *et al.* (2015), the Chinese species are allocated among 23 families within three suborders. The former family Megapodagrionidae was recently reclassified. But relationships of several genera are uncertain and they are technically designated "*incertae sedis*". This mixed group includes over 30 Chinese species in six genera. For convenience in the main text these genera are still kept under the family name "Calopterygoidea *incertae sedis*".

蜻蜓目 Order Odonata Fabricius, 1793

差翅亚目 Suborder Anisoptera Selys, 1854

蜓总科 Superfamily Aeshnoidea Leach, 1815

蜓科 Family Aeshnidae Leach, 1815

春蜓总科 Superfamily Gomphoidea Rambur, 1842

春蜓科 Family Gomphidae Rambur, 1842

大蜓总科 Superfamily Cordulegastroidea Hagen, 1875

裂唇蜓科 Family Chlorogomphidae Needham, 1903

大蜓科 Family Cordulegastridae Hagen, 1875

蜻总科 Superfamily Libelluloidea Leach, 1815

伪蜻科 Family Corduliidae Selys, 1850

大伪蜻科 Family Macromiidae Needham, 1903

综蜻科 Family Synthemistidae Tillyard, 1911

蜻科 Family Libellulidae Leach, 1815

间翅亚目 Suborder Anisozygoptera Handlirsch, 1906

昔蜓总科 Superfamily Epiophlebioidea Muttkowski, 1910

昔蜓科 Family Epiophlebiidae Muttkowski, 1910

束翅亚目 Suborder Zygoptera Selys, 1854

色蟌总科 Superfamily Calopterygoidea Selys, 1850

丽蟌科 Family Devadattidae Dijkstra & al., 2014

色蟌科 Family Calopterygidae Selys, 1850

鼻蟌科 Family Chlorocyphidae Cowley, 1937

溪蟌科 Family Euphaeidae Yakobson& Bianchi, 1905

大溪蟌科 Family Philogangidae Kennedy, 1920

黑山蟌科 Family Philosinidae Kennedy, 1925

野山蟌科 Family Argiolestidae Fraser, 1957

色蟌总科待定科 Family "Calopterygoidea *incertae sedis*"

拟丝蟌科 Family Pseudolestidae Fraser, 1957

丝蟌总科 Superfamily Lestoidea Calvert, 1901

丝蟌科 Family Lestidae Calvert, 1901

综蟌科 Family Synlestidae Tillyard, 1917

蟌总科 Superfamily Coenagrionoidea Kirby, 1890

扇蟌科 Family Platycnemididae Yakobson& Bianchi, 1905

蟌科 Family Coenagrionidae Kirby, 1890

扁蟌总科 Superfamily Platystictoidea Kennedy, 1920

扁蟌科 Family Platystictidae Kennedy, 1920

中国蜻蜓多样性数据 Diversity of dragonflies from China

根据作者的蜻蜓收藏和历史资料，此处汇总了一份“中国蜻蜓名录”。这份名录包括中国蜻蜓目昆虫830个确定种（含7个存疑种）、25个亚种（如果1个种内包含多个亚种，第1个亚种被统计为“种数”，其余亚种被统计为“亚种数”）和128个待定种，总计3亚目23科175属983种（含亚种数）。此外还包含模式标本产自中国的114个异名，由Matti Hämäläinen厘定。束翅亚目包括14科66属344种（含4个亚种和46个待定种），间翅亚目1科1属3种（含1个待定种），差翅亚目8科108属636种（含21个亚种和81个待定种）。在855个确定种中（含亚种数），有462种根据产自中国的模式标本发表。

本书收录了中国蜻蜓3亚目22科172属820种（含亚种数），占中国已知蜻蜓总数的83.42%；其中束翅亚目13科65属293种，间翅亚目1科1属3种（以概述的形式描述），差翅亚目8科106属524种。叶山蟌属、似沼蜓属和曙春蜓属未做描述。

Based on my collection and historical records, a new checklist of Odonata from China is included below. The checklist includes 830 named full species (including seven dubious species), 25 additional subspecies (for each species only one ssp. is included in the full species accounts, the others are “additional”) and 128 species identified to genus level only, a total of 983 species-group taxa belonging to 175 genera in 23 families in three suborders. In addition it includes 114 synonymic names of species and subspecies originally described on type specimens from China; these synonymic names were contributed by Matti Hämäläinen. In suborder Zygoptera a total of 344 species-group taxa (including 4 additional subspecies and 46 taxa unidentified to the species level) are listed; they belong to 66 genera in 14 families. In suborder Anisozygoptera three species (one unidentified) are listed representing one genus and one family. In suborder Anisoptera 636 species-group taxa (including 21 additional subspecies and 81 taxa unidentified to the species level) are listed; they belonging to 108 genera in 8 families. Of the valid 855 full species listed in the checklist, a total of 462 species have been originally described from specimens from China.

In this book, a total of 820 species or subspecies, representing three suborders, 22 families and 172 genera are described and illustrated, 83.42% of the known species-group taxa from China. Of these the Zygoptera includes 293 species in 65 genera and 13 families, Anisozygoptera includes three species (in general introduction), and the Anisoptera includes 524 species in 106 genera and 8 families. The genera *Podolestes* Selys, 1862, *Oligoaeschna* Selys, 1889 and *Eogomphus* Needham, 1941 are not included in this volume.

分类学位置的变动 Taxonomic changes incorporated in this book

1. 赵氏佩蜓之前被放在头蜓属，根据雌性产卵管的形态将其移入佩蜓属。

Periaeschna chaoi (Asahina, 1982), previously placed in genus *Cephalaeschna*, is moved to genus *Periaeschna*, based on the morphology of the female ovipositor.

2. 赵氏多棘蜓作为红褐多棘蜓的异名处理。

Polycanthagyna chaoi Yang & Li, 1994 is considered a junior synonym of *Polycanthagyna erythromelas* (McLachlan, 1896).

3. 福建小叶春蜓之前被认为是并纹小叶春蜓的亚种，根据雄性后钩片及雌性下生殖板的构造将其提升至种。

Gomphidia fukienensis Chao, 1955 has previously been regarded as a subspecies of *Gomphidia kruegeri* Martin, 1904, but based on the structure of male posterior hamule and female vulvar lamina it is raised to species level here.

4. 高山环尾春蜓作为黄尾环尾春蜓的异名处理。

Lamelligomphus laetus Yang & Davies, 1993 is considered a junior synonym of *Lamelligomphus biforceps* (Selys, 1878).

5. 泰国长足春蜓之前被放在异春蜓属，本属根据雄性肛附器特征将其移入长足春蜓属。

Merogomphus pinratani (Hämäläinen, 1991), previously placed in genus *Anisogomphus*, is moved to genus *Merogomphus*, based on morphology of the male caudal appendages.

6. 黎氏日春蜓作为汤氏日春蜓的异名处理。

Nihonogomphus lieftincki Chao, 1954 is considered a junior synonym of *Nihonogomphus thomassoni* (Kirby, 1900).

7. 赵氏显春蜓已经作为细尾显春蜓的异名处理，根据雌性下生殖板构造上的差异，本书恢复其有效种的地位。

Phaenandrogomphus chaoi Zhu & Liang, 1994, previously ranked by Wilson & Xu (2009) as a junior synonym of *Phaenandrogomphus tonkinicus* (Fraser, 1926), is here retained as a valid species, based on the structure of female vulvar lamina.

8. 侗族裂唇蜓作为铃木裂唇蜓的异名处理。

Chlorogomphus tunti Needham, 1930 is considered a junior synonym of *Chlorogomphus suzukii* (Oguma, 1926).

9. 朴氏裂唇蜓之前被放在山裂唇蜓亚属，此处将其移入华裂唇蜓亚属。

Chlorogomphus (*Sinorogomphus*) *piaoacensis* Karube, 2013, previously placed in subgenus *Orogomphus*, is moved to subgenus *Sinorogomphus*.

10. 高翔裂唇蜓之前被放在凹尾裂唇蜓属，此处将其移入楔尾裂唇蜓属。

Watanabeopetalia (*Matsumotopetalia*) *soarer* (Wilson, 2002), previously placed in genus *Chloropetalia*, is moved to genus *Watanabeopetalia*.

11. 中国的大蜓属种类都被移入角臀大蜓属。

All Chinese species of *Cordulegaster* are moved to genus *Neallogaster*.

12. 版纳丽大伪蜻作为黄斑丽大伪蜻的异名处理。

Epophthalmia bannaensis Zha & Jiang, 2010 is considered a junior synonym of *Epophthalmia frontalis* Selys, 1871.

13. 蓝额疏脉蜻之前被认为是褐胸疏脉蜻的亚种，本书将其提升到种。

Brachydiplax flavovittata Ris, 1911, previously considered a subspecies of *Brachydiplax chalybea* Brauer, 1868, is raised to species level.

14. 喜马赤蜻作为条斑赤蜻的亚种处理。

Sympetrum commixtum (Selys, 1884) is considered a subspecies of *Sympetrum striolatum* (Charpentier, 1840).

种类描述的信息说明 Explanation of species accounts

栖息环境 Habitat

大多数物种的栖息地概括是根据作者本人的野外调查，少见种的信息综合了文献的记录。

The information provided on habitats for most species is based on my own observations or is obtained from the literature, especially for poorly known species.

分布 Distribution

本书所包含物种在国内的分布信息以省级单位列出，云南省除外。物种在中国云南的分布信息提供到保护区或者市县级。物种在中国以外的分布信息以分号"；"与其在国内的分布分隔，以国家级单位或者一般性的概括列出。

对于一些广布种，有时用地理区划来简要概括。本书中这些地理区划所代表的省份如下：

东北：黑龙江、辽宁、吉林；

华北：北京、天津、河北、山西、内蒙古；

西北：陕西、甘肃、青海、宁夏、新疆；

华中：山东、河南、湖北、湖南、上海、江苏、浙江、安徽、江西、福建、台湾；

西南：西藏、云南、贵州、四川、重庆；

华南：广西、广东、海南、香港、澳门。

Distribution within China is listed at the level of province, with the exception of Yunnan Province. For Yunnan the autonomous prefecture, city or nature reserve are also given. Distribution outside China follows after a semicolon, either by country level or in more general terms.

For some widespread species, the distribution range is summarized by the geographical regions. The provinces included in each regions are as follow:

Northeast: Heilongjing, Liaoning, Jilin;

North: Beijing, Tianjin, Hebei, Shanxi, Inner Mongolia;

Northwest: Shaanxi, Gansu, Qinghai, Ningxia, Xinjiang;

Central: Shandong, Henan, Hubei, Hunan, Shanghai, Jiangsu, Zhejiang, Anhui, Jiangxi, Fujian, Taiwan;

Southwest: Tibet, Yunnan, Guizhou, Sichuan, Chongqing;

South: Guangxi, Guangdong, Hainan, Hong Kong, Macau.

飞行期 Flight season

飞行期是根据个人野外考察并结合历史信息，给出物种的最长飞行期。对于广布种而言，其飞行期从南至北逐渐缩短，北方出现月份滞后于南方。

Flight season data are based both on my own observations and published data, encompassing the greatest range for each species. It should be kept in mind that in widespread species the flight season usually starts later and is shorter in the more northerly part of their range.

照片 Photos

大多数的彩色生态照为作者野外工作所获。所有照片均给出拍摄地点信息，作者本人拍摄的照片省略了姓名标注，其他来源的照片标注了摄影师姓名。所有的分类特征彩色图版由作者根据个人收藏的实体标本拍摄所获。

为本书提供照片的国内外摄影师：Adolfo Cordero-Rivera（西班牙）、Graham Reels（英国）、Oleg E.Kosterin（俄罗斯）、Matti Hämäläinen（芬兰）、Sami Karjalainen（芬兰）、酒井正次（日本）、Tom Kompier（荷兰）、Jörg Arlt（德国）、莫善濂、吴宏道、宋黎明、宋睿斌、金洪光、秦彧、吴超、陈尽、陈炜、张巍巍、计云、齐天博、安起迪、钟彦葵、蒋先兰、袁屏、黄海燕、刘辉、梁嘉景、吕非、祁麟峰、苏毅雄、张智民和嘎嘎。

Most photographs of living adults were taken by me during my field work. Locality data accompany all photos. Unless otherwise stated, all photographs were taken by me. All color plates showing the structural details for identification were taken from the specimens in my collection.

Photographers who contributed nice photos to the book: Adolfo Cordero-Rivera (Spain), Graham Reels (United Kingdom), Oleg E.Kosterin (Russia), Matti Hämäläinen (Finland), Sami Karjalainen (Finland), Shoji Sakai (Japan), Tom Kompier (the Netherlands), Jörg Arlt (Germany), Shanlian Mo, Hongdao Wu, Liming Song, Ruibin Song, Hongguang Jin, Yu Qin, Chao Wu, Jin Chen, Wei Chen, Weiwei Zhang, Yun Ji, Tianbo Qi, Qidi An, Yankui Zhong, Xianlan Jiang, Ping Yuan, Haiyan Huang, Hui Liu, Kenneth Leung, Fei Lv, Mahler Ka, Samson So, CheungChi Man and Gaga.

体长 Measurements

绝大多数物种的体长信息是根据作者本人的蜻蜓收藏实体标本测量所得。若某物种的标本未在作者本人的收藏中，则以历史资料为准。体长的范围包括雌雄两性。

All measurements are taken from specimens in the author's Odonata collection. Species absent in the collection are based on published data. Range of measurements includes both sexes.

大小的定义 Size

体长在40 mm以下为小型，体长在40～70 mm为中型，体长在70～100 mm为大型，体长超过100 mm为巨型。

Total body lengths shorter than 40 mm are considered "small-sized", those between 40-70 mm "medium-sized", those between 70-100 mm "large-sized", and those exceeding 100 mm "huge".

缩写 Abbreviations

腹部第1～10节。

S1-10 abdominal segments 1-10.

工作小组和个人收藏 Working team and collection

蜻蜓工作小组 Working team

蜻蜓工作小组的大力协助使图鉴更全面，他们是来自全国各地的昆虫摄影师：莫善濂、吴宏道及其夫人严翠珍、宋黎明及其子宋睿斌和金洪光先生。

从1998年至今，工作小组已先后对中国黑龙江、辽宁、吉林、内蒙古、北京、天津、河北、山东、安徽、浙江、江苏、湖南、湖北、广东、广西、海南、四川、重庆、云南、贵州等20余省市进行了野外考察。考察期间拍摄了大量的生态照片，并获得了许多有价值的标本。

Members of the working team who helped make this volume more complete came are from all over China: Mr. Shanlian Mo, Mr. Hongdao Wu and his wife Mrs. Cuizhen Yan, Mr. Liming Song and his son Mr. Ruibin Song, and Mr. Hongguang Jin.

From 1998 we conducted surveys in provinces or provincial administrative regions, including: Heilongjiang, Jilin,

Liaoning, Inner Mongolia, Beijing, Tianjin, Hebei, Shandong, Anhui, Zhejiang, Jiangsu, Hunan, Hubei, Guangdong, Guangxi, Hainan, Sichuan, Chongqing, Yunnan, Guizhou, etc. We were able to obtain many nice photos as well as valuable specimens.

张浩淼的世界蜻蜓收藏 Haomiao Zhang's World Odonata Collection

目前作者收藏了来自世界各地的蜻蜓标本近2000种，超过2万号，其中中国地区共计800余种，是世界最齐全的中国蜻蜓标本收藏，对于研究蜻蜓目的分类学、生态学和系统发育有重要价值。

The Haomiao Zhang's World Odonata Collection contains about 2000 species, over 20000 specimens from all over the world with over 800 species from China, and ranks as one of the most complete collections of Chinese Odonata in the world and is a valuable resource for future studies involving the taxonomy, ecology and phylogeny of the order.

来自张浩淼世界蜻蜓收藏的豆娘标本
Damselfly specimens from Haomiao Zhang's World Odonata Collection

中国蜻蜓名录 A checklist of Odonata from China

本书未包含的种类以红色标记并以省级单位列出其在中国的分布。本名录所包含的异名仅为模式标本产自中国的物种，异名的字号略小于有效名。身份存疑的物种被列于名单的最后。

Species not included in this book are marked with red and prorinces in China where recorded listed. Only those synonymic names are included, which are based on type specimens collected in China, the synonymic names are a bit smaller than the valid names. Dubious species from China are listed in the end.

束翅亚目（豆娘） Suborder Zygoptera (Damselfly)
（14科66属337种4亚种 14 families 66 genera 337 species 4 subspecies）

德丽蟌科 Family Devadattidae （1属1种 1 genus 1 species）

德丽蟌属 Genus *Devadatta* Kirby, 1890

褐顶德丽蟌 *Devadatta ducatrix* Lieftinck, 1969

色蟌科 Family Calopterygidae （12属44种2亚种 12 genera 44 species 2 subspecies）

基色蟌属 Genus *Archineura* Kirby, 1894

霜基色蟌 *Archineura hetaerinoides* (Fraser, 1933)

赤基色蟌 *Archineura incarnata* (Karsch, 1892)

Syn. *Archineura basilactea* Kirby, 1894

暗色蟌属 Genus *Atrocalopteryx* Dumont, Vanfleteren, De Jonckheere & Weekers, 2005

黑暗色蟌 *Atrocalopteryx atrata* (Selys, 1853)

Syn. *Calopteryx grandaeva* Selys, 1853

Syn. *Calopteryx smaragdina* Selys, 1853

Syn. *Vestalis tristis* Navás, 1932

黑蓝暗色蟌 *Atrocalopteryx atrocyana* (Fraser, 1935)

越南暗色蟌 *Atrocalopteryx coomani* (Fraser, 1935)

褐带暗色蟌 *Atrocalopteryx fasciata* Yang, Hämäläinen & Zhang, 2014

黑顶暗色蟌指名亚种 *Atrocalopteryx melli melli* Ris, 1912

黑顶暗色蟌海南亚种 *Atrocalopteryx melli orohainani* Guan, Han & Dumont, 2012

透顶暗色蟌 *Atrocalopteryx oberthueri* (McLachlan, 1894)

Syn. *Agrion grahami* Needham, 1930

闪色蟌属 Genus *Caliphaea* Hagen, 1859

昂卡闪色蟌 *Caliphaea angka* Hämäläinen, 2003

绿闪色蟌 *Caliphaea confusa* Hagen, 1859

紫闪色蟌 *Caliphaea consimilis* McLachlan, 1894

亮闪色蟌 *Caliphaea nitens* Navás, 1934

泰国闪色蟌 *Caliphaea thailandica* Asahina, 1976

闪色蟌属待定种 *Caliphaea* sp.

色蟌属 Genus *Calopteryx* Leach, 1815

日本色蟌 *Calopteryx japonica* Selys, 1869

续表

华丽色蟌 *Calopteryx splendens* (Harris, 1780)
亮翅色蟌属 Genus *Echo* Selys, 1853 白背亮翅色蟌 *Echo candens* Zhang, Hämäläinen & Cai, 2015 黑顶亮翅色蟌 *Echo margarita* Selys, 1853 华丽亮翅色蟌 *Echo perornata* Yu & Hämäläinen, 2012
单脉色蟌属 Genus *Matrona* Selys, 1853 安妮单脉色蟌 *Matrona annina* Zhang & Hämäläinen, 2012 透顶单脉色蟌 *Matrona basilaris* Selys, 1853 Syn. *Matrona kricheldorffi* Karsch, 1892 褐单脉色蟌 *Matrona corephaea* Hämäläinen, Yu & Zhang, 2011 台湾单脉色蟌 *Matrona cyanoptera* Hämäläinen &Yeh, 2000 妈祖单脉色蟌 *Matrona mazu* Yu, Xue & Hämäläinen, 2015 黑单脉色蟌 *Matrona nigripectus* Selys, 1879 神女单脉色蟌 *Matrona oreades* Hämäläinen, Yu & Zhang, 2011
绿色蟌属 Genus *Mnais* Selys, 1853 安氏绿色蟌 *Mnais andersoni* McLachlan, 1873 黑带绿色蟌 *Mnais gregoryi* Fraser, 1924 Syn. *Mnais maclachlani* Fraser, 1924 Syn. *Mnais semiopaca* May, 1935 烟翅绿色蟌 *Mnais mneme* Ris, 1916 Syn. *Mnais earnshawi thoracicus* May, 1935 黄翅绿色蟌 *Mnais tenuis* Oguma, 1913 ?Syn. *Mnais decolorata* Bartenev, 1913 ?Syn. *Mnais auripennis* Needham, 1930 ?Syn. *Mnais pieli* Navás, 1936 绿色蟌属待定种1 *Mnais* sp. 1 绿色蟌属待定种2 *Mnais* sp. 2 绿色蟌属待定种3 *Mnais* sp. 3 绿色蟌属待定种4 *Mnais* sp. 4 绿色蟌属待定种5 *Mnais* sp. 5
艳色蟌属 Genus *Neurobasis* Selys, 1853 安氏艳色蟌 *Neurobasis anderssoni* Sjöstedt, 1926 华艳色蟌 *Neurobasis chinensis* (Linnaeus, 1758)
爱色蟌属 Genus *Noguchiphaea* Asahina, 1976 美子爱色蟌 *Noguchiphaea yoshikoae* Asahina, 1976
褐顶色蟌属 Genus *Psolodesmus* McLachlan, 1870 褐顶色蟌南台亚种 *Psolodesmus mandarinus dorothea* Williamson, 1904 褐顶色蟌指名亚种 *Psolodesmus mandarinus mandarinus* McLachlan, 1870

续表

黄细色蟌属 Genus *Vestalaria* May, 1935
苗黄细色蟌 ***Vestalaria miao*** (Wilson & Reels, 2001)
透翅黄细色蟌 ***Vestalaria smaragdina*** (Selys, 1879)
褐翅黄细色蟌 ***Vestalaria velata*** (Ris, 1912)
Syn. *Vestalis virens* Needham, 1930
黑角黄细色蟌 ***Vestalaria venusta*** (Hämäläinen, 2004)
细色蟌属 Genus *Vestalis* Selys, 1853
多横细色蟌 ***Vestalis gracilis*** (Rambur, 1842)
鼻蟌科 Family Chlorocyphidae (6属23种 6 genera 23 species)
圣鼻蟌属 Genus *Aristocypha* Laidlaw, 1950
蓝脊圣鼻蟌 ***Aristocypha aino*** Hämäläinen, Reels & Zhang, 2009
簾格圣鼻蟌 *Aristocypha baibarana* (Matsumura, 1931) [台湾 Taiwan]
赵氏圣鼻蟌 ***Aristocypha chaoi*** (Wilson, 2004)
西藏圣鼻蟌 ***Aristocypha cuneata*** (Selys, 1853)
黄脊圣鼻蟌 ***Aristocypha fenestrella*** (Rambur, 1842)
蓝纹圣鼻蟌 ***Aristocypha iridea*** (Selys, 1891)
四斑圣鼻蟌 ***Aristocypha quadrimaculata*** (Selys, 1853)
阳鼻蟌属 Genus *Heliocypha* Fraser, 1949
月斑阳鼻蟌 ***Heliocypha biforata*** (Selys, 1859)
三斑阳鼻蟌 ***Heliocypha perforata*** (Percheron, 1835)
Syn. *Rhinocypha whiteheadi* Kirby, 1900
Syn. *Rhynocyyha* (*Sic!*) *14-maculata* Oguma, 1913
Syn. *Rhinocypha maculata* Matsumura, 1931
Syn. *Heliocypha yunnanensis* Zhou & Zhou, 2004
隐鼻蟌属 Genus *Heterocypha* Laidlaw, 1950
印度隐鼻蟌 ***Heterocypha vitrinella*** (Fraser, 1935)
印鼻蟌属 Genus *Indocypha* Fraser, 1949
显著印鼻蟌 ***Indocypha catopta*** Zhang, Hämäläinen & Tong, 2010
蓝尾印鼻蟌 ***Indocypha cyanicauda*** Zhang & Hämäläinen, 2018
卡萨印鼻蟌 ***Indocypha katharina*** (Needham, 1930)
Syn. *Indocypha chishuiensis* Zhou & Zhou, 2006
红尾印鼻蟌 ***Indocypha silbergliedi*** Asahina, 1988
四川印鼻蟌 *Indocypha svenhedini* (Sjöstedt, 1932) [四川 Sichuan]
黑白印鼻蟌 ***Indocypha vittata*** (Selys, 1891)
隼蟌属 Genus *Libellago* Selys, 1840
点斑隼蟌 ***Libellago lineata*** (Burmeister, 1839)
鼻蟌属 Genus *Rhinocypha* Rambur, 1842
黄侧鼻蟌 ***Rhinocypha arguta*** Hämäläinen & Divasiri, 1997

续表

线纹鼻蟌 *Rhinocypha drusilla* Needham, 1930
Syn. *Indocypha maolanensis* Zhou & Bao, 2002
华氏鼻蟌 *Rhinocypha huai* (Zhou & Zhou, 2006)
翠顶鼻蟌 *Rhinocypha orea* Hämäläinen & Karube, 2001
台湾鼻蟌 *Rhinocypha taiwana* Wang & Chang, 2013
三纹鼻蟌 *Rhinocypha trimaculata* Selys, 1853 [西藏 Tibet]
溪蟌科 Family Euphaeidae (5属33种 5 genera 33 species)
异翅溪蟌属 Genus *Anisopleura* Selys, 1853
蓝斑异翅溪蟌 *Anisopleura furcata* Selys, 1891
斧尾异翅溪蟌 *Anisopleura pelecyphora* Zhang, Hämäläinen & Cai, 2014
庆元异翅溪蟌 *Anisopleura qingyuanensis* Zhou, 1982
三彩异翅溪蟌 *Anisopleura subplatystyla* Fraser, 1927
云南异翅溪蟌 *Anisopleura yunnanensis* Zhu & Zhou, 1999
郑氏异翅溪蟌 *Anisopleura zhengi* Yang, 1996 [陕西 Shaanxi]
尾溪蟌属 Genus *Bayadera* Selys, 1853
二齿尾溪蟌 *Bayadera bidentata* Needham, 1930
短尾尾溪蟌 *Bayadera brevicauda* Fraser, 1928
大陆尾溪蟌 *Bayadera continentalis* Asahina, 1973
斑翅尾溪蟌 *Bayadera fasciata* Sjöstedt, 1932 [四川 Sichuan]
钳尾溪蟌 *Bayadera forcipata* Needham, 1930 [四川 Sichuan]
墨端尾溪蟌 *Bayadera hatvan* Hämäläinen & Kompier, 2015
透翅尾溪蟌 *Bayadera hyalina* Selys, 1879
科氏尾溪蟌 *Bayadera kirbyi* Wilson & Reels, 2001
巨齿尾溪蟌 *Bayadera melanopteryx* Ris, 1912
Syn. *Bayadera melania* Navás, 1934
褐翅尾溪蟌 *Bayadera nephelopennis* Davies & Yang, 1996
锯突尾溪蟌 *Bayadera serrata* Davies & Yang, 1996
条斑尾溪蟌 *Bayadera strigata* Davies & Yang, 1996
隐溪蟌属 Genus *Cryptophaea* Hämäläinen, 2003
优雅隐溪蟌 *Cryptophaea saukra* Hämäläinen, 2003
越南隐溪蟌 *Cryptophaea vietnamensis* (Van Tol & Rozendaal, 1995)
云南隐溪蟌 *Cryptophaea yunnanensis* (Davies & Yang, 1996)
暗溪蟌属 Genus *Dysphaea* Selys, 1853
黑斑暗溪蟌 *Dysphaea basitincta* Martin, 1904
华丽暗溪蟌 *Dysphaea gloriosa* Fraser, 1938
浩淼暗溪蟌 *Dysphaea haomiao* Hämäläinen, 2012
溪蟌属 Genus *Euphaea* Selys, 1840
方带溪蟌 *Euphaea decorata* Hagen, 1853

续表

台湾溪蟌 *Euphaea formosa* Hagen, 1869

Syn. *Euphaea compar* McLachlan, 1870

绿翅溪蟌 *Euphaea guerini* Rambur, 1842

透顶溪蟌 *Euphaea masoni* Selys, 1879

黄翅溪蟌 *Euphaea ochracea* Selys, 1859

褐翅溪蟌 *Euphaea opaca* Selys, 1853

宽带溪蟌 *Euphaea ornata* (Campion, 1924)

华丽溪蟌 *Euphaea superba* Kimmins, 1936

溪蟌属待定种 *Euphaea* sp.

大溪蟌科 Family Philogangidae (**1属2种1亚种** 1 genus 2 species 1 subspecies)

大溪蟌属 Genus *Philoganga* Kirby, 1890

壮大溪蟌瑛凤亚种 *Philoganga robusta infantua* Yang & Li, 1994 [陕西 Shaanxi]

壮大溪蟌指名亚种 *Philoganga robusta robusta* Navás, 1936

大溪蟌 *Philoganga vetusta* Ris, 1912

黑山蟌科 Family Philosinidae (**2属3种** 2 genera 3 species)

黑山蟌属 Genus *Philosina* Ris, 1917

覆雪黑山蟌 *Philosina alba* Wilson, 1999

红尾黑山蟌 *Philosina buchi* Ris, 1917

鲨山蟌属 Genus *Rhinagrion* Calvert, 1913

海南鲨山蟌 *Rhinagrion hainanense* Wilson & Reels, 2001

野山蟌科Family Argiolestidae Fraser, 1957 (**1属1种** 1 genus 1 species)

叶山蟌属 Genus *Podolestes* Selys, 1862

露兜叶山蟌 *Podolestes pandanus* Wilson & Reels, 2001 [海南 Hainan]

色蟌总科待定科 Family "Calopterygoidea *incertae sedis*" (**6属33种1亚种** 6 genera 33 species 1 subspecies)

野蟌属 Genus *Agriomorpha* May, 1933

白尾野蟌 *Agriomorpha fusca* May, 1933

兴隆野蟌 *Agriomorpha xinglongensis* (Wilson & Reels, 2001)

野蟌属待定种 *Agriomorpha* sp.

缅山蟌属 Genus *Burmargiolestes* Kennedy, 1925

黑缅山蟌 *Burmargiolestes melanothorax* (Selys, 1891)

凸尾山蟌属 Genus *Mesopodagrion* McLachlan, 1896

藏凸尾山蟌南方亚种 *Mesopodagrion tibetanum australe* Yu & Bu, 2009

藏凸尾山蟌指名亚种 *Mesopodagrion tibetanum tibetanum* McLachlan, 1896

雅州凸尾山蟌 *Mesopodagrion yachowense* Chao, 1953

凸尾山蟌属待定种 *Mesopodagrion* sp.

古山蟌属 Genus *Priscagrion* Zhou & Wilson, 2001

克氏古山蟌 *Priscagrion kiautai* Zhou & Wilson, 2001

续表

宾黑古山蟌 *Priscagrion pinheyi* Zhou & Wilson, 2001
古山蟌属待定种 *Priscagrion* sp.
扇山蟌属 Genus *Rhipidolestes* Ris, 1912
棘扇山蟌 *Rhipidolestes aculeatus* Ris, 1912
艾伦扇山蟌 *Rhipidolestes alleni* Wilson, 2000
尖扇山蟌 *Rhipidolestes apicatus* Navás, 1934 [浙江 Zhejiang]
巴斯扇山蟌 *Rhipidolestes bastiaani* Zhu & Yang, 1998 [陕西 Shaanxi, 四川 Sichuan]
二齿扇山蟌 *Rhipidolestes bidens* Schmidt, 1931 [浙江 Zhejiang]
赵氏扇山蟌 *Rhipidolestes chaoi* Wilson, 2004 [湖南 Hunan]
黄蓝扇山蟌 *Rhipidolestes cyanoflavus* Wilson, 2000
褐带扇山蟌 *Rhipidolestes fascia* Zhou, 2003
珍妮扇山蟌 *Rhipidolestes janetae* Wilson, 1997
愉快扇山蟌 *Rhipidolestes jucundus* Lieftinck, 1948
劳氏扇山蟌 *Rhipidolestes laui* Wilson & Reels, 2003 [广西 Guangxi]
李氏扇山蟌 *Rhipidolestes lii* Zhou, 2003
水鬼扇山蟌 *Rhipidolestes nectans* (Needham, 1929)
黄白扇山蟌 *Rhipidolestes owadai* Asahina, 1997
红足扇山蟌 *Rhipidolestes rubripes* (Navás, 1936) [江西 Jiangxi]
褐顶扇山蟌 *Rhipidolestes truncatidens* Schmidt, 1931 Syn. *Lestomima flavostigma* May, 1933
杨冰扇山蟌 *Rhipidolestes yangbingi* Davies, 1998 [四川 Sichuan]
扇山蟌属待定种1 *Rhipidolestes* sp. 1
扇山蟌属待定种2 *Rhipidolestes* sp. 2
扇山蟌属待定种3 *Rhipidolestes* sp. 3
扇山蟌属待定种4 *Rhipidolestes* sp. 4
华山蟌属 Genus *Sinocnemis* Wilson & Zhou, 2000
杜氏华山蟌 *Sinocnemis dumonti* Wilson & Zhou, 2000 [贵州 Guizhou]
杨氏华山蟌 *Sinocnemis yangbingi* Wilson & Zhou, 2000 Syn. *Sinocnemis henanese* Wang, 2003
拟丝蟌科 Family Pseudolestidae (1属1种 1 genus 1 species)
拟丝蟌属 Genus *Pseudolestes* Kirby, 1900
丽拟丝蟌 *Pseudolestes mirabilis* Kirby, 1900
丝蟌科 Family Lestidae (4属21种 4 genera 21 species)
印丝蟌属 Genus *Indolestes* Fraser, 1922
黄面印丝蟌 *Indolestes assamicus* Fraser, 1930
蓝印丝蟌 *Indolestes cyaneus* (Selys, 1862)
贵州印丝蟌 *Indolestes guizhouensis* Zhou & Zhou, 2005 [贵州 Guizhou]
斑脊印丝蟌 *Indolestes gracilis* (Hagen, 1862) [云南 Yunnan]

续表

奇印丝蟌 *Indolestes peregrinus* (Ris, 1916) ?Syn. *Indolestes coeruleus* (Fraser, 1924) Syn. *Lestes extranea* Needham, 1930 Syn. *Lestes monteili* Navás, 1935
丝蟌属 Genus *Lestes* Leach, 1815 刀尾丝蟌 *Lestes barbarus* (Fabricius, 1798) [新疆 Xinjiang, 内蒙古 Inner Mongolia, 吉林 Jilin, 甘肃 Gansu] 整齐丝蟌 *Lestes concinnus* Hagen, 1862 多罗丝蟌 *Lestes dorothea* Fraser, 1924 足尾丝蟌 *Lestes dryas* Kirby, 1890 高丝蟌 *Lestes elatus* Hagen, 1862 [海南 Hainan] 日本丝蟌 *Lestes japonicus* Selys, 1883 大痣丝蟌 *Lestes macrostigma* (Eversmann, 1836) 蕾尾丝蟌 *Lestes nodalis* Selys, 1891 舟尾丝蟌 *Lestes praemorsus* Hagen, 1862 桨尾丝蟌 *Lestes sponsa* (Hansemann, 1823) 蓝绿丝蟌 *Lestes temporalis* Selys, 1883 锯尾丝蟌 *Lestes umbrinus* Selys, 1891 [广西 Guangxi, 海南 Hainan, 云南 Yunnan] 绿丝蟌 *Lestes virens* (Charpentier, 1825) 丝蟌属待定种 *Lestes* sp.
长痣丝蟌属 Genus *Orolestes* McLachlan, 1895 长痣丝蟌 *Orolestes selysi* McLachlan, 1895 Syn. *Megalestes mirabilis* Matsumura, 1913 Syn. *Orolestes koxingai* Chen, 1950
黄丝蟌属 Genus *Sympecma* Burmeister, 1839 三叶黄丝蟌 *Sympecma paedisca* (Brauer, 1877)
综蟌科 Family Synlestidae (2属13种 2 genera 13 species)
绿综蟌属 Genus *Megalestes* Selys, 1862 褐腹绿综蟌 *Megalestes chengi* Chao, 1947 [福建 Fujian] 盘绿综蟌 *Megalestes discus* Wilson, 2004 褐尾绿综蟌 *Megalestes distans* Needham, 1930 郝氏绿综蟌 *Megalestes haui* Wilson & Reels, 2003 黄腹绿综蟌 *Megalestes heros* Needham, 1930 Syn. *Megalestes suensoni* Asahina, 1956 泰国绿综蟌 *Megalestes kurahashii* Asahina, 1985 大黄尾绿综蟌 *Megalestes maai* Chen, 1947 细腹绿综蟌 *Megalestes micans* Needham, 1930 峨眉绿综蟌 *Megalestes omeiensis* Chao, 1965 铲形绿综蟌 *Megalestes palaceus* Zhou & Zhou, 2008 [贵州 Guizhou]

续表

白尾绿综蟌 ***Megalestes riccii*** Navás, 1935
狼牙绿综蟌 *Megalestes tuska* Wilson & Reels, 2003 [广西 Guangxi]
华综蟌属 Genus *Sinolestes* Needham, 1930
黄肩华综蟌 ***Sinolestes editus*** Needham, 1930
Syn. *Sinolestes ornata* Needham, 1930
Syn. *Sinolestes truncata* Needham, 1930
扇蟌科 Family Platycnemididae (7属58种 7 genera 58 species)
丽扇蟌属 Genus *Calicnemia* Strand, 1928
赵氏丽扇蟌 ***Calicnemia chaoi*** Wilson, 2004
赭腹丽扇蟌 ***Calicnemia erythromelas*** (Selys, 1891)
朱腹丽扇蟌 ***Calicnemia eximia*** (Selys, 1863)
古蔺丽扇蟌 ***Calicnemia gulinensis*** Yu & Bu, 2008
黑丽扇蟌 ***Calicnemia haksik*** Wilson & Reels, 2003
灰丽扇蟌 ***Calicnemia imitans*** Lieftinck, 1948
迈尔丽扇蟌 ***Calicnemia miles*** (Laidlaw, 1917)
黑尾丽扇蟌 *Calicnemia miniata* (Selys, 1886) [西藏 Tibet]
中脊丽扇蟌 *Calicnemia porcata* Yu & Bu, 2008 [四川 Sichuan]
华丽扇蟌 ***Calicnemia sinensis*** Lieftinck, 1984
黑袜丽扇蟌 ***Calicnemia soccifera*** Yu & Chen, 2013
朱氏丽扇蟌 *Calicnemia zhuae* Zhang & Yang, 2008 [陕西 Shaanxi]
丽扇蟌属待定种 ***Calicnemia*** sp.
长腹扇蟌属 Genus *Coeliccia* Kirby, 1890
金脊长腹扇蟌 ***Coeliccia chromothorax*** (Selys, 1891)
黄纹长腹扇蟌 ***Coeliccia cyanomelas*** Ris, 1912
四斑长腹扇蟌 ***Coeliccia didyma*** (Selys, 1863)
黄尾长腹扇蟌 ***Coeliccia flavicauda*** Ris, 1912
蓝黑长腹扇蟌 ***Coeliccia furcata*** Hämäläinen, 1986
黄绿长腹扇蟌 *Coeliccia galbina* Wilson & Reels, 2003 [广西 Guangxi]
海南长腹扇蟌 ***Coeliccia hainanense*** Laidlaw, 1932
蓝斑长腹扇蟌 ***Coeliccia loogali*** Fraser, 1932
明溪长腹扇蟌 *Coeliccia mingxiensis* Xu, 2006 [福建 Fujian]
蓝脊长腹扇蟌 ***Coeliccia poungyi*** Fraser, 1924
黄蓝长腹扇蟌 ***Coeliccia pyriformis*** Laidlaw, 1932
佐藤长腹扇蟌 ***Coeliccia satoi*** Asahina, 1997
截斑长腹扇蟌 ***Coeliccia scutellum*** Laidlaw, 1932
六斑长腹扇蟌 *Coeliccia sexmaculata* Wang, 1994 [河南 Henan]
双色长腹扇蟌 ***Coeliccia svihleri*** Asahina, 1970
韦氏长腹扇蟌 *Coeliccia wilsoni* Zhang & Yang, 2011 [陕西 Shaanxi, 四川 Sichuan]

续表

长腹扇蟌属待定种1 *Coeliccia* sp. 1

长腹扇蟌属待定种2 *Coeliccia* sp. 2

长腹扇蟌属待定种3 Coeliccia sp. 3

长腹扇蟌属待定种4 *Coeliccia* sp. 4

长腹扇蟌属待定种5 *Coeliccia* sp. 5

长腹扇蟌属待定种6 *Coeliccia* sp. 6

狭扇蟌属 Genus *Copera* Kirby, 1890

白狭扇蟌 ***Copera annulata*** (Selys, 1863)

毛狭扇蟌 ***Copera ciliata*** (Selys, 1863)

黄狭扇蟌 ***Copera marginipes*** (Rambur, 1842)

Syn. *Platycnemis pierrati* Navás, 1935

黑狭扇蟌 ***Copera tokyoensis*** Asahina, 1948

?Syn. *Copera rubripes* Navás, 1934

褐狭扇蟌 ***Copera vittata*** (Selys, 1863)

印扇蟌属 Genus *Indocnemis* Laidlaw, 1917

黑背印扇蟌 ***Indocnemis ambigua*** (Asahina, 1997)

印扇蟌 ***Indocnemis orang*** (Förster, 1907)

同痣蟌属 Genus *Onychargia* Selys, 1865

毛面同痣蟌 ***Onychargia atrocyana*** Selys, 1865

扇蟌属 Genus *Platycnemis* Burmeister, 1839

白扇蟌 ***Platycnemis foliacea*** Selys, 1886

Syn. *Platycnemis foliosa* Navás, 1932

叶足扇蟌 ***Platycnemis phyllopoda*** Djakonov, 1926

?Syn. *Platycnemis ulmifolia* Ris, 1930

?Syn. *Platycnemis hummeli* Sjöstedt, 1932

微桥原蟌属 Genus *Prodasineura* Cowley, 1934

金脊微桥原蟌 *Prodasineura auricolor* (Fraser, 1927) [云南 Yunnan]

乌微桥原蟌 ***Prodasineura autumnalis*** (Fraser, 1922)

Syn. *Indoneura dolorosa* Needham, 1930

朱背微桥原蟌 ***Prodasineura croconota*** (Ris, 1916)

福建微桥原蟌 ***Prodasineura fujianensis*** Xu, 2006

汉中微桥原蟌 *Prodasineura hanzhongensis* Yang & Li, 1995 [陕西 Shaanxi]

华氏微桥原蟌 *Prodasineura huai* Zhou & Zhou, 2007 [广东 Guangdong]

龙井微桥原蟌 *Prodasineura longjingensis* (Zhou, 1981) [浙江 Zhejiang]

黑微桥原蟌 *Prodasineura nigra* (Fraser, 1922) [云南 Yunnan]

黄条微桥原蟌 *Prodasineura sita* (Kirby, 1893) [云南 Yunnan, 福建 Fujian, 海南 Hainan, 广东 Guangdong]

微桥原蟌属待定种1 *Prodasineura* sp.1

微桥原蟌属待定种2 *Prodasineura* sp. 2

续表

微桥原蟌属待定种3 *Prodasineura* sp. 3
微桥原蟌属待定种4 *Prodasineura* sp. 4
蟌科 Family Coenagrionidae (14属72种 14 genera 72 species)
狭翅蟌属 Genus *Aciagrion* Selys, 1891
霜蓝狭翅蟌 *Aciagrion approximans* (Selys, 1876)
华安狭翅蟌 *Aciagrion huaanensis* Xu, 2005 [福建 Fujian]
针尾狭翅蟌 *Aciagrion migratum* (Selys,1876)
蓝尾狭翅蟌 *Aciagrion olympicum* Laidlaw, 1919
森狭翅蟌 *Aciagrion pallidum* Selys, 1891
小蟌属 Genus *Agriocnemis* Selys, 1877
眼斑小蟌 *Agriocnemis dabreui* Fraser, 1919 [云南 Yunnan]
蓝斑小蟌 *Agriocnemis clauseni* Fraser, 1922
杯斑小蟌 *Agriocnemis femina* (Brauer, 1868)
白腹小蟌 *Agriocnemis lacteola* Selys, 1877
黑尾小蟌 *Agriocnemis naia* Fraser, 1923 [云南 Yunnan]
樽斑小蟌 *Agriocnemis nana* Laidlaw, 1914
黄尾小蟌 *Agriocnemis pygmaea* (Rambur, 1842)
Syn. *Agrion kagiensis* Matsumura, 1911
印度小蟌 *Agriocnemis splendidissima* Laidlaw, 1919
安蟌属 Genus *Amphiallagma* Kennedy, 1920
天蓝安蟌 *Amphiallagma parvum* (Selys, 1876)
黑蟌属 Genus *Argiocnemis* Selys, 1877
蓝唇黑蟌 *Argiocnemis rubescens* Selys, 1877
黄蟌属 Genus *Ceriagrion* Selys, 1876
翠胸黄蟌 *Ceriagrion auranticum ryukyuanum* Asahina, 1967
天蓝黄蟌 *Ceriagrion azureum* (Selys, 1891)
橙黄蟌 *Ceriagrion chaoi* Schmidt, 1964
长尾黄蟌 *Ceriagrion fallax* Ris, 1914
柠檬黄蟌 *Ceriagrion indochinense* Asahina, 1967
短尾黄蟌 *Ceriagrion melanurum* Selys, 1876
赤黄蟌 *Ceriagrion nipponicum* Asahina, 1967
钩尾黄蟌 *Ceriagrion olivaceum* Laidlaw, 1914
中华黄蟌 *Ceriagrion sinense* Asahina, 1967
黄蟌属待定种1 *Ceriagrion* sp. 1
黄蟌属待定种2 *Ceriagrion* sp. 2
黄蟌属待定种3 *Ceriagrion* sp. 3
黄蟌属待定种4 *Ceriagrion* sp. 4
蟌属 Genus *Coenagrion* Kirby, 1890

续表

多棘蟌 *Coenagrion aculeatum* Yu & Bu, 2007
盃纹蟌 *Coenagrion ecornutum* (Selys, 1872)
腾冲蟌 *Coenagrion exclamationis* (Fraser, 1919)
Syn. *Coenagrion tengchongensis* Yu & Bu, 2007
框纹蟌 *Coenagrion glaciale* (Selys, 1872) [黑龙江 Heilongjiang]
三纹蟌 *Coenagrion hastulatum* (Charpentier, 1825)
叶纹蟌 *Coenagrion holdereri* (Förster, 1900) [西藏 Tibet]
黑格蟌 *Coenagrion hylas* (Trybom, 1889)
纤腹蟌 *Coenagrion johanssoni* (Wallengren, 1894)
Syn. *Agrion convalescens* Bartenev, 1914
Syn. *Coenagrion bifurcatum* Zhu & Ou-yan, 2000
矛斑蟌 *Coenagrion lanceolatum* (Selys, 1872)
月斑蟌 *Coenagrion lunulatum* (Charpentier, 1840)
绿蟌属 Genus *Enallagma* Charpentier, 1840
心斑绿蟌 *Enallagma cyathigerum* (Charpentier, 1840)
红眼蟌属 Genus *Erythromma* Charpentier, 1840
红眼蟌 *Erythromma najas* (Hansemann, 1823)
火蟌属 Genus *Huosoma* Guan, Dumont, Yu, Han & Vierstraete, 2013
阔叶火蟌 *Huosoma latiloba* (Yu, Yang & Bu, 2008)
窄叶火蟌 *Huosoma tinctipenne* (McLachlan, 1894) [四川 Sichuan, 云南 Yunnan]
异痣蟌属 Genus *Ischnura* Charpentier, 1840
东亚异痣蟌 *Ischnura asiatica* (Brauer, 1865)
Syn. *Ischnura orientalis* Selys, 1876
Syn. *Ischnura lobata* Needham, 1930
Syn. *Ischnura formosanus* Chujo, 1931
?Syn. *Coenagrion needhami* Navás, 1933
黄腹异痣蟌 *Ischnura aurora* (Brauer, 1865)
Syn. *Agriocnemis amelia* Needham, 1930
长叶异痣蟌 *Ischnura elegans* (Vander Linden, 1820)
蓝斑异痣蟌 *Ischnura evansi* Morton, 1919 [内蒙古 Inner Mongolia]
水源异痣蟌 *Ischnura fountaineae* Morton, 1905 [西藏 Tibet]
蓝壮异痣蟌 *Ischnura pumilio* (Charpentier, 1825)
赤斑异痣蟌 *Ischnura rufostigma* Selys, 1876
褐斑异痣蟌 *Ischnura senegalensis* (Rambur, 1842)
异痣蟌属待定种 *Ischnura* sp.
妹蟌属 Genus *Mortonagrion* Fraser, 1920
蓝尾妹蟌 *Mortonagrion aborense* (Laidlaw, 1914)
广濑妹蟌 *Mortonagrion hirosei* Asahina, 1972
钩斑妹蟌 *Mortonagrion selenion* (Ris, 1916)

续表

绿背蟌属 Genus *Nehalennia* Selys, 1850
黑面绿背蟌 ***Nehalennia speciosa*** (Charpentier, 1840)
尾蟌属 Genus *Paracercion* Weekers & Dumont, 2004
挫齿尾蟌 ***Paracercion barbatum*** (Needham,1930)
蓝纹尾蟌 ***Paracercion calamorum*** (Ris, 1916)
钳尾蟌 ***Paracercion dorothea*** (Fraser, 1924)
Syn. *Coenagrion impar* Needham, 1930
Syn. *Cercion yunnanensis* Zhu & Han, 2000
隼尾蟌 ***Paracercion hieroglyphicum*** (Brauer, 1865)
Syn. *Nehalennia atrinuchalis* Selys, 1876
Syn. *Coenagrion chusanicum* Navás, 1933
黑背尾蟌 ***Paracercion melanotum*** (Selys, 1876)
Syn. *Coenagrion admirationis* Navás, 1933
Syn. *Coenagrion trilineatum* Navás, 1933
七条尾蟌 ***Paracercion plagiosum*** (Needham, 1930)
钱博尾蟌 *Paracercion sieboldii* (Selys, 1876) [台湾 Taiwan]
Syn. *Agrion sauteri* Ris, 1916
捷尾蟌 ***Paracercion v-nigrum*** (Needham, 1930)
尾蟌属待定种1 ***Paracercion*** sp. 1
尾蟌属待定种2 ***Paracercion*** sp. 2
斑蟌属 Genus *Pseudagrion* Selys, 1876
亚澳斑蟌 ***Pseudagrion australasiae*** Selys, 1876
大盘山斑蟌 *Pseudagrion daponshanensis* Zhou & Zhou, 2007 [浙江 Zhejiang]
绿斑蟌 ***Pseudagrion microcephalum*** (Rambur, 1842)
红玉斑蟌 ***Pseudagrion pilidorsum*** (Brauer, 1868)
赤斑蟌 ***Pseudagrion pruinosum*** (Burmeister, 1839)
Syn. *Pseudagrion elongatum* Needham, 1930
丹顶斑蟌 ***Pseudagrion rubriceps*** Selys, 1876
褐斑蟌 ***Pseudagrion spencei*** Fraser, 1922
扁蟌科 Family Platystictidae (4属32种 4 genera 32 species)
镰扁蟌属 Genus *Drepanosticta* Laidlaw, 1917
白尾镰扁蟌 ***Drepanosticta brownelli*** (Tinkham, 1938)
修长镰扁蟌 ***Drepanosticta elongata*** Wilson & Reels, 2001
香港镰扁蟌 ***Drepanosticta hongkongensis*** Wilson, 1996
巨镰扁蟌 ***Drepanosticta magna*** Wilson & Reels, 2003
周氏镰扁蟌 ***Drepanosticta zhoui*** Wilson & Reels, 2001
镰扁蟌属待定种1 ***Drepanosticta*** sp. 1
镰扁蟌属待定种2 ***Drepanosticta*** sp. 2
镰扁蟌属待定种3 ***Drepanosticta*** sp. 3

续表

镰扁蟌属待定种4 *Drepanosticta* sp. 4
镰扁蟌属待定种5 *Drepanosticta* sp. 5
原扁蟌属 Genus *Protosticta* Selys, 1885
黄颈原扁蟌 *Protosticta beaumonti* Wilson, 1997
卡罗原扁蟌 *Protosticta caroli* Van Tol, 2008
奇异原扁蟌 *Protosticta curiosa* Fraser, 1934
Syn. *Protosticta zhengi* Yu & Bu, 2009
暗色原扁蟌 *Protosticta grandis* Asahina, 1985
泰国原扁蟌 *Protosticta khaosoidaoensis* Asahina, 1984
克氏原扁蟌 *Protosticta kiautai* Zhou, 1986 [浙江 Zhejiang]
黑胸原扁蟌 *Protosticta nigra* Kompier, 2016
白瑞原扁蟌 *Protosticta taipokauensis* Asahina & Dudgeon, 1987
原扁蟌属待定种1 *Protosticta* sp. 1
原扁蟌属待定种2 *Protosticta* sp. 2
原扁蟌属待定种3 *Protosticta* sp. 3
原扁蟌属待定种4 *Protosticta* sp. 4
原扁蟌属待定种5 *Protosticta* sp. 5
原扁蟌属待定种6 *Protosticta* sp. 6
华扁蟌属 Genus *Sinosticta* Wilson, 1997
戴波华扁蟌 *Sinosticta debra* Wilson & Xu, 2007
海南华扁蟌 *Sinosticta hainanense* Wilson & Reels, 2001
绪方华扁蟌 *Sinosticta ogatai* (Matsuki & Saito, 1996)
深林华扁蟌 *Sinosticta sylvatica* Yu & Bu, 2009
华扁蟌属待定种 *Sinosticta* sp.
云扁蟌属 Genus *Yunnanosticta* Dow & Zhang, 2018
蓝颈云扁蟌 *Yunnanosticta cyaneocollaris* Dow & Zhang, 2018
韦氏云扁蟌 *Yunnanosticta wilsoni* Dow & Zhang, 2018
云扁蟌属待定种 *Yunnanosticta* sp.
间翅亚目 Suborder Anisozygoptera **(1科1属3种 1 family 1 genus 3 species)**
昔蜓科 Family Epiophlebiidae (1属3种 1 genus 3 species)
昔蜓属 Genus *Epiophlebia* Calvert, 1903
川昔蜓 *Epiophlebia diana* Carle, 2012
华昔蜓 *Epiophlebia sinensis* Li & Nel, 2012
昔蜓属待定种 *Epiophlebia* sp.
差翅亚目（蜻蜓） Suborder Anisoptera (Dragonfly) **(8科108属611种21亚种 8 families 108 genera 611 species 21 subspecies)**

续表

蜓科 Family Aeshnidae (15属107种2亚种 15 genera 107 species 2 subspecies)
绿蜓属 Genus *Aeschnophlebia* Selys, 1883 黑纹绿蜓 ***Aeschnophlebia anisoptera*** Selys, 1883 长痣绿蜓 ***Aeschnophlebia longistigma*** Selys, 1883 Syn. *Aeschnophlebia kolthoffi* Sjöstedt, 1925
蜓属 Genus *Aeshna* Fabricius, 1775 硕斑蜓 ***Aeshna affinis*** Vander Linden, 1820 琉璃蜓 ***Aeshna crenata*** Hagen, 1856 峻蜓 ***Aeshna juncea*** (Linnaeus, 1758) 混合蜓 ***Aeshna mixta*** Latreille, 1805 蝶斑蜓指名亚种 ***Aeshna petalura petalura*** Martin, 1909 蝶斑蜓台湾亚种 ***Aeshna petalura taiyal*** Asahina, 1938 神农蜓 ***Aeshna shennong*** Zhang & Cai, 2014 极北蜓 ***Aeshna subarctica*** Walker, 1908
翠蜓属 Genus *Anaciaeschna* Selys, 1878 碧翠蜓 ***Anaciaeschna jaspidea*** (Burmeister, 1839) 褐翠蜓 ***Anaciaeschna martini*** (Selys, 1897)
伟蜓属 Genus *Anax* Leach, 1815 斑伟蜓 ***Anax guttatus*** (Burmeister, 1839) 黄伟蜓 ***Anax immaculifrons*** Rambur, 1842 印度伟蜓 ***Anax indicus*** Lieftinck, 1942 黑纹伟蜓 ***Anax nigrofasciatus*** Oguma, 1915 东亚伟蜓 ***Anax panybeus*** Hagen, 1867 碧伟蜓灰胸亚种 ***Anax parthenope parthenope*** (Selys, 1839) 碧伟蜓东亚亚种 ***Anax parthenope julius*** Brauer, 1865
细腰蜓属 Genus *Boyeria* McLachlan, 1895 褐面细腰蜓 ***Boyeria karubei*** Yokoi, 2002 华细腰蜓 *Boyeria sinensis* Asahina, 1978 [四川 Sichuan]
头蜓属 Genus *Cephalaeschna* Selys, 1883 长角头蜓 ***Cephalaeschna cornifrons*** Zhang & Cai, 2013 鼎湖头蜓 ***Cephalaeschna dinghuensis*** Wilson, 1999 异色头蜓 ***Cephalaeschna discolor*** Zhang, Cai & Liao, 2013 克氏头蜓 ***Cephalaeschna klotsi*** Asahina, 1982 Syn. *Periaeschna rotunda* Wilson, 2005 马蒂头蜓 ***Cephalaeschna mattii*** Zhang, Cai & Liao, 2013 尼氏头蜓 ***Cephalaeschna needhami*** Asahina, 1981 暗色头蜓 ***Cephalaeschna obversa*** Needham, 1930 蝶斑头蜓 ***Cephalaeschna ordopapiliones*** Zhang & Cai, 2013

续表

长者头蜓 *Cephalaeschna patrorum* Needham, 1930
李氏头蜓 *Cephalaeschna risi* Asahina, 1981
邵武头蜓 *Cephalaeschna shaowuensis* Xu, 2006
独行头蜓 *Cephalaeschna solitaria* Zhang, Cai & Liao, 2013
西乡头蜓 *Cephalaeschna xixiangensis* Zhang, 2013 [陕西 Shaanxi]
头蜓属待定种1 *Cephalaeschna* sp. 1
头蜓属待定种2 *Cephalaeschna* sp. 2
头蜓属待定种3 *Cephalaeschna* sp. 3
头蜓属待定种4 *Cephalaeschna* sp. 4
头蜓属待定种5 *Cephalaeschna* sp. 5
长尾蜓属 Genus *Gynacantha* Rambur, 1842
无纹长尾蜓 *Gynacantha bayadera* Selys, 1891
透翅长尾蜓 *Gynacantha hyalina* Selys, 1882
基凹长尾蜓 *Gynacantha incisura* Fraser, 1935 [云南 Yunnan]
日本长尾蜓 *Gynacantha japonica* Bartenev, 1909
琉球长尾蜓 *Gynacantha ryukyuensis* Asahina, 1962
跳长尾蜓 *Gynacantha saltatrix* Martin, 1909
细腰长尾蜓 *Gynacantha subinterrupta* Rambur, 1842
长尾蜓属待定种1 *Gynacantha* sp. 1
长尾蜓属待定种2 *Gynacantha* sp. 2
棘蜓属 Genus *Gynacanthaeschna* Fraser, 1921
锡金棘蜓 *Gynacanthaeschna sikkima* (Karsch, 1891)
似沼蜓属 Genus *Oligoaeschna* Selys, 1889
广西似沼蜓 *Oligoaeschna aquilonaris* Wilson, 2005 [广西 Guangxi]
海南似沼蜓 *Oligoaeschna petalura* Lieftinck, 1968 [海南 Hainan]
佩蜓属 Genus *Periaeschna* Martin, 1908
赵氏佩蜓 *Periaeschna chaoi* (Asahina, 1982)
福临佩蜓 *Periaeschna flinti* Asahina, 1978
黄脊佩蜓 *Periaeschna gerrhon* (Wilson, 2005) [广西 Guangxi, 广东 Guangdong]
狭痣佩蜓 *Periaeschna magdalena* Martin, 1909
奇异佩蜓 *Periaeschna mira* Navás, 1936 [江西 Jiangxi]
浅色佩蜓 *Periaeschna nocturnalis* Fraser, 1927
雅珍佩蜓 *Periaeschna yazhenae* Xu, 2012
漳州佩蜓 *Periaeschna zhangzhouensis* Xu, 2007
佩蜓属待定种1 *Periaeschna* sp. 1
佩蜓属待定种2 *Periaeschna* sp. 2
佩蜓属待定种3 *Periaeschna* sp. 3

续表

叶蜓属 Genus *Petaliaeschna* Fraser, 1927
科氏叶蜓 *Petaliaeschna corneliae* Asahina, 1982
黄纹叶蜓 *Petaliaeschna flavipes* Karube, 1999
黎氏叶蜓 *Petaliaeschna lieftincki* Asahina, 1982 [陕西 Shaanxi]
叶蜓属待定种1 *Petaliaeschna* sp. 1
叶蜓属待定种2 *Petaliaeschna* sp. 2
黑额蜓属 Genus *Planaeschna* McLachlan, 1896
棘尾黑额蜓 *Planaeschna caudispina* Zhang & Cai, 2013
希里黑额蜓 *Planaeschna celia* Wilson & Reels, 2001 [海南 Hainan]
清迈黑额蜓 *Planaeschna chiengmaiensis* Asahina, 1981
联纹黑额蜓 *Planaeschna gressitti* Karube, 2002
郝氏黑额蜓 *Planaeschna haui* Wilson & Xu, 2008
石垣黑额蜓 *Planaeschna ishigakiana flavostria* Yeh, 1996
崂山黑额蜓 *Planaeschna laoshanensis* Zhang, Yeh & Tong, 2010
刘氏黑额蜓 *Planaeschna liui* Xu, Chen & Qiu, 2009
角斑黑额蜓 *Planaeschna maculifrons* Zhang & Cai, 2013
茂兰黑额蜓 *Planaeschna maolanensis* Zhou & Bao, 2002
高山黑额蜓 *Planaeschna monticola* Zhang & Cai, 2013
南昆山黑额蜓 *Planaeschna nankunshanensis* Zhang, Yeh & Tong, 2010
南岭黑额蜓 *Planaeschna nanlingensis* Wilson & Xu, 2008
褐面黑额蜓 *Planaeschna owadai* Karube, 2002
李氏黑额蜓 *Planaeschna risi* Asahina, 1964
粗壮黑额蜓 *Planaeschna robusta* Zhang & Cai, 2013
山西黑额蜓 *Planaeschna shanxiensis* Zhu & Zhang, 2001
幽灵黑额蜓 *Planaeschna skiaperipola* Wilson & Xu, 2008
遂昌黑额蜓 *Planaeschna suichangensis* Zhou & Wei, 1980
台湾黑额蜓 *Planaeschna taiwana* Asahina, 1951
黑额蜓属待定种1 *Planaeschna* sp. 1
黑额蜓属待定种2 *Planaeschna* sp. 2
黑额蜓属待定种3 *Planaeschna* sp. 3
黑额蜓属待定种4 *Planaeschna* sp. 4
黑额蜓属待定种5 *Planaeschna* sp. 5
黑额蜓属待定种6 *Planaeschna* sp. 6
黑额蜓属待定种7 *Planaeschna* sp. 7
多棘蜓属 Genus *Polycanthagyna* Fraser, 1933
红褐多棘蜓 *Polycanthagyna erythromelas* (McLachlan, 1896)
Syn. *Polycanthagyna erythromelas paiwan* Asahina, 1951
Syn.n. *Polycanthagyna chaoi* Yang & Li, 1994

续表

黄绿多棘蜓 *Polycanthagyna melanictera* (Selys, 1883)
蓝黑多棘蜓 *Polycanthagyna ornithocephala* (McLachlan, 1896)
多棘蜓属待定种1 *Polycanthagyna* sp. 1
多棘蜓属待定种2 *Polycanthagyna* sp. 2

沼蜓属 Genus *Sarasaeschna* Karube & Yeh, 2001
台湾沼蜓 *Sarasaeschna chiangchinlii* Chen & Yeh, 2014 [台湾 Taiwan]
连氏沼蜓 *Sarasaeschna lieni* (Yeh & Chen, 2000)
高峰沼蜓 *Sarasaeschna gaofengensis* Yeh & Kiyoshi, 2015
曹氏沼蜓 *Sarasaeschna kaoi* Yeh, Lee & Wong, 2015 [台湾 Taiwan]
尼氏沼蜓 *Sarasaeschna niisatoi* (Karube, 1998)
源垭沼蜓 *Sarasaeschna pyanan* (Asahina, 1951)
刃尾沼蜓 *Sarasaeschna sabre* (Wilson & Reels, 2001)
锋刀沼蜓 *Sarasaeschna tsaopiensis* (Yeh & Chen, 2000)
朱氏沼蜓 *Sarasaeschna zhuae* Xu, 2008 [福建 Fujian]

短痣蜓属 Genus *Tetracanthagyna* Selys, 1883
沃氏短痣蜓 *Tetracanthagyna waterhousei* McLachlan, 1898

春蜓科 Family Gomphidae (**37属241种4亚种** 37 genera 241 species 4 subspecies)

安春蜓属 Genus *Amphigomphus* Chao, 1954
汉森安春蜓 *Amphigomphus hansoni* Chao, 1954
娜卡安春蜓 *Amphigomphus nakamurai* Karube, 2001
索氏安春蜓 *Amphigomphus somnuki* Hämäläinen, 1996
安春蜓属待定种 *Amphigomphus* sp.

异春蜓属 Genus *Anisogomphus* Selys, 1858
安氏异春蜓 *Anisogomphus anderi* Lieftinck, 1948
双条异春蜓 *Anisogomphus bivittatus* (Selys, 1854)
镰尾异春蜓 *Anisogomphus caudalis* Fraser, 1926
赵氏异春蜓 *Anisogomphus chaoi* Liu, 1991 [福建 Fujian]
黄脸异春蜓 *Anisogomphus flavifacies* Klots, 1947
福氏异春蜓 *Anisogomphus forresti* (Morton, 1928)
福建异春蜓 *Anisogomphus fujianensis* Zhou & Wu, 1992 [福建 Fujian]
井冈山异春蜓 *Anisogomphus jinggangshanus* Liu, 1991 [江西 Jiangxi]
国姓异春蜓 *Anisogomphus koxingai* Chao, 1954
马奇异春蜓 *Anisogomphus maacki* (Selys, 1872)
Syn. *Gomphus m-flavum* Selys, 1878
高山异春蜓 *Anisogomphus nitidus* Yang & Davies, 1993
欢乐异春蜓 *Anisogomphus resortus* Yang & Davies, 1996 [四川 Sichuan]
五指山异春蜓 *Anisogomphus wuzhishanus* Chao, 1982 [海南 Hainan]

续表

云南异春蜓 *Anisogomphus yunnanensis* Zhou & Wu, 1992
异春蜓属待定种 *Anisogomphus* sp.

亚春蜓属 Genus *Asiagomphus* Asahina, 1985
黄基亚春蜓 *Asiagomphus acco* Asahina, 1996
金斑亚春蜓 *Asiagomphus auricolor* (Fraser, 1926)
云南亚春蜓 *Asiagomphus corniger* (Morton, 1928) [云南 Yunnan]
长角亚春蜓 *Asiagomphus cuneatus* (Needham, 1930) [江西 Jiangxi, 浙江 Zhejiang, 福建 Fujian]
短角亚春蜓 *Asiagomphus giza* Wilson, 2005 [广西 Guangxi, 贵州 Guizhou]
贡山亚春蜓 *Asiagomphus gongshanensis* Yang, Mao & Zhang, 2006 [云南 Yunnan]
海南亚春蜓 *Asiagomphus hainanensis* (Chao, 1953)
西南亚春蜓 *Asiagomphus hesperius* (Chao, 1953)
墨脱亚春蜓 *Asiagomphus motuoensis* Liu & Chao, 1990 [西藏 Tibet]
安定亚春蜓 *Asiagomphus pacatus* (Chao, 1953) [四川 Sichuan]
和平亚春蜓 *Asiagomphus pacificus* (Chao, 1953)
三角亚春蜓 *Asiagomphus perlaetus* (Chao, 1953)
面具亚春蜓 *Asiagomphus personatus* (Selys, 1873)
凹缘亚春蜓 *Asiagomphus septimus* (Needham, 1930)
卧佛亚春蜓 *Asiagomphus somnolens* (Needham, 1930) [北京 Beijing]
凸缘亚春蜓 *Asiagomphus xanthenatus* (Williamson, 1907)
亚春蜓属待定种1 *Asiagomphus* sp. 1
亚春蜓属待定种2 *Asiagomphus* sp. 2
亚春蜓属待定种3 *Asiagomphus* sp. 3

缅春蜓属 Genus *Burmagomphus* Williamson, 1907
太阳缅春蜓 *Burmagomphus apricus* Zhang, Kosterin & Cai, 2015
朝比奈缅春蜓 *Burmagomphus asahinai* Kosterin, Makbun & Dawwrueng, 2012
双纹缅春蜓 *Burmagomphus arvalis* (Needham, 1930) [江苏 Jiangsu]
巴山缅春蜓 *Burmagomphus bashanensis* Yang & Li, 1994 [四川 Sichuan]
领纹缅春蜓 *Burmagomphus collaris* (Needham, 1930)
Syn. *Gomphus campestris* Needham, 1930
齿尾缅春蜓 *Burmagomphus dentatus* Zhang, Kosterin & Cai, 2015
歧角缅春蜓 *Burmagomphus divaricatus* Lieftinck, 1964
欢乐缅春蜓 *Burmagomphus gratiosus* Chao, 1954
溪居缅春蜓 *Burmagomphus intinctus* (Needham, 1930)
林神缅春蜓 *Burmagomphus latescens* Zhang, Kosterin & Cai, 2015
巨缅春蜓 *Burmagomphus magnus* Zhang, Kosterin & Cai, 2015
索氏缅春蜓 *Burmagomphus sowerbyi* (Needham, 1930)
联纹缅春蜓 *Burmagomphus vermicularis* (Martin, 1904)

续表

威廉缅春蜓 ***Burmagomphus williamsoni*** **Förster, 1914**
戴春蜓属 Genus *Davidius* Selys, 1878 双角戴春蜓 ***Davidius bicornutus*** **Selys, 1878** 赵氏戴春蜓 *Davidius chaoi* Cao & Zheng, 1988 [陕西 Shaanxi] Syn. *Davidius miaotaiziensis* Zhu, Yan & Li, 1988 戴氏戴春蜓指名亚种 *Davidius davidii davidii* Selys, 1878 [四川 Sichuan] 戴氏戴春蜓陕西亚种 *Davidius davidii shaanxiensis* Zhu, Yan & Li, 1988 [陕西 Shaanxi] 戴氏戴春蜓云南亚种 *Davidius davidii yunnanensis* Yang & Davies, 1996 [云南 Yunnan] 细纹戴春蜓 ***Davidius delineatus*** **Fraser, 1926** 弗鲁戴春蜓 ***Davidius fruhstorferi*** **Martin, 1904** Syn. *Davidius fruhstorferi guizhoensis* Chao & Liu, 1990 Syn. *Davidius fruhstorferi simaoensis* Zhou & Wu, 1992 新月戴春蜓 *Davidius lunatus* (Bartenev, 1914) [吉林 Jilin, 辽宁 Liaoning] 秦岭戴春蜓 *Davidius qinlingensis* Cao & Zheng, 1989 [陕西 Shaanxi] 方钩戴春蜓 ***Davidius squarrosus*** **Zhu, 1991** 三角戴春蜓 *Davidius triangularis* Chao & Yang, 1995 [陕西 Shaanxi] 黑尾戴春蜓 ***Davidius trox*** **Needham, 1931** 平截戴春蜓 *Davidius truncus* Chao, 1995 [福建 Fujian, 浙江 Zhejiang, 江西 Jiangxi] 元坝戴春蜓 *Davidius yuanbaensis* Zhu, Yan & Li, 1988 [陕西 Shaanxi] 扎洛克戴春蜓 ***Davidius zallorensis*** **Hagen, 1878** 周氏戴春蜓 ***Davidius zhoui*** **Chao, 1995** 戴春蜓属待定种1 ***Davidius*** **sp. 1** 戴春蜓属待定种2 ***Davidius*** **sp. 2** 戴春蜓属待定种3 ***Davidius*** **sp. 3**
曙春蜓属 Genus *Eogomphus* Needham, 1941 忽视曙春蜓 *Eogomphus neglectus* (Needham, 1930) [四川 Sichuan]
闽春蜓属 Genus *Fukienogomphus* Chao, 1954 赛芳闽春蜓 ***Fukienogomphus choifongae*** **Wilson & Tam, 2006** 深山闽春蜓 ***Fukienogomphus prometheus*** **(Lieftinck, 1939)** 显著闽春蜓 ***Fukienogomphus promineus*** **Chao, 1954** Syn. *Fukienogomphus margarita* Chao, 1954
长腹春蜓属 Genus *Gastrogomphus* Needham, 1941 长腹春蜓 ***Gastrogomphus abdominalis*** **(McLachlan, 1884)**
小叶春蜓属 Genus *Gomphidia* Selys, 1854 黄纹小叶春蜓 ***Gomphidia abbotti*** **Williamson, 1907** 联纹小叶春蜓 ***Gomphidia confluens*** **Selys, 1878** 福建小叶春蜓 ***Gomphidia fukienensis*** **Chao, 1955** 克氏小叶春蜓 ***Gomphidia kelloggi*** **Needham, 1930**

续表

并纹小叶春蜓 ***Gomphidia kruegeri*** **Martin, 1904**
小叶春蜓属待定种1 ***Gomphidia*** **sp. 1**
小叶春蜓属待定种2 ***Gomphidia*** **sp. 2**
类春蜓属 Genus *Gomphidictinus* Fraser, 1942
黄纹类春蜓 ***Gomphidictinus perakensis*** **(Laidlaw, 1902)**
Syn. *Gomphidia interruptstria* Zha, Zhang & Zheng, 2005
童氏类春蜓 ***Gomphidictinus tongi*** **Zhang, Guan & Wang, 2017**
曦春蜓属 Genus *Heliogomphus* Laidlaw, 1922
扭尾曦春蜓 ***Heliogomphus retroflexus*** **(Ris, 1912)**
独角曦春蜓 ***Heliogomphus scorpio*** **(Ris, 1912)**
Syn. *Davidius unicornis* Needham, 1930
赛丽曦春蜓 ***Heliogomphus selysi*** **Fraser, 1925**
曦春蜓属待定种 ***Heliogomphus*** **sp.**
叶春蜓属 Genus *Ictinogomphus* Cowley, 1934
华饰叶春蜓 ***Ictinogomphus decoratus*** **(Selys, 1854)**
霸王叶春蜓 ***Ictinogomphus pertinax*** **(Hagen, 1854)**
猛春蜓属 Genus *Labrogomphus* Needham, 1931
凶猛春蜓 ***Labrogomphus torvus*** **Needham, 1931**
环尾春蜓属 Genus *Lamelligomphus* Fraser, 1922
安娜环尾春蜓 ***Lamelligomphus annakarlorum*** **Zhang, Yang & Cai, 2016**
黄尾环尾春蜓 ***Lamelligomphus biforceps*** **(Selys, 1878)**
Syn.n. *Lamelligomphus laetus* Yang & Davies, 1993
Syn. *Lamelligomphus parvulus* Zhou & Li, 2000
驼峰环尾春蜓 ***Lamelligomphus camelus*** **(Martin, 1904)**
赵氏环纹春蜓 *Lamelligomphus chaoi* Zhu, 1999 [云南 Yunnan]
周氏环尾春蜓指名亚种 ***Lamelligomphus choui choui*** **Chao & Liu, 1989**
周氏环尾春蜓天府亚种 *Lamelligomphus choui tienfuensis* (Chao, 1995) [四川 Sichuan]
台湾环尾春蜓 ***Lamelligomphus formosanus*** **(Matsumura, 1926)**
Syn. *Onychogomphus micans* Needham, 1930
Syn. *Lamelligomphus jiuquensis* Liu, 1993
海南环尾春蜓 ***Lamelligomphus hainanensis*** **(Chao, 1954)**
Syn.*Lamelligomphus hongkongensis* Wilson, 1995
汉中环尾春蜓 *Lamelligomphus hanzhongensis* Yang & Zhu, 2001 [陕西 Shaanxi]
墨脱环尾春蜓 ***Lamelligomphus motuoensis*** **Chao, 1983**
环纹环尾春蜓 ***Lamelligomphus ringens*** **(Needham, 1930)**
Syn. *Onychogomphus ridens* Needham, 1930
李氏环尾春蜓 ***Lamelligomphus risi*** **(Fraser, 1922)**
脊纹环尾春蜓 *Lamelligomphus trinus* (Navás, 1936) [江西 Jiangxi]
双髻环尾春蜓 ***Lamelligomphus tutulus*** **Liu & Chao, 1990**

续表

环尾春蜓属待定种1 *Lamelligomphus* sp. 1

环尾春蜓属待定种2 *Lamelligomphus* sp. 2

环尾春蜓属待定种3 *Lamelligomphus* sp. 3

环尾春蜓属待定种4 *Lamelligomphus* sp. 4

纤春蜓属 Genus *Leptogomphus* Selys, 1878

欢庆纤春蜓 *Leptogomphus celebratus* Chao, 1982

Syn. *Leptogomphus hainanensis* Chao, 1984

歧角纤春蜓 *Leptogomphus divaricatus* Chao, 1984

优美纤春蜓 *Leptogomphus elegans* Lieftinck, 1948

尖尾纤春蜓 *Leptogomphus gestroi* Selys, 1891

香港纤春蜓 *Leptogomphus hongkongensis* Asahina, 1988

居间纤春蜓 *Leptogomphus intermedius* Chao, 1982

圆腔纤春蜓 *Leptogomphus perforatus* Ris, 1912

苏氏纤春蜓台湾亚种 *Leptogomphus sauteri formosanus* Matsumura, 1926

苏氏纤春蜓指名亚种 *Leptogomphus sauteri sauteri* Ris, 1912

三道纤春蜓 *Leptogomphus tamdaoensis* Karube, 2014

羚角纤春蜓 *Leptogomphus uenoi* Asahina, 1996

纤春蜓属待定种 *Leptogomphus* sp.

大春蜓属 Genus *Macrogomphus* Selys, 1858

桂林大春蜓 *Macrogomphus guilinensis* Chao, 1982 [广西 Guangxi]

黄绿大春蜓 *Macrogomphus matsukii* Asahina, 1986

粗壮大春蜓 *Macrogomphus robustus* (Selys, 1854) [西藏 Tibet]

大春蜓属待定种 *Macrogomphus* sp.

硕春蜓属 Genus *Megalogomphus* Campion, 1923

萨默硕春蜓 *Megalogomphus sommeri* (Selys, 1854)

弯尾春蜓属 Genus *Melligomphus* Chao, 1990

双峰弯尾春蜓 *Melligomphus ardens* (Needham, 1930)

瀑布弯尾春蜓 *Melligomphus cataractus* Chao & Liu, 1990 [贵州 Guizhou]

罗城弯尾春蜓 *Melligomphus dolus* (Needham, 1930) [广西 Guangxi]

广东弯尾春蜓 *Melligomphus guangdongensis* (Chao, 1994)

Syn. *Mellligomphus moluami* Wilson, 1995

无峰弯尾春蜓 *Melligomphus ludens* (Needham, 1930)

长足春蜓属 Genus *Merogomphus* Martin, 1904

赵氏长足春蜓 *Merogomphus chaoi* Yang & Davies, 1993 [云南 Yunnan]

马丁长足春蜓 *Merogomphus martini* (Fraser, 1922) [云南 Yunnan]

帕维长足春蜓 *Merogomphus pavici* Martin, 1904

Syn. *Anisogomphus pieli* Navás, 1932

Syn. *Merogomphus chui* Asahina, 1968

续表

Syn. *Merogomphus lingyinensis* Zhu & Wu, 1985

泰国长足春蜓 ***Merogomphus pinratani*** (Hämäläinen, 1991)

越南长足春蜓 ***Merogomphus tamdaoensis*** Karube, 2001

小长足春蜓 *Merogomphus torpens* (Needham, 1930) [四川 Sichuan]

江浙长足春蜓 ***Merogomphus vandykei*** Needham, 1930

四川长足春蜓 ***Merogomphus vespertinus*** Chao, 1999

小春蜓属 Genus *Microgomphus* Selys, 1858

越南小春蜓 ***Microgomphus jurzitzai*** Karube, 2000

小春蜓属待定种 ***Microgomphus*** sp.

内春蜓属 Genus *Nepogomphus* Fraser, 1934

优雅内春蜓 *Nepogomphus modestus* (Selys, 1878) [西藏 Tibet]

沃尔内春蜓 ***Nepogomphus walli*** (Fraser, 1924)

内春蜓属待定种 ***Nepogomphus*** sp.

日春蜓属 Genus *Nihonogomphus* Oguma, 1926

贝氏日春蜓 ***Nihonogomphus bequaerti*** Chao, 1954

短翅日春蜓 ***Nihonogomphus brevipennis*** (Needham, 1930)

Syn. *Nihonogomphus hummeli* Sjöstedt, 1932

赵氏日春蜓 *Nihonogomphus chaoi* Zhou & Wu, 1992 [浙江 Zhejiang]

刀日春蜓 ***Nihonogomphus cultratus*** Chao & Wang, 1990

浅黄日春蜓 *Nihonogomphus gilvus* Chao, 1954 [福建 Fujian]

黄沙日春蜓 ***Nihonogomphus huangshaensis*** Chao & Zhu, 1999

黄侧日春蜓 ***Nihonogomphus luteolatus*** Chao & Liu, 1990

山日春蜓 *Nihonogomphus montanus* Zhou & Wu, 1992 [浙江 Zhejiang]

臼齿日春蜓 ***Nihonogomphus ruptus*** (Selys, 1858)

长钩日春蜓 ***Nihonogomphus semanticus*** Chao, 1954

邵武日春蜓 *Nihonogomphus shaowuensis* Chao, 1954 [浙江 Zhejiang, 福建 Fujian]

Syn. *Nihonogomphus zhejiangensis* Chao & Zhou, 1990

浙江日春蜓 *Nihonogomphus silvanus* Zhou & Wu, 1992 [浙江 Zhejiang]

相似日春蜓 *Nihonogomphus simillimus* Chao, 1982 [福建 Fujian]

汤氏日春蜓 ***Nihonogomphus thomassoni*** (Kirby, 1900)

Syn.n. *Nihonogomphus lieftincki* Chao, 1954

日春蜓属待定种 ***Nihonogomphus*** sp.

奈春蜓属 Genus *Nychogomphus* Carle, 1986

双齿奈春蜓 *Nychogomphus bidentatus* Yang, Mao & Zhang, 2010 [云南 Yunnan]

基齿奈春蜓 ***Nychogomphus duaricus*** (Fraser, 1924)

黄尾奈春蜓 ***Nychogomphus flavicaudus*** (Chao, 1982)

卢氏奈春蜓 ***Nychogomphus lui*** Zhou, Zhou & Lu, 2005

双条奈春蜓 *Nychogomphus striatus* (Fraser, 1924) [云南 Yunnan]

续表

杨氏奈春蜓 *Nychogomphus yangi* Zhang, 2014
钩尾春蜓属 Genus *Onychogomphus* Selys, 1854 豹纹钩尾春蜓 *Onychogomphus forcipatus* (Linnaeus, 1758)
蛇纹春蜓属 Genus *Ophiogomphus* Selys, 1854 越南长钩春蜓 *Ophiogomphus longihamulus* Karube, 2014 暗色蛇纹春蜓 *Ophiogomphus obscurus* Bartenev, 1909 中华长钩春蜓 *Ophiogomphus sinicus* (Chao, 1954) 棘角蛇纹春蜓 *Ophiogomphus spinicornis* Selys, 1878
东方春蜓属 Genus *Orientogomphus* Chao & Xu, 1987 具突东方春蜓 *Orientogomphus armatus* Chao & Xu, 1987
副春蜓属 Genus *Paragomphus* Cowley, 1934 钩尾副春蜓 *Paragomphus capricornis* (Förster, 1914) 贺副春蜓 *Paragomphus hoffmanni* (Needham, 1931) [海南 Hainan] 豹纹副春蜓 *Paragomphus pardalinus* Needham, 1942 五指山副春蜓 *Paragomphus wuzhishanensis* Liu, 1988 [海南 Hainan] 副春蜓属待定种 *Paragomphus* sp.
奇春蜓属 Genus *Perissogomphus* Laidlaw, 1922 朝氏奇春蜓 *Perissogomphus asahinai* Zhu, Yang & Wu, 2007 [云南 Yunnan] 史蒂奇春蜓 *Perissogomphus stevensi* Laidlaw, 1922
显春蜓属 Genus *Phaenandrogomphus* Lieftinck, 1964 金黄显春蜓 *Phaenandrogomphus aureus* (Laidlaw, 1922) 赵氏显春蜓 *Phaenandrogomphus chaoi* Zhu & Liang, 1994 细尾显春蜓 *Phaenandrogomphus tonkinicus* (Fraser, 1926) 云南显春蜓 *Phaenandrogomphus yunnanensis* Zhou, 1999 [云南 Yunnan]
刀春蜓属 Genus *Scalmogomphus* Chao, 1990 黄条刀春蜓 *Scalmogomphus bistrigatus* (Hagen, 1854) 丁格刀春蜓 *Scalmogomphus dingavani* (Fraser, 1924) 镰状刀春蜓 *Scalmogomphus falcatus* Chao, 1990 [四川 Sichuan] 贵州刀春蜓 *Scalmogomphus guizhouensis* Zhou & Li, 2000 [贵州 Guizhou] 文山刀春蜓 *Scalmogomphus wenshanensis* Zhou, Zhou & Lu, 2005 刀春蜓属待定种1 *Scalmogomphus* sp. 1 刀春蜓属待定种2 *Scalmogomphus* sp. 2
邵春蜓属 Genus *Shaogomphus* Chao, 1984 黎氏邵春蜓 *Shaogomphus lieftincki* Chao, 1984 [福建 Fujian] 寒冷邵春蜓 *Shaogomphus postocularis epophthalmus* (Selys, 1872) 施氏邵春蜓 *Shaogomphus schmidti* (Asahina, 1956)
施春蜓属 Genus *Sieboldius* Selys, 1854

续表

艾氏施春蜓 ***Sieboldius albardae*** **Selys, 1886** 亚力施春蜓 ***Sieboldius alexanderi*** **(Chao, 1955)** 折尾施春蜓 ***Sieboldius deflexus*** **(Chao, 1955)** 环纹施春蜓 *Sieboldius herculeus* Needham, 1930 [福建 Fujian] 马氏施春蜓 *Sieboldius maai* Chao, 1990 [浙江 Zhejiang, 福建 Fujian] 黑纹施春蜓 ***Sieboldius nigricolor*** **(Fraser, 1924)**
新叶春蜓属 Genus *Sinictinogomphus* Fraser, 1939 大团扇春蜓 ***Sinictinogomphus clavatus*** **(Fabricius, 1775)** Syn. *Ictinus phaleratus* Selys, 1854
华春蜓属 Genus *Sinogomphus* May, 1935 朝比奈华春蜓 *Sinogomphus asahinai* Chao, 1984 [广西 Guangxi] 台湾华春蜓 ***Sinogomphus formosanus*** **Asahina, 1951** 细尾华春蜓 *Sinogomphus leptocercus* Chao, 1983 [西藏 Tibet] 三尖华春蜓 *Sinogomphus orestes* (Lieftinck, 1939) [福建 Fujian] 黄侧华春蜓 *Sinogomphus peleus* (Lieftinck, 1939) [浙江 Zhejiang, 福建 Fujian] 无裂华春蜓 *Sinogomphus pylades* (Lieftinck, 1939) [福建 Fujian] 长角华春蜓 ***Sinogomphus scissus*** **(McLachlan, 1896)** Syn. *Sinogomphus edax* Needham, 1930 Syn. *Sinogomphus nigrofasciatus* May, 1935 Syn. *Sinogomphus shennongjianus* Liu, 1989 修氏华春蜓 ***Sinogomphus suensoni*** **(Lieftinck, 1939)** 双纹华春蜓 *Sinogomphus telamon* (Lieftinck, 1939) [福建 Fujian] 华春蜓属待定种 ***Sinogomphus*** **sp.**
尖尾春蜓属 Genus *Stylogomphus* Fraser, 1922 越中尖尾春蜓 ***Stylogomphus annamensis*** **Kompier, 2017** 张氏尖尾春蜓 ***Stylogomphus changi*** **Asahina, 1968** 纯鎏尖尾春蜓 ***Stylogomphus chunliuae*** **Chao, 1954** 英格尖尾春蜓 ***Stylogomphus inglisi*** **Fraser, 1922** 劳伦斯尖尾春蜓 ***Stylogomphus lawrenceae*** **Yang & Davies, 1996** 肖小尖尾春蜓 *Stylogomphus lutantus* Chao, 1983 [西藏 Tibet] 台湾尖尾春蜓 ***Stylogomphus shirozui*** **Asahina, 1966** 小尖尾春蜓 ***Stylogomphus tantulus*** **Chao, 1954** 尖尾春蜓属待定种1 ***Stylogomphus*** **sp. 1** 尖尾春蜓属待定种2 ***Stylogomphus*** **sp. 2**
扩腹春蜓属 Genus *Stylurus* Needham, 1897 长节扩腹春蜓 ***Stylurus amicus*** **(Needham, 1930)** Syn. *Gomphus szechuanicus* Chao, 1953 黑面扩腹春蜓 *Stylurus clathratus* (Needham, 1930) [湖北 Hubei, 四川 Sichuan, 福建 Fujian, 广东 Guangdong]

续表

恩迪扩腹春蜓 *Stylurus endicotti* (Needham, 1930) [四川 Sichuan] 竖角扩腹春蜓 *Stylurus erectocornus* Liu & Chao, 1990 [广西 Guangxi] 黄角扩腹春蜓 *Stylurus flavicornis* (Needham, 1931) 黄足扩腹春蜓 *Stylurus flavipes* (Charpentier, 1825) 愉快扩腹春蜓 *Stylurus gaudens* (Chao, 1953) [四川 Sichuan] 双斑扩腹春蜓 *Stylurus gideon* (Needham, 1941) [湖北 Hubei, 四川 Sichuan] 克雷扩腹春蜓 *Stylurus kreyenbergi* (Ris, 1928) 南宁扩腹春蜓 *Stylurus nanningensis* Liu, 1985 高尚扩腹春蜓 *Stylurus nobilis* Liu & Chao, 1990 [宁夏 Ningxia] 奇特扩腹春蜓 *Stylurus occultus* (Selys, 1878) [吉林 Jilin, 天津 Tianjin, 甘肃 Gansu, 河北 Hebei, 河南 Henan, 江西 Jiangxi] 文雅扩腹春蜓 *Stylurus placidus* Liu & Chao, 1990 [四川 Sichuan] 深山扩腹春蜓 *Stylurus takashii* (Asahina, 1966) [台湾 Taiwan] 铜仁扩腹春蜓 *Stylurus tongrensis* Liu, 1991 [贵州 Guizhou]
棘尾春蜓属 Genus *Trigomphus* Bartenev, 1911 野居棘尾春蜓 *Trigomphus agricola* (Ris, 1916) 黄唇棘尾春蜓 *Trigomphus beatus* Chao, 1954 亲棘尾春蜓 *Trigomphus carus* Chao, 1954 [浙江 Zhejiang, 福建 Fujian] 吉林棘尾春蜓 *Trigomphus citimus* (Needham, 1931) 海南棘尾春蜓 *Trigomphus hainanensis* Zhang & Tong, 2009 净棘尾春蜓 *Trigomphus lautus* (Needham, 1931) 黑足棘尾春蜓 *Trigomphus nigripes* (Selys, 1887) 斜纹棘尾春蜓 *Trigomphus succumbens* (Needham, 1930) [吉林 Jilin] 斯氏棘尾春蜓 *Trigomphus svenhedini* Sjöstedt, 1932 云南棘尾春蜓 *Trigomphus yunnanensis* Zhou & Wu, 1992 [云南 Yunnan]
裂唇蜓科 Family Chlorogomphidae (3属32种1亚种 3 genera 32 species 1 subspecies)
裂唇蜓属 Genus *Chlorogomphus* Selys, 1854
蝶裂唇蜓亚属 Subgenus *Aurorachlorus* Carle, 1995 蝴蝶裂唇蜓 *Chlorogomphus* (*Aurorachlorus*) *papilio* Ris, 1927
金翅裂唇蜓亚属 Subgenus *Neorogomphus* Carle, 1995 黄翅裂唇蜓 *Chlorogomphus* (*Neorogomphus*) *auratus* Martin, 1910 金翅裂唇蜓 *Chlorogomphus* (*Neorogomphus*) *canhvang* Kompier & Karube, 2018 戴维裂唇蜓 *Chlorogomphus* (*Neorogomphus*) *daviesi* Karube, 2001 褐基裂唇蜓 *Chlorogomphus* (*Neorogomphus*) *yokoii* Karube, 1995 金翅裂唇蜓亚属待定种 *Chlorogomphus* (*Neorogomphus*) sp.
山裂唇蜓亚属 Subgenus *Orogomphus* Selys, 1878 金翼裂唇蜓 *Chlorogomphus* (*Orogomphus*) *auripennis* Zhang & Cai, 2014

续表

短痣裂唇蜓 *Chlorogomphus* (*Orogomphus*) *brevistigma* Oguma, 1926 李氏裂唇蜓 *Chlorogomphus* (*Orogomphus*) *risi* Chen, 1950 阔翅裂唇蜓 *Chlorogomphus* (*Orogomphus*) *splendidus* (Selys, 1878) [台湾 Taiwan] 斑翅裂唇蜓 *Chlorogomphus* (*Orogomphus*) *usudai* Ishida, 1996 Syn. *Chlorogomphus icarus* Wilson & Reels, 2001
华裂唇蜓亚属 Subgenus *Sinorogomphus* Carle, 1995 细腹裂唇蜓 *Chlorogomphus* (*Sinorogomphus*) *gracilis* Wilson & Reels, 2001 老挝裂唇蜓 *Chlorogomphus* (*Sinorogomphus*) *hiten* (Sasamoto, Yokoi & Teramoto, 2011) 褐翅裂唇蜓 *Chlorogomphus* (*Sinorogomphus*) *infuscatus* Needham, 1930 Syn. *Chlorogomphus infuscatus holophaea* Navás, 1936 长腹裂唇蜓 *Chlorogomphus* (*Sinorogomphus*) *kitawakii* Karube, 1995 武夷裂唇蜓 *Chlorogomphus* (*Sinorogomphus*) *montanus* (Chao, 1999) 长鼻裂唇蜓指名亚种 *Chlorogomphus* (*Sinorogomphus*) *nasutus nasutus* Needham, 1930 长鼻裂唇蜓越南亚种 *Chlorogomphus* (*Sinorogomphus*) *nasutus satoi* Asahina, 1995 朴氏裂唇蜓 *Chlorogomphus* (*Sinorogomphus*) *piaoacensis* Karube, 2013 中越裂唇蜓 *Chlorogomphus* (*Sinorogomphus*) *sachiyoae* Karube, 1995 山裂唇蜓 *Chlorogomphus* (*Sinorogomphus*) *shanicus* Wilson, 2002 铃木裂唇蜓 *Chlorogomphus* (*Sinorogomphus*) *suzukii* (Oguma, 1926) Syn.n. *Chlorogomphus tunti* Needham, 1930 叶尾裂唇蜓 *Chlorogomphus* (*Sinorogomphus*) *urolobatus* Chen, 1950 华裂唇蜓亚属待定种1 *Chlorogomphus* (*Sinorogomphus*) sp. 1 华裂唇蜓亚属待定种2 *Chlorogomphus* (*Sinorogomphus*) sp. 2 华裂唇蜓亚属待定种3 *Chlorogomphus* (*Sinorogomphus*) sp. 3 华裂唇蜓亚属待定种4 *Chlorogomphus* (*Sinorogomphus*) sp. 4
花裂唇蜓亚属 Subgenus *Petaliorogomphus* Karube, 2013 黄唇裂唇蜓 *Chlorogomphus* (*Petaliorogomphus*) *miyashitai* Karube, 1995 花裂唇蜓亚属待定种 *Chlorogomphus* (*Petaliorogomphus*) sp.
凹尾裂唇蜓属 Genus *Chloropetalia* Carle, 1995 赛丽裂唇蜓 *Chloropetalia selysi* (Fraser, 1929)
楔尾裂唇蜓属 Genus *Watanabeopetalia* Karube, 2002 高翔裂唇蜓 *Watanabeopetalia* (*Matsumotopetalia*) *soarer* (Wilson, 2002) U纹裂唇蜓 *Watanabeopetalia* (*Matsumotopetalia*) *usignata* (Chao, 1999) 楔尾裂唇蜓属待定种 *Watanabeopetalia* (*Matsumotopetalia*) sp.
大蜓科 Family Cordulegastridae (2属28种 2 genera 28 species)
圆臀大蜓属 Genus *Anotogaster* Selys, 1854 黄肩圆臀大蜓 *Anotogaster antehumeralis* Lohmann, 1993 [新疆 Xinjiang] 赵氏圆臀大蜓 *Anotogaster chaoi* Zhou, 1998 长角圆臀大蜓 *Anotogaster cornutifrons* Lohmann, 1993 [陕西 Shaanxi]

续表

黑额圆臀大蜓 *Anotogaster gigantica* Fraser, 1924
格氏圆臀大蜓 *Anotogaster gregoryi* Fraser, 1924
金斑圆臀大蜓 *Anotogaster klossi* Fraser, 1919
? Syn. *Anotogaster flaveola* Lohman, 1993
双斑圆臀大蜓 *Anotogaster kuchenbeiseri* (Förster, 1899)
细纹圆臀大蜓 *Anotogaster myosa* Needham, 1930
褐面圆臀大蜓 *Anotogaster nipalensis* (Selys, 1854)
清六圆臀大蜓 *Anotogaster sakaii* Zhou, 1988
萨帕圆臀大蜓 *Anotogaster sapaensis* Karube, 2012
巨圆臀大蜓 *Anotogaster sieboldii* (Selys, 1854)
圆臀大蜓属待定种1 *Anotogaster* sp. 1
圆臀大蜓属待定种2 *Anotogaster* sp. 2
圆臀大蜓属待定种3 *Anotogaster* sp. 3
圆臀大蜓属待定种4 *Anotogaster* sp. 4
圆臀大蜓属待定种5 *Anotogaster* sp. 5
圆臀大蜓属待定种6 *Anotogaster* sp. 6
圆臀大蜓属待定种7 *Anotogaster* sp. 7
圆臀大蜓属待定种8 *Anotogaster* sp. 8
角臀大蜓属 Genus *Neallogaster* Cowley, 1934
云南角臀大蜓 *Neallogaster annandalei* (Fraser, 1924)
周氏角臀大蜓 *Neallogaster choui* Yang & Li, 1994 [陕西 Shaanxi]
浅色角臀大蜓 *Neallogaster hermionae* (Fraser, 1927)
晋角臀大蜓 *Neallogaster jinensis* (Zhu & Han, 1992) [山西 Shanxi]
Syn. *Neallogaster lieftincki* Lohmann, 1993
褐面角臀大蜓 *Neallogaster latifrons* (Selys, 1878)
月纹角臀大蜓 *Neallogaster lunifera* (Selys, 1878) [四川 Sichuan]
东方角臀大蜓 *Neallogaster orientalis* (Van Pelt, 1994) [山东 Shandong]
北京角臀大蜓 *Neallogaster pekinensis* (Selys, 1886)
伪蜻科 Family Corduliidae (5属17种 5 genera 17 species)
伪蜻属 Genus *Cordulia* Leach, 1815
青铜伪蜻 *Cordulia amurensis* Selys, 1887
毛伪蜻属 Genus *Epitheca* Burmeister, 1839
虎斑毛伪蜻 *Epitheca bimaculata* (Charpentier, 1825)
缘斑毛伪蜻 *Epitheca marginata* (Selys, 1883)
半伪蜻属 Genus *Hemicordulia* Selys, 1870
高山半伪蜻 *Hemicordulia edai* Karube & Katatani, 2012
岷峨半伪蜻 *Hemicordulia mindana nipponica* Asahina,1980
半伪蜻属待定种 *Hemicordulia* sp.

续表

褐伪蜻属 Genus *Procordulia* Martin, 1907
朝比奈褐伪蜻 *Procordulia asahinai* Karube, 2007
金光伪蜻属 Genus *Somatochlora* Selys, 1871
高地金光伪蜻 *Somatochlora alpestris* (Selys, 1840)
北极金光伪蜻 *Somatochlora arctica* (Zetterstedt, 1840)
绿金光伪蜻 *Somatochlora dido* Needham, 1930
日本金光伪蜻 *Somatochlora exuberata* Bartenev, 1910
格氏金光伪蜻 *Somatochlora graeseri* Selys, 1887
灵隐金光伪蜻 *Somatochlora lingyinensis* Zhou & Wei, 1979 [浙江 Zhejiang]
凝翠金光伪蜻 *Somatochlora metallica* (Vander Linden, 1825)
山西金光伪蜻 *Somatochlora shanxiensis* Zhu & Zhang, 1999
神农金光伪蜻 *Somatochlora shennong* Zhang, Vogt & Cai, 2014
台湾金光伪蜻 *Somatochlora taiwana* Inoue & Yokota, 2001
大伪蜻科 Family Macromiidae (2属30种1亚种 2 genera 30 species 1 subspecies)
丽大伪蜻属 Genus *Epophthalmia* Burmeister, 1839
闪蓝丽大伪蜻 *Epophthalmia elegans* (Brauer, 1865)
黄斑丽大伪蜻 *Epophthalmia frontalis* Selys, 1871
Syn.n. *Epophthalmia bannaensis* Zha & Jiang, 2010
大伪蜻属 Genus *Macromia* Rambur, 1842
圆大伪蜻 *Macromia amphigena* Selys, 1871
北京大伪蜻 *Macromia beijingensis* Zhu & Chen, 2005
伯兰大伪蜻 *Macromia berlandi* Lieftinck, 1941
笛尾大伪蜻 *Macromia calliope* Ris, 1916
泰国大伪蜻 *Macromia chaiyaphumensis* Hämäläinen, 1986
海神大伪蜻 *Macromia clio* Ris, 1916
Syn. *Macromia hamifera* Lieftinck, 1955
褐蓝大伪蜻 *Macromia cupricincta* Fraser, 1924
大斑大伪蜻 *Macromia daimoji* Okumura, 1949
Syn. *Macromia chui* Asahina, 1968
黄斑大伪蜻 *Macromia flavocolorata* Fraser, 1922
亮面大伪蜻 *Macromia fulgidifrons* Wilson, 1998
锤钩大伪蜻 *Macromia hamata* Zhou, 2003
广东大伪蜻 *Macromia icterica* Lieftinck, 1929 [广东 Guangdong]
天使大伪蜻 *Macromia katae* Wilson, 1993
克氏大伪蜻 *Macromia kiautai* Zhou, Wang, Shuai & Liu, 1994 [云南 Yunnan]
斑点大伪蜻 *Macromia macula* Zhou, Wang, Shuai & Liu, 1994 [浙江 Zhejiang]
福建大伪蜻 *Macromia malleifera* Lieftinck, 1955
东北大伪蜻 *Macromia manchurica* Asahina, 1964

续表

莫氏大伪蜻指名亚种 *Macromia moorei moorei* Selys, 1874

莫氏大伪蜻马来亚种 *Macromia moorei malayana* Laidlaw, 1928

褐面大伪蜻 *Macromia pinratani vietnamica* Asahina, 1996

沙天马大伪蜻 *Macromia septima* Martin, 1904

弯钩大伪蜻 *Macromia unca* Wilson, 2004

天王大伪蜻 *Macromia urania* Ris, 1916

万荣大伪蜻 *Macromia vangviengensis* Yokoi & Mitamura, 2002

云南大伪蜻 *Macromia yunnanensis* Zhou, Luo, Hu & Wu, 1993 [云南 Yunnan]

大伪蜻属待定种1 *Macromia* sp. 1

大伪蜻属待定种2 *Macromia* sp. 2

大伪蜻属待定种3 *Macromia* sp. 3

大伪蜻属待定种4 *Macromia* sp. 4

综蜻科 Family Synthemistidae (2属19种 2 genera 19 species)

异伪蜻属 Genus *Idionyx* Hagen, 1867

朝比奈异伪蜻 *Idionyx asahinai* Karube, 2011

长角异伪蜻 *Idionyx carinata* Fraser, 1926

Syn. *Idionyx lieftincki* Zhou, 1984

郁异伪蜻 *Idionyx claudia* Ris, 1912

三角异伪蜻 *Idionyx optata* Selys, 1878

伪威异伪蜻 *Idionyx pseudovictor* Xu, 2013 [福建 Fujian]

赛丽异伪蜻 *Idionyx selysi* Fraser, 1926

黄面异伪蜻 *Idionyx stevensi* Fraser, 1924

具角异伪蜻 *Idionyx unguiculata* Fraser, 1926 [广西 Guangxi]

威异伪蜻 *Idionyx victor* Hämäläinen, 1991

云南异伪蜻 *Idionyx yunnanensis* Zhou, Wang, Shuai & Liu, 1994

异伪蜻属待定种1 *Idionyx* sp. 1

异伪蜻属待定种2 *Idionyx* sp. 2

中伪蜻属 Genus *Macromidia* Martin, 1907

伊中伪蜻 *Macromidia ellenae* Wilson, 1996

黑尾中伪蜻 *Macromidia genialis* Laidlaw, 1923

黄尾中伪蜻 *Macromidia ishidai* Asahina, 1964

克氏中伪蜻 *Macromidia kelloggi* Asahina, 1978

Syn. *Macromidia hangzhouensis* Zhou & Wei, 1979

飓中伪蜻 *Macromidia rapida* Martin, 1907

Syn. *Macromia cantonensis* Tinkham, 1936

谢氏中伪蜻 *Macromidia shiehae* Jiang, Li & Yu, 2008 [江西 Jiangxi]

中伪蜻属待定种 *Macromidia* sp.

蜻科 Family Libellulidae (42属137种13亚种 42 genera 137 species 13 subspecies)

续表

续表
锥腹蜻属 Genus *Acisoma* Rambur, 1842
锥腹蜻 ***Acisoma panorpoides*** Rambur, 1842
异蜻属 Genus *Aethriamanta* Kirby, 1889
褐基异蜻 ***Aethriamanta aethra*** Ris, 1912
红腹异蜻 ***Aethriamanta brevipennis*** (Rambur, 1842)
霜蓝异蜻 ***Aethriamanta gracilis*** (Brauer, 1878)
豹纹蜻属 Genus *Agrionoptera* Brauer, 1864
豹纹蜻指名亚种 ***Agrionoptera insignis insignis*** (Rambur, 1842)
豹纹蜻台湾亚种 ***Agrionoptera insignis similis*** Selys, 1879
安蜻属 Genus *Amphithemis* Selys, 1891
红安蜻 ***Amphithemis curvistyla*** Selys, 1891
长腹安蜻 ***Amphithemis vacillans*** Selys, 1891
黑斑蜻属 Genus *Atratothemis* Wilson, 2005
黑斑蜻 ***Atratothemis reelsi*** Wilson, 2005
疏脉蜻属 Genus *Brachydiplax* Brauer, 1868
褐胸疏脉蜻 ***Brachydiplax chalybea*** Brauer, 1868
霜白疏脉蜻 ***Brachydiplax farinosa*** Krüger, 1902
蓝额疏脉蜻 ***Brachydiplax flavovittata*** Ris,1911
暗色疏脉蜻 ***Brachydiplax sobrina*** (Rambur, 1842)
云南疏脉蜻 *Brachydiplax yunnanensis* Fraser, 1924 [云南 Yunnan]
疏脉蜻属待定种 ***Brachydiplax*** sp.
黄翅蜻属 Genus *Brachythemis* Brauer, 1868
黄翅蜻 ***Brachythemis contaminata*** (Fabricius, 1793) Syn. *Sympetrum aureolum* Navás, 1932
岩蜻属 Genus *Bradinopyga* Kirby, 1893
赭岩蜻 ***Bradinopyga geminata*** (Rambur, 1842)
巨蜻属 Genus *Camacinia* Kirby, 1889
亚洲巨蜻 ***Camacinia gigantea*** (Brauer, 1867)
森林巨蜻 ***Camacinia harterti*** Karsch, 1890
林蜻属 Genus *Cratilla* Kirby, 1900
线纹林蜻台湾亚种 ***Cratilla lineata assidua*** Lieftinck, 1953
线纹林蜻指名亚种 ***Cratilla lineata lineata*** (Brauer, 1878)
红蜻属 Genus *Crocothemis* Brauer, 1868
长尾红蜻 ***Crocothemis erythraea*** (Brullé, 1832)
红蜻古北亚种 ***Crocothemis servilia mariannae*** Kiauta, 1983
红蜻指名亚种 ***Crocothemis servilia*** servilia (Drury, 1773) ?Syn. *Rhodothemis flavostigma* Navás, 1932
多纹蜻属 Genus *Deielia* Kirby, 1889

续表

异色多纹蜻 *Deielia phaon* (Selys, 1883) Syn. *Deielia phaon brevistigma* Oguma, 1915 Syn. *Leucorrhinia nebulifera* Navás, 1935
蓝小蜻属 Genus *Diplacodes* Kirby, 1889 斑蓝小蜻 ***Diplacodes nebulosa*** (Fabricius, 1793) 纹蓝小蜻 ***Diplacodes trivialis*** (Rambur, 1842)
楔翅蜻属 Genus *Hydrobasileus* Kirby, 1889 臀斑楔翅蜻 ***Hydrobasileus croceus*** (Brauer, 1867)
沼蜻属 Genus *Hylaeothemis* Ris, 1909 雨林沼蜻 ***Hylaeothemis clementia*** Ris, 1909
印蜻属 Genus *Indothemis* Ris, 1909 深蓝印蜻 ***Indothemis carnatica*** (Fabricius, 1798) 蓝黑印蜻 ***Indothemis limbata*** (Selys, 1891)
秘蜻属 Genus *Lathrecista* Kirby, 1889 亚洲秘蜻 ***Lathrecista asiatica*** (Fabricius, 1798)
白颜蜻属 Genus *Leucorrhinia* Brittinger, 1850 短斑白颜蜻 ***Leucorrhinia dubia*** (Vander Linden, 1825) 居间白颜蜻 ***Leucorrhinia intermedia*** Bartenev, 1912
蜻属 Genus *Libellula* Linnaeus, 1758 低斑蜻 ***Libellula angelina*** Selys, 1883 高斑蜻 ***Libellula basilinea*** McLachlan, 1894 基斑蜻 ***Libellula depressa*** Linnaeus, 1758 米尔蜻 ***Libellula melli*** Schmidt, 1948 小斑蜻 ***Libellula quadrimaculata*** Linnaeus, 1758
宽腹蜻属 Genus *Lyriothemis* Brauer, 1868 西藏宽腹蜻 *Lyriothemis acigastra* (Selys, 1878) [西藏 Tibet] 双纹宽腹蜻 ***Lyriothemis bivittata*** (Rambur, 1842) 华丽宽腹蜻 ***Lyriothemis elegantissima*** Selys, 1883 金黄宽腹蜻 ***Lyriothemis flava*** Oguma, 1915 Syn. *Lyriothemis tricolor* Ris, 1916 卡米宽腹蜻 ***Lyriothemis kameliyae*** Kompier, 2017 闪绿宽腹蜻 ***Lyriothemis pachygastra*** (Selys, 1878)
漭蜻属 Genus *Macrodiplax* Brauer, 1868 高翔漭蜻 ***Macrodiplax cora*** (Brauer, 1867)
红小蜻属 Genus *Nannophya* Rambur, 1842 侏红小蜻 ***Nannophya pygmaea*** Rambur, 1842 Syn. *Nannodiplax yutsehongi* Navás, 1935
斑小蜻属 Genus *Nannophyopsis* Lieftinck, 1935

续表

膨腹斑小蜻 *Nannophyopsis clara* (Needham, 1930)
脉蜻属 Genus *Neurothemis* Brauer, 1867 月斑脉蜻 *Neurothemis fluctuans* (Fabricius, 1793) 网脉蜻 *Neurothemis fulvia* (Drury, 1773) 褐基脉蜻 *Neurothemis intermedia* (Rambur, 1842) 台湾脉蜻 *Neurothemis taiwanensis* Seehausen & Dow, 2016 截斑脉蜻 *Neurothemis tullia* (Drury, 1773) 脉蜻属待定种 *Neurothemis* sp.
爪蜻属 Genus *Onychothemis* Brauer, 1868 红腹爪蜻 *Onychothemis culminicola* Förster, 1904 雨林爪蜻 *Onychothemis testacea* Laidlaw, 1902 海湾爪蜻 *Onychothemis tonkinensis* Martin, 1904
灰蜻属 Genus *Orthetrum* Newman, 1833 白尾灰蜻 *Orthetrum albistylum* Selys, 1848 天蓝灰蜻 *Orthetrum brunneum* (Fonscolombe, 1837) 粗灰蜻 *Orthetrum cancellatum* (Linnaeus, 1758) 华丽灰蜻 *Orthetrum chrysis* (Selys, 1891) 黑尾灰蜻 *Orthetrum glaucum* (Brauer, 1865) 褐肩灰蜻 *Orthetrum internum* McLachlan, 1894 线痣灰蜻 *Orthetrum lineostigma* (Selys, 1886) 吕宋灰蜻 *Orthetrum luzonicum* (Brauer, 1868) Syn. *Orthetrum devium* Needham, 1930 异色灰蜻 *Orthetrum melania melania* (Selys, 1883) 斑灰蜻 *Orthetrum poecilops* Ris, 1919 赤褐灰蜻西里亚种 *Orthetrum pruinosum clelia* (Selys, 1878) 赤褐灰蜻中印亚种 *Orthetrum pruinosum neglectum* (Rambur, 1842) Syn. *Libellula petalura* Brauer, 1865 狭腹灰蜻 *Orthetrum sabina* (Drury, 1773) 鼎脉灰蜻 *Orthetrum triangulare* (Selys, 1878) 灰蜻属待定种1 *Orthetrum* sp. 1 灰蜻属待定种2 *Orthetrum* sp. 2 灰蜻属待定种3 *Orthetrum* sp. 3 灰蜻属待定种4 *Orthetrum* sp. 4
曲缘蜻属 Genus *Palpopleura* Rambur, 1842 六斑曲缘蜻 *Palpopleura sexmaculata* (Fabricius, 1787) Syn. *Aeshna minuta* Fabricius, 1787
黄蜻属 Genus *Pantala* Hagen, 1861 黄蜻 *Pantala flavescens* (Fabricius, 1798)

续表

长足蜻属 Genus *Phyllothemis* Fraser, 1935 沼长足蜻 ***Phyllothemis eltoni*** Fraser, 1935
狭翅蜻属 Genus *Potamarcha* Karsch, 1890 湿地狭翅蜻 ***Potamarcha congener*** (Rambur, 1842) 黄面狭翅蜻 *Potamarcha puella* Needham, 1930 [江苏 Jiangsu]
玉带蜻属 Genus *Pseudothemis* Kirby, 1889 玉带蜻 ***Pseudothemis zonata*** (Burmeister, 1839) Syn. *Orthetrum zonale* Navás, 1932
红胭蜻属 Genus *Rhodothemis* Ris, 1909 红胭蜻 ***Rhodothemis rufa*** (Rambur, 1842)
丽翅蜻属 Genus *Rhyothemis* Hagen, 1867 黑丽翅蜻 ***Rhyothemis fuliginosa*** Selys, 1883 青铜丽翅蜻 ***Rhyothemis obsolescens*** Kirby, 1889 臀斑丽翅蜻 ***Rhyothemis phyllis*** (Sulzer, 1776) 曜丽翅蜻 ***Rhyothemis plutonia*** Selys, 1883 灰黑丽翅蜻 ***Rhyothemis regia*** (Brauer, 1867) 赛琳丽翅蜻 ***Rhyothemis severini*** Ris, 1913 三角丽翅蜻 ***Rhyothemis triangularis*** Kirby, 1889 斑丽翅蜻多斑亚种 ***Rhyothemis variegata arria*** Drury, 1773 Syn. *Libellula splendida* Rambur, 1842 斑丽翅蜻指名亚种 ***Rhyothemis variegata variegata*** (Linnaeus, 1763) 丽翅蜻属待定种 ***Rhyothemis*** sp.
赛丽蜻属 Genus *Selysiothemis* Ris, 1897 黑赛丽蜻 ***Selysiothemis nigra*** (Vander Linden, 1825)
赤蜻属 Genus *Sympetrum* Newman, 1833 暗赤蜻 *Sympetrum anomalum* Needham, 1930 [江苏 Jiangsu] 大赤蜻指名亚种 ***Sympetrum baccha baccha*** (Selys, 1884) 大赤蜻褐顶亚种 ***Sympetrum baccha matutinum*** Ris, 1911 长尾赤蜻 ***Sympetrum cordulegaster*** (Selys, 1883) 半黄赤蜻 ***Sympetrum croceolum*** Selys, 1883 大理赤蜻 ***Sympetrum daliensis*** Zhu, 1999 黑赤蜻 ***Sympetrum danae*** (Sulzer, 1776) 夏赤蜻 ***Sympetrum darwinianum*** Selys, 1883 Syn. *Diplax sinensis* Selys, 1883 扁腹赤蜻 ***Sympetrum depressiusculum*** (Selys, 1841) 竖眉赤蜻指名亚种 ***Sympetrum eroticum eroticum*** (Selys, 1883) Syn. *Sympetrum ignotum* Needham, 1930 竖眉赤蜻多纹亚种 ***Sympetrum eroticum ardens*** (McLachlan, 1894)

续表

黄斑赤蜻 *Sympetrum flaveolum* (Linnaeus, 1758)
方氏赤蜻 *Sympetrum fonscolombii* (Selys, 1840)
秋赤蜻 *Sympetrum frequens* (Selys, 1883)
旭光赤蜻 *Sympetrum hypomelas* (Selys, 1884)
褐顶赤蜻 *Sympetrum infuscatum* (Selys, 1883)
小黄赤蜻 *Sympetrum kunckeli* (Selys, 1884)
Syn. *Sympetrum flavicauda* Navás, 1932
南投赤蜻 *Sympetrum nantouensis* Tang, Yeh & Chen, 2013
牧赤蜻 *Sympetrum nomurai* Asahina, 1997 [四川 Sichuan]
东方赤蜻 *Sympetrum orientale* (Selys, 1883) [中国 China]
姬赤蜻 *Sympetrum parvulum* (Bartenev, 1913)
褐带赤蜻 *Sympetrum pedemontanum* (Müller, 1766)
李氏赤蜻 *Sympetrum risi risi* Bartenev, 1914
双横赤蜻 *Sympetrum ruptum* Needham, 1930 [四川 Sichuan]
血红赤蜻 *Sympetrum sanguineum* (Müller, 1764)
黄基赤蜻微斑亚种 *Sympetrum speciosum haematoneura* Fraser, 1924
黄基赤蜻指名亚种 *Sympetrum speciosum speciosum* Oguma, 1915
黄基赤蜻台湾亚种 *Sympetrum speciosum taiwanum* Asahina,1951
条斑赤蜻喜马亚种 *Sympetrum striolatum commixtum* (Selys, 1884)
条斑赤蜻指名亚种 *Sympetrum striolatum striolatum* (Charpentier, 1840)
Syn. *Sympetrum striolatum pallidum* Selys, 1887
黄足赤蜻 *Sympetrum tibiale* (Ris, 1897) [新疆 Xinjiang]
大黄赤蜻 *Sympetrum uniforme* (Selys, 1883)
Syn. *Sympetrum fatigans* Needham, 1930
普赤蜻 *Sympetrum vulgatum* (Linnaeus, 1758)
Syn. *Diplax imitans* Selys, 1886
肖氏赤蜻 *Sympetrum xiaoi* Han & Zhu, 1997
赤蜻属待定种1 *Sympetrum* sp. 1
赤蜻属待定种2 *Sympetrum* sp. 2
赤蜻属待定种3 *Sympetrum* sp. 3
赤蜻属待定种4 *Sympetrum* sp. 4

方蜻属 Genus *Tetrathemis* Brauer, 1868
钩尾方蜻 *Tetrathemis irregularis* Brauer, 1868
宽翅方蜻 *Tetrathemis platyptera* Selys, 1878

云斑蜻属 Genus *Tholymis* Hagen, 1867
云斑蜻 *Tholymis tillarga* (Fabricius, 1798)

斜痣蜻属 Genus *Tramea* Hagen, 1861
浅色斜痣蜻 *Tramea basilaris burmeisteri* Kirby,1889

续表

缘环斜痣蜻 *Tramea limbata similata* Rambur, 1842
海神斜痣蜻微斑亚种 *Tramea transmarina euryale* (Selys, 1878)
海神斜痣蜻粗斑亚种 *Tramea transmarina propinqua* Lieftinck, 1942
华斜痣蜻 *Tramea virginia* (Rambur, 1842)
褐蜻属 Genus *Trithemis* Brauer, 1868
晓褐蜻 *Trithemis aurora* (Burmeister, 1839)
庆褐蜻 *Trithemis festiva* (Rambur, 1842)
灰脉褐蜻 *Trithemis pallidinervis* (Kirby, 1889)
曲钩脉蜻属 Genus *Urothemis* Brauer, 1868
赤斑曲钩脉蜻微斑亚种 *Urothemis signata insignata* (Selys, 1872)
赤斑曲钩脉蜻指名亚种 *Urothemis signata signata* (Rambur, 1842)
赤斑曲钩脉蜻台湾亚种 *Urothemis signata yiei* Asahina, 1972
虹蜻属 Genus *Zygonyx* Hagen, 1867
朝比奈虹蜻 *Zygonyx asahinai* Matsuki & Saito, 1995
彩虹蜻 *Zygonyx iris insignis* Kirby, 1900
高砂虹蜻 *Zygonyx takasago* Asahina, 1966
开臀蜻属 Genus *Zyxomma* Rambur, 1842
霜白开臀蜻 *Zyxomma obtusum* Albarda, 1881
细腹开臀蜻 Zyxomma *petiolatum* Rambur, 1842
存疑种 Dubious species from China
土黄小蟌 *Agriocnemis luteola* Navás, 1936 [浙江 Zhejiang]
皮尔小蟌 *Agriocnemis pieli* Navás, 1933 [浙江 Zhejiang]
疑绿蟌 *Enallagma ambiguum* Navás, 1936 [江西 Jiangxi]
阿赛蜓 *Aeshna athalia* Needham, 1930 [浙江 Zhejiang]
福罗蜓 *Aeshna frontalis* Navás, 1936 [江西 Jiangxi]
光明蜓 *Aeshna lucia* Needham, 1930 [湖北 Hubei]
管氏丽大伪蜻 *Epophthalmia kuani* Jiang, 1998 [江苏 Jiangsu]

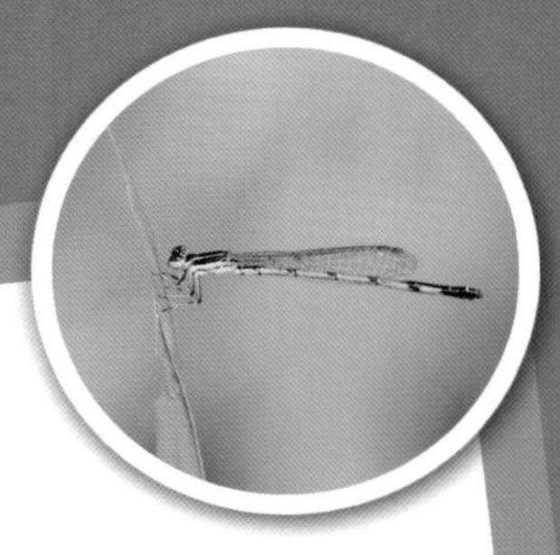

中国蜻蜓图鉴

DRAGONFLIES AND DAMSELFLIES OF CHINA

中国蜻蜓大图鉴
DRAGONFLIES AND DAMSELFLIES OF CHINA

差翅亚目

SUBORDER ANISOPTERA

1 蜓科 Family Aeshnidae

本科世界性分布，全球已知54属近500种，中国已知14属约100种。蜓科包含很多珍稀物种，例如头蜓属和黑额蜓属，其中多数种类仅有模式标本的记录。本科蜻蜓是体型中至大型的飞行高手，身体黑色或褐色并具丰富的条纹和斑点。复眼发达，在头顶交汇处呈一条直线，多数面部窄而长；胸部粗壮，绝大多数种类翅透明，少数染有色斑；腹部较长。雄性的肛附器、阳茎构造以及雌性的产卵器和尾毛长度是重要的辨识特征。

本科蜻蜓的栖息环境包括各种静水水域的池塘、湖泊和沼泽地，以及清澈的山区小溪。一些最常见的蜓科种类，比如拥有绿体色的伟蜓属种类喜欢栖息于静水环境，在繁华的大都市中亦可以见到。许多稀有物种则要到茂盛的森林或者特殊的环境才有机会见到。溪栖的蜓科物种是山区溪流中的幽灵，大多数种类惧怕阳光，白天很难发现，却在清晨和黄昏时非常活跃，它们高超的飞行技能可以轻松躲避采集人的捕虫网。

The dragonfly family Aeshnidae is distributed worldwide with about 500 species assigned to 54 genera. 14 genera and about 100 species are known from China. The family contains many rare, or seldom encountered, species, such as species of genera *Cephalaeschna* and *Planaeschna*. A majority of them are known only from type material. Species of Aeshnidae are medium to large strong-flying dragonflies. The body is fundamentally brown or black and often striped (thorax) or spotted (abdomen). The eyes are large and broadly confluent above, and their face is relatively narrow and long. The thorax is robust and the wings hyaline in most species but a few have coloured wings. The abdomen is long. Many species are similar in appearance and the male anal appendages, penis and female ovipositor and cerci are key diagnostic characters.

Species of the family are found at both standing water habitats (ponds, lakes and marshes) and clear montane streams. Some of the most familiar aeshnids, such as some brilliant green *Anax* species, prefer standing water, and adults can be easily seen even in the metropolis of many large cities. Many rare species can only be seen in dense forests or in specialized habitats. Most species at running water habitats are the "ghosts" and only active in early morning and twilight, their excellent flying capabilities coupled with their unpredictable path allows them to easily elude the collector's net.

东亚伟蜓 雄

Anax panybeus, male

绿蜓属 Genus *Aeschnophlebia* Selys, 1883

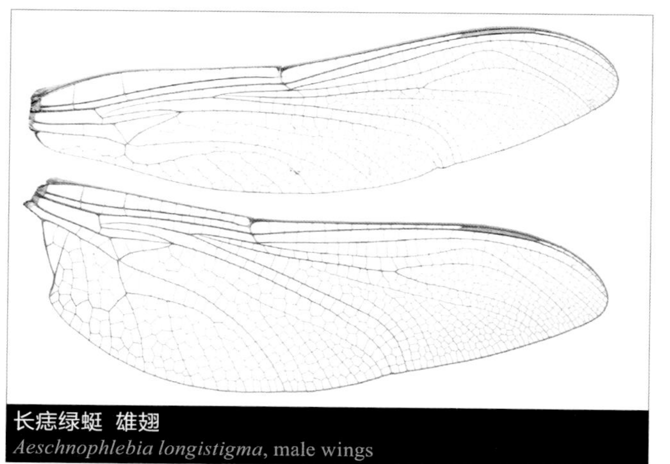

长痣绿蜓 雄翅
Aeschnophlebia longistigma, male wings

本属全世界已知4种，中国2种。其中长痣绿蜓在东北和华北地区较常见，而黑纹绿蜓仅有1个标本的记录。本属体中型至大型，雄性上肛附器以及雌性尾毛甚长，翅痣较长，基室无横脉，R3在近翅端显著弯曲，IR3叉状，臀三角室3室。

本属蜻蜓主要栖息环境是较低海拔的静水池塘以及流速缓慢的开阔溪流。成虫喜欢在芦苇丛中游荡，并且会经常停落在芦苇丛中。雌性会在芦苇的茎干中产卵。

This genus contains four species with two recorded from China. *Aeschnophlebia longistigma* is a common species in North and Northeast China, but *A. anisoptera* has only a single specimen record. Species of the genus are medium to large sized, male superior appendages and female cerci very long, pterostigma long, median space without crossveins, R3 strongly curved near the wing tips, IR3 forked, anal triangle 3-celled.

Aeschnophlebia species inhabit ponds and slow flowing streams in lowland. Adults usually fly in reeds, and sometimes perch. Females lay their eggs into the stalks of reed.

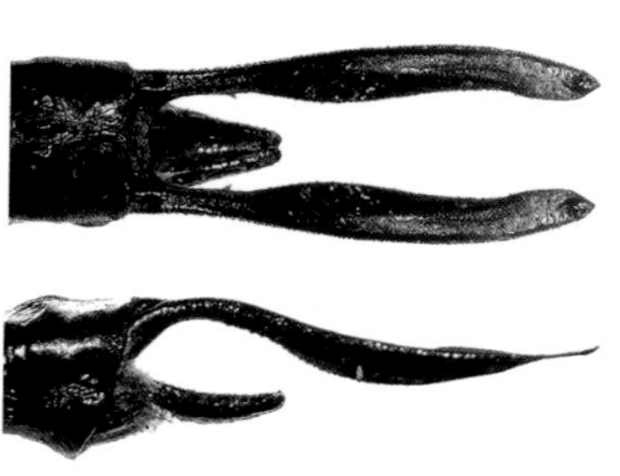

黑纹绿蜓
Aeschnophlebia anisoptera

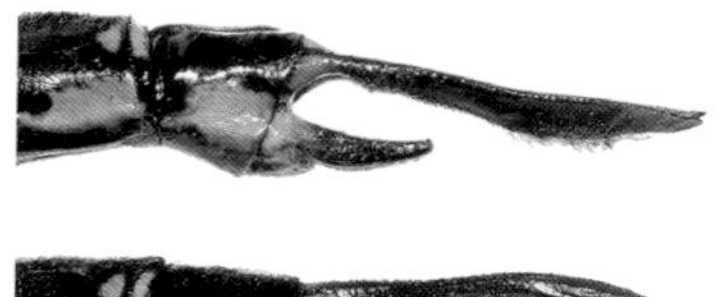

长痣绿蜓
Aeschnophlebia longistigma

绿蜓属 雄性肛附器
Genus *Aeschnophlebia*, male anal appenages

长痣绿蜓 交尾 | 金洪光 摄
Aeschnophlebia longistigma, mating pair | Photo by Hongguang Jin

黑纹绿蜓 *Aeschnophlebia anisoptera* Selys, 1883

【形态特征】雄性复眼深蓝色，面部黄色和黑色，上唇黑色具1条黄条纹，前唇基和后唇基下方黑色，前额具1个黑色斑，上额具黑色"T"形斑；合胸黑色，肩条纹甚阔，肩前下点甚小，侧面具2条宽阔的黄绿色条纹，后胸前侧板具2个小黄斑，足黑色，翅染有淡褐色；腹部黑色，第1～9节具黄色斑纹，上肛附器甚长。【长度】体长 75～88 mm，腹长 49～63 mm，后翅 45～55 mm。【栖息环境】挺水植物茂盛的湿地。【分布】浙江；朝鲜半岛、日本。【飞行期】不详。

黑纹绿蜓 雄，日本
Aeschnophlebia anisoptera, male from Japan

黑纹绿蜓 雌，日本
Aeschnophlebia anisoptera, female from Japan

[Identification] Male eyes dark blue, face yellow and black, labrum black with a yellow stripe, anteclypeus and lower margin of postclypeus black, antefrons with a black spot, top of frons with black T-mark. Thorax black with broad antehumeral stripes and small antehumeral spots, sides with two broad yellowish green stripes and two small yellow spots on metepisternum, legs black, wings tinted with light brown. Abdomen black, S1-S9 with yellow markings, superior appendages very long. **[Measurements]** Total length 75-88 mm, abdomen 49-63 mm, hind wing 45-55 mm. **[Habitat]** Wetlands with numerous emerging plants. **[Distribution]** Zhejiang; Korean peninsula, Japan. **[Flight Season]** Unknown.

长痣绿蜓 *Aeschnophlebia longistigma* Selys, 1883

【形态特征】复眼绿色带蓝色斑，面部黄绿色，上额具黑色"T"形斑；合胸绿色，脊黑色，肩条纹甚阔，黑色，足黑色具黄色条纹；腹部绿色，各节脊两侧具2条较长的黑带，雄性肛附器和雌性尾毛甚长。【长度】体长 67～69 mm，腹长 50～51 mm，后翅 41～45 mm。【栖息环境】海拔 500 m以下挺水植物茂盛的湿地。【分布】黑龙江、吉林、辽宁、河北、北京、山东、江苏；日本、朝鲜半岛、俄罗斯远东。【飞行期】5—7月。

[Identification] Eyes green with blue spots, face yellowish green, top of frons with black T-mark. Thorax green, dorsal carina black, the black humeral stripes broad, legs black with yellow stripes. Abdomen green with two long bands besides the dorsal carina. Male anal appendages and female cerci long. **[Measurements]** Total length 67-69 mm, abdomen 50-51 mm, hind wing 41-45 mm. **[Habitat]** Wetlands with numerous emerging plants below 500 m elevation. **[Distribution]** Heilongjiang, Jilin, Liaoning, Hebei, Beijing, Shandong, Jiangsu; Japan, Korean peninsula, Russian Far East. **[Flight Season]** May to July.

长痣绿蜓 雄，吉林 | 金洪光 摄
Aeschnophlebia longistigma, male from Jilin | Photo by Honguang Jin

长痣绿蜓 交尾，吉林 | 金洪光 摄
Aeschnophlebia longistigma, mating pair from Jilin | Photo by Hongguang Jin

长痣绿蜓 雌，北京 | 陈炜 摄
Aeschnophlebia longistigma, female from Beijing | Photo by Wei Chen

蜓属 Genus *Aeshna* Fabricius, 1775

本属全世界已知30余种，主要分布在北美洲和欧洲的温带区域，中国已知约10种，新疆地区可能会有一些欧洲广布种的新记录，而中部至南部的高山地区可能会有特殊的新物种等待发现，现已在湖北神农架的高山湿地发现本属较特殊的一种——神农蜓。本属蜻蜓，多是以蓝黄搭配，即雄性体色蓝色为主，雌性黄色为主。有些种类雌性

多型，分为黄色型和蓝色型。翅透明，雌性翅稍微染有琥珀色，基室无横脉，R3在翅痣下方弯曲，IR3叉状，臀三角室2～4室，雄性后翅臀角通常为角状，腹部第2节具耳状突。

本属蜻蜓多栖息于静水环境，偏爱挺水植物茂盛的池塘、湖泊和沼泽地。雄性具领域行为，通常在水面上小范围内来回飞行或者定点低空悬停。雌性在水草上产卵。本属成虫通常发生期较晚，在夏末和秋季开始繁殖。

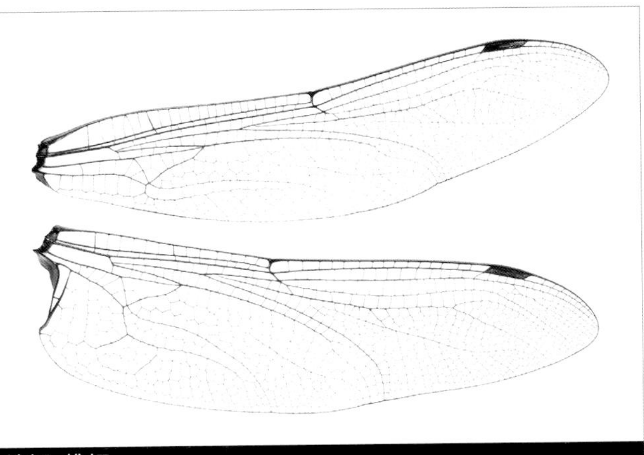

峻蜓 雄翅
Aeshna juncea, male wings

The genus contains over 30 species, most of them found in the temperate areas of North America and Europe. About ten species are recorded from China. Some species widespread in Europe and eastwards may be found in Xinjiang province and become new national records, and there are some potential undescribed species in the high mountains in the central and southern part of China. *Aeshna shennong*, a special species has been recently described from the high mountains in Shennongjia Mountains, Hubei Province. Most members of the genus are blue and

混合蜓 雄
Aeshna mixta, male

spotted yellow, males are usually spotted blue and females yellow. Female of many species are polymorphic, including the yellow morph and the blue morph. Wings hyaline, slightly tinted with amber in some females, median space without crossveins, R3 curved at the level of pterostigma, IR3 forked, anal triangle 2- to 4-celled, male tornus usually angled, S2 with auricles.

Aeshna species prefer standing water habitats, especially ponds, lakes ard swampy areas with plenty of emergent aquatic plants. Males hold the territory by patrolling a small area or hovering. Females lay their eggs into plants. Most members of the genus begin to fly in late summer and autumn.

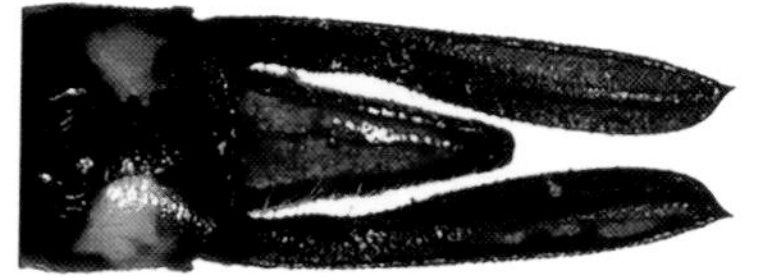

硕斑蜓
Aeshna affinis

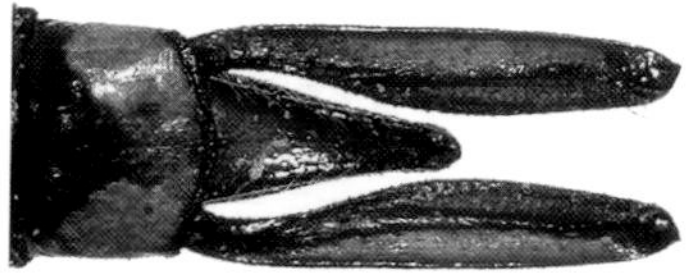

琉璃蜓
Aeshna crenata

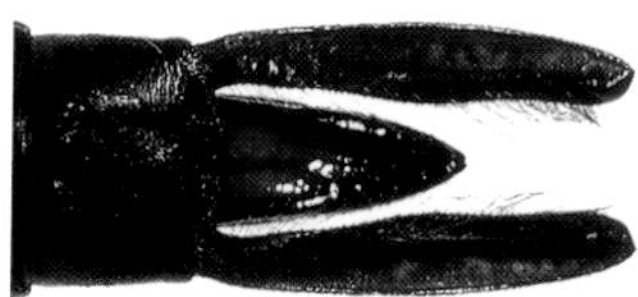

峻蜓
Aeshna juncea

混合蜓
Aeshna mixta

神农蜓
Aeshna shennong

蝶斑蜓指名亚种
Aeshna petalura petalura

极北蜓
Aeshna subarctica

蜓属 雄性肛附器
Genus *Aeshna*, male anal appenages

硕斑蜓 *Aeshna affinis* Vander Linden, 1820

【形态特征】雄性复眼天蓝色，面部白色，上额具1个黑色的"T"形斑；胸部大面积黄色，合胸背面褐色，足黑色，翅透明；腹部黑色具发达的蓝色斑。雌性多型，蓝色型与雄性相似，黄色型身体褐色具黄色条纹。【长度】体长 57～66 mm，腹长 39～49 mm，后翅 37～42 mm。【栖息环境】海拔较低、水草茂盛的静水环境。【分布】新疆；地中海至蒙古、非洲北部。【飞行期】6—8月。

[Identification] Male eyes sky-blue, face white, top of frons with a black T-mark. Thorax largely yellow with dorsal part brown, legs black, wings hyaline. Abdomen black with numerous blue markings. Female polymorphic, the blue morph similar to male and the yellow morph brown with yellow markings. [Measurements] Total length 57-66 mm, abdomen 39-49 mm, hind wing 37-42 mm. [Habitat] Standing water with numerous emerging plants in lowland. [Distribution] Xinjiang; Mediterranean to Mongolia, North Africa. [Flight Season] June to August.

硕斑蜓 雄，保加利亚 | Sami Karjalainen 摄
Aeshna affinis, male from Bulgaria | Photo by Sami Karjalainen

硕斑蜓 雌雄连结，俄罗斯 | Oleg E. Kosterin 摄
Aeshna affinis, pair in tandem from Russia | Photo by Oleg E. Kosterin

琉璃蜓 *Aeshna crenata* Hagen, 1856

【形态特征】雄性复眼深蓝色，面部白色，上额具1个黑色的"T"形斑；胸部大面积褐色，具较宽阔的肩前条纹，侧面具2条宽阔的蓝黄色条纹，足黑色，翅透明；腹部黑色具蓝色斑纹。雌性多型，蓝色型与雄性色彩相似，黄色型身体褐色具黄色条纹，翅端半部染有琥珀色。【长度】体长 71~86 mm，腹长 53~67 mm，后翅 44~60 mm。【栖息环境】海拔较低、水草茂盛的静水环境。【分布】新疆、内蒙古、黑龙江、吉林、辽宁；从欧洲东北部经西伯利亚至朝鲜半岛和日本。【飞行期】6—9月。

[Identification] Male eyes dark blue, face white, top of frons with a black T-mark. Thorax largely brown with broad antehumeral sripes, sides with two broad bluish yellow stripes, legs black, wings hyaline. Abdomen black with numerous blue markings. Female polymorphic, the blue morph similar to male and the yellow morph brown with yellow markings, wings with amber tint in the apical half. **[Measurements]** Total length 71-86 mm, abdomen 53-67 mm, hind wing 44-60 mm. **[Habitat]** Standing water with abundant emerging plants in lowland. **[Distribution]** Xinjiang, Inner Mongolia, Heilongjiang, Jilin, Liaoning; From northeastern Europe throughout Siberia to Korean peninsula and Japan. **[Flight Season]** June to September.

琉璃蜓 雌，产卵，吉林 | 张巍巍 摄
Aeshna crenata, female, laying eggs from Jilin | Photo by Weiwei Zhang

琉璃蜓 雌，产卵，黑龙江
Aeshna crenata, female, laying eggs from Heilongjiang

琉璃蜓 雄，黑龙江 | 莫善濂 摄
Aeshna crenata, male from Heilongjiang | Photo by Shanlian Mo

峻蜓 *Aeshna juncea* (Linnaeus, 1758)

【形态特征】雄性复眼深蓝色，面部淡黄色，上额具1个黑色的“T”形斑；胸部大面积褐色，具较宽阔的肩前条纹，侧面具2条宽阔的黄色条纹，足黑色，翅透明；腹部黑色具丰富的蓝色斑纹。雌性多型，蓝色型与雄性色彩相似，黄色型身体褐色具黄色条纹。【长度】体长 65~80 mm，腹长 50~59 mm，后翅 40~48 mm。【栖息环境】海拔 1500 m以上水草茂盛的静水环境。【分布】本种在中国的分布记录需要重新修订，许多省为错误记录，目前确定的分布地是内蒙古和湖北；北美洲、从欧洲北部经西伯利亚至朝鲜半岛和日本。【飞行期】6—9月。

峻蜓 雄，湖北
Aeshna juncea, male from Hubei

[Identification] Male eyes dark blue, face pale yellow, top of frons with a black T-mark. Thorax largely brown with broad antehumeral sripes, sides with two broad yellow stripes, legs black, wings hyaline. Abdomen black with numerous blue markings. Female polymorphic, the blue morph similar to male and the yellow morph brown with yellow markings. **[Measurements]** Total length 65-80 mm, abdomen 50-59 mm, hind wing 40-48 mm. **[Habitat]** Standing water with abundant emergent plants above 1500 m elevation. **[Distribution]** The distribution of this species in China is very poorly known, the published records from many provinces are misidentified, currently there are confirmed records only from Inner Mongolia and Hubei; North America, from the north of Europe throughout Siberia to Korean peninsula and Japan. **[Flight Season]** June to September.

峻蜓 雄，湖北
Aeshna juncea, male from Hubei

峻蜓 雌，黄色型，湖北
Aeshna juncea, female, the yellow morph from Hubei

峻蜓 雌，蓝色型，湖北
Aeshna juncea, female, the blue morph from Hubei

峻蜓 雌，黄色型，湖北
Aeshna juncea, female, the yellow morph from Hubei

峻蜓 雌，蓝色型，湖北
Aeshna juncea, female, the blue morph from Hubei

混合蜓 *Aeshna mixta* Latreille, 1805

混合蜓 雄，黑龙江
Aeshna mixta, male from Heilongjiang

混合蜓 雌，蓝色型，吉林 |金洪光 摄
Aeshna mixta, female, the blue morph from Jilin | Photo by Hongguang Jin

混合蜓 交尾，吉林 |金洪光 摄
Aeshna mixta, mating pair from Jilin | Photo by Hongguang Jin

【形态特征】雄性复眼天蓝色，面部白色，上额具1个黑色的"T"形斑；胸部大面积黄褐色，具甚短的肩前条纹，侧面具2条宽阔的黄条纹，足大面积黑色，基方红褐色，翅透明；腹部黑色具发达的蓝色斑。雌性多型，蓝色型与雄性色彩相似，黄色型身体褐色具黄色条纹。**【长度】**体长 56～64 mm，腹长 43～54 mm，后翅 37～42 mm。**【栖息环境】**海拔较低、水草茂盛的静水环境，在芦苇塘尤为常见。**【分布】**新疆、内蒙古、黑龙江、吉林、辽宁、河北、北京、山东；从欧洲分布至亚洲北部东至日本，非洲北部。**【飞行期】**7—10月。

[Identification] Male eyes sky-blue, face white, top of frons with a black T-mark. Thorax largely yellowish brown with very short antehumeral stripes, sides with two broad yellow stripes, legs largely black with bases reddish brown, wings hyaline. Abdomen black with numerous blue markings. Female polymorphic, the blue morph similar to male and the yellow morph brown with yellow markings. **[Measurements]** Total length 56-64 mm, abdomen 43-54 mm, hind wing 37-42 mm. **[Habitat]** Standing water with abundant emergent plants in lowland, especially the reed ponds. **[Distribution]** Xinjiang, Inner Mongolia, Heilongjiang, Jilin, Liaoning, Hebei, Beijing, Shandong; From Europe to the north of Asia and eastwards to Japan, the north of Africa. **[Flight Season]** July to October.

蝶斑蜓指名亚种 *Aeshna petalura petalura* Martin, 1909

【形态特征】雄性复眼深蓝色，面部黄色，前额上半部具1个黑色斑，上额具1个黑色的“T”形斑；胸部黑褐色，肩前条纹宽阔，侧面具2条宽阔的黄绿色条纹，足黑色，翅透明；腹部黑色具丰富的蓝色和黄色斑点，其中第3～7节背面具蝶形斑点。雌性腹部短而粗壮，产卵器较发达。【长度】体长 69～74 mm，腹长 52～56 mm，后翅 45～47 mm。【栖息环境】海拔 1000～3000 m的高山湿地和小型水塘。【分布】四川、贵州、广西；不丹、尼泊尔、印度。【飞行期】6—10月。

[Identification] Male eyes dark blue, face yellow, antefrons with a black spot in the upper half, top of frons with a black T-mark. Thorax blackish brown with broad antehumeral stripes, sides with two broad yellowish green stripes, legs black, wings hyaline. Abdomen black with numerous blue and yellow markings, dorsum of S3-S7 with butterfly-shaped

蝶斑蜓指名亚种 雄，四川
Aeshna petalura petalura, male from Sichuan

spots. Female abdomen short and stout, ovipositor developed. **[Measurements]** Total length 69-74 mm, abdomen 52-56 mm, hind wing 45-47 mm. **[Habitat]** Marshes and ponds in mountains at 1000-3000 m elevation. **[Distribution]** Sichuan, Guizhou, Guangxi; Bhutan, Nepal, India. **[Flight Season]** June to October.

蝶斑蜓指名亚种 雌，四川
Aeshna petalura petalura, female from Sichuan

蝶斑蜓台湾亚种 *Aeshna petalura taiyal* Asahina, 1938

【形态特征】雄性复眼深蓝色，面部黄色，前额上半部具1个黑色斑，上额具1个黑色的“T”形斑；胸部黑褐色，肩前条纹宽阔，侧面具2条宽阔的黄绿色条纹，足黑色，翅透明；腹部黑色具丰富的蓝色和黄色斑点，其中第3~7节背面具蝶形斑点。雌性腹部短而粗壮，产卵器较发达。【长度】体长 67~72 mm，腹长 50~54 mm，后翅 44~46 mm。【栖息环境】海拔 1000~3000 m的高山湿地和小型水塘。【分布】中国台湾特有。【飞行期】4—12月。

[Identification] Male eyes dark blue, face yellow, antefrons with a black spot in the upper half, top of frons with a black T-mark. Thorax blackish brown with broad antehumeral stripes, sides with two broad yellowish green stripes, legs black, wings hyaline. Abdomen black with numerous blue and yellow markings, dorsum of S3-S7 with butterfly-shaped spots. Female abdomen short and stout, ovipositor developed. **[Measurements]** Total length 67-72 mm, abdomen 50-54 mm, hind wing 44-46 mm. **[Habitat]** Marshes and ponds in mountains at 1000-3000 m elevation. **[Distribution]** Endemic to Taiwan of China. **[Flight Season]** April to December.

蝶斑蜓台湾亚种 雄，台湾 | 嘎嘎 摄
Aeshna petalura taiyal, male from Taiwan | Photo by Gaga

极北蜓 *Aeshna subarctica* Walker, 1908

【形态特征】雄性复眼深蓝色，面部淡黄色，上额具1个黑色“T”形斑；胸部大面积褐色，具较宽阔的肩前条纹，侧面具2条宽阔的黄色条纹，足黑色，翅透明；腹部黑色具丰富的蓝色斑纹。雌性多型，蓝色型与雄性色彩相似，黄色型身体褐色具黄色条纹。【长度】体长 70～76 mm，腹长 47～57 mm，后翅 39～46 mm。【栖息环境】海拔较低、水草茂盛的静水环境，在芦苇塘尤为常见。【分布】黑龙江、吉林、北京；北美洲、从欧洲分布至亚洲北部东至日本。【飞行期】6—10月。

[Identification] Male eyes dark blue, face pale yellow, top of frons with a black T-mark. Thorax largely brown with broad antehumeral stripes, sides with two broad yellow stripes, legs black, wings hyaline. Abdomen black with numerous blue markings. Female polymorphic, the blue morph similar to male and the yellow morph brown with yellow markings. [Measurements] Total length 70-76 mm, abdomen 47-57 mm, hind wing 39-46 mm. [Habitat] Standing water with abundant emergent plants in lowland, especially reed ponds. [Distribution] Heilongjiang, Jilin, Beijing; North America, from Europe to the north of Asia and eastwards to Japan. [Flight Season] June to October.

极北蜓 雄，黑龙江
Aeshna subarctica, male from Heilongjiang

极北蜓 雌，黑龙江
Aeshna subarctica, female from Heilongjiang

神农蜓 *Aeshna shennong* Zhang & Cai, 2014

【形态特征】雄性复眼深蓝色，面部黄褐色，前额上半部具1个黑色斑，上额具1个黑色的“T”形斑；胸部黑褐色，肩前条纹中央间断，肩条纹仅为下方的1个小圆点，侧面大面积黄色，足大面积黑色，翅透明；腹部黑色具丰富的蓝色和黄色斑点。雌性体色较淡，全身具黄色条纹，腹部短而粗壮，产卵管短。【长度】体长 65~74 mm，腹长 48~56 mm，后翅 46~49 mm。【栖息环境】海拔 1800 m左右挺水植物茂盛的静水环境。【分布】中国湖北特有。【飞行期】6—9月。

[Identification] Male eyes dark blue, face yellowish brown, antefrons with a black spot in the upper half, top of frons with a black T-mark. Thorax blackish brown, dorsal part of synthorax with antehumeral stripes interrupted medially and an oval spot in posterior lower corner, sides largely yellow, legs largely black, wings hyaline. Abdomen black with numerous blue and yellow markings. Female paler with yellow markings, abdomen short and stout, ovipositor

神农蜓 雄，湖北
Aeshna shennong, male from Hubei

short. **[Measurements]** Total length 65-74 mm, abdomen 48-56 mm, hind wing 46-49 mm. **[Habitat]** Standing water with abundant emerging plants at about 1800 m elevation. **[Distribution]** Endemic to Hubei of China. **[Flight Season]** June to September.

神农蜓 雄，湖北
Aeshna shennong, male from Hubei

神农蜓 雌，湖北
Aeshna shennong, female from Hubei

翠蜓属 Genus *Anaciaeschna* Selys, 1878

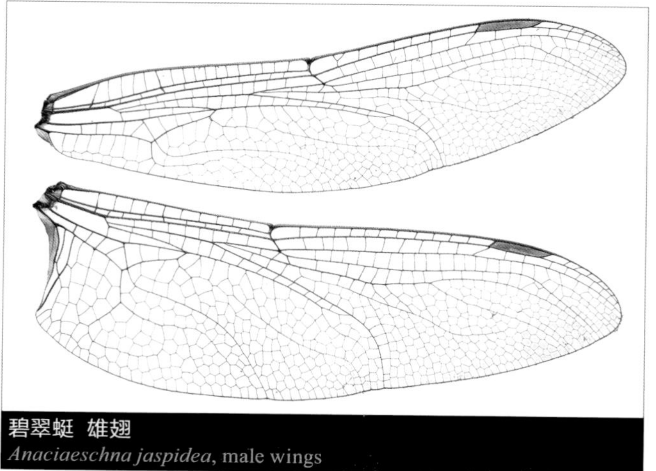

碧翠蜓 雄翅
Anaciaeschna jaspidea, male wings

本属全球已知约10种，中国已知2种，主要在中国南方分布。本属蜻蜓体中型；雄性复眼为天蓝色或者深蓝色，与一些北方分布的蜓属蜻蜓相似；翅染有淡褐色，基室无横脉，R3在近翅端显著弯曲，IR3叉状，臀三角室3～5室；腹部第2节具较小的耳状突。

本属蜻蜓栖息于杂草丛生的湿地和水稻田，白天很难遇见，在茂盛的森林中躲避阳光。黄昏时较活跃，集群捕食。雌性多在下午3至5点以半身潜水的方式产卵。

The genus contains about ten species with two recorded from China, mainly found in the southern region. Species of the genus are medium-sized. Eyes of males are sky-blue or dark blue, similar to some *Aeshna* species from the north. Wings slightly tinted with brown, median space without crossveins, R3 strongly curved near wing tip, IR3 forked, anal triangle 3-5 celled. Abdomen with small auricles on S2.

Anaciaeschna species inhabit grassy wetlands and paddy fields, they are seldom seen in the daytime, usually perching in dense forest to escape the sunshine. They are active at twilight when they are on forage in groups. Females lay eggs by submerging half the body between 3 pm and 5 pm.

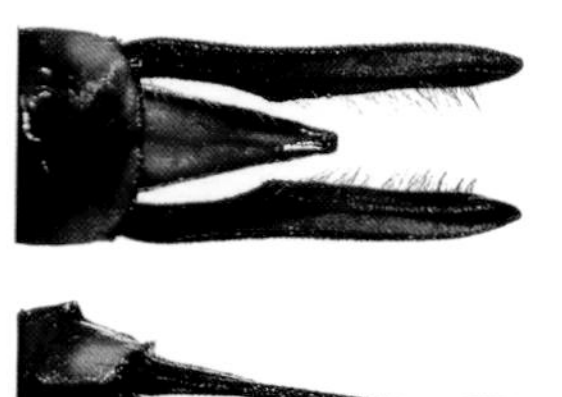

碧翠蜓
Anaciaeschna jaspidea

褐翠蜓
Anaciaeschna martini

翠蜓属 雄性肛附器
Genus *Anaciaeschna*, male anal appenages

褐翠蜓 雄
Anaciaeschna martini, male

碧翠蜓 *Anaciaeschna jaspidea* (Burmeister, 1839)

【形态特征】雄性具有十分显著的天蓝色复眼，面部黄色，具黑色的额横纹；胸部浅褐色，侧面具2条甚阔的黄色条纹，足黑色，基方红褐色，翅透明，翅痣褐色；腹部红褐色具黄白色的小斑点。雌性多型，色彩多变，复眼和身体已发现有土黄色、绿色和浅蓝色3种色型。【长度】体长61~64 mm，腹长 46~48 mm，后翅 41~43 mm。【栖息环境】海拔 1500 m以下杂草丛生的浅水池塘、沼泽以及暂时性水塘。【分布】广东、广西、海南、福建、云南、香港、台湾；广布于南亚、东南亚、澳新界和大洋洲。【飞行期】3—12月。

碧翠蜓 雄，云南（西双版纳）
Anaciaeschna jaspidea, male from Yunnan (Xishuangbanna)

[Identification] Male with very distinct sky-blue eyes, face yellow with a black stripe on top of frons. Thorax light brown with two broad lateral yellow stripes, legs black with reddish brown bases, wings hyaline with brown pterostigma. Abdomen reddish brown with small yellowish white spots. Female polymorphic, color variable, three morphs present, eyes and body markings khaki yellow, green and pale blue. **[Measurements]** Total length 61-64 mm, abdomen 46-48 mm, hind wing 41-43 mm. **[Habitat]** Grassy and shallow ponds, marshes and temporary pools below 1500 m elevation. **[Distribution]** Guangdong, Guangxi, Hainan, Fujian,Yunnan, Hong Kong, Taiwan; Widespread in South and Southeast Asia, Australasia and Oceania. **[Flight Season]** March to December.

碧翠蜓 雄，云南（西双版纳）
Anaciaeschna jaspidea, male from Yunnan (Xishuangbanna)

碧翠蜓 雌，广东 | 宋睿斌 摄
Anaciaeschna jaspidea, female from Guangdong | Photo by Ruibin Song

褐翠蜓 *Anaciaeschna martini* (Selys, 1897)

【形态特征】雄性具深蓝色的复眼，面部浅蓝色；胸部褐色，侧面具2条宽阔的浅蓝色条纹；足黑色，基方稍染红棕色；翅染有淡淡的褐色；腹部棕色，第2节和第3节侧面具蓝色斑点。雌性复眼黄褐色，面部黄色；合胸侧面具2条黄色条纹，翅琥珀色，基方具黑色斑；腹部红褐色。【长度】体长 68~76 mm，腹长 51~57 mm，后翅 42~48 mm。【栖息环境】海拔 2500 m以下杂草丛生的湿地。【分布】湖北、贵州、重庆、江西、广东、云南、台湾；泰国、老挝、越南、朝鲜半岛、日本。【飞行期】5—10月。

[Identification] Male eyes dark blue, face pale blue. Thorax brown, sides with two broad light blue stripes, legs black, bases slightly tinted with reddish brown, wings tinted with light brown. Abdomen brown, S2-S3 with lateral blue spots. Female eyes yellowish brown, face yellow. Thorax with two yellow stripes, wings amber, bases with black spots. Abdomen reddish brown. [Measurements] Total length 68-76 mm, abdomen 51-57 mm, hind wing 42-48 mm. [Habitat] Grassy marshes below 2500 m elevation. [Distribution] Hubei, Guizhou, Chongqing, Jiangxi, Guangdong, Yunnan, Taiwan; Thailand, Laos, Vietnam, Korean peninsula, Japan. [Flight Season] May to October.

褐翠蜓 雄，贵州
Anaciaeschna martini, male from Guizhou

褐翠蜓 雄，贵州
Anaciaeschna martini, male from Guizhou

褐翠蜓 雌，广东 | 宋睿斌 摄
Anaciaeschna martini, female from Guangdong | Photo by Ruibin Song

伟蜓属 Genus *Anax* Leach, 1815

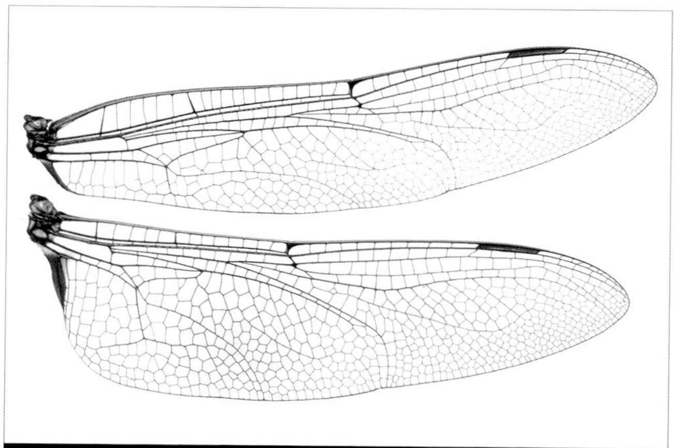

斑伟蜓 雄翅
Anax guttatus, male wings

本属世界性分布，全球已知约30种，中国已知7种及亚种。本属蜻蜓胸部色彩均一，有时侧缝具黑色条纹；翅大面积透明，有些种类具有琥珀色斑，基室无横脉，IR3有时呈叉状，雄性无臀三角室，臀角圆弧形；雄性腹部第2节没有耳状突，肛附器粗壮。雌性产卵管较短。伟蜓属种类经常穿梭于城市中的池塘和水渠，也是公园中最常见的蜻蜓，是少数能飞进公众视线而被认识的蜻蜓成员。

本属蜻蜓主要栖息于静水环境和流速缓慢的溪流。成虫白天活动，雄性在池塘中来回飞行，巡视并具领域行为，两雄相遇会展开争斗，不同种类的雄性也会进行激战。交尾的时间较长，通常停落在附近的灌木丛或者树丛上。许多种类有连结产卵的习性。

碧伟蜓东亚亚种 连结产卵 | 莫善濂 摄
Anax parthenope julius, oviposting in tandem | Photo by Shanlian Mo

This genus is distributed worldwide with over 30 recognized species, seven species and subspecies are recorded from China. Species of the genus usually possess uniform color in thorax, sutures black in some species. Wings largely hyaline, amber spots sometimes present, median space without crossveins, IR3 sometimes forked, male anal triangle absent, tornus rounded. Male without auricle on S2, anal appendages robust. Female ovipositor short. *Anax* species are strong-flying dragonflies and can be seen in the parks of cities, being among the few dragonflies recognized by city dwellers.

Anax species inhabit standing water and slow flowing streams. Males patrol along the ponds and defend territories, regularly fighting with other males of the same species and sometimes also with males of an other *Anax* species. Mating is protracted, with pairs perching on trees. Many species oviposit in tandem.

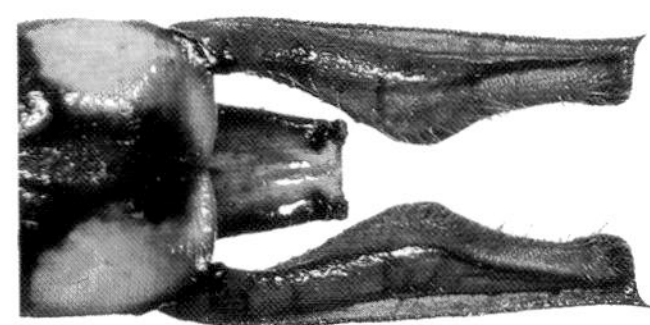

印度伟蜓
Anax indicus

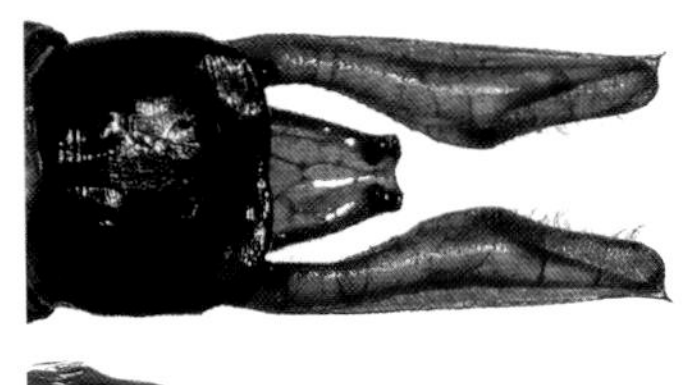

斑伟蜓
Anax guttatus

黄伟蜓
Anax immaculifrons

碧伟蜓东亚亚种
Anax parthenope julius

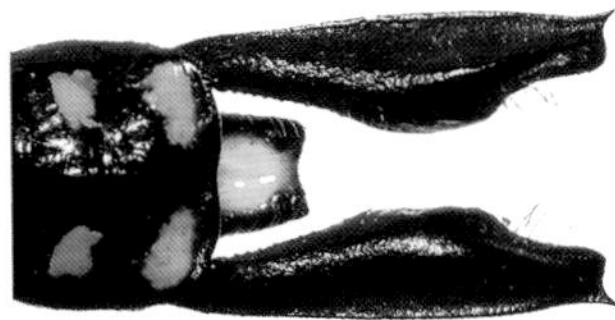

黑纹伟蜓
Anax nigrofasciatus

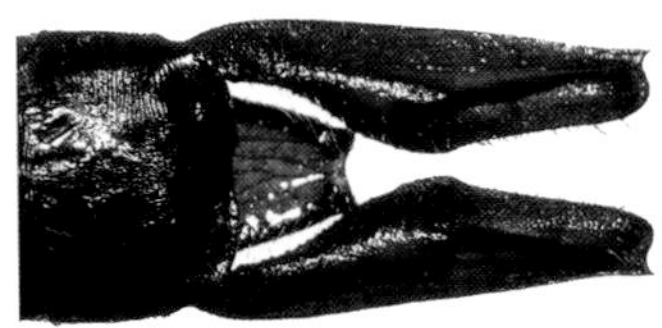

东亚伟蜓
Anax panybeus

碧伟蜓灰胸亚种
Anax parthenope parthenope

伟蜓属 雄性肛附器
Genus *Anax*, male anal appenages

斑伟蜓 *Anax guttatus* (Burmeister, 1839)

【形态特征】复眼绿色，面部黄色，额无显著的“T”形斑；胸部绿色，足黑色，基方红褐色，翅透明，雄性后翅亚基部具琥珀色斑；腹部黑色具黄白色斑点，雄性第2节主要蓝色，雌性此节色彩变异较大，蓝色、深绿色或者黄绿色。【长度】体长 78~86 mm，腹长 58~64 mm，后翅 52~55 mm。【栖息环境】海拔 1500 m以下的池塘、沼泽和溪流中流速缓慢的宽阔水域。【分布】华中、华南和西南地区广布，辽宁和山东也有零星记录；广布于南亚、东南亚、澳新界和大洋洲。【飞行期】全年可见。

[Identification] Eyes green, face yellow, frons without clear T-mark. Thorax green, legs black with reddish brown bases, wings hyaline, male hind wings with amber spots sub-basally. Abdomen black with yellowish white spots, male with S2 largely blue, female with S2 blue, dark green or yellowish green. **[Measurements]** Total length 78-86 mm, abdomen 58-64 mm, hind wing 52-55 mm. **[Habitat]** Ponds, marshes and slow flowing streams below 1500 m elevation. **[Distribution]** Widespread in Central, South and Southwest China, and a few records from Liaoning and Shandong; Widespread in South and Southeast Asia, Australasia, Oceania. **[Flight Season]** Throughout the year.

斑伟蜓 雄，云南（西双版纳）
Anax guttatus, male from Yunnan (Xishuangbanna)

斑伟蜓 雌，云南（西双版纳）
Anax guttatus, female from Yunnan (Xishuangbanna)

斑伟蜓 雄，云南（西双版纳）
Anax guttatus, male from Yunnan (Xishuangbanna)

斑伟蜓 雌，云南（西双版纳）
Anax guttatus, female from Yunnan (Xishuangbanna)

斑伟蜓 交尾，云南（西双版纳）
Anax guttatus, mating pair from Yunnan (Xishuangbanna)

黄伟蜓 *Anax immaculifrons* Rambur, 1842

【形态特征】雄性复眼蓝绿色，雌性复眼深绿色，面部黄色，额无"T"形斑；胸部主要黄绿色，中胸和后胸侧缝具宽阔的黑色条纹，足黑色，翅金黄色；腹部主要橙红色。【长度】体长 78~84 mm，腹长 57~61 mm，后翅 54~60 mm。【栖息环境】海拔 1000 m以下阴暗的池塘、溪流中流速缓慢的宽阔水域。【分布】福建、广东、广西、海南、香港；零散分布于欧洲地中海地区、土耳其、阿富汗，广布于印度、斯里兰卡、中南半岛。【飞行期】2—10月。

[Identification] Male eyes bluish green, female eyes dark green, face yellow without T-mark on top of frons. Thorax mainly yellowish green, humeral suture and metapleural suture with broad black stripes, legs black, wings golden. Abdomen mainly orange red. **[Measurements]** Total length 78-84 mm, abdomen 57-61 mm, hind wing 54-

黄伟蜓 雄，广东 | 莫善濂 摄
Anax immaculifrons, male from Guangdong | Photo by Shanlian Mo

60 mm. **[Habitat]** Shady ponds and slow flowing streams below 1000 m elevation. **[Distribution]** Fujian, Guangdong, Guangxi, Hainan, Hong Kong; Scattered distribution in the Eastern Mediterranean countries in Europe, Turkey, Afganistan, and widespread in India, Sri Lanka, Indochina. **[Flight Season]** February to Ocotber.

黄伟蜓 雄，广东 | 莫善濂 摄
Anax immaculifrons, male from Guangdong | Photo by Shanlian Mo

黄伟蜓 雌，广东 | 宋睿斌 摄
Anax immaculifrons, female from Guangdong | Photo by Ruibin Song

印度伟蜓 *Anax indicus* Lieftinck, 1942

【形态特征】雄性复眼绿色，面部黄色，上额无显著的“T”形斑；胸部绿色，足黑色，基方红褐色，翅透明，雄性后翅亚基部具琥珀色斑；腹部黑色具黄白色斑点，第2节主要蓝色。**【长度】**体长 75~81 mm，腹长 56~60 mm，后翅 48~51 mm。**【栖息环境】**海拔 1000 m以下水草茂盛的沼泽和季节性池塘。**【分布】**云南（德宏、西双版纳、临沧）、香港；印度、尼泊尔、巴基斯坦、斯里兰卡、泰国、越南。**【飞行期】**4—11月。

[Identification] Male eyes green, face yellow, frons without clear T-mark. Thorax green, legs black, bases reddish brown, wings hyaline, hind wings with amber spots subbasally. Abdomen black with yellowish white spots, S2 mainly blue. **[Measurements]** Total length 75-81 mm, abdomen 56-60 mm, hind wing 48-51 mm. **[Habitat]** Marshes with numerous aquatic plants and seasonal ponds below 1000 m elevation. **[Distribution]** Yunnan (Dehong, Xishuangbanna, Lincang), Hong Kong; India, Nepal, Pakistan, Sri Lanka, Thailand, Vietnam. **[Flight Season]** April to November.

印度伟蜓 雄，云南（德宏）
Anax indicus, male from Yunnan (Dehong)

印度伟蜓 雄，云南（德宏）
Anax indicus, male from Yunnan (Dehong)

黑纹伟蜓 *Anax nigrofasciatus* Oguma, 1915

黑纹伟蜓 雌，黄色型，产卵，贵州 | 莫善濂 摄
Anax nigrofasciatus, female, the yellow morph, laying eggs from Guizhou | Photo by Shanlian Mo

黑纹伟蜓 雌，蓝色型，产卵，湖北
Anax nigrofasciatus, female, the blue morph, laying eggs from Hubei

黑纹伟蜓 雄，湖北
Anax nigrofasciatus, male from Hubei

【形态特征】雄性复眼蓝色，面部黄色，额具1个显著的“T”形斑；胸部绿色，中胸和后胸侧缝具黑色条纹，足黑色，翅透明；腹部黑色具蓝色斑点，第2节主要蓝色。雌性多型，按腹部第2节色彩可分为蓝色型、绿色型和黄色型。【长度】体长 75~80 mm，腹长 55~58 mm，后翅 50~52 mm。【栖息环境】海拔 3000 m以下的各类湿地和溪流中流速缓慢的宽阔水域。【分布】除西北地区和海南外广布于全国；不丹、印度、尼泊尔、泰国、老挝、越南、朝鲜半岛、日本。【飞行期】2—12月。

[Identification] Male eyes blue, face yellow, frons with a clear T-mark. Thorax green, humeral suture and metapleural suture with black stripes, legs black, wings hyaline. Abdomen black with blue spots, S2 mainly blue. Female

黑纹伟蜓 雄，广东 | 宋睿斌 摄
Anax nigrofasciatus, male from Guangdong | Photo by Ruibin Song

polymorphic according to color of S2, including the blue, green and yellow morphs. **[Measurements]** Total length 75-80 mm, abdomen 55-58 mm, hind wing 50-52 mm. **[Habitat]** Wetlands and slow flowing streams below 3000 m elevation. **[Distribution]** Widespread in China except the Northwest and Hainan; Bhutan, India, Nepal, Thailand, Laos, Vietnam, Korean peninsula, Japan **[Flight Season]** February to December.

东亚伟蜓 *Anax panybeus* Hagen, 1867

【形态特征】复眼深绿色，面部黄色，额具1个显著的“T”形斑；胸部绿色，足黑色，基方红褐色，翅大面积透明，雄性后翅亚基部具琥珀色斑；腹部黑色具黄白色斑点，第2节主要蓝色。【长度】体长 84~92 mm，腹长 63~65 mm，后翅 52~54 mm。【栖息环境】海拔 2000 m以下的山区水潭和挺水植物匮乏具宽阔水面的静水环境。【分布】云南（德宏、西双版纳）、台湾；广布于东南亚，东至印度尼西亚帝纹岛，北至日本南部岛屿。【飞行期】全年可见。

[Identification] Eyes dark green, face yellow, frons with a clear T-mark. Thorax green, legs black, bases reddish brown, wings largely hyaline, male hind wings with amber spots subbasally. Abdomen black with yellowish white spots, S2 mainly blue. **[Measurements]** Total length 84-92 mm, abdomen 63-65 mm, hind wing 52-54 mm. **[Habitat]**

Montane ponds and larger water boby lacking emergent plants below 2000 m elevation. **[Distribution]** Yunnan (Dehong, Xishuangbanna), Taiwan; Widespread in Southeast Asia, ranging to Timor of Indonesia in east and to the southern islands of Japan in north. **[Flight Season]** Throughout the year.

东亚伟蜓 雄，云南（德宏）
Anax panybeus, male from Yunnan (Dehong)

碧伟蜓灰胸亚种 *Anax parthenope parthenope* (Selys, 1839)

【形态特征】复眼绿色，面部浅黄色，具黑色的额横纹；胸部灰褐色，足黑色，腿节红褐色，翅透明，稍染黄褐色；腹部第2节大面积蓝色，其余各节雄性大面积灰白色或灰褐色，并沿腹部中脊具1条甚阔的黑色条纹，雌性腹部大面积灰白色或淡蓝色，【长度】体长 62~75 mm，腹长 46~53 mm，后翅 44~51 mm。【栖息环境】大型水塘和湖泊等静水环境。【分布】新疆；从欧洲和非洲北部分布至阿拉伯半岛、西伯利亚、印度、日本。【飞行期】5—8月。

[Identification] Eyes green, face light yellow, top of frons with a black stripe. Thorax greyish brown, legs black, bases reddish brown, wings slightly tinted with yellowish brown. Abdomen S2 largely blue, male abdomen largely greyish white or greyish brown with a broad black stripe along the carina, female abdomen large greyish white or pale

碧伟蜓灰胸亚种 雄，希腊 | Jörg Arlt 摄
Anax parthenope parthenope, male from Greece | Photo by Jörg Arlt

碧伟蜓灰胸亚种 雌，希腊 | Jörg Arlt 摄
Anax parthenope parthenope, female from Greece | Photo by Jörg Arlt

blue. **[Measurements]** Total length 62-75 mm, abdomen 46-53 mm, hind wing 44-51 mm. **[Habitat]** Standing water including large ponds and lakes. **[Distribution]** Xinjiang; From Europe and North Africa to the Arabian Peninsula, Siberia, India, Japan. **[Flight Season]** May to August.

碧伟蜓东亚亚种 *Anax parthenope julius* Brauer, 1865

【形态特征】复眼绿色，面部浅黄色，具黑色的额横纹；胸部绿色，足黑色，腿节红褐色，翅透明，稍染黄褐色；雄性腹部第2节蓝色，其余各节灰白色，沿腹部中脊具1条甚阔的黑色条纹；雌性多型，按腹部第2节色彩可分为蓝色型、黄色型和绿色型，腹部中脊的宽阔条纹红褐色。【长度】体长 68~76 mm，腹长 49~55 mm，后翅 50~51 mm。【栖息环境】海拔 2500 m以下的湿地、水库等静水环境以及溪流中流速缓慢的宽阔水域。【分布】除新疆外全国广布；缅甸、越南、朝鲜半岛、日本。【飞行期】全年可见。

[Identification] Eyes green, face light yellow, top of frons with a black stripe. Thorax green, legs black, bases reddish brown, wings hyaline, slightly tinted with yellowish brown. Male abdomen largely grey with a broad black stripe along the carina, S2 largely blue. Female with S2 blue, yellow or green, the stripe along the carina reddish brown. **[Measurements]** Total length 68-76 mm, abdomen 49-55 mm, hind wing 50-51 mm. **[Habitat]** Standing water including reservoirs and slow flowing streams below 2500 m elevation. **[Distribution]** Widespread throughout China except Xinjiang; Myanmar, Vietnam, Korean peninsula, Japan. **[Flight Season]** Throughout the year.

碧伟蜓东亚亚种　雄，云南（西双版纳）
Anax parthenope julius, male from Yunnan (Xishuangbanna)

碧伟蜓东亚亚种　雌，蓝色型，山东
Anax parthenope julius, female, the blue morph from Shandong

碧伟蜓东亚亚种　交尾，黑龙江 | 莫善濂 摄
Anax parthenope julius, mating pair from Heilongjiang | Photo by Shanlian Mo

碧伟蜓东亚亚种　雌，绿色型，湖北
Anax parthenope julius, female, the green morph from Hubei

细腰蜓属 Genus *Boyeria* McLachlan, 1895

本属全球已知8种，分布于北美洲、欧洲和亚洲。中国已知2种，最早的记录是中华细腰蜓，根据采自四川的一头未熟雄性描述。之后褐面细腰蜓记录于中国华南地区，但两者的关系尚未完全明确。本属蜻蜓体型较大，色彩较暗。面部很窄，复眼非常发达；胸部和腹部深褐色具黄色条纹；翅透明，基室具横脉，翅痣较长，前缘具支持脉，R3在翅痣前开始弯曲，IR3未分叉，臀三角室3室，臀角略呈角状。

本属蜻蜓栖息于茂盛森林中的小溪和沟渠。成虫白天很难遇见，通常躲在茂盛的森林中，而在天色近黑时活跃。雌性曾被发现于下午4至5点在小溪边缘产卵。

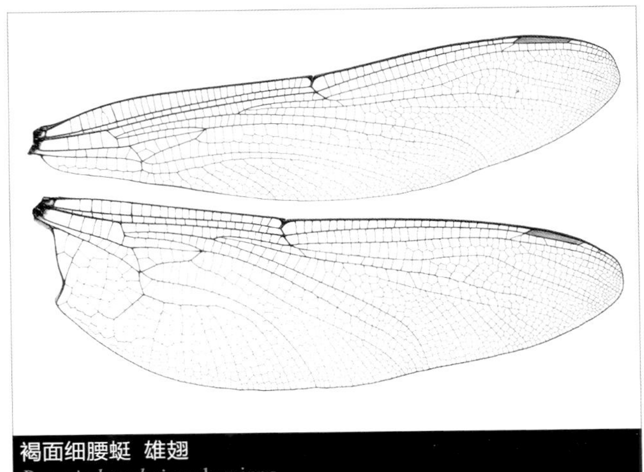

褐面细腰蜓 雄翅
Boyeria karubei, male wings

This genus contains eight species, distributed in North America, Europe and Asia. Two species are recorded from China, the first record, *Boyeria sinensis*, was described based on an immature male specimen from Sichuan. *B. karubei* was later recorded from South China, but the relationship of the two species is not yet clear. Species of the genus are large and dark species. Face narrow, eyes very large. Thorax and abdomen dark brown with yellow stripes. Wings hyaline, median space with crossveins, pterostigma long and well braced, R3 begins to curve before level of of pterostigma, IR3 not forked, anal triangle 3-celled, tornus slightly angled.

Boyeria species inhabit streams and ditches in forests. The adults are difficult to see during the day time, when they perch in dense foliage, but are active at twilight. Females have been observed to lay eggs in the edge of small streams between 4 pm and 5 pm.

褐面细腰蜓 雄
Boyeria karubei, male

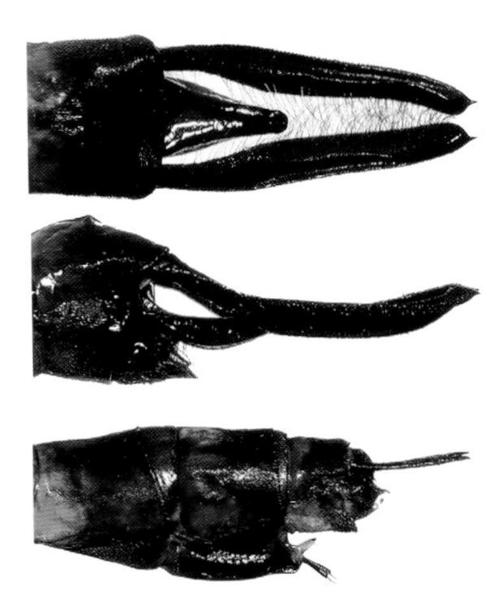

褐面细腰蜓
Boyeria karubei

细腰蜓属 雄性肛附器和雌性产卵器
Genus *Boyeria*, Male anal appendages and female ovipositor

褐面细腰蜓 *Boyeria karubei* Yokoi, 2002

【形态特征】复眼深绿色，面部褐色；合胸深褐色，具黄绿色肩前条纹，合胸侧面具2条甚阔的黄色条纹，足大面积黑色，基方红褐色，翅透明，甚阔，翅痣长；腹部黑褐色并具黄色环纹。【长度】体长 78～83 mm，腹长 61～64 mm，后翅 51～60 mm。【栖息环境】海拔 1000 m以下的山区小溪。【分布】福建、广东、广西、海南、云南（西双版纳）；老挝、越南。【飞行期】5—9月。

[Identification] Eyes dark green, face brown. Thorax dark brown with yellowish green anterhumeral stripes, sides with two broad yellow stripes, legs largely black with bases reddish brown, wings hyaline and broad, pterostigma long. Abdomen blackish brown with yellow rings. [Measurements] Total length 78-83 mm, abdomen 61-64 mm, hind wing 51-60 mm. [Habitat] Montane streams below 1000 m elevation. [Distribution] Fujian, Guangdong, Guangxi, Hainan, Yunnan (Xishuangbanna); Laos, Vietnam. [Flight Season] May to September.

褐面细腰蜓 雄，广东
Boyeria karubei, male from Guangdong

褐面细腰蜓 雄，广东
Boyeria karubei, male from Guangdong

褐面细腰蜓 雌，广东
Boyeria karubei, female from Guangdong

褐面细腰蜓 雌，广东
Boyeria karubei, female from Guangdong

头蜓属 Genus *Cephalaeschna* Selys, 1883

本属仅在亚洲分布，全世界已知约25种，中国已知约10种。本属是一类体中型至大型的溪栖蜓科物种，它们身体黑褐色并具绿色和黄色条纹。一些种类面部很窄，另一些则很宽阔。翅透明，基室具横脉，翅痣较短。主要辨识特征包括面部的宽窄、肛附器和阳茎的构造、产卵器的长度、臀三角室的翅室数量等。

本属蜻蜓生活在茂盛森林中的林荫小溪、沟渠和开阔溪流，多数种类生活在具有一定海拔高度的林区，有些可以在3000 m以上的高山生活，是一类较少见的蜻蜓。雄性在下午活跃，在有树荫遮蔽的小溪上方，以低空定点悬停占据领地。雌性在小溪边缘的土坡和苔藓上产卵。

This genus is confined to Asia and contains about 25 species, more than ten species have been recorded from China. Species of the genus are medium to large sized, body blackish brown with green and yellow stripes. Some species have a narrow face and others a very broad face. Wings hyaline, median space with crossveins, pterostigma short. The diagnostic characters for identification include width of face, male anal appendages and penis, female ovipositor, cell number of anal triangle.

Cephalaeschna species live in shady streams, ditches or exposed streams in forest. Most species prefer moderate altitude, but some species live in high mountains above 3000 m elevation and are difficult to find. Males are active in the afternoon, hovering very low above the water in the shady part of streams. Females lay eggs in the earthen slopes of stream margin.

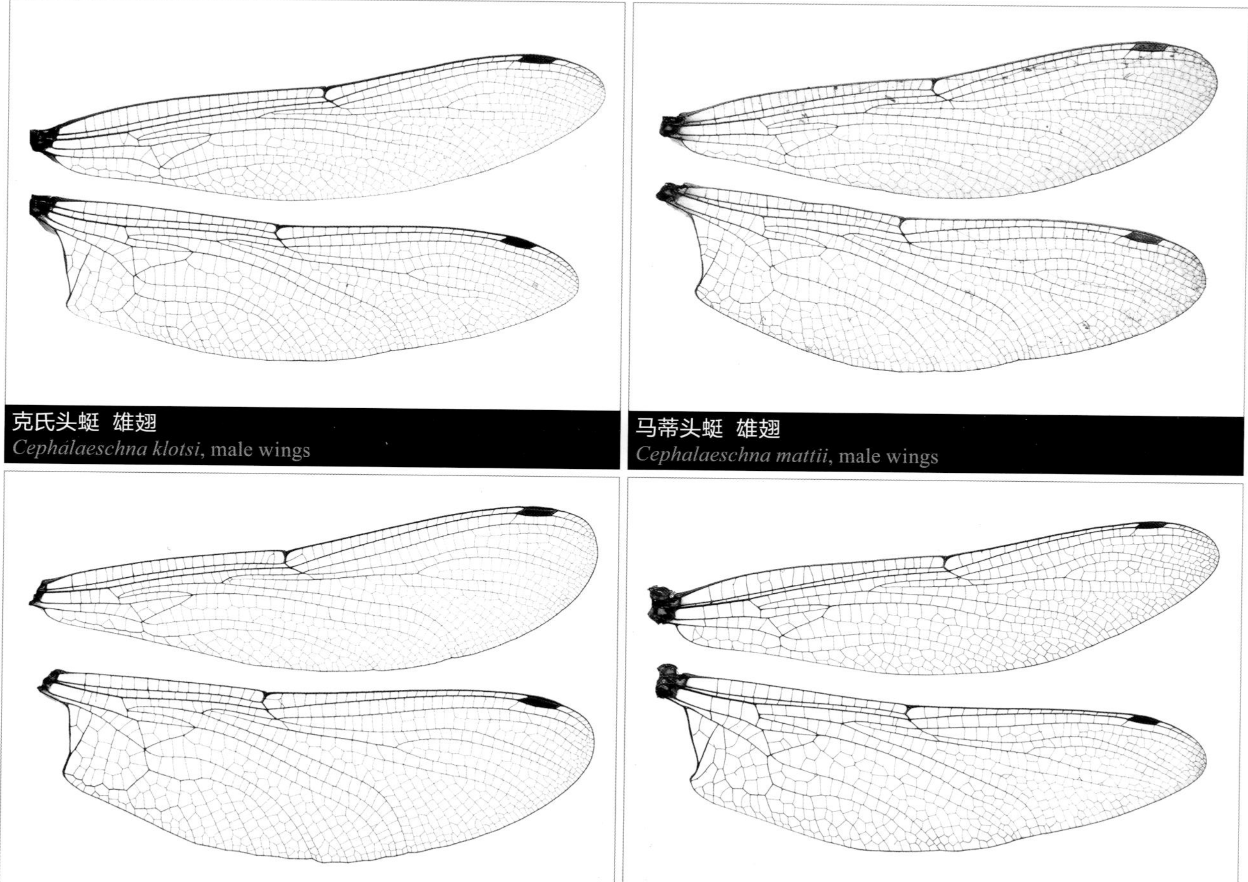

克氏头蜓 雄翅
Cephalaeschna klotsi, male wings

马蒂头蜓 雄翅
Cephalaeschna mattii, male wings

鼎湖头蜓 雄翅
Cephalaeschna dinghuensis, male wings

长角头蜓 雄翅
Cephalaeschna cornifrons, male wings

尼氏头蜓 雄
Cephalaeschna needhami, male

长角头蜓
Cephalaeschna cornifrons

鼎湖头蜓
Cephalaeschna dinghuensis

异色头蜓
Cephalaeschna discolor

头蜓属 雄性肛附器和雌性产卵器
Genus *Cephalaeschna*, male anal appendages and female ovipositor

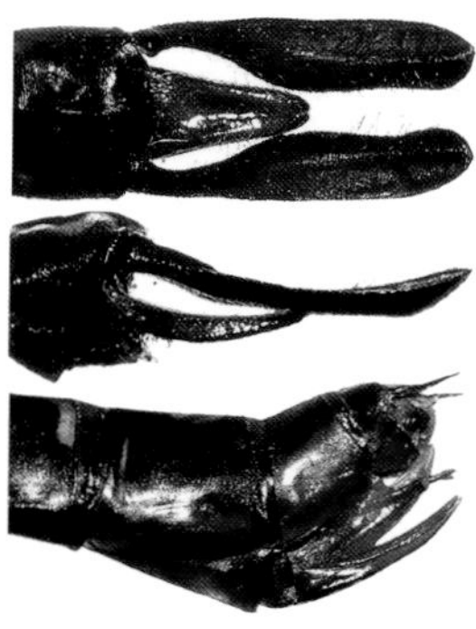

克氏头蜓
Cephalaeschna klotsi

马蒂头蜓
Cephalaeschna mattii

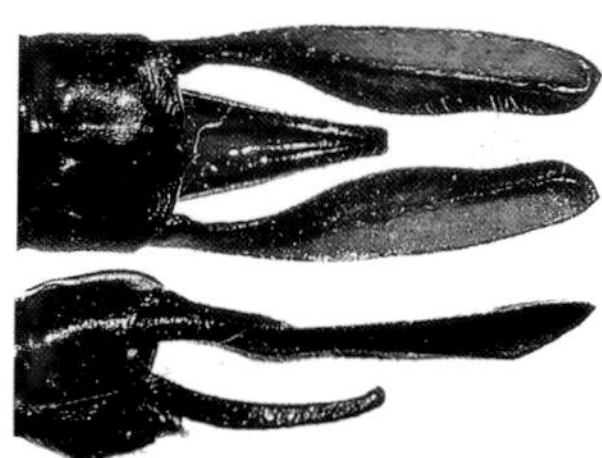

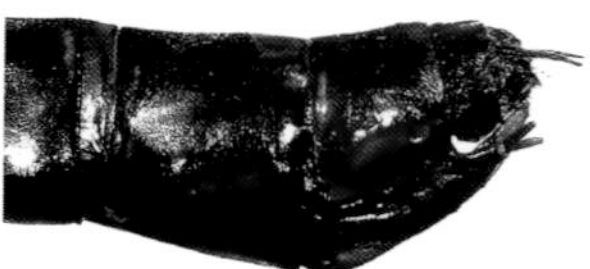

尼氏头蜓
Cephalaeschna needhami

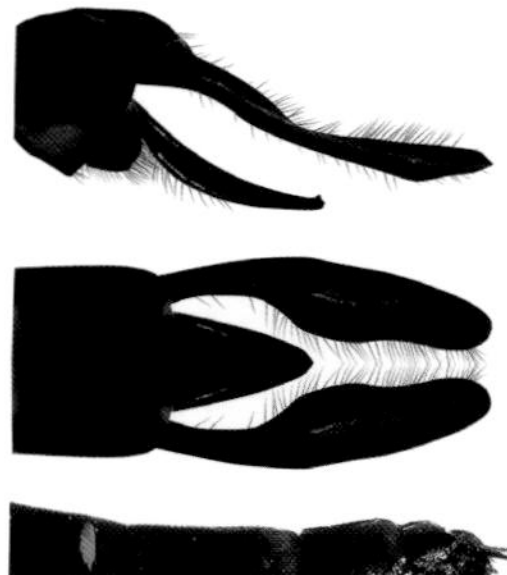

暗色头蜓
Cephalaeschna obversa

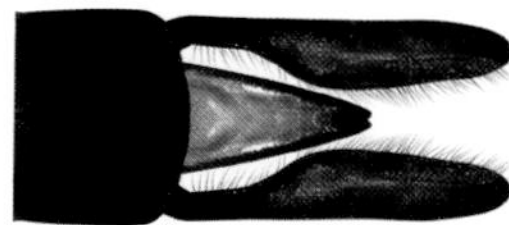

蝶斑头蜓
Cephalaeschna ordopapiliones

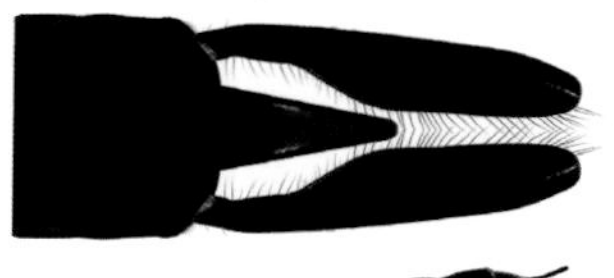

长者头蜓
Cephalaeschna patrorum

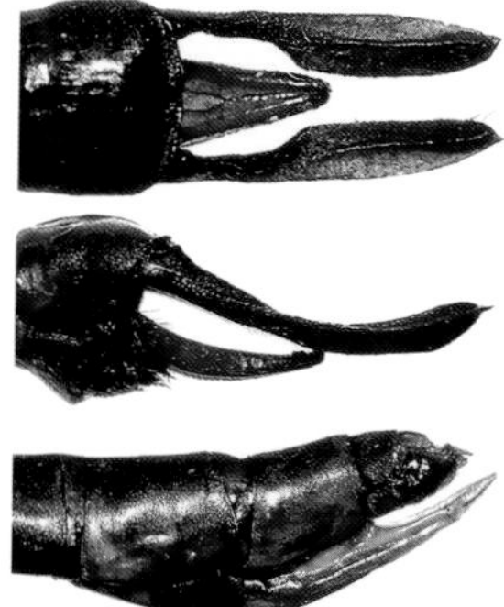

李氏头蜓
Cephalaeschna risi

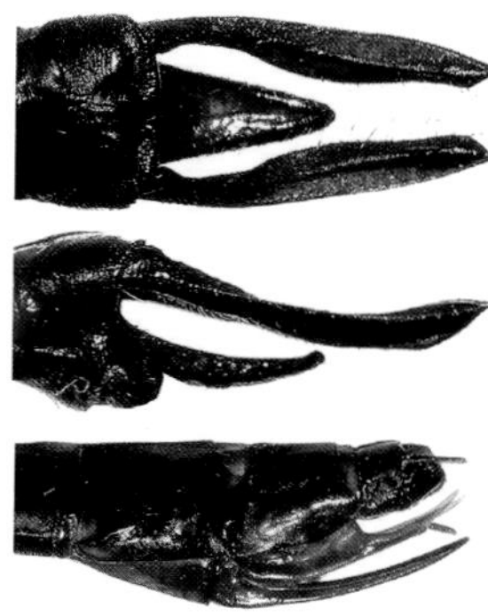

邵武头蜓
Cephalaeschna shaowuensis

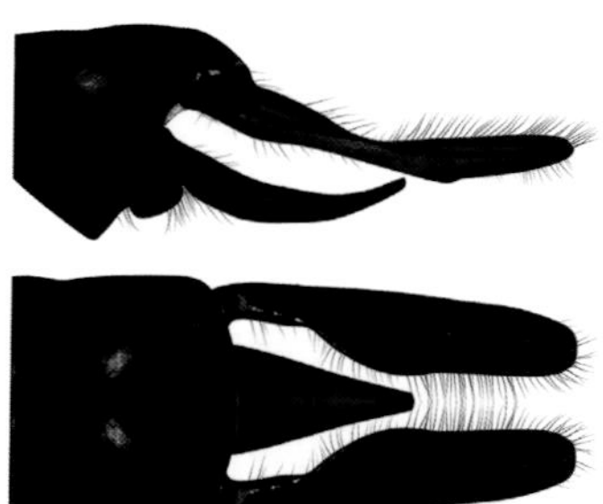

独行头蜓
Cephalaeschna solitaria

头蜓属 雄性肛附器和雌性产卵器
Genus *Cephalaeschna*, male anal appendages and female ovipositor

长角头蜓 *Cephalaeschna cornifrons* Zhang & Cai, 2013

【形态特征】雄性复眼深蓝色，面部大面积黄绿色，前额褐色，上额具1个“T”形斑，额上缘中央具1个角状突起；胸部褐色，具黄绿色肩前条纹，合胸侧面具2条甚阔的黄色条纹，足深褐色，翅透明；腹部黑褐色，各节具丰富的黄色斑点。雌性较粗壮，色彩与雄性相似，额上的角状突起更长，第10节腹板延长。【长度】体长 60~66 mm，腹长 47~51 mm，后翅 40~43 mm。【栖息环境】海拔 2000~3000 m的开阔溪流。【分布】中国云南（大理）特有。【飞行期】6—12月。

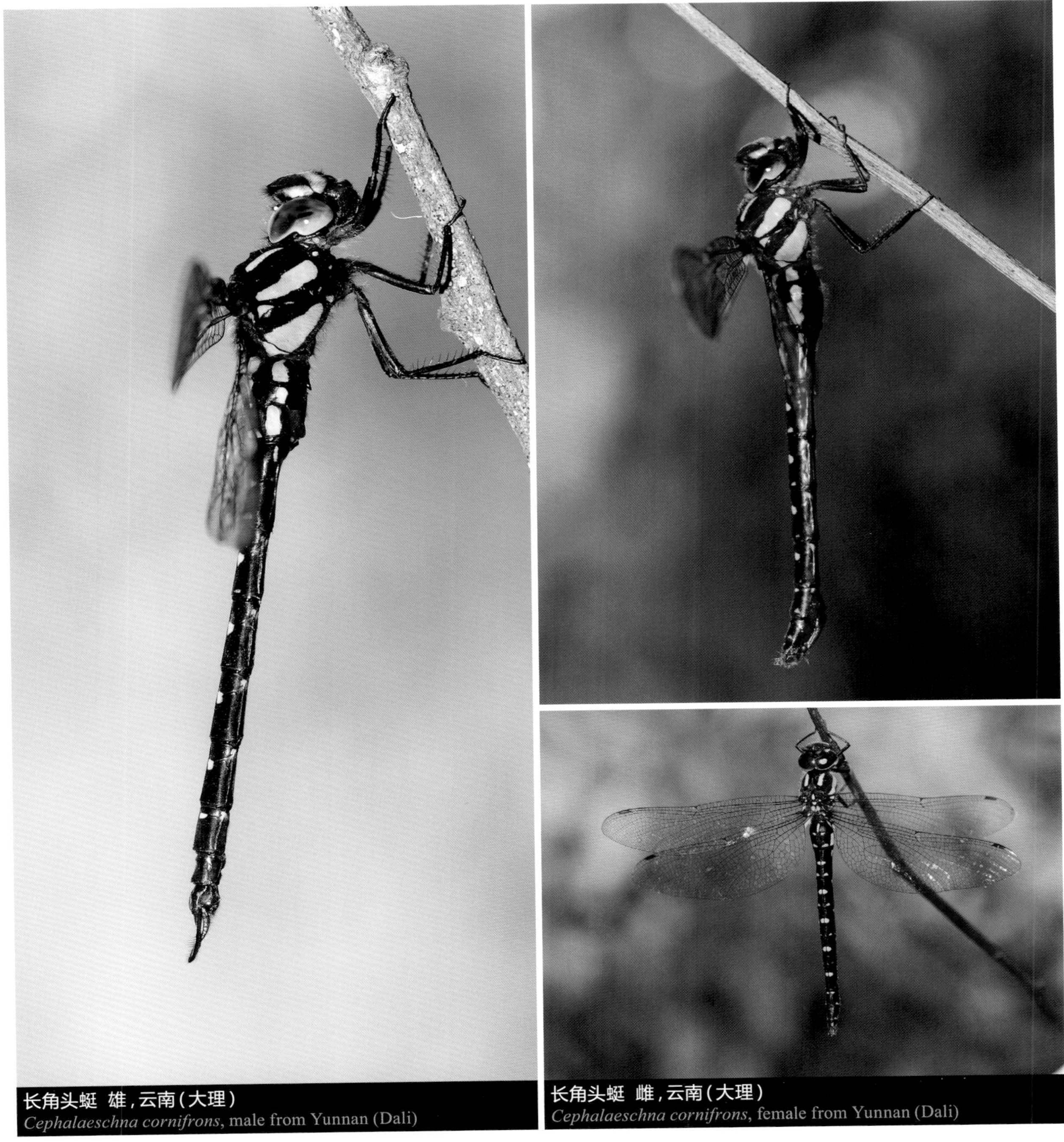

长角头蜓 雄，云南（大理）
Cephalaeschna cornifrons, male from Yunnan (Dali)

长角头蜓 雌，云南（大理）
Cephalaeschna cornifrons, female from Yunnan (Dali)

[Identification] Male eyes dark blue, face largely yellowish green, antefrons brown, tope of frons with a T-mark, the anterior ridge of frons with a median horn. Thorax brown with yellowish green antehumeral stripes, sides with two broad yellow stripes, legs dark brown, wings hyaline. Abdomen blackish brown with numerous yellow spots on S1-S10. Female stouter, body marking similar to male, the horn on the frons longer, sternite of S10 elongated. **[Measurements]** Total length 60-66 mm, abdomen 47-51 mm, hind wing 40-43 mm. **[Habitat]** Exposed montane streams at 2000-3000 m elevation. **[Distribution]** Endemic to Yunnan (Dali) of China. **[Flight Season]** June to December.

长角头蜓 雄，云南（大理）
Cephalaeschna cornifrons, male from Yunnan (Dali)

鼎湖头蜓 *Cephalaeschna dinghuensis* Wilson, 1999

【形态特征】雄性复眼蓝色，面部大面积黄色，前额具1个较大黑斑；胸部褐色，具甚细的黄色肩前条纹，合胸侧面具2条甚阔的黄色条纹，2个条纹之间具1条黄色细纹，足基方黄色，翅透明，端部甚阔；腹部黑褐色，具非常细小的黄色斑点。雌性与雄性相似，第10节腹板稍微延长。【长度】体长 62～69 mm，腹长 47～54 mm，后翅 41～47 mm。【栖息环境】海拔 500～1000 m的林荫狭窄小溪和渗流地。【分布】中国广东特有。【飞行期】6—8月。

[Identification] Male eyes blue, face largely yellow, antefrons with a large black spot. Thorax brown with narrow yellow antehumeral stripes, lateral sides with two broad yellow stripes and a fine yellow linear stripe between them, legs with bases yellow, wings hyaline, broad to the tips. Abdomen blackish brown with small yellow spots. Female similar to male, the sternite of S10 slightly elongated. **[Measurements]** Total length 62-69 mm, abdomen 47-54 mm, hind wing 41-47 mm. **[Habitat]** Shady and narrow montane streams and seepages at 500-1000 m elevation. **[Distribution]** Endemic to Guangdong of China. **[Flight Season]** June to August.

鼎湖头蜓 雄，广东
Cephalaeschna dinghuensis, male from Guangdong

鼎湖头蜓 雌，广东
Cephalaeschna dinghuensis, female from Guangdong

异色头蜓 *Cephalaeschna discolour* Zhang, Cai & Liao, 2013

【形态特征】雄性复眼深绿色，面部大面积黄绿色，前额具1个甚大的褐色斑，上额具“T”形斑；胸部深褐色，具黄绿色肩前条纹，合胸侧面具3条黄绿色条纹，足黑色具红褐色条纹，翅透明；腹部黑褐色，具甚细的黄色条纹和甚小的斑点。雌性较粗壮，色彩与雄性相似，产卵器十分发达，产卵管末端超出第10节末端。【长度】体长65～67 mm，腹长 51～53 mm，后翅 45～50 mm。【栖息环境】海拔 1000～1500 m的开阔溪流。【分布】中国湖北特有。【飞行期】8—9月。

[Identification] Male eyes dark green, face largely yellowish green, antefrons with a large brown spot, top of frons with T-mark. Thorax dark brown with yellowish green antehumeral stripes, sides with three broad yellow and green stripes, legs largely black with reddish brown stripes, wings hyaline. Abdomen dark brown with tiny yellow markings. Female body marking similar to male but stouter, ovipositor very long, exceeding the end of S10. **[Measurements]** Total length 65-67 mm, abdomen 51-53 mm, hind wing 45-50 mm. **[Habitat]** Exposed streams in forest at 1000-1500 m elevation. **[Distribution]** Endemic to Hubei of China. **[Flight Season]** August to September.

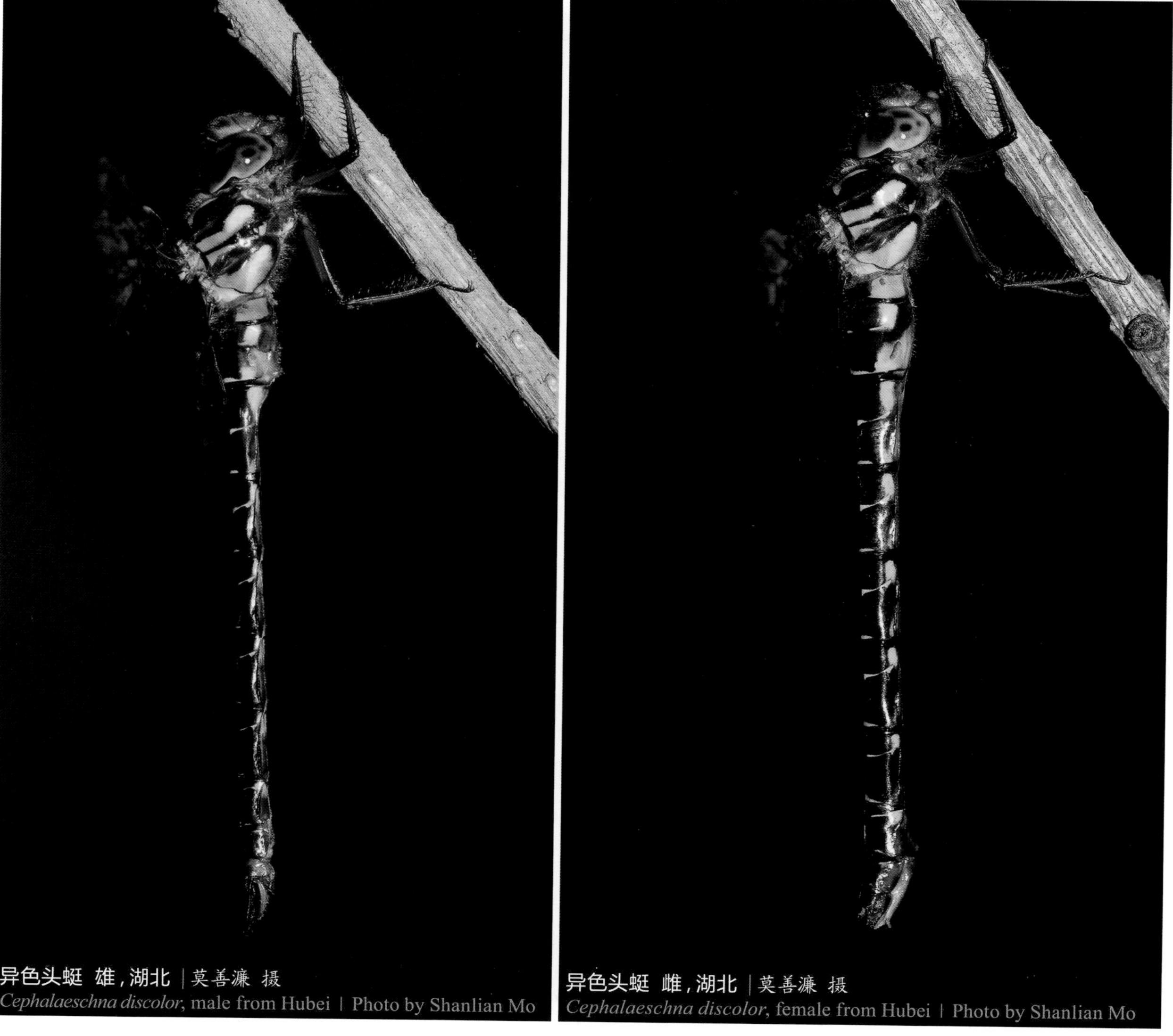

异色头蜓 雄，湖北 | 莫善濂 摄
Cephalaeschna discolor, male from Hubei | Photo by Shanlian Mo

异色头蜓 雌，湖北 | 莫善濂 摄
Cephalaeschna discolor, female from Hubei | Photo by Shanlian Mo

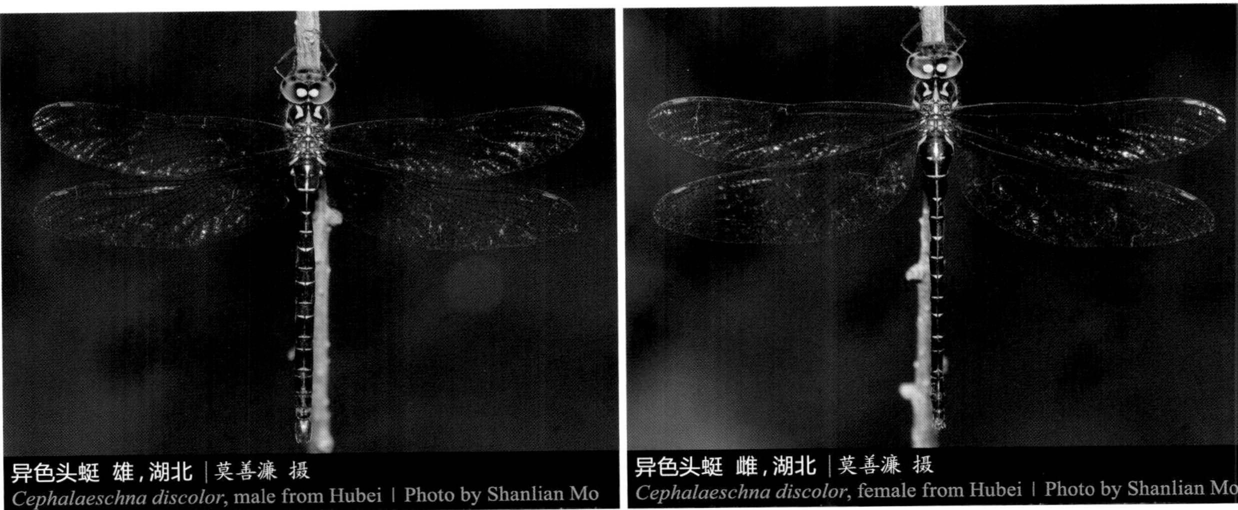

异色头蜓 雄，湖北 | 莫善濂 摄
Cephalaeschna discolor, male from Hubei | Photo by Shanlian Mo

异色头蜓 雌，湖北 | 莫善濂 摄
Cephalaeschna discolor, female from Hubei | Photo by Shanlian Mo

克氏头蜓 *Cephalaeschna klotsi* Asahina, 1982

【形态特征】雄性复眼深绿色，面部黑褐色；胸部黑色，具绿色肩前条纹，合胸侧面具2条宽阔的绿色条纹，足黑褐色，翅透明；腹部黑色，第1~8节具甚细小的黄色斑纹。雌性与雄性相似，但条纹色彩较浅，产卵器不发达，产卵管未超出第10节末端。【长度】体长 70~77 mm，腹长 54~61 mm，后翅 46~49 mm。【栖息环境】海拔 500~2000 m的小型瀑布和渗流石壁。【分布】中国特有，分布于湖北、福建、浙江、广东、香港。【飞行期】5—10月。

克氏头蜓 雄，浙江
Cephalaeschna klotsi, male from Zhejiang

[Identification] Male eyes dark green, face largely blackish brown. Thorax black with green antehumeral stripes, sides with two broad green stripes, legs blackish brown, wings hyaline. Abdomen black, S1-S8 with tiny yellow markings. Female body marking similar to male but paler, ovipositor short, not exceeding the end of S10. **[Measurements]** Total length 70-77 mm, abdomen 54-61 mm, hind wing 46-49 mm. **[Habitat]** Small waterfalls and precipices with trickles in forest at 500-2000 m elevation. **[Distribution]** Endemic to China, recorded from Hubei, Fujian, Zhejiang, Guangdong, Hong Kong. **[Flight Season]** May to October.

克氏头蜓 雄，浙江
Cephalaeschna klotsi, male from Zhejiang

克氏头蜓 雌，浙江
Cephalaeschna klotsi, female from Zhejiang

马蒂头蜓 *Cephalaeschna mattii* Zhang, Cai & Liao, 2013

【形态特征】雄性复眼深绿色，面部大面积黄褐色，前额具1个甚大的褐色斑，上额具“T”形斑；胸部深褐色，具黄绿色肩前条纹，合胸侧面具3条黄绿色条纹，足黑色具黄色条纹，翅透明，翅痣黑褐色；腹部黑褐色，具不清晰的黄色条纹和斑点。雌性与雄性相似，翅痣色彩黄褐色，产卵器十分发达，产卵管末端超出第10节末端。【长度】体长68～69 mm，腹长 53～55 mm，后翅 46～49 mm。【栖息环境】海拔 500～1500 m的开阔溪流。【分布】中国特有，分布于湖北、四川。【飞行期】7—9月。

[Identification] Male eyes dark green, face largely yellowish brown, antefrons with a large brown spot, top of frons with T-mark. Thorax dark brown with yellowish green antehumeral stripes, sides with three broad yellow and green stripes, legs largely black with yellow stripes, wings hyaline, pterostigma blackish brown. Abdomen blackish brown with unclear yellow markings. Female body marking similar to male, pterostigma yellowish brown, ovipositor very long, exceeding the end of S10. **[Measurements]** Total length 68-69 mm, abdomen 53-55 mm, hind wing 46-49 mm. **[Habitat]** Exposed montane streams at 500-1500 m elevation. **[Distribution]** Endemic to China, recorded from Hubei, Sichuan. **[Flight Season]** July to September.

马蒂头蜓 雄，湖北
Cephalaeschna mattii, male from Hubei

马蒂头蜓 雄，湖北
Cephalaeschna mattii, male from Hubei

马蒂头蜓 雌，湖北 | 莫善濂 摄
Cephalaeschna mattii, female from Hubei | Photo by Shanlian Mo

尼氏头蜓 *Cephalaeschna needhami* Asahina, 1981

【形态特征】雄性复眼深绿色，面部黑褐色；胸部黑色，具黄绿色肩前条纹，合胸侧面具2条宽阔的黄绿色条纹，2个条纹之间具细纹，足黑褐色，翅透明，翅痣深褐色；腹部黑色，第1～7节具甚细小的黄色斑纹。雌性与雄性相似，但条纹色彩较浅，产卵器不发达，产卵管未超出第10节末端。【长度】体长 61～67 mm，腹长 47～52 mm，后翅 40～45 mm。【栖息环境】海拔 500～1500 m的小型瀑布和渗流石壁。【分布】中国特有，分布于湖北、江西、福建、广西、广东。【飞行期】7—10月。

尼氏头蜓 雄，湖北
Cephalaeschna needhami, male from Hubei

尼氏头蜓 雌，湖北
Cephalaeschna needhami, female from Hubei

[Identification] Male eyes dark green, face largely blackish brown. Thorax black with yellowish green antehumeral stripes, sides with two broad yellowish green stripes and a linear stripe between them, legs blackish brown, wings hyaline. Abdomen black, S1-S7 with fine yellow markings. Female body marking similar to male but paler, ovipositor short, not exceeding the end of S10. **[Measurements]** Total length 61-67 mm, abdomen 47-52 mm, hind wing 40-45 mm. **[Habitat]** Small waterfalls and precipies with trickles in forest at 500-1500 m elevation. **[Distribution]** Endemic to China, recorded from Hubei, Jiangxi, Fujian, Guangxi, Guangdong. **[Flight Season]** July to October.

尼氏头蜓 雄，湖北
Cephalaeschna needhami, male from Hubei

暗色头蜓 *Cephalaeschna obversa* Needham, 1930

【形态特征】雄性复眼深绿色，面部黄绿色，额黑色；胸部深褐色，具黄绿色肩前条纹，合胸侧面具2条绿色条纹，足黑褐色，翅透明；腹部黑色，第1～9节具黄色斑点。雌性与雄性相似，但条纹色彩较浅，产卵器不发达，产卵管未超出第10节末端。【长度】体长 63～67 mm，腹长 49～52 mm，后翅 40～42 mm。【栖息环境】海拔1000～1500 m的开阔溪流。【分布】中国特有，分布于湖北、四川、贵州。【飞行期】7—10月。

[Identification] Male eyes dark green, face largely yellowish green, frons black. Thorax dark brown with yellowish green antehumeral stripes, lateral sides with two green stripes, legs dark brown, wings hyaline. Abdomen black, S1-S9 with yellow markings. Female body marking similar to male but paler, ovipositor short, not exceeding the end of S10.

[Measurements] Total length 63-67 mm, abdomen 49-52 mm, hind wing 40-42 mm. **[Habitat]** Exposed montane streams at 1000-1500 m elevation. **[Distribution]** Endemic to China, recorded from Hubei, Sichuan, Guizhou. **[Flight Season]** July to October.

暗色头蜓 雄，贵州
Cephalaeschna obversa, male from Guizhou

暗色头蜓 雌，贵州
Cephalaeschna obversa, female from Guizhou

蝶斑头蜓 *Cephalaeschna ordopapiliones* Zhang & Cai, 2013

蝶斑头蜓 雄，云南（大理）
Cephalaeschna ordopapiliones, male from Yunnan (Dali)

【形态特征】雄性复眼深蓝色，面部大面积黄绿色，前额上缘褐色；合胸深褐色，具黄绿色肩前条纹，合胸侧面具2条甚阔的黄绿色条纹，足深褐色，翅透明；腹部黑褐色，第1～9节具黄色斑点，其中第4～7节背面具蝶形斑。【长度】体长 60～65 mm，腹长 46～49 mm，后翅 40～42 mm。【栖息环境】海拔 2000～3000 m的狭窄小溪。【分布】中国云南（大理）特有。【飞行期】9—11月。

[Identification] Male eyes dark blue, face largely yellowish green, antefrons with upper margin brown. Thorax dark brown with yellowish green antehumeral stripes, sides with two broad yellowish green stripes, legs dark brown, wings hyaline. Abdomen blackish brown with yellow spots on S1-S9, the spots on S4-S7 butterfly-shaped. [Measurements] Total length 60-65 mm, abdomen 46-49 mm, hind wing 40-42 mm. [Habitat] Narrow montane streams at 2000-3000 m elevation. [Distribution] Endemic to Yunnan (Dali) of China. [Flight Season] September to November.

长者头蜓 *Cephalaeschna patrorum* Needham, 1930

【形态特征】雄性复眼深蓝色，面部大面积黄绿色，前额具1个褐色斑，上额具1个“T”形斑；胸部黑色，具黄绿色肩前条纹，合胸侧面具2条甚阔的黄绿色条纹，足深褐色，翅透明；腹部黑色，第1～8节具黄色斑纹。雌性较粗壮，色彩与雄性相似，第10节腹板延长。【长度】体长 69~71 mm，腹长 51~54 mm，后翅 45~51 mm。【栖息环境】海拔 500~1500 m的开阔溪流。【分布】中国特有，分布于北京、河南、陕西、山西、四川。【飞行期】6—10月。

[Identification] Male eyes dark blue, face largely yellowish green, antefrons with a brown spot, top of frons with a T-mark. Thorax black with yellowish green antehumeral stripes, sides with two broad yellowish green stripes, legs dark brown, wings hyaline. Abdomen black with yellow stripes on S1-S8. Female stouter, body marking similar to male, sternite of S10 elongated. **[Measurements]** Total length 69-71 mm, abdomen 51-54 mm, hind wing 45-51 mm. **[Habitat]** Exposed montane streams at 500-1500 m elevation. **[Distribution]** Endemic to China, recorded from Beijing, Henan, Shaanxi, Shanxi, Sichuan. **[Flight Season]** June to October.

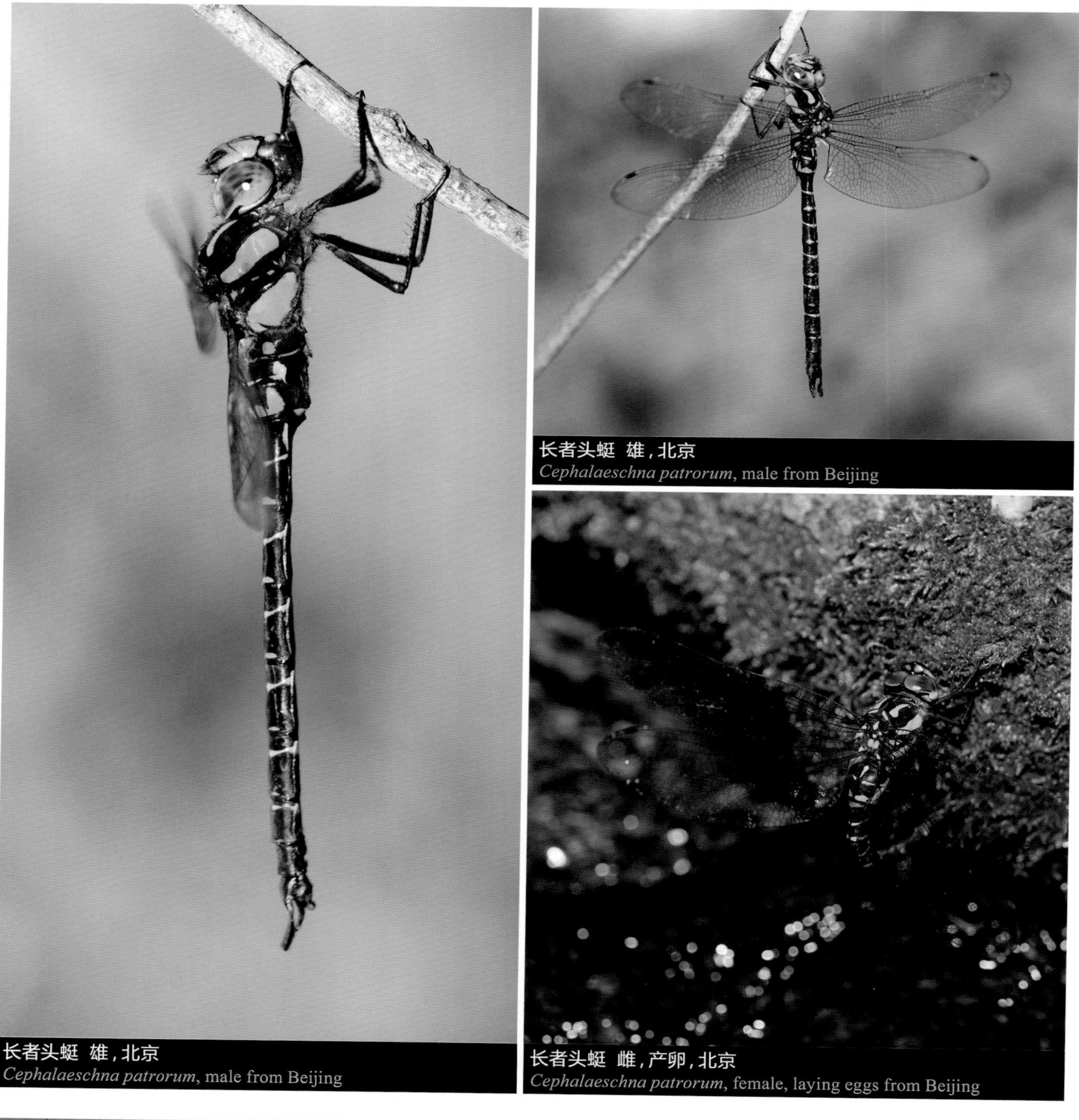

长者头蜓 雄，北京
Cephalaeschna patrorum, male from Beijing

长者头蜓 雄，北京
Cephalaeschna patrorum, male from Beijing

长者头蜓 雌，产卵，北京
Cephalaeschna patrorum, female, laying eggs from Beijing

李氏头蜓 *Cephalaeschna risi* Asahina, 1981

【形态特征】雄性复眼深绿色，面部大面积黄色，额具1个褐色斑；胸部深褐色，脊黄色，肩前条纹甚细，合胸侧面具2条黄绿色条纹，足黑色具黄色条纹，翅透明，翅痣褐色；腹部黑色，第2～8节背面具甚细的黄色条纹。雌性与雄性相似，翅痣黄色，产卵管末端稍微超出第10节末端。【长度】体长 66～67 mm，腹长 52～53 mm，后翅 45～49 mm。【栖息环境】海拔 500～2000 m的林荫小溪。【分布】中国特有，分布于浙江、福建、广东、台湾。【飞行期】6—9月。

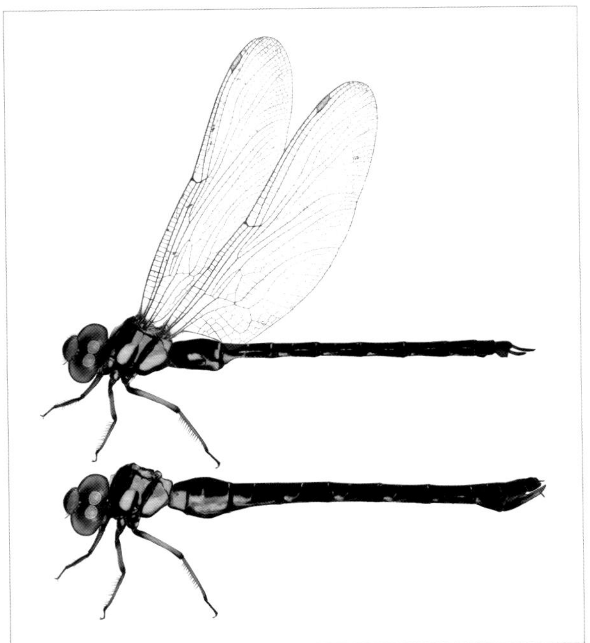
李氏头蜓 上雄下雌，浙江
Cephalaeschna risi, male (above) and female (below) from Zhejiang

[Identification] Male eyes dark green, face largely yellow, frons with a brown spot. Thorax dark brown with yellow dorsal carina and tiny yellow antehumeral stripes, sides with two yellow and green stripes, legs largely black with yellow stripes, wings hyaline, pterostigma brown. Abdomen black with tiny yellow markings on S2-S8. Female similar to male, pterostigma yellow, ovipositor slightly exceeding the end of S10. **[Measurements]** Total length 66-67 mm, abdomen 52-53 mm, hind wing 45-49 mm. **[Habitat]** Shady montane streams at 500-2000 m elevation. **[Distribution]** Endemic to China, recorded from Zhejiang, Fujian, Guangdong, Taiwan. **[Flight Season]** June to September.

邵武头蜓 *Cephalaeschna shaowuensis* Xu, 2006

邵武头蜓 雌，广东
Cephalaeschna shaowuensis, female from Guangdong

【形态特征】雄性复眼深绿色，面部大面积黄色，额具1个褐色斑；胸部深褐色，脊黄色，肩前条纹甚细小，合胸侧面具2条黄绿色条纹，足黑色具黄色条纹，翅透明，翅痣深褐色；腹部大面积黑色，第1~3节侧面具黄斑，第2节具背中条纹。雌性与雄性相似，翅痣黄褐色，产卵管末端稍微超出第10节末端。【长度】体长65~70 mm，腹长 52~55 mm，后翅 45~52 mm。【栖息环境】海拔 1000~1500 m的开阔溪流。【分布】中国特有，分布于福建、广东。【飞行期】7—10月。

[Identification] Male eyes dark green, face largely yellow, frons with a brown spot. Thorax dark brown with yellow dorsal carina and tiny yellow antehumeral stripes, sides with two yellow and green stripes, legs largely black with yellow stripes, wings hyaline, pterostigma dark brown. Abdomen largely black, S1-S3 with lateral yellow spots, S2 with median dorsal stripe. Female similar to male, pterostigma yellowish brown, ovipositor long, slightly exceeding the end of S10. **[Measurements]** Total length 65-70 mm, abdomen 52-55 mm, hind wing 45-52 mm. **[Habitat]** Exposed montane streams at 1000-1500 m elevation. **[Distribution]** Endemic to China, recorded from Fujian, Guangdong. **[Flight Season]** July to October.

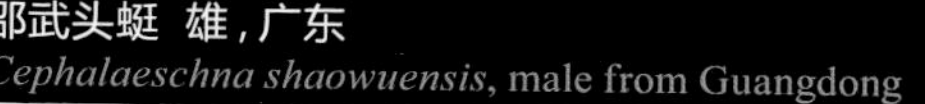

邵武头蜓 雄，广东
Cephalaeschna shaowuensis, male from Guangdong

邵武头蜓 雌，广东
Cephalaeschna shaowuensis, female from Guangdong

独行头蜓 *Cephalaeschna solitaria* Zhang, Cai & Liao, 2013

独行头蜓 雄，湖北
Cephalaeschna solitaria, male from Hubei

【形态特征】雄性复眼深蓝色，面部大面积黄绿色，前额具1个褐色斑，上额具“T”形斑；合胸黑色，具黄绿色肩前条纹，合胸侧面具2条甚阔的黄绿色条纹，足深褐色，翅透明；腹部黑色，第1～9节具黄色斑纹。【长度】体长66～67 mm，腹长 50～51 mm，后翅 42～43 mm。【栖息环境】海拔 2000～3000 m的开阔溪流。【分布】中国湖北特有。【飞行期】6—10月。

[Identification] Male eyes dark blue, face largely yellowish green, antefrons with a brown spot, top of frons with a T-mark. Thorax black with yellowish green antehumeral stripes, sides with two broad yellowish green stripes, legs dark brown, wings hyaline. Abdomen black with yellow stripes on S1-S9. **[Measurements]** Total length 66-67 mm, abdomen 50-51 mm, hind wing 42-43 mm. **[Habitat]** Exposed montane streams at 2000-3000 m elevation. **[Distribution]** Endemic to Hubei of China. **[Flight Season]** June to October.

头蜓属待定种1 *Cephalaeschna* sp. 1

【形态特征】雄性复眼深蓝色，面部黄褐色，额甚阔，褐色；胸部深褐色，具黄色肩前条纹，合胸侧面具2条黄绿色条纹，中胸的黄条纹中央间断，足黑色具褐色条纹，翅透明；腹部黑褐色，各节具黄色条纹和甚小的黄色斑点。【长度】体长 61~65 mm，腹长 47~51 mm，后翅 40~43 mm。【栖息环境】海拔 2000~3000 m的开阔溪流。【分布】云南（大理）。【飞行期】7—10月。

[Identification] Male eyes dark blue, face yellowish brown, frons brown and very broad. Thorax dark brown with yellow antehumeral stripes, sides with two yellowish green stripes and the first one interrupted centrally, legs largely black with brown stripes, wings hyaline. Abdomen blackish brown with yellow stripes and small yellow spots. **[Measurements]** Total length 61-65 mm, abdomen 47-51 mm, hind wing 40-43 mm. **[Habitat]** Exposed montane streams at 2000-3000 m elevation. **[Distribution]** Yunnan (Dali). **[Flight Season]** July to October.

头蜓属待定种1 雄，云南（大理）
Cephalaeschna sp. 1, male from Yunnan (Dali)

头蜓属待定种2 *Cephalaeschna* sp. 2

【形态特征】雄性复眼深绿色，面部黑褐色；胸部黑色，具绿色肩前条纹，合胸侧面具2条宽阔的绿色条纹，足黑褐色，翅透明；腹部黑色，第1~8节具甚细小的黄色斑纹。雌性与雄性相似，但条纹色彩较浅，产卵管未超出第10节末端。【长度】体长 68~69 mm，腹长 52~54 mm，后翅 43~48 mm。【栖息环境】海拔 1000~1500 m的开阔溪流。【分布】贵州。【飞行期】7—9月。

[Identification] Male eyes dark green, face largely blackish brown. Thorax black with green antehumeral stripes, sides with two broad green stripes, legs blackish brown, wings hyaline. Abdomen black, S1-S8 with fine yellow markings. Female body marking similar to male but paler, ovipositor not exceeding the end of S10. [Measurements] Total length 68-69 mm, abdomen 52-54 mm, hind wing 43-48 mm. [Habitat] Exposed montane streams at 1000-1500 m elevation. [Distribution] Guizhou. [Flight Season] July to September.

头蜓属待定种2 雄，贵州
Cephalaeschna sp. 2, male from Guizhou

头蜓属待定种2 雌，贵州
Cephalaeschna sp. 2, female from Guizhou

头蜓属待定种3 *Cephalaeschna* sp. 3

头蜓属待定种3 雄，贵州
Cephalaeschna sp. 3, male from Guizhou

【形态特征】雄性复眼深绿色，面部黑褐色；胸部黑色，具绿色肩前条纹，合胸侧面具2条宽阔的绿色条纹，足黑褐色，翅透明；腹部黑色，第1～8节具甚细小的黄色斑纹。雌性与雄性相似，但条纹色彩较浅，产卵管未超出第10节末端。【长度】体长 66～73 mm，腹长 51～58 mm，后翅 42～48 mm。【栖息环境】海拔 500～1500 m的开阔溪流。【分布】湖北、贵州。【飞行期】7—10月。

[Identification] Male eyes dark green, face largely blackish brown. Thorax black with green antehumeral stripes, sides with two broad green stripes, legs blackish brown, wings hyaline. Abdomen black, S1-S8 with fine yellow markings. Female body marking similar to male but paler, ovipositor not exceeding the end of S10. **[Measurements]** Total length 66-73 mm, abdomen 51-58 mm, hind wing 42-48 mm. **[Habitat]** Exposed montane streams at 500-1500 m elevation. **[Distribution]** Hubei, Guizhou. **[Flight Season]** July to October.

头蜓属待定种3 雄，贵州
Cephalaeschna sp. 3, male from Guizhou

头蜓属待定种3 雌，贵州
Cephalaeschna sp. 3, female from Guizhou

头蜓属待定种3 雌，贵州
Cephalaeschna sp. 3, female from Guizhou

头蜓属待定种4 *Cephalaeschna* sp. 4

【形态特征】雄性复眼深绿色，面部大面积黄色，上额具1个“T”形斑；胸部深褐色，具绿色的肩前条纹，合胸侧面具2条黄绿色条纹，足黑色具红褐色条纹，翅透明；腹部大面积黑色，第1~7节具较细小的黄色斑纹。【长度】雄性体长 68 mm，腹长 52 mm，后翅 47 mm。【栖息环境】山区溪流。【分布】西藏。【飞行期】7—9月。

[Identification] Male eyes dark green, face largely yellow, top of frons with a T-mark. Thorax dark brown with green antehumeral stripes, sides with two yellow and green stripes, legs largely black with reddish brown stripes, wings hyaline. Abdomen largely black, S1-S7 with small yellow spots. **[Measurements]** Male total length 68 mm, abdomen 52 mm, hind wing 47 mm. **[Habitat]** Montane streams. **[Distribution]** Tibet. **[Flight Season]** July to September.

头蜓属待定种4 雄，西藏 | 吴超 摄
Cephalaeschna sp. 4, male from Tibet | Photo by Chao Wu

头蜓属待定种5 *Cephalaeschna* sp. 5

【形态特征】雄性复眼深蓝色，面部大面积黄褐色；胸部深褐色，具黄色肩前条纹，合胸侧面具3条弯曲的黄条纹，足大面积深褐色具红褐色条纹，翅透明；腹部黑色，第1~9节具黄色斑纹。【长度】雄性体长 63~66 mm，腹长 50~52 mm，后翅 42~43 mm。【栖息环境】山区溪流。【分布】西藏。【飞行期】7—9月。

[Identification] Male eyes dark blue, face largely yellowish brown. Thorax dark brown with yellow antehumeral stripes, sides with three curved yellow stripes, legs largely dark brown with reddish brown stripes, wings hyaline. Abdomen black, S1-S9 with yellow stripes. **[Measurements]** Male total length 63-66 mm, abdomen 50-52 mm, hind wing 42-43 mm. **[Habitat]** Montane streams. **[Distribution]** Tibet. **[Flight Season]** July to September.

头蜓属待定种5 雄，西藏 | 吴超 摄
Cephalaeschna sp. 5, male from Tibet | Photo by Chao Wu

长尾蜓属 Genus *Gynacantha* Rambur, 1842

本属已知超过90种，是本科多样性最高的一个属。本属在热带种类较多，包括非洲和大洋洲炎热干旱的区域。中国约有10种，主要分布在华南和西南等地。本属蜻蜓体型中型至大型，面部较窄，复眼发达；翅透明，基室无横脉，IR3叉状，臀角略呈角状；腹部较细长，雄性的肛附器和雌性的尾毛非常长。

日本长尾蜓 雄翅
Gynacantha japonica, male wings

本属蜻蜓主要栖息于水稻田、森林中的沼泽地和泥潭以及暂时性的积雨潭。雄性偶见领域行为，有时在林荫处低空定点悬停，并经常移动位置。在热带地区，很多长尾蜓全年可见。它们以成虫的形式度过旱季。雌性可以将卵产在低洼地的泥土中，等待雨季到来以后，卵才开始孵化。稚虫速生，可以在短时间内迅速发育成熟。本属成虫在白天通常栖息于茂盛的森林中，黄昏时活跃，在天色近黑时开始捕食，是一天中最晚活动的蜻蜓。

This genus contains over 90 species, the most speciose genus in the family. Most species live in the tropical zone, including the dry areas of Africa and Oceania. About ten species are recorded from China, mainly found in the South and Southwest regions. Species of the genus are medium to large dragonflies. Face narrow and eyes large. Wings hyaline, median space without crossveins, IR3 forked, tornus slightly angled. Abdomen long and slim, male anal appendages and female cerci very long,

Gynacantha species inhabit paddy fields, marshes and temporary pools. Male territorial behavior is occasionally seen, when the male hovers near the ground but he rarely defends the same area for long. In lowland tropical habitats they are on the wing throughout the year. They survive the dry season as adults. Females lay eggs into mud in low lying places and eggs hatch when the rainy season comes. The larva can grow very fast. The adults usually perch in dense trees during day and become active at twilight, they are on forage in very late time.

跳长尾蜓 雄 | 宋睿斌 摄
Gynacantha saltatrix, male | Photo by Ruibin Song

无纹长尾蜓
Gynacantha bayadera

透翅长尾蜓
Gynacantha hyalina

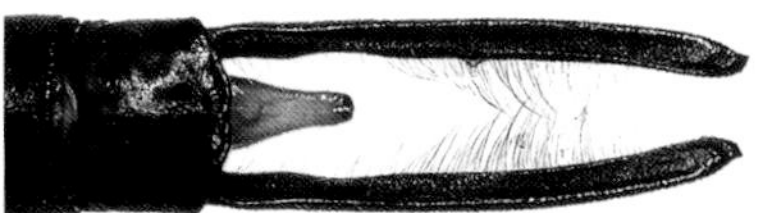

日本长尾蜓
Gynacantha japonica

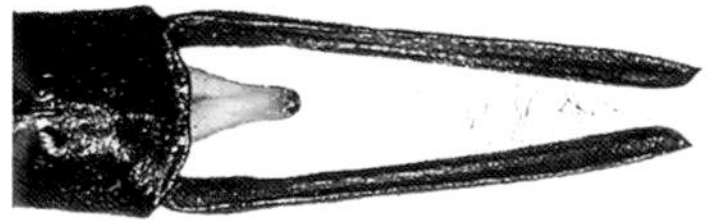

琉球长尾蜓
Gynacantha ryukyuensis

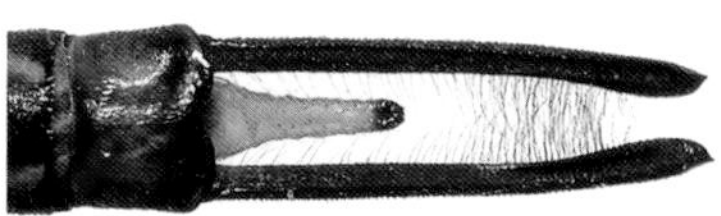

跳长尾蜓
Gynacantha saltatrix

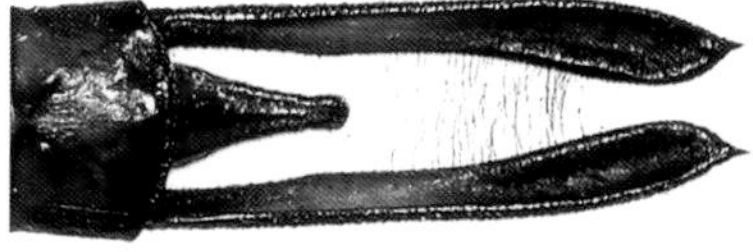

细腰长尾蜓
Gynacantha subinterrupta

长尾蜓属 雄性肛附器
Genus *Gynacantha*, male anal appenages

无纹长尾蜓 *Gynacantha bayadera* Selys, 1891

无纹长尾蜓 雄，海南
Gynacantha bayadera, male from Hainan

无纹长尾蜓 雌，云南（德宏）
Gynacantha bayadera, female from Yunnan (Dehong)

【形态特征】雄性复眼深蓝色，雌性灰色或灰蓝色，面部黄绿色；胸部绿色，足浅褐色，翅透明；腹部灰褐色，具黄绿色斑点。【长度】体长 59~61 mm，腹长 45~47 mm，后翅 40~42 mm。【栖息环境】海拔 1500 m以下的浅水池塘和季节性水塘。【分布】广东、海南、云南、台湾；广布于南亚、东南亚，东至新几内亚。【飞行期】全年可见。

[Identification] Male eyes dark blue, female eyes grey or greyish blue, face yellowish green. Thorax green, legs light brown, wings hyaline. Abdomen greyish brown with yellowish green spots. **[Measurements]** Total length 59-61 mm, abdomen 45-47 mm, hind wing 40-42 mm. **[Habitat]** Shallow and seasonal ponds below 1500 m elevation. **[Distribution]** Guangdong, Hainan, Yunnan, Taiwan; Widespread in South and Southeast Asia reaching New Guinea in the east. **[Flight Season]** Throughout the year.

透翅长尾蜓 *Gynacantha hyalina* Selys, 1882

【形态特征】雄性复眼蓝绿色，面部黄褐色，额具1个“T”形斑；胸部绿色，足红褐色，翅透明；腹部褐色，雄性第2节具蓝绿相间的斑纹，第3～10节具黄绿色斑点。雌性色彩变异较大，复眼蓝绿色或褐色；胸部褐色或绿色，足浅褐色。【长度】体长 67～70 mm，腹长 52～55 mm，后翅 43～45 mm。【栖息环境】海拔 500 m以下的浅水池塘。【分布】中国台湾；菲律宾。亚洲大陆地区的记录待确定。【飞行期】全年可见。

[Identification] Male eyes bluish green, face yellowish brown, frons with a T-mark. Thorax green, legs reddish brown, wings hyaline. Abdomen brown, S2 with blue and green markings, S3-S10 with yellowish green spots. Female body color variable, eyes bluish green or brown. Thorax brown or green, legs light brown. **[Measurements]** Total length 67-70 mm, abdomen 52-55 mm, hind wing 43-45 mm. **[Habitat]** Shallow ponds below 500 m elevation. **[Distribution]** Taiwan of China; Philippines. Published records from continental Asia are in need of confirmation. **[Flight Season]** Throughout the year.

透翅长尾蜓 雄，台湾
Gynacantha hyalina, male from Taiwan

日本长尾蜓 *Gynacantha japonica* Bartenev, 1909

【形态特征】雄性复眼深蓝色，面部黄褐色，额具1个“T”形斑；胸部绿色，足基方至腿节中部红褐色，其余各节黑褐色，翅透明；腹部黑褐色，第2节具蓝绿相间的条纹和斑点，第3～8节具黄绿色条纹，上肛附器甚长。雌性多型，绿色型复眼绿色，腹部第2～3节无蓝色斑点；蓝色型复眼蓝绿色，腹部第2～3节具蓝色斑点。【长度】体长

68~76 mm，腹长 55~59 mm，后翅 46~48 mm。【栖息环境】海拔 1500 m以下的浅水池塘、季节性水塘、水稻田和狭窄的林荫小溪。【分布】湖北、湖南、安徽、浙江、福建、江西、贵州、广东、广西、香港、台湾；朝鲜半岛、日本。【飞行期】5—11月。

[Identification] Male eyes dark blue, face yellowish brown, frons with a T-mark. Thorax green, legs reddish brown from bases to the mid femora, the rest segments blackish brown, wings hyaline. Abdomen dark brown, S2 with blue and green markings, S3-S8 with yellowish green stripes. Female polymorphic, the green morph with eyes green, S2-S3 without blue spots. The blue morph with eyes bluish green, S2-S3 with blue spots. [Measurements] Total length 68-76 mm, abdomen 55-59 mm, hind wing 46-48 mm. [Habitat] Shallow or seasonal ponds, paddy fields and shady narrow streams below 1500 m elevation. [Distribution] Hubei, Hunan, Anhui, Zhejiang, Fujian, Jiangxi, Guizhou, Guangdong, Guangxi, Hong Kong, Taiwan; Korean peninsula, Japan. [Flight Season] May to November.

日本长尾蜓 雌，广东 | 宋睿斌 摄
Gynacantha japonica, female from Guangdong | Photo by Ruibin Song

日本长尾蜓 雄，贵州
Gynacantha japonica, male from Guizhou

日本长尾蜓 雄，广东 | 宋睿斌 摄
Gynacantha japonica, male from Guangdong | Photo by Ruibin Song

琉球长尾蜓 *Gynacantha ryukyuensis* Asahina, 1962

【形态特征】雄性复眼橄榄绿色，面部黄色，额具1个“T”形斑；胸部黄绿色，后胸稍染天蓝色，足大面积黑褐色，基部红褐色，翅透明，基部稍染黑色；腹部黑色，第2节具黄色和蓝色相间的条纹，第3～8节具黄绿色斑。雌性与雄性相似，翅染有浅褐色，且基方具黑斑。【长度】体长 64～66 mm，腹长 49～51 mm，后翅 43～45 mm。【栖息环境】海拔 1500 m以下的浅水池塘、季节性水塘和狭窄的林荫小溪。【分布】广西、海南、香港、台湾；日本。【飞行期】1—10月。

[Identification] Male eyes olive green, face yellow, frons with a T-mark. Thorax yellowish green, metipimeron tinted with blue, legs largely blackish brown, bases reddish brown, wings hyaline, bases slightly tinted with black. Abdomen black, S2 with blue and yellow markings, S3-S8 with basal yellowish green spots laterally. Female similar to male, wings tinted with light brown and bases with black spots. [Measurements] Total length 64-66 mm, abdomen 49-51 mm, hind wing 43-45 mm. [Habitat] Shallow and seasonal ponds, paddy fields and shady narrow streams below 1500 m elevation. [Distribution] Guangxi, Hainan, Hong Kong, Taiwan; Japan. [Flight Season] January to October.

琉球长尾蜓 雄，海南
Gynacantha ryukyuensis, male from Hainan

琉球长尾蜓 雌，海南
Gynacantha ryukyuensis, female from Hainan

琉球长尾蜓 雄，海南
Gynacantha ryukyuensis, male from Hainan

跳长尾蜓 *Gynacantha saltatrix* Martin, 1909

跳长尾蜓 雄，广东 | 宋睿斌 摄
Gynacantha saltatrix, male from Guangdong | Photo by Ruibin Song

【形态特征】雄性复眼黄褐色并具天蓝色眼斑，面部黄褐色，额具1个“T”形斑；胸部浅绿色，足浅褐色，翅透明；腹部黑褐色，第2节具蓝绿相间的条纹和斑点，第3～9节具黄绿色条纹。雌性与雄性相似，第2节无蓝色斑点。【长度】体长 62～66 mm，腹长 49～51 mm，后翅 39～41 mm。【栖息环境】海拔 1000 m以下的浅水池塘、季节性水塘和狭窄的林荫小溪。【分布】广东、广西、海南、香港、台湾；缅甸、泰国、柬埔寨、老挝、越南。【飞行期】4—11月。

[Identification] Male eyes yellowish brown with blue spots, face yellowish brown, frons with a T-mark. Thorax light green, legs light brown, wings hyaline. Abdomen blackish brown, S2 with blue and green markings, S3-S9 with yellowish green stripes. Female body color similar to male, S2 without blue markings. **[Measurements]** Total length

62-66 mm, abdomen 49-51 mm, hind wing 39-41 mm. **[Habitat]** Shallow or seasonal ponds, paddy fields and shady narrow streams below 1000 m elevation. **[Distribution]** Guangdong, Guangxi, Hainan, Hong Kong, Taiwan; Myanmar, Thailand, Cambodia, Laos, Vietnam. **[Flight Season]** April to November.

跳长尾蜓 雄，广东 | 宋睿斌 摄
Gynacantha saltatrix, male from Guangdong | Photo by Ruibin Song

跳长尾蜓 雌，广东 | 宋睿斌 摄
Gynacantha saltatrix, female from Guangdong | Photo by Ruibin Song

细腰长尾蜓 *Gynacantha subinterrupta* Rambur, 1842

【形态特征】雄性复眼蓝绿色，面部黄褐色，额具1个“T”形斑；胸部绿色，足红褐色，翅透明；腹部褐色，第2节具蓝绿相间的条纹和斑点，第3～10节具黄绿色条纹和斑点。雌性色彩变异较大，复眼蓝绿色或褐色；胸部褐色或绿色，足浅褐色。未熟体为浅褐色。【长度】体长 61～70 mm，腹长 48～54 mm，后翅 43～48 mm。【栖息环境】海拔 1500 m以下的浅水池塘、季节性水塘和狭窄的林荫小溪。【分布】云南、贵州、湖南、福建、广东、广西、海南、香港；印度、尼泊尔、柬埔寨、老挝、越南、泰国、马来半岛、新加坡、印度尼西亚。【飞行期】全年可见。

[Identification] Male eyes bluish green, face yellowish brown, frons with a T-mark. Thorax green, legs reddish brown, wings hyaline. Abdomen brown, S2 with blue and green markings, S3-S10 with yellowish green spots. Female body color variable, eyes bluish green or brown. Thorax brown or green, legs light brown. Immature adult largely light brown. **[Measurements]** Total length 61-70 mm, abdomen 48-54 mm, hind wing 43-48 mm. **[Habitat]** Shallow or seasonal ponds, paddy fields and shady narrow streams below 1500 m elevation. **[Distribution]** Yunnan, Guizhou, Hunan, Fujian, Guangdong, Guangxi, Hainan, Hong Kong; India, Nepal, Cambodia, Laos, Vietnam, Thailand, Peninsular Malaysia, Singapore, Indonesia. **[Flight Season]** Throughout the year.

细腰长尾蜓 雌，广东 | 宋睿斌 摄
Gynacantha subinterrupta, female from Guangdong | Photo by Ruibin Song

细腰长尾蜓 雄，海南
Gynacantha subinterrupta, male from Hainan

细腰长尾蜓 雄，广东 | 宋睿斌 摄
Gynacantha subinterrupta, male from Guangdong | Photo by Ruibin Song

长尾蜓属待定种1 *Gynacantha* sp. 1

【形态特征】雄性复眼深蓝色，面部黄褐色，额具1个“T”形斑；胸部绿色，后胸苍白色，足红色，翅透明；腹部黑褐色，第2节具蓝绿相间的条纹和斑点，第3~7节具黄绿色条纹。雌性复眼蓝绿色，胸部黄绿色，腹部第2节的蓝色较淡。【长度】体长 65~68 mm，腹长 50~52 mm，后翅 40~45 mm。【栖息环境】海拔 600~1300 m的山区狭窄林荫小溪。【分布】云南（德宏、临沧、普洱、西双版纳）。【飞行期】9—12月。

[Identification] Male eyes dark blue, face yellowish brown, frons with a T-mark. Thorax green, metepimeron grey, legs red, wings hyaline. Abdomen blackish brown, S2 with blue and green markings, S3-S7 with yellowish green

长尾蜓属待定种1 雌，云南（德宏）
Gynacantha sp.1, female from Yunnan (Dehong)

长尾蜓属待定种1 雄，云南（德宏）
Gynacantha sp.1, male from Yunnan (Dehong)

长尾蜓属待定种1 雄，云南（德宏）
Gynacantha sp.1, male from Yunnan (Dehong)

stripes. Female eyes bluish green, thorax yellowish green, S2 with pale blue markings. **[Measurements]** Total length 65-68 mm, abdomen 50-52 mm, hind wing 40-45 mm. **[Habitat]** Shady narrow streams at 600-1300 m elevation. **[Distribution]** Yunnan (Dehong, Lincang, Pu'er, Xishuangbanna). **[Flight Season]** September to December.

长尾蜓属待定种2 *Gynacantha* sp. 2

【形态特征】雄性复眼蓝色，面部黄褐色，额具1个"T"形斑；胸部绿色和红色，足红色，翅透明；腹部黑褐色，第2节具蓝绿相间的条纹和斑点，第3～7节具黄绿色条纹。雌性复眼褐色；胸部和腹部大面积红褐色，第2节具绿色条纹。【长度】体长 57～64 mm，腹长 43～49 mm，后翅 40～44 mm。【栖息环境】海拔 1500～2500 m的水稻田。【分布】云南（昆明、大理）。【飞行期】8—11月。

[Identification] Male eyes blue, face yellowish brown, frons with a T-mark. Thorax green and red, legs red, wings hyaline. Abdomen blackish brown, S2 with blue and green markings, S3-S7 with yellowish green stripes. Female eyes brown. Thorax and abdomen largely reddish brown, S2 with green markings. **[Measurements]** Total length 57-64 mm, abdomen 43-49 mm, hind wing 40-44 mm. **[Habitat]** Paddy fields at 1500-2500 m elevation **[Distribution]** Yunnan (Kunming, Dali). **[Flight Season]** August and November.

长尾蜓属待定种2 雄，云南（昆明）
Gynacantha sp. 2, male from Yunnan (Kunming)

长尾蜓属待定种2 雌，云南（昆明）
Gynacantha sp. 2, female from Yunnan (Kunming)

棘蜓属 Genus *Gynacanthaeschna* Fraser, 1921

本属全球仅知1种，分布于喜马拉雅和中国西南的横断山地区。本属蜻蜓体中型；面部甚阔，其宽度大于头部宽度的1/2；翅透明，基室具横脉，翅痣较短，具支持脉，IR3叉状，Rspl和Mspl之上仅有1列翅室，臀三角室3室。

本属蜻蜓栖息于茂盛森林中的溪流，成虫在下午4至5点期间靠近水面，或者沿着山路飞行，并时而悬停。云南西部的个体在10月以后才可见，是晚季节飞行的种类。

锡金棘蜓 雄翅
Gynacanthaeschna sikkima, male wings

This genus contains just a single species from the Himalayas and Hengduan Mountains in Southwest China. The species is medium-sized. Face broad, the width of face more than half width of head. Wings hyaline, median space with crossveins, pterostigma short and well braced, IR3 forked, a single row of cells above Rspl and Mspl, anal triangle 3-celled.

Gynacanthaeschna species inhabits forest streams. The adults are active between 4 pm and 5 pm, when they approach water, or fly along paths, sometimes hovering. Individuals from the west of Yunnan begin to fly in October, a late season species.

锡金棘蜓 雄
Gynacanthaeschna sikkima, male

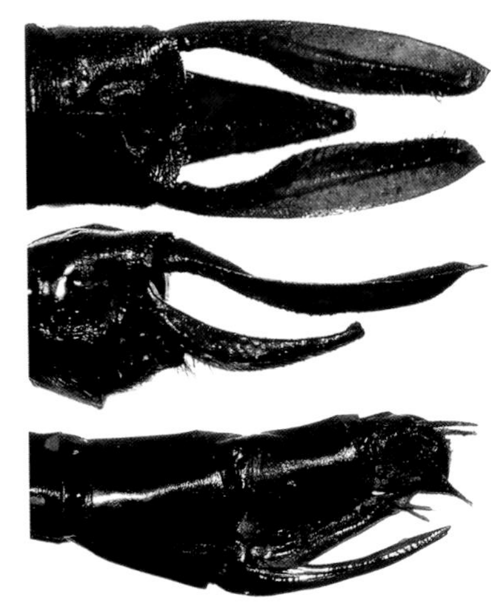

锡金棘蜓
Gynacanthaeschna sikkima

棘蜓属 雄性肛附器和雌性产卵器
Genus *Gynacanthaeschna*, male anal appendages and female ovipositor

锡金棘蜓 *Gynacanthaeschna sikkima* (Karsch, 1891)

【形态特征】雄性复眼绿色，面部黄褐色，前额具1个甚大的黑色斑；胸部黑褐色，肩前条纹较宽阔，合胸侧面具2条苹果绿色条纹，后胸前侧板具1个三角形斑，足大面积黑色，翅透明；腹部黑褐色具黄绿色斑。雌性与雄性相似，色彩稍淡。【长度】体长 57~60 mm，腹长 44~47 mm，后翅 38~39 mm。【栖息环境】海拔 1000~1500 m植被茂盛的山区小溪。【分布】云南（德宏）、西藏；不丹、尼泊尔、印度。【飞行期】10—12月。

[Identification] Male eyes green, face yellowish brown, antefrons with a large black spot. Thorax blackish brown with broad antehumeral stripes, sides with two apple green stripes, metepisternum with a triangular spot, legs largely black, wings hyaline. Abdomen blackish brown with yellowish green spots. Female similar to male but paler. **[Measurements]** Total length 57-60 mm, abdomen 44-47 mm, hind wing 38-39 mm. **[Habitat]** Montane streams in dense forest at 1000-1500 m elevation. **[Distribution]** Yunnan (Dehong), Tibet; Bhutan, Nepal, India. **[Flight Season]** October to December.

锡金棘蜓 雄，云南（德宏）
Gynacanthaeschna sikkima, male from Yunnan (Dehong)

锡金棘蜓 雌，云南（德宏）
Gynacanthaeschna sikkima, female from Yunnan (Dehong)

锡金棘蜓 雄，云南（德宏）
Gynacanthaeschna sikkima, male from Yunnan (Dehong)

锡金棘蜓 雌，云南（德宏）
Gynacanthaeschna sikkima, female from Yunnan (Dehong)

佩蜓属 Genus *Periaeschna* Martin, 1908

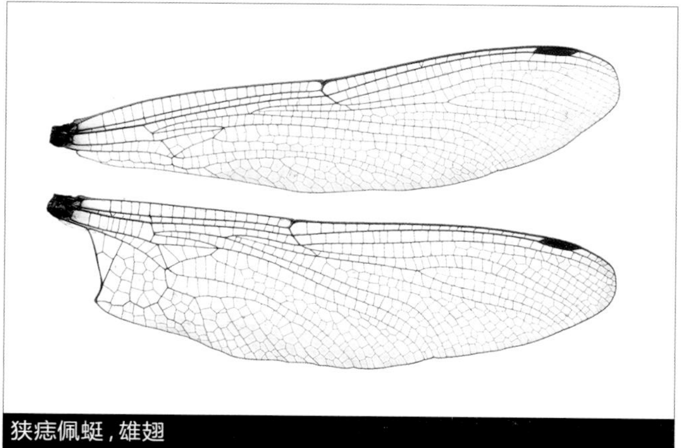
狭痣佩蜓，雄翅
Periaeschna magdalena, male wings

本属全球已知11种，中国已发现8种。本属蜻蜓是一类体中型的溪栖蜓科物种，身体黑褐色具有较发达的绿色和黄色条纹。复眼通常绿色，面部很窄；翅透明，基室具横脉，翅痣较短，IR3叉状，Rspl和Mspl之上仅有1列翅室，臀三角室3室。

本属蜻蜓广泛分布于中国南方，在茂盛森林中较狭窄的林荫小溪和沟渠中较容易遇见。雄性在下午活跃，在有树荫遮蔽的小溪上方，以低空定点悬停来占据领地，两雄相遇会展开激战。雌性在小溪边缘的土坡上产卵。

This genus contains 11 species, eight species are recorded from China. Species of the genus are medium-sized dragonflies, their bodies blackish brown with green or yellow stripes. Eyes often green, face narrow, wings hyaline, median space with crossveins, pterostigma short, a single row of cells above Rspl and Mspl, anal triangle 3-celled.

Periaeschna species are widespread in the south of China, found in forested streams or ditches. Males are active in the afternoon, hovering low above water in the shade, males fight fiercely for territory. Females lay eggs in the earth slope of the stream margin.

福临佩蜓 雄
Periaeschna flinti, male

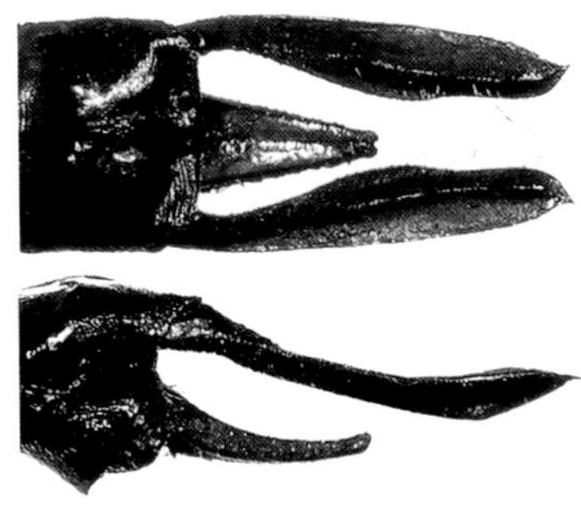

赵氏佩蜓
Periaeschna chaoi

福临佩蜓
Periaeschna flinti

狭痣佩蜓
Periaeschna magdalena

浅色佩蜓
Periaeschna nocturnalis

雅珍佩蜓
Periaeschna yazhenae

漳州佩蜓
Periaeschna zhangzhouensis

佩蜓属 雄性肛附器
Genus *Periaeschna*, male anal appenages

福临佩蜓
Periaeschna flinti

赵氏佩蜓
Periaeschna chaoi

佩蜓属 雌性产卵器
Genus *Periaeschna*, female ovipositor

赵氏佩蜓 *Periaeschna chaoi* (Asahina, 1982)

【形态特征】雄性复眼绿色，面部黄褐色，前额具1个甚大的黑色斑；合胸黑色，具黄绿色肩前条纹，合胸侧面具2条甚阔的黄绿色条纹，足深褐色，翅透明；腹部黑色，第1～2节背面和侧面具黄绿色斑，第3～7节背面后缘具三角形斑点。【长度】雄性体长 69 mm，腹长 55 mm，后翅 46 mm。【栖息环境】海拔 1000 m左右的山区小溪。【分布】中国特有，分布于湖北、重庆、江西、福建、广东。【飞行期】7—9月。

[Identification] Male eyes green, face yellowish brown, antefrons with a large black spot. Thorax black with yellowish green antehumeral stripes, sides with two broad yellowish green stripes, legs dark brown, wings hyaline. Abdomen black, S1-S2 with dorsal and lateral yellowish green stripes, S3-S7 with triangular dorsal spots apically. **[Measurements]** Male total length 69 mm, abdomen 55 mm, hind wing 46 mm. **[Habitat]** Montane streams at about 1000 m elevation. **[Distribution]** Endemic to China, recorded from Hubei, Chongqing, Jiangxi, Fujian, Guangdong. **[Flight Season]** July to September.

赵氏佩蜓 雄，重庆
Periaeschna chaoi, male from Chongqing

赵氏佩蜓 雄，重庆
Periaeschna chaoi, male from Chongqing

赵氏佩蜓 雌，重庆
Periaeschna chaoi, female from Chongqing

福临佩蜓 *Periaeschna flinti* Asahina, 1978

【形态特征】雄性复眼绿色，面部黄褐色，前额具1个甚大的黑色斑；胸部黑褐色，肩前条纹较宽阔，合胸侧面具2条苹果绿色条纹，足黑色，翅透明；腹部黑褐色，第3～9节末端具半圆形的背斑。雌性色彩较淡，条纹为黄色。【长度】体长 63～65 mm，腹长 48～50 mm，后翅 39～44 mm。【栖息环境】海拔 1500 m以下植被茂盛的山区小溪和沟渠。【分布】湖北、湖南、四川、云南、贵州、安徽、浙江、福建、广东、广西；印度。【飞行期】5—9月。

[Identification] Male eyes green, face yellowish brown, antefrons with a large black spot. Thorax blackish brown with broad antehumeral stripes, sides with two apple green stripes, legs black, wings hyaline. Abdomen blackish brown, S3-S9 with dorsal semicircular spots apically. Female paler with yellow markings. **[Measurements]** Total length 63-65 mm, abdomen 48-50 mm, hind wing 39-44 mm. **[Habitat]** Montane streams and ditches in dense forest below 1500 m elevation. **[Distribution]** Hubei, Hunan, Sichuan, Yunnan, Guizhou, Anhui, Zhejiang, Fujian, Guangdong, Guangxi; India. **[Flight Season]** May to September.

福临佩蜓 雌，产卵，贵州
Periaeschna flinti, female, laying eggs from Guizhou

福临佩蜓 雌，浙江 | 莫善濂 摄
Periaeschna flinti, female from Zhejiang | Photo by Shanlian Mo

福临佩蜓 雄，贵州
Periaeschna flinti, male from Guizhou

狭痣佩蜓 *Periaeschna magdalena* Martin, 1909

【形态特征】雄性复眼黄绿色，面部黄褐色，额具1个不清晰的"T"形斑；胸部黑褐色，肩前条纹甚阔，合胸侧面具2条黄色条纹，翅稍染褐色；腹部黑褐色，具不发达的黄色斑点，有时第9~10节具甚大的黄斑。雌性与雄性相似，但体型粗壮。【长度】体长 65~74 mm，腹长 50~57 mm，后翅 43~50 mm。【栖息环境】海拔 500~1500 m植被茂盛的山区小溪。【分布】陕西、湖北、湖南、四川、云南、贵州、重庆、江苏、安徽、浙江、福建、江西、广东、广西、台湾；不丹、印度、老挝、越南。【飞行期】4—8月。

[Identification] Male eyes yellowish green, face yellowish brown, frons with an unclear T-mark. Thorax blackish brown with broad antehumeral stripes, sides with two yellow stripes, wings tinted with light brown. Abdomen blackish brown with fine yellow spots, sometimes S9-S10 with large yellow spots. Female similar to male but stouter. [Measurements] Total length 65-74 mm, abdomen 50-57 mm, hind wing 43-50 mm. [Habitat] Montane streams in dense forest at 500-1500 m elevation. [Distribution] Shaanxi, Hubei, Hunan, Sichuan, Yunnan, Guizhou, Chongqing, Jiangsu, Anhui, Zhejiang, Fujian, Jiangxi, Guangdong, Guangxi, Taiwan; Bhutan, India, Laos, Vietnam. [Flight Season] April to August.

狭痣佩蜓 雄，广东 | 莫善濂 摄
Periaeschna magdalena, male from Guangdong | Photo by Shanlian Mo

狭痣佩蜓 雌，产卵，广东 | 宋睿斌 摄
Periaeschna magdalena, female laying eggs from Guangdong | Photo by Ruibin Song

狭痣佩蜓 交尾，湖北
Periaeschna magdalena, mating pair from Hubei

浅色佩蜓 *Periaeschna nocturnalis* Fraser, 1927

浅色佩蜓 雄，云南（红河）| 莫善濂 摄
Periaeschna nocturnalis, male from Yunnan (Honghe) | Photo by Shanlian Mo

浅色佩蜓 雄，云南（临沧）
Periaeschna nocturnalis, male from Yunnan (Lincang)

浅色佩蜓 雌，云南（红河）| 莫善濂 摄
Periaeschna nocturnalis, female from Yunnan (Honghe) | Photo by Shanlian Mo

【形态特征】雄性复眼蓝灰色，面部黄褐色；胸部褐色，肩前条纹甚阔，合胸侧面具2条黄色条纹，翅透明；腹部主要褐色，具黄色条纹，有时第8～10节具甚大的黄褐色斑。雌性与雄性相似。【长度】体长 63～68 mm，腹长 49～53 mm，后翅 40～45 mm。【栖息环境】海拔 500～2500 m植被茂盛的山区小溪。【分布】云南；印度、老挝、泰国。【飞行期】4—7月。

[Identification] Male eyes bluish grey, face yellowish brown. Thorax brown with broad antehumeral stripes, sides with two yellow stripes, wings hyaline. Abdomen mainly brown with yellow stripes, sometimes S8-S10 with large yellowish brown spots. Female similar to male. **[Measurements]** Total length 63-68 mm, abdomen 49-53 mm, hind wing 40-45 mm. **[Habitat]** Montane streams in dense forest at 500-2500 m elevation. **[Distribution]** Yunnan; India, Laos, Thailand. **[Flight Season]** April to July.

雅珍佩蜓 *Periaeschna yazhenae* Xu, 2012

【形态特征】雄性复眼蓝绿色，面部褐色；胸部黑褐色，肩前条纹甚阔，合胸侧面具2条黄色条纹，足黑色，翅透明但基部具黑色斑；腹部黑色，基部3节具不发达的条纹。雌性与雄性相似，翅基方的黑色斑更显著。【长度】体长 62~63 mm，腹长 48~49 mm，后翅 42~45 mm。【栖息环境】海拔 1000 m以下植被茂盛的山区小溪和渗流地。【分布】中国特有，分布于福建、广东、广西。【飞行期】5—8月。

雅珍佩蜓 雄，广西
Periaeschna yazhenae, male from Guangxi

雅珍佩蜓 雌，广西
Periaeschna yazhenae, female from Guangxi

[Identification] Male eyes bluish green, face brown. Thorax blackish brown with broad antehumeral stripes, sides with two yellow stripes, legs black, wings hyaline, bases with black spots. Abdomen black with reduced stripes on basal three segments. Female similar to male but wing bases with larger black markings. **[Measurements]** Total length 62-63 mm, abdomen 48-49 mm, hind wing 42-45 mm. **[Habitat]** Montane streams and ditches in dense forest below 1000 m elevation. **[Distribution]** Endemic to China, recorded from Fujian, Guangdong, Guangxi. **[Flight Season]** May to August.

雅珍佩蜓 雄，广西
Periaeschna yazhenae, male from Guangxi

漳州佩蜓 *Periaeschna zhangzhouensis* Xu, 2007

【形态特征】雄性复眼绿色，面部褐色，额黑色；胸部黑褐色，肩前条纹较宽阔，合胸侧面具2条苹果绿色条纹，足黑褐色，翅透明；腹部黑褐色，具苹果绿色斑点。雌性色彩较淡，条纹为黄绿色。【长度】体长 62~67 mm，腹长 48~51 mm，后翅 41~45 mm。【栖息环境】海拔 500~1500 m植被茂盛的山区小溪。【分布】中国特有，分布于四川、贵州、福建、湖南、江西、广东。【飞行期】5—9月。

[Identification] Male eyes green, face brown, frons black. Thorax blackish brown with broad antehumeral stripes, sides with two apple green stripes, legs blackish brown, wings hyaline. Abdomen blackish brown with apple green stripes. Female paler with yellowish green markings. **[Measurements]** Total length 62-67 mm, abdomen 48-51 mm, hind wing 41-45 mm. **[Habitat]** Montane streams in dense forest at 500-1500 m elevation. **[Distribution]** Endemic to China, recorded form Sichuan, Guizhou, Fujian, Hunan, Jiangxi, Guangdong. **[Flight Season]** May to September.

漳州佩蜓 雄，贵州
Periaeschna zhangzhouensis, male from Guizhou

漳州佩蜓 雌，贵州
Periaeschna zhangzhouensis, female from Guizhou

佩蜓属待定种1 *Periaeschna* sp. 1

佩蜓属待定种1 雄，海南
Periaeschna sp. 1, male from Hainan

【形态特征】雄性复眼绿色，面部黄褐色；胸部黑褐色，具黄绿色的肩前条纹，合胸侧面具2条黄绿色条纹，2条纹间具1条细纹，翅稍染褐色；腹部主要黑褐色，第2～8节侧缘具黄色斑点，有时第9～10节具背面具甚大黄斑。雌性与雄性相似，但体型更粗壮。【长度】体长 69～72 mm，腹长 54～56 mm，后翅 44～48 mm。【栖息环境】海拔500～1000 m植被茂盛的山区林荫小溪。【分布】海南。【飞行期】4—6月。

[Identification] Male eyes green, face yellowish brown. Thorax blackish brown with yellowish green antehumeral stripes, sides with two yellowish green stripes and a linear stripe between them, wings tinted with light brown. Abdomen blackish brown, S2-S8 with lateral yellow spots, sometimes S9-S10 with large yellow spots dorsally. Female similar to male but stouter. **[Measurements]** Total length 69-72 mm, abdomen 54-56 mm, hind wing 44-48 mm. **[Habitat]** Shady montane streams in dense forest at 500-1000 m elevation. **[Distribution]** Hainan. **[Flight Season]** April to June.

佩蜓属待定种1 雄，海南
Periaeschna sp. 1, male from Hainan

佩蜓属待定种1 雌，海南
Periaeschna sp. 1, female from Hainan

佩蜓属待定种1 雌，海南
Periaeschna sp. 1, female from Hainan

佩蜓属待定种2 *Periaeschna* sp. 2

【形态特征】雄性复眼绿色，面部褐色；胸部黑褐色，具黄绿色的肩前条纹，合胸侧面具2条黄绿色条纹，翅稍染褐色；腹部主要黑褐色。雌性与雄性相似，但腹部侧缘具较发达的黄斑。【长度】体长 67~70 mm，腹长 53~56 mm，后翅 44~47 mm。【栖息环境】海拔 500~1000 m植被茂盛的山区林荫小溪。【分布】海南。【飞行期】4—7月。

佩蜓属待定种2 雄，海南
Periaeschna sp. 2, male from Hainan

佩蜓属待定种2 雌，海南
Periaeschna sp. 2, female from Hainan

[Identification] Male eyes green, face brown. Thorax blackish brown with yellowish green antehumeral stripes, sides with two yellowish green stripes, wings tinted with light brown. Abdomen largely blackish brown. Female similar to male but abdomen with lateral yellow spots. **[Measurements]** Total length 67-70 mm, abdomen 53-56 mm, hind wing 44-47 mm. **[Habitat]** Shady montane streams in dense forest at 500-1000 m elevation. **[Distribution]** Hainan. **[Flight Season]** April to July.

佩蜓属待定种2 雄，海南
Periaeschna sp. 2, male from Hainan

佩蜓属待定种3 *Periaeschna* sp. 3

【形态特征】雄性复眼绿色，面部黄绿色，上唇具1对苹果绿色斑点，额具1个甚大的黑色斑；胸部黑褐色，肩前条纹较宽阔，合胸侧面具3条苹果绿色条纹，足大部分深褐色，足基方至腿节红褐色，翅透明；腹部黑褐色具较发达的苹果绿色斑点。【长度】体长 55~62 mm，腹长 42~48 mm，后翅 39~40 mm。【栖息环境】海拔 1000~1500 m植被茂盛的山区开阔小溪。【分布】云南（德宏、普洱）。【飞行期】5—9月。

[Identification] Male eyes green, face yellowish green, labrum with a pair of apple green spots, frons with a large black spot. Thorax blackish brown with broad antehumeral stripes, sides with three apple green stripes, legs largely dark brown, from bases to mid femora reddish brown, wings hyaline. Abdomen blackish brown, with numerous apple green stripes. **[Measurements]** Total length 55-62 mm, abdomen 42-48 mm, hind wing 39-40 mm. **[Habitat]** Exposed montane streams in dense forest at 1000-1500 m elevation. **[Distribution]** Yunnan (Dehong, Pu'er). **[Flight Season]** May to September.

佩蜓属待定种3 雌，云南（德宏）
Periaeschna sp. 3, female from Yunnan (Dehong)

佩蜓属待定种3 雄，云南（德宏）
Periaeschna sp. 3, male from Yunnan (Dehong)

叶蜓属 Genus *Petaliaeschna* Fraser, 1927

本属全球已知5种，分布于亚洲。中国已知3种。本属是一类体型较小且罕见的溪栖蜓科物种。面部黄色；足黄色，翅透明，基室具横脉，IR3叉状，臀三角室3室；雄性下肛附器黄色。

本属蜻蜓主要栖息于茂盛森林中较狭窄的林荫小溪和沟渠。雄性下午活跃，在有树荫遮蔽的小溪上方，以定点低空悬停的方式占据领地。

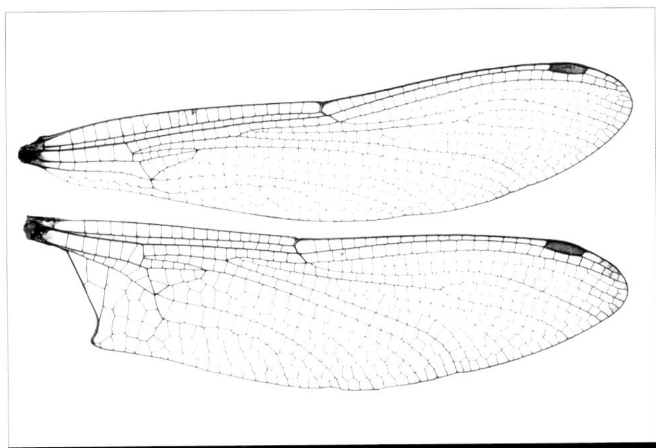
科氏叶蜓 雄翅
Petaliaeschna corneliae, male wings

This genus contains five Asian species with three recorded from China. Species of the genus are relatively small for the family, rarely encountered. Face yellow. Legs yellow, wings hyaline, median space with crossveins, IR3 forked, anal triangle 3-celled. Inferior appendage yellow.

Petaliaeschna species inhabit shady streams and ditches in dense forests. Males are active in the afternoon, hovering above the streams where they defend territories.

黄纹叶蜓
Petaliaeschna flavipes

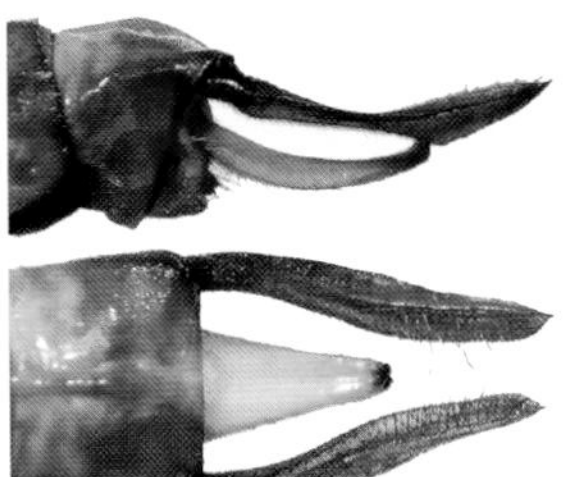
叶蜓属待定种2
Petaliaeschna sp. 2

叶蜓属 雄性肛附器
Genus *Petaliaeschna*, male anal appenages

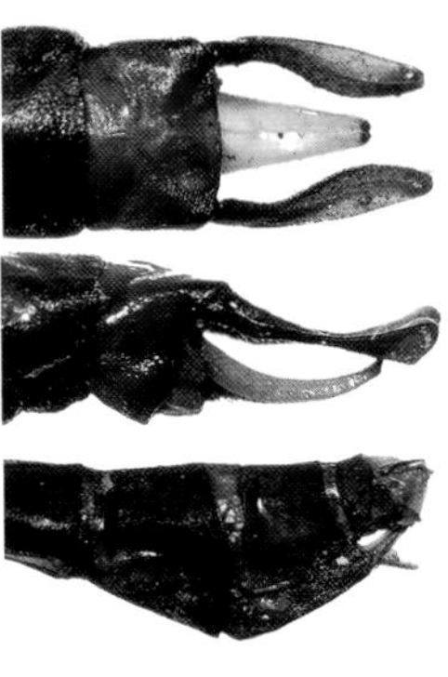
科氏叶蜓
Petaliaeschna corneliae

叶蜓属 雄性肛附器和雌性产卵器
Genus *Petaliaeschna*, male anal appendages and female ovipositor

黄纹叶蜓 雄
Petaliaeschna flavipes, male

科氏叶蜓 *Petaliaeschna corneliae* Asahina, 1982

【形态特征】雄性复眼深蓝色，面部黄色，额浅褐色；胸部黑色，合胸脊黄色，侧面具2条宽阔的黄色条纹，足黄色，翅透明；腹部主要黑色，第1～3节侧面和第2节背面具黄色斑点。雌性与雄性相似，腹部第1～7节侧面具发达的黄斑。【长度】体长 55～57 mm，腹长 43～46 mm，后翅 34～38 mm。【栖息环境】海拔 500～1000 m植被茂盛的山区小溪和渗流地。【分布】中国特有，分布于福建、广东。【飞行期】4—6月。

[Identification] Male eyes dark blue, face yellow, frons light brown. Thorax black, dorsal carina yellow, sides with two broad yellow stripes, legs yellow, wings hyaline. Abdomen mainly black, sides of S1-S3 and dorsum of S2 with yellow spots. Female similar to male, S1-S7 with lateral yellow spots. **[Measurements]** Total length 55-57 mm, abdomen 43-46 mm, hind wing 34-38 mm. **[Habitat]** Montane streams and seepages in forest at 500-1000 m elevation. **[Distribution]** Endemic to China, recorded from Fujian, Guangdong. **[Flight Season]** April to June.

科氏叶蜓 雄，广东 | 吴宏道 摄
Petaliaeschna corneliae, male from Guangdong | Photo by Hongdao Wu

科氏叶蜓 雌，广东 | 吴宏道 摄
Petaliaeschna corneliae, female from Guangdong | Photo by Hongdao Wu

黄纹叶蜓 *Petaliaeschna flavipes* Karube, 1999

【形态特征】雄性复眼深褐色，面部黄色，额浅褐色；胸部黑色，合胸脊黄色，侧面具2条宽阔的黄色条纹，足黄色，翅透明；腹部主要黑色，第1～2节侧面和第2节背面具黄色斑点。雌性与雄性相似，但体型更粗壮。【长度】雄性体长 57 mm，腹长 45 mm，后翅 37 mm。【栖息环境】海拔 1000～2000 m植被茂盛的山区小溪和渗流地。【分布】云南（红河）；泰国、老挝、越南。【飞行期】4—6月。

[Identification] Male eyes dark brown, face yellow, frons light brown. Thorax black, dorsal carina yellow, sides with two broad yellow stripes, legs yellow, wings hyaline. Abdomen mainly black, sides of S1-S2 and dorsum of S2 with yellow spots. Female similar to male but stouter. **[Measurements]** Male total length 57 mm, abdomen 45 mm, hind wing 37 mm. **[Habitat]** Montane streams and seepages in forest at 1000-2000 m elevation. **[Distribution]** Yunnan (Honghe); Thailand, Laos, Vietnam. **[Flight Season]** April to June.

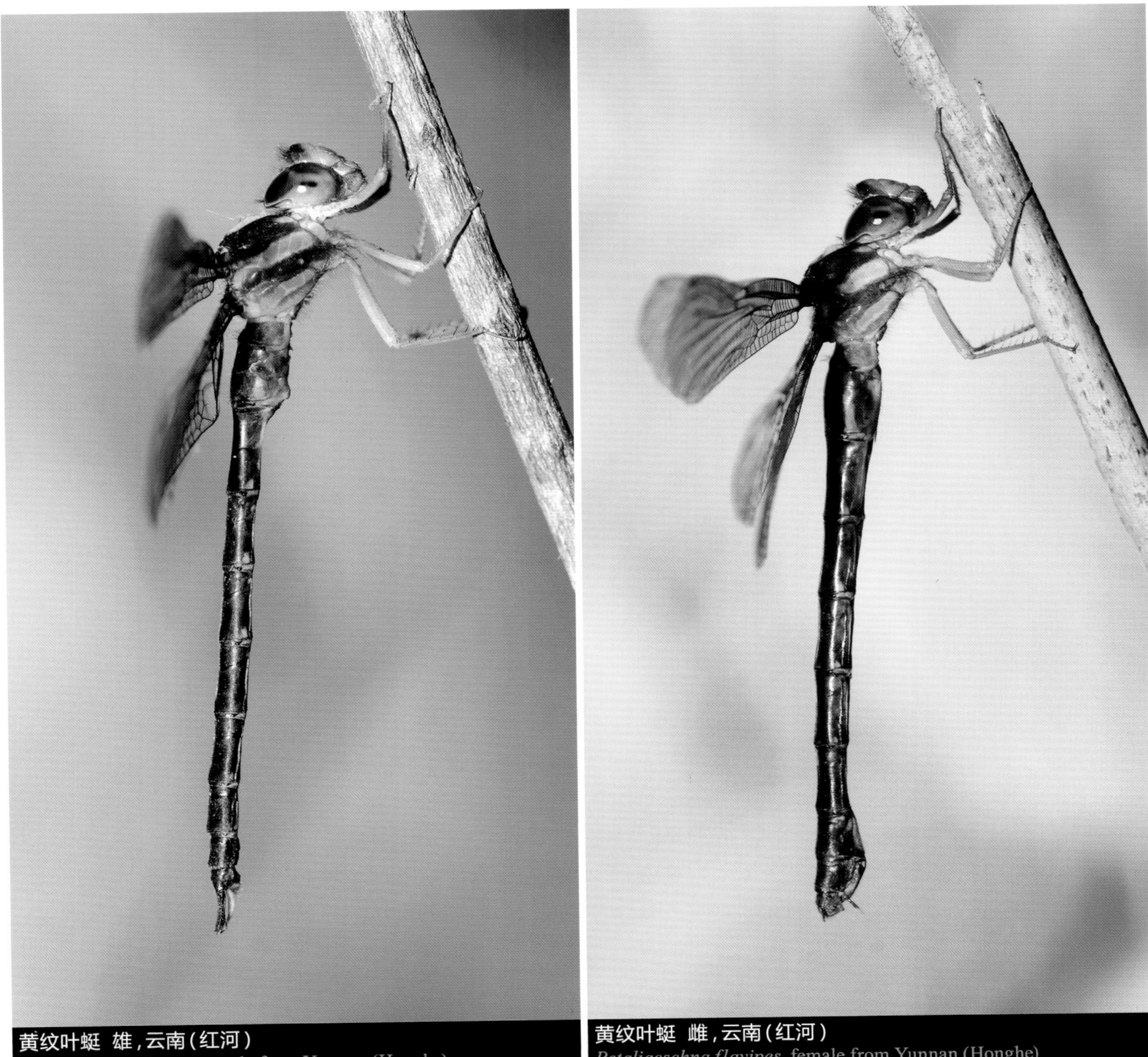

黄纹叶蜓 雄，云南（红河）
Petaliaeschna flavipes, male from Yunnan (Honghe)

黄纹叶蜓 雌，云南（红河）
Petaliaeschna flavipes, female from Yunnan (Honghe)

黄纹叶蜓 雄，云南（红河）
Petaliaeschna flavipes, male from Yunnan (Honghe)

黄纹叶蜓 雌，云南（红河）
Petaliaeschna flavipes, female from Yunnan (Honghe)

叶蜓属待定种1 雌，云南（大理）
Petaliaeschna sp. 1, female from Yunnan (Dali)

叶蜓属待定种1 *Petaliaeschna* sp. 1

【形态特征】雌性复眼深褐色，面部黄色，具黑色的额横纹；胸部褐色，合胸脊黄色，具黄色的肩前下点，侧面具2条宽阔的黄白色条纹，足浅褐色，翅透明；腹部浅褐色，具不甚清晰的黄斑。【长度】雌性体长 60 mm，腹长 47 mm，后翅 40 mm。【栖息环境】海拔 2000~2500 m的山区开阔溪流。【分布】云南（大理）。【飞行期】5—6月。

[Identification] Female eyes dark brown, face yellow, frons with black stripe. Thorax brown, dorsal carina yellow, with yellow antehumeral spots, sides with two broad yellowish white stripes, legs light brown, wings hyaline. Abdomen light brown with obscure yellow spots. **[Measurements]** Female total length 60 mm, abdomen 47 mm, hind wing 40 mm. **[Habitat]** Exposed montane streams at 2000-2500 m elevation. **[Distribution]** Yunnan (Dali). **[Flight Season]** May to June.

叶蜓属待定种2 *Petaliaeschna* sp. 2

【形态特征】雄性复眼褐色，面部黄色，额浅褐色；胸部褐色，合胸脊黄色，侧面具2条宽阔的黄色条纹，足黄色，翅透明；腹部主要褐色，第1～2节侧面和第2节背面具黄色斑点。【长度】雄性体长 66 mm，腹长 52 mm，后翅 42 mm。【栖息环境】海拔 1000 m植被茂盛的山区小溪。【分布】云南（普洱）。【飞行期】5—6月。

[Identification] Male eyes brown, face yellow, frons light brown. Thorax brown, dorsal carina yellow, sides with two broad yellow stripes, legs yellow, wings hyaline. Abdomen mainly brown, sides of S1-S2 and dorsum of S2 with yellow spots. **[Measurements]** Male total length 66 mm, abdomen 52 mm, hind wing 42 mm. **[Habitat]** Montane streams in forest at 1000 m elevation. **[Distribution]** Yunnan (Pu'er). **[Flight Season]** May to June.

叶蜓属待定种2 雄，云南（普洱）
Petaliaeschna sp. 2, male from Yunnan (Pu'er)

黑额蜓属 Genus *Planaeschna* McLachlan, 1896

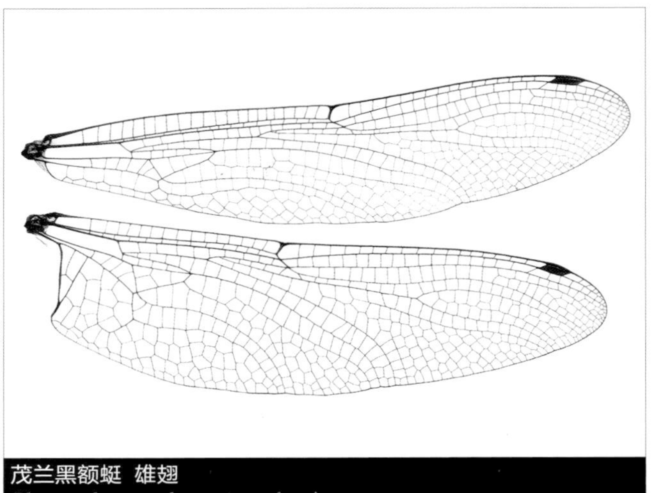
茂兰黑额蜓 雄翅
Planaeschna maolanensis, male wings

本属仅在亚洲分布，已知约30种，中国已经发现20余种。本属蜻蜓体型中型至大型，体黑色或褐色具黄色或绿色条纹，外观上与头蜓属和佩蜓属种类近似，但本属的基室没有横脉，可以与头蜓、佩蜓区分。本属蜻蜓雄性可以通过身体条纹结合肛附器和阳茎构造来区分，雌性可以根据体色、产卵器和尾毛长度区分。

本属蜻蜓栖息于山区溪流环境，在植被茂盛的森林深处可以发现。多数种类在晚季节发生，有些飞行期可以持续到冬季。白天通常不活跃，隐蔽在树林中，在黄昏时活跃，或者集群捕食，或者四处游荡。雄性具领域行为，在小溪上方低空定点悬停飞行，时而更换位置。雌性产卵于溪流边缘的泥土或者朽木中。

This genus is confined to Asia and contains about 30 species, among them over 20 species have been recorded from China. Species of the genus are medium to large sized dragonflies, body black or brown with yellow or green markings, their appearance is quite similar to species of genera *Cephalaeschna* and *Periaeschna*, but wings of *Planaeschna* species have no crossveins in the median space. Males of the genus can be separated by the body color pattern, anal appendages and penis, females can be distinguished by the body color pattern, ovipositor and length of cerci.

Planaeschna species inhabit montane streams in dense forests. Most species are on the wing late in the season, and some species can fly until winter. They usually hide in the forest during the heat of the day but become active at twilight, forage in small groups or patrolling in the forest. Males display territorial behavior, hovering very low above streams, sometimes changing position. Females lay eggs into mud or deadwood.

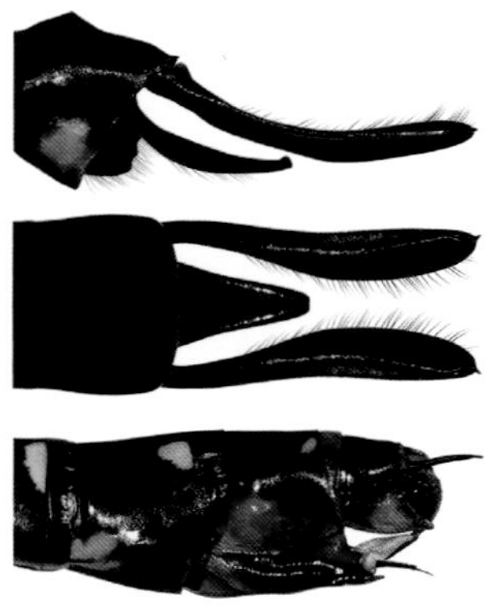
棘尾黑额蜓
Planaeschna caudispina

黑额蜓属 雄性肛附器和雌性产卵器
Genus *Planaeschna*, male anal appendages and female ovipositor

遂昌黑额蜓 雄 | 宋睿斌 摄
Planaeschna suichangensis, male | Photo by Ruibin Song

郝氏黑额蜓
Planaeschna haui

石垣黑额蜓
Planaeschna ishigakiana flavostria

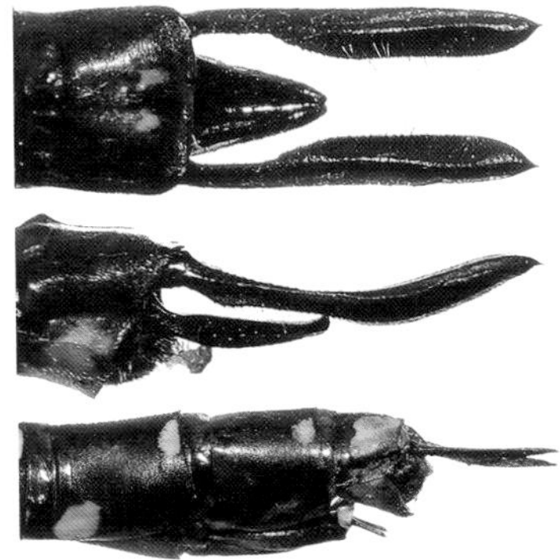

崂山黑额蜓
Planaeschna laoshanensis

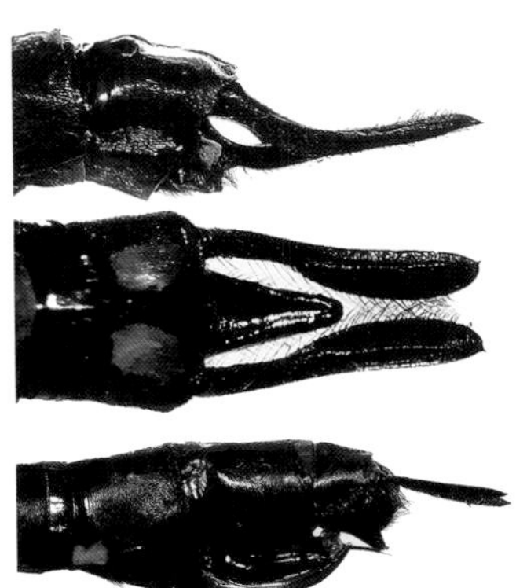

刘氏黑额蜓
Planaeschna liui

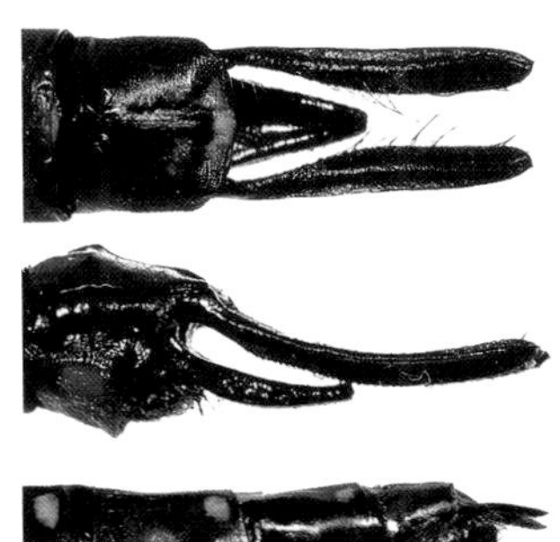

茂兰黑额蜓
Planaeschna maolanensis

高山黑额蜓
Planaeschna monticola

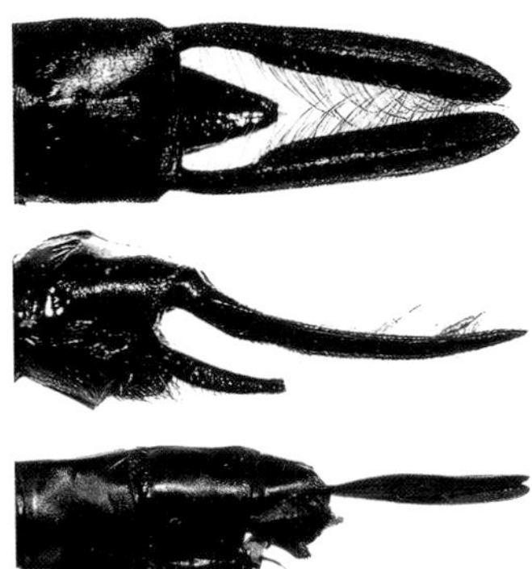

南昆山黑额蜓
Planaeschna nankunshanensis

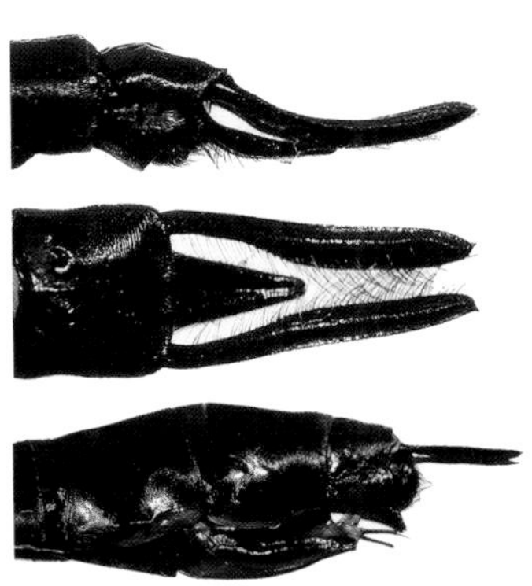

南岭黑额蜓
Planaeschna nanlingensis

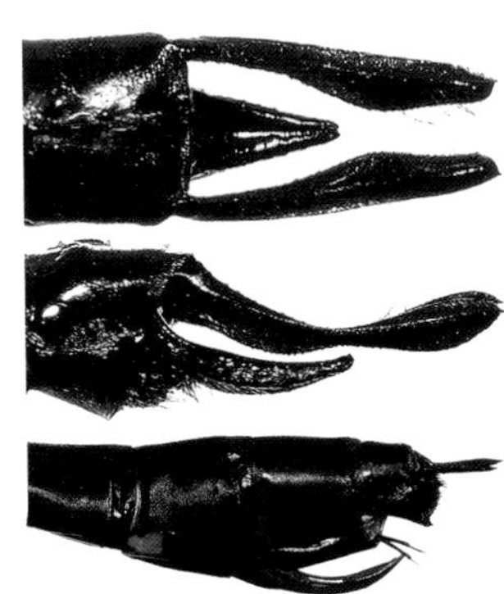

褐面黑额蜓
Planaeschna owadai

黑额蜓属 雄性肛附器和雌性产卵器
Genus *Planaeschna*, male anal appendages and female ovipositor

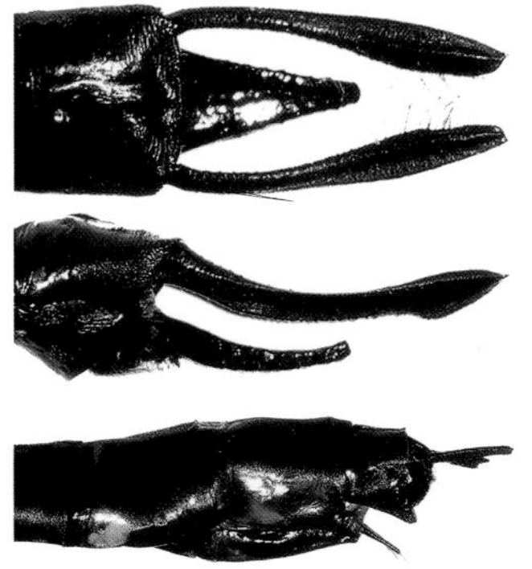

李氏黑额蜓
Planaeschna risi

粗壮黑额蜓
Planaeschna robusta

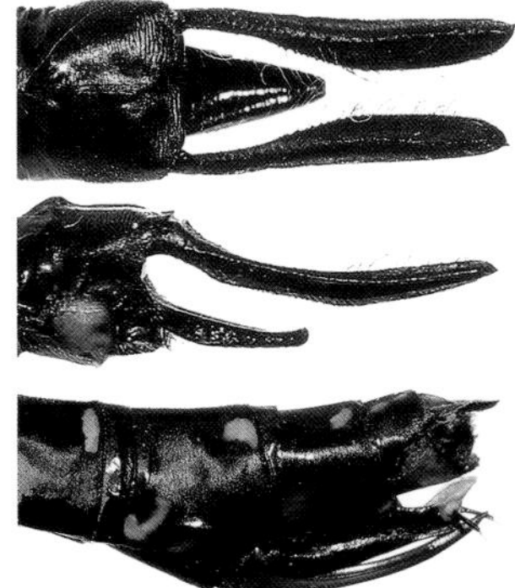

山西黑额蜓
Planaeschna shanxiensis

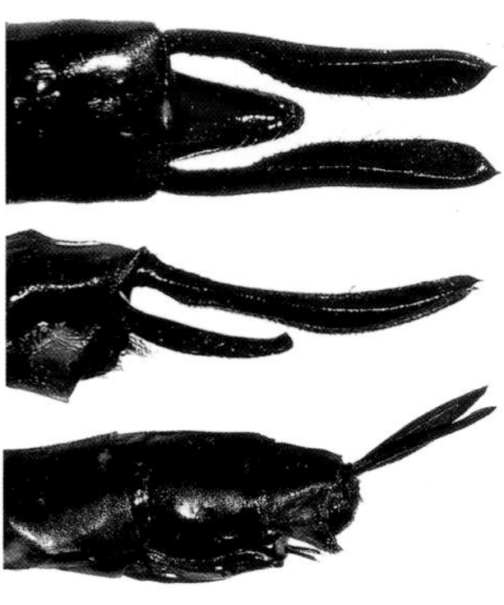

幽灵黑额蜓
Planaeschna skiaperipola

遂昌黑额蜓
Planaeschna suichangensis

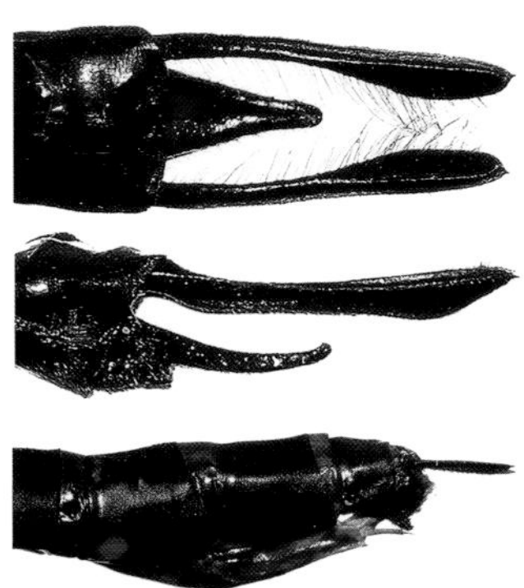

台湾黑额蜓
Planaeschna taiwana

黑额蜓属 雄性肛附器和雌性产卵器
Genus *Planaeschna*, male anal appendages and female ovipositor

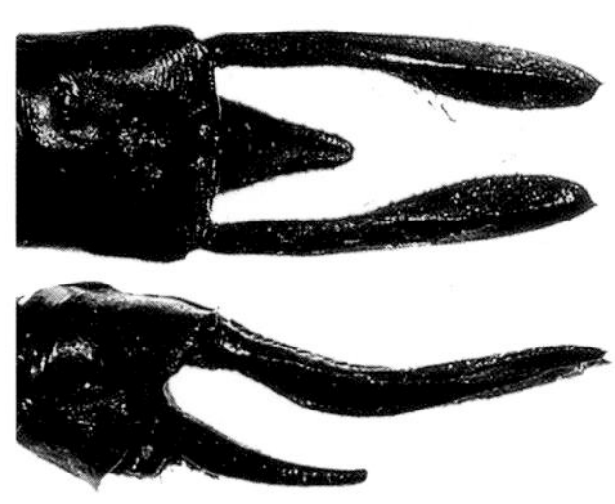

清迈黑额蜓
Planaeschna chiengmaiensis

联纹黑额蜓
Planaeschna gressitti

角斑黑额蜓
Planaeschna maculifrons

黑额蜓属 雄性肛附器
Genus *Planaeschna*, male anal appenages

棘尾黑额蜓 *Planaeschna caudispina* Zhang & Cai, 2013

【形态特征】雄性复眼墨绿色具蓝色斑，面部黄绿色，无显著的黑色斑；胸部黑色，具肩前条纹和肩前下点，合胸侧面具3条黄绿色条纹，足黑色，翅透明；腹部黑色具发达的黄绿色斑点，第10节具锥状突起。雌性尾毛较短，和第10节等长。【长度】体长 65～71 mm，腹长 49～54 mm，后翅 43～47 mm。【栖息环境】海拔 1000～1500 m的山区小溪。【分布】中国四川特有。【飞行期】7—9月。

[Identification] Male eyes dark green with blue spots, face yellowish green without black marking. Thorax black with antehumeral stripes and spots, sides with three yellowish green stripes, legs black, wings hyaline. Abdomen black with numerous yellowish green markings, S10 with a pyramidal prominence. Female cerci short, same length as S10. **[Measurements]** Total length 65-71 mm, abdomen 49-54 mm, hind wing 43-47 mm. **[Habitat]** Montane streams at 1000-1500 m elevation. **[Distribution]** Endemic to Sichuan of China. **[Flight Season]** July to September.

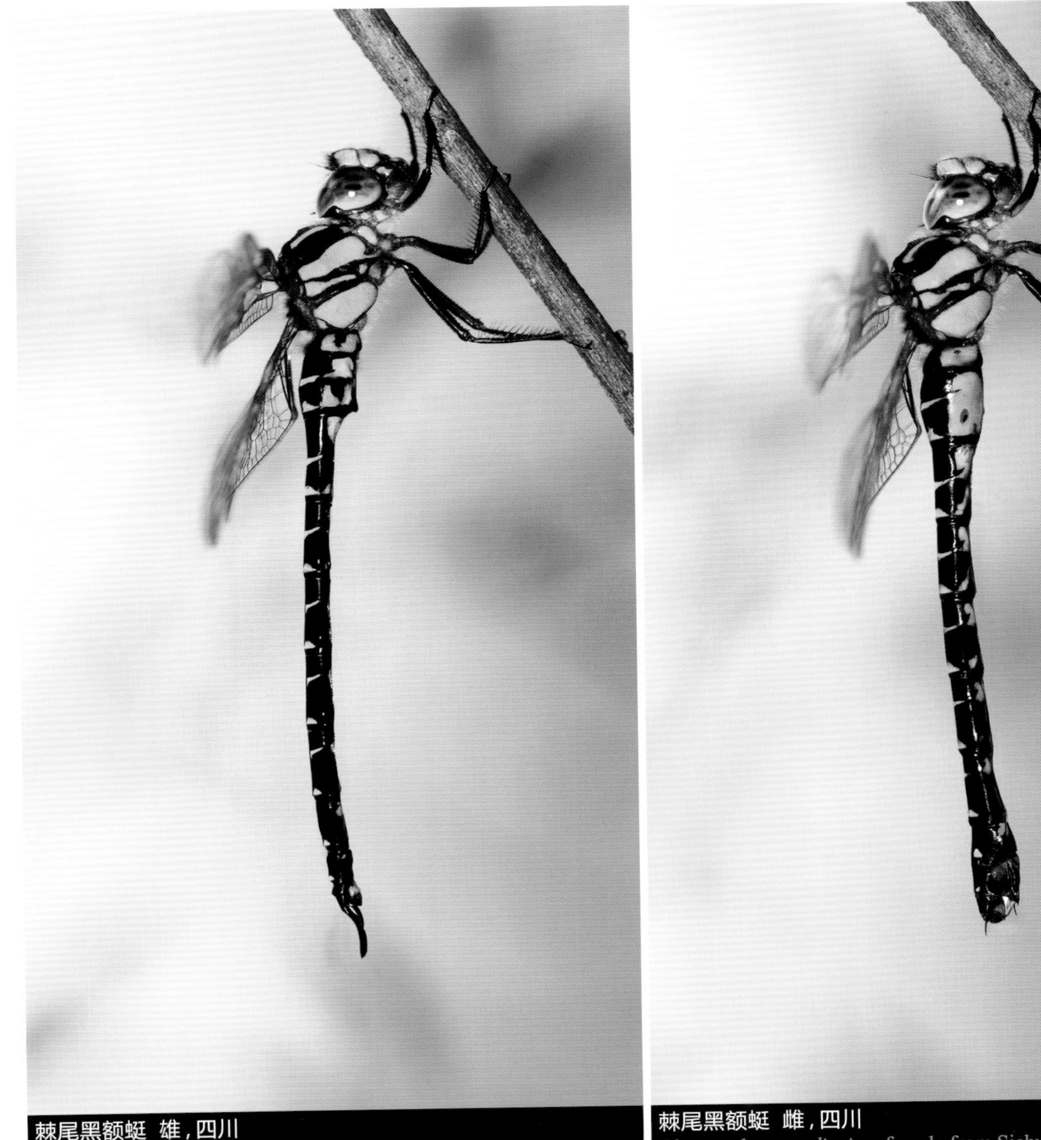

棘尾黑额蜓 雄，四川
Planaeschna caudispina, male from Sichuan

棘尾黑额蜓 雌，四川
Planaeschna caudispina, female from Sichuan

清迈黑额蜓 *Planaeschna chiengmaiensis* Asahina, 1981

【形态特征】雄性复眼绿色，面部黄褐色；胸部褐色，肩前条纹甚阔，合胸侧面具2条宽阔的黄色条纹，后胸前侧板具2个黄色斑点，足大面积黑色，翅透明；腹部黑色具黄色斑点。雌性翅基方染有较大的黑褐色斑，尾毛长度约为第10节的1.5倍。【长度】体长 54～60 mm，腹长 43～47 mm，后翅 35～41 mm。【栖息环境】海拔 500～1000 m的山区林荫小溪和沟渠。【分布】云南（西双版纳）；泰国。【飞行期】4—6月。

[Identification] Male eyes green, face yellowish brown. Thorax brown with broad antehumeral stripes, sides with two broad yellow stripes, metepisternum with two yellow spots, legs largely black, wings hyaline. Abdomen black with yellow markings. Female wing bases with blackish brown markings, cerci about one and a half times as long as S10. **[Measurements]** Total length 54-60 mm, abdomen 43-47 mm, hind wing 35-41 mm. **[Habitat]** Shady streams and ditches at 500-1000 m elevation. **[Distribution]** Yunnan (Xishuangbanna); Thailand. **[Flight Season]** April to June.

清迈黑额蜓 雄，云南（西双版纳）
Planaeschna chiengmaiensis, male from Yunnan (Xishuangbanna)

清迈黑额蜓 雌，云南（西双版纳）
Planaeschna chiengmaiensis, female from Yunnan (Xishuangbanna)

清迈黑额蜓 雄，云南（西双版纳）
Planaeschna chiengmaiensis, male from Yunnan (Xishuangbanna)

联纹黑额蜓 *Planaeschna gressitti* Karube, 2002

【形态特征】雄性复眼绿色，面部黄色，前额具1个甚大的黑色斑；胸部黑色，肩前条纹甚阔，合胸侧面具2条宽阔的黄色条纹，足大面积黑色，翅透明；腹部黑色具黄色斑点。雌性与雄性相似，尾毛长度约为第10节的1.5倍。本种与遂昌黑额蜓相似，但阳茎的构造不同。【长度】体长 64~66 mm，腹长 49~50 mm，后翅 42~43 mm。【栖息环境】海拔 500~1500 m的山区小溪。【分布】中国特有，分布于四川、贵州、广东。【飞行期】7—11月。

[Identification] Male eyes green, face yellow, antefrons with a large black spot. Thorax black with broad antehumeral stripes, sides with two broad yellow stripes, legs largely black, wings hyaline. Abdomen black with yellow markings. Female similar to male, cerci about one and a half times as long as S10. The species is similar to *P. suichangensis*, but the penis is of a different shape. **[Measurements]** Total length 64-66 mm, abdomen 49-50 mm, hind wing 42-43 mm. **[Habitat]** Montane streams at 500-1500 m elevation. **[Distribution]** Endemic to China, recorded from Sichuan, Guizhou, Guangdong. **[Flight Season]** July to November.

联纹黑额蜓 雄，四川
Planaeschna gressitti, male from Sichuan

联纹黑额蜓 雌，四川
Planaeschna gressitti, female from Sichuan

郝氏黑额蜓 *Planaeschna haui* Wilson & Xu, 2008

【形态特征】雄性复眼绿色，面部黄色和黑色，上唇黑色具1个黄色斑，前唇基和后唇基下缘黑色，前额具1个甚大的黑色斑；胸部黑色，肩前条纹绿色，肩前下点甚小，合胸侧面具2条宽阔的黄绿色条纹，后胸前侧板具1条绿色细条纹，足大面积黑色，基方黄褐色，翅透明；腹部黑色具黄色和绿色斑点。雌性与雄性相似，尾毛约与第10节等长。【长度】腹长 52~57 mm，后翅 44~50 mm。【栖息环境】海拔 500 m以下的山区溪流。【分布】中国特有，分布于广东、广西。【飞行期】6—8月。

[Identification] Male eyes green, face yellow and black, labrum black with a yellow spot, anteclypeus and lower edge of postclypeus black, antefrons with a large black spot. Thorax black with green antehumeral stripes and small antehumeral spots, sides with two broad yellow and green stripes, metepisternum with a narrow green stripe, legs largely black with bases yellowish brown, wings hyaline. Abdomen black with yellow and green markings. Female similar to male, cerci about same length as S10. **[Measurements]** Abdomen 52-57 mm, hind wing 44-50 mm. **[Habitat]** Montane streams below 500 m elevation. **[Distribution]** Endemic to China, recorded from Guangdong, Guangxi. **[Flight Season]** June to August.

郝氏黑额蜓 雄，广东 | 吴宏道 摄
Planaeschna haui, male from Guangdong | Photo by Hongdao Wu

郝氏黑额蜓 雌，广东 | 吴宏道 摄
Planaeschna haui, female from Guangdong | Photo by Hongdao Wu

石垣黑额蜓 *Planaeschna ishigakiana flavostria* Yeh, 1996

【形态特征】雄性复眼蓝绿色，面部黄绿色，前唇基黑色，前额具1个甚大的黑色斑；胸部黑色，肩前条纹甚阔，合胸侧面具2条宽阔的黄绿色条纹，后胸前侧板具1个短条纹，足大面积黑色，翅透明；腹部黑色具发达的黄色斑点。雌性与雄性相似，尾毛较长。【长度】体长 69~71 mm，腹长 53~55 mm，后翅 46~50 mm。【栖息环境】海拔1000 m以下森林中的小溪。【分布】中国台湾特有。【飞行期】5—10月。

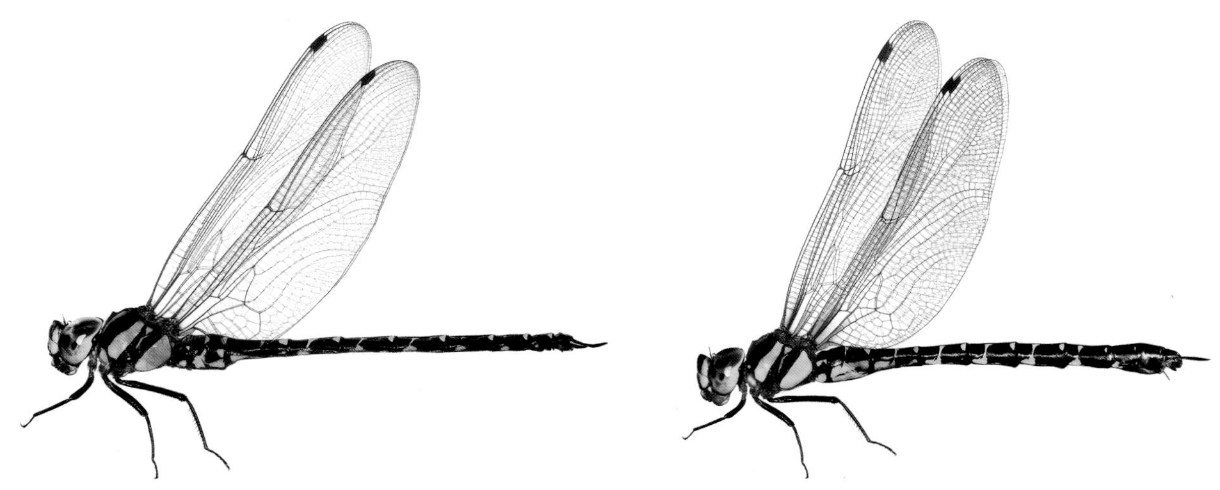

石垣黑额蜓 雄，台湾
Planaeschna ishigakiana flavostria, male from Taiwan

石垣黑额蜓 雌，台湾
Planaeschna ishigakiana flavostria, female from Taiwan

[Identification] Male eyes bluish green, face yellowish green, anteclypeus black, antefrons with a large black spot. Thorax black with broad antehumeral stripes, sides with two broad yellowish green stripes, metepisternum with a short stripe, legs largely black, wings hyaline. Abdomen black with numerous yellow markings. Female similar to male, cerci long. **[Measurements]** Total length 69-71 mm, abdomen 53-55 mm, hind wing 46-50 mm. **[Habitat]** Montane streams below 1000 m elevation. **[Distribution]** Endemic to Taiwan of China. **[Flight Season]** May to October.

崂山黑额蜓 *Planaeschna laoshanensis* Zhang, Yeh & Tong, 2010

【形态特征】雄性复眼蓝绿色，面部黄绿色，前额具1个甚大的黑色斑，上额具1个“T”形斑，后唇基中央具褐色条纹，有时缺如；胸部黑色，肩前条纹甚阔，合胸侧面具2条宽阔的黄绿色条纹，后胸前侧板具1个三角形小黄斑，足大面积黑色，翅透明；腹部黑色具发达的黄色斑点。雌性与雄性相似，尾毛较长，长度约为第10节的2倍。【长度】体长 67~72 mm，腹长 52~55 mm，后翅 43~47 mm。【栖息环境】海拔 300~800 m的山区林荫小溪。【分布】中国山东特有。【飞行期】8—10月。

[Identification] Male eyes bluish green, face yellowish green, antefrons with a large black spot, top of frons with a T-mark, postclypeus with a brown making centrally or absent. Thorax black with broad antehumeral stripes, sides with two broad yellowish green stripes, metepisternum with a small triangular yellow spot, legs largely black, wings hyaline. Abdomen black with yellow markings. Female similar to male, cerci about twice as long as S10. **[Measurements]** Total length 67-72 mm, abdomen 52-55 mm, hind wing 43-47 mm. **[Habitat]** Shady montane streams at 300-800 m elevation. **[Distribution]** Endemic to Shandong of China. **[Flight Season]** August to October.

崂山黑额蜓 雌，山东
Planaeschna laoshansensis, female from Shandong

崂山黑额蜓 雄，山东
Planaeschna laoshansensis, male from Shandong

崂山黑额蜓 雄，山东
Planaeschna laoshansensis, male from Shandong

刘氏黑额蜓 *Planaeschna liui* Xu, Chen & Qiu, 2009

【形态特征】雄性复眼绿色，面部黑色和黄色，前额具1个甚大的黑色斑；胸部黑色，肩前条甚阔，合胸侧面具2条宽阔的黄条纹，后胸前侧板具1个甚小黄斑，足大面积黑色，翅透明；腹部黑色具黄色斑点，雄性肛附器长度约为第10节的1.5倍。雌性尾毛长度约为第10节的1.5倍。本种与遂昌黑额蜓相似，但雄性肛附器和阳茎构造不同。【长度】体长 61~68 mm，腹长 46~53 mm，后翅 44~47 mm。【栖息环境】海拔 1000~1500 m的山区小溪。【分布】中国特有，分布于福建、广东。【飞行期】7—11月。

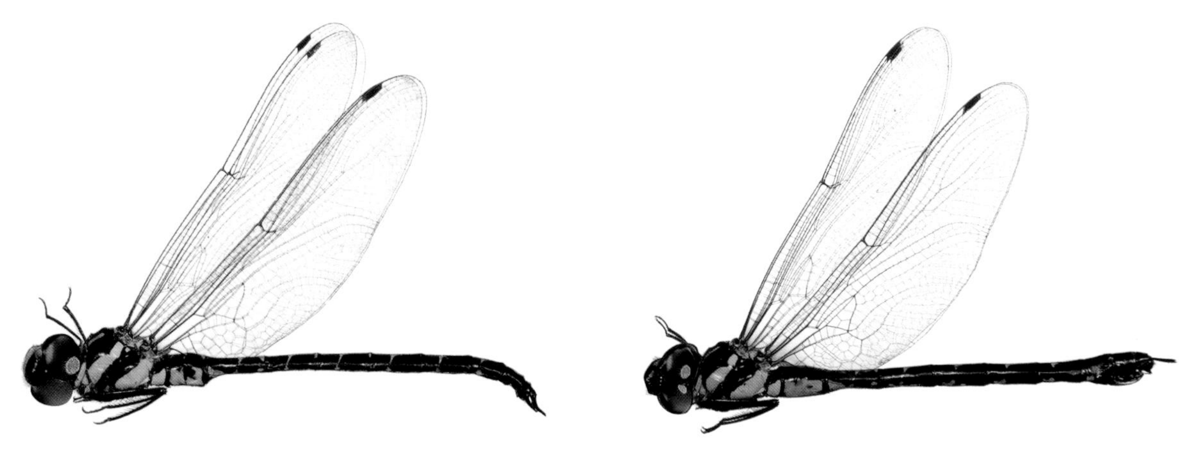

刘氏黑额蜓 雄，广东
Planaeschna liui, male from Guangdong

刘氏黑额蜓 雌，广东
Planaeschna liui, female from Guangdong

[Identification] Male eyes green, face black and yellow, antefrons with a large black spot. Thorax black with broad antehumeral stripes, sides with two broad yellow stripes, metepisternum with a small spot, legs largely black, wings hyaline. Abdomen black with yellow markings, anal appendages about one and a half times as long as S10. Female similar to male, cerci about one and a half times as long as S10. The species is similar to *P. suichangensis* but the structure of the male anal appendages and penis are different. **[Measurements]** Total length 61-68 mm, abdomen 46-53 mm, hind wing 44-47 mm. **[Habitat]** Montane streams at 1000-1500 m elevation. **[Distribution]** Endemic to China, recorded from Fujian, Guangdong. **[Flight Season]** July to November.

角斑黑额蜓 *Planaeschna maculifrons* Zhang & Cai, 2013

【形态特征】雄性复眼蓝绿色，面部黄绿色，前额具1个甚大的三角形黑色斑；胸部黑色，具肩前条纹和肩前下点，合胸侧面具2条宽阔的黄绿色条纹，后胸前侧板具3个大小不等的绿色斑点，足黑色，腿节具黄斑，翅透明；腹部黑色具黄绿色斑点。【长度】雄性体长 72 mm，腹长 56 mm，后翅 47 mm。【栖息环境】海拔 1000 m左右的山区林荫小溪。【分布】中国四川特有。【飞行期】6—8月。

[Identification] Male eyes bluish green, face yellowish green, antefrons with a large triangular black spot. Thorax black with antehumeral stripes and spots, sides with two broad yellowish green stripes, metepisternum with three green spots, legs black with yellow markings on femora, wings hyaline. Abdomen black with yellowish green markings. **[Measurements]** Male total length 72 mm, abdomen 56 mm, hind wing 47 mm. **[Habitat]** Shady montane streams at about 1000 m elevation. **[Distribution]** Endemic to Sichuan of China. **[Flight Season]** June to August.

角斑黑额蜓 雄，四川
Planaeschna maculifrons, male from Sichuan

茂兰黑额蜓 *Planaeschna maolanensis* Zhou & Bao, 2002

茂兰黑额蜓 雄，贵州
Planaeschna maolanensis, male from Guizhou

茂兰黑额蜓 雌，贵州
Planaeschna maolanensis, female from Guizhou

【形态特征】雄性复眼蓝绿色，面部黄褐色，前额具1个褐色斑；胸部黑色，肩前条纹甚阔，具甚小的肩前下点，合胸侧面具2条宽阔的黄绿色条纹，后胸前侧板具1个三角形小黄斑，足大面积黑色，翅透明；腹部黑色具发达的黄色斑点。雌性复眼黄褐色，尾毛短，约与第10节等长。【长度】体长 62~66 mm，腹长 47~51 mm，后翅 45~47 mm。【栖息环境】海拔 800~1500 m的山区开阔小溪。【分布】中国特有，分布于贵州、重庆。【飞行期】7—10月。

茂兰黑额蜓 雄，贵州
Planaeschna maolanensis, male from Guizhou

[Identification] Male eyes bluish green, face yellowish brown, antefrons with a brown spot. Thorax black with broad antehumeral stripes and small antehumeral spots, sides with two broad yellowish green stripes, metepisternum with a small triangular spot, legs largely black, wings hyaline. Abdomen black with numerous yellow markings. Female eyes yellowish brown, cerci short, about same length as S10. **[Measurements]** Total length 62-66 mm, abdomen 47-51 mm, hind wing 45-47 mm. **[Habitat]** Exposed montane streams at 800-1500 m elevation. **[Distribution]** Endemic to China, recorded from Guizhou, Chongqing. **[Flight Season]** July to October.

高山黑额蜓 *Planaeschna monticola* Zhang & Cai, 2013

【形态特征】雄性复眼深蓝色，面部棕黄色，前额具1个黑褐色斑；胸部褐色，肩前条纹甚阔，合胸侧面具2条宽阔的黄色条纹，后胸前侧板具1个三角形小黄斑，足大面积黑色，基方至腿节中部红褐色，其余各节黑色，翅透明；腹部褐色具发达的黄色斑点。雌性复眼黄褐色，尾毛较短，与第10节等长。**【长度】**体长 63～67 mm，腹长 49～52 mm，后翅 42～44 mm。**【栖息环境】**海拔 1800～2500 m的开阔小溪。**【分布】**中国云南（昆明、大理）特有。**【飞行期】**9—12月。

[Identification] Male eyes dark blue, face brownish yellow, antefrons with a blackish brown spot. Thorax brown with broad antehumeral stripes, sides with two broad yellow stripes, metepisternum with a small triangular yellow spot, legs largely black, from bases to mid femora reddish brown, wings hyaline. Abdomen brown with numerous yellow markings. Female eyes yellowish brown, cerci short, same length as S10. **[Measurements]** Total length 63-67 mm, abdomen 49-52 mm, hind wing 42-44 mm. **[Habitat]** Exposed montane streams at 1800-2500 m elevation. **[Distribution]** Endemic to Yunnan (Kunming, Dali) of China. **[Flight Season]** September to December.

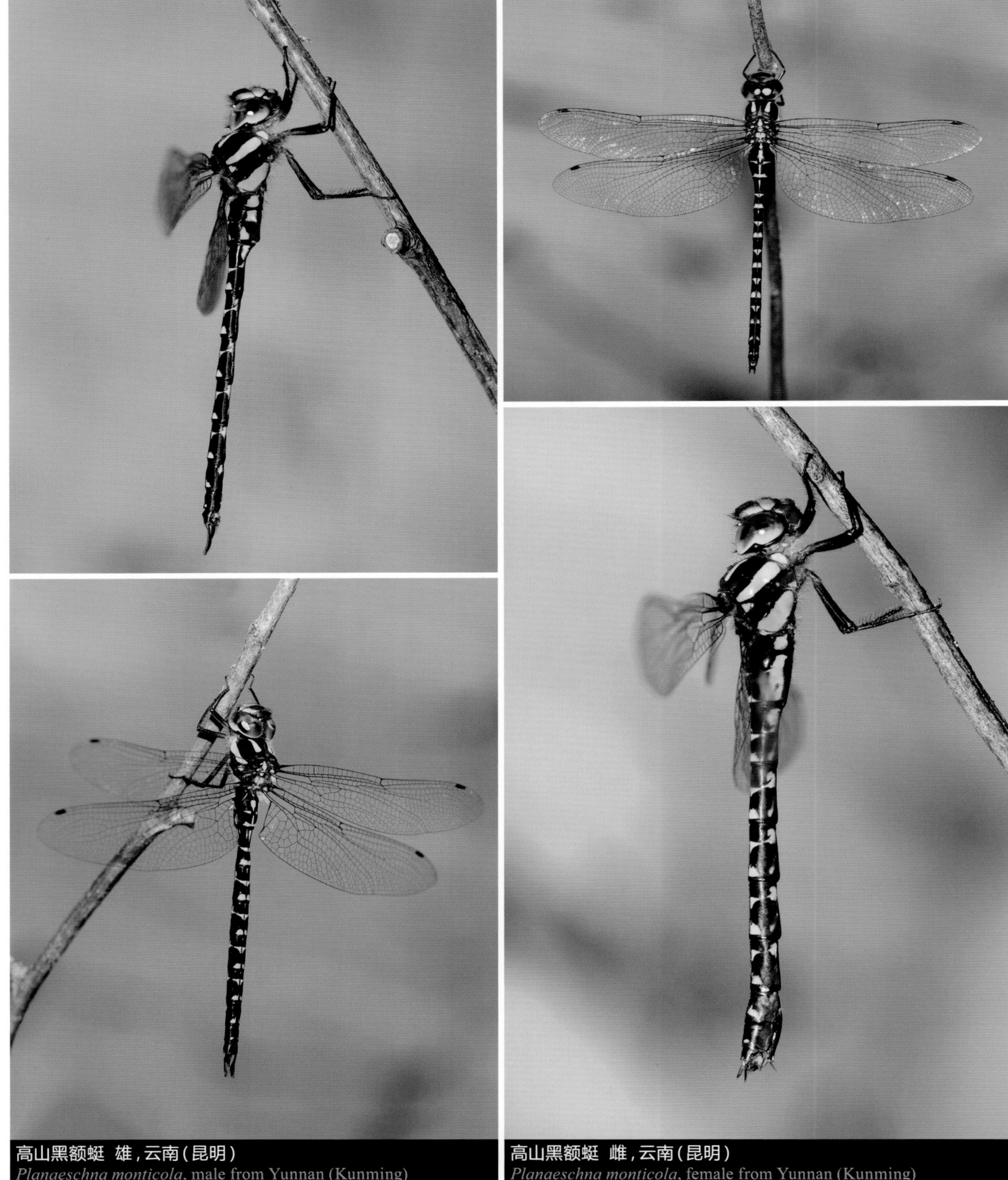

高山黑额蜓 雄，云南（昆明）
Planaeschna monticola, male from Yunnan (Kunming)

高山黑额蜓 雌，云南（昆明）
Planaeschna monticola, female from Yunnan (Kunming)

南昆山黑额蜓 *Planaeschna nankunshanensis* Zhang, Yeh & Tong, 2010

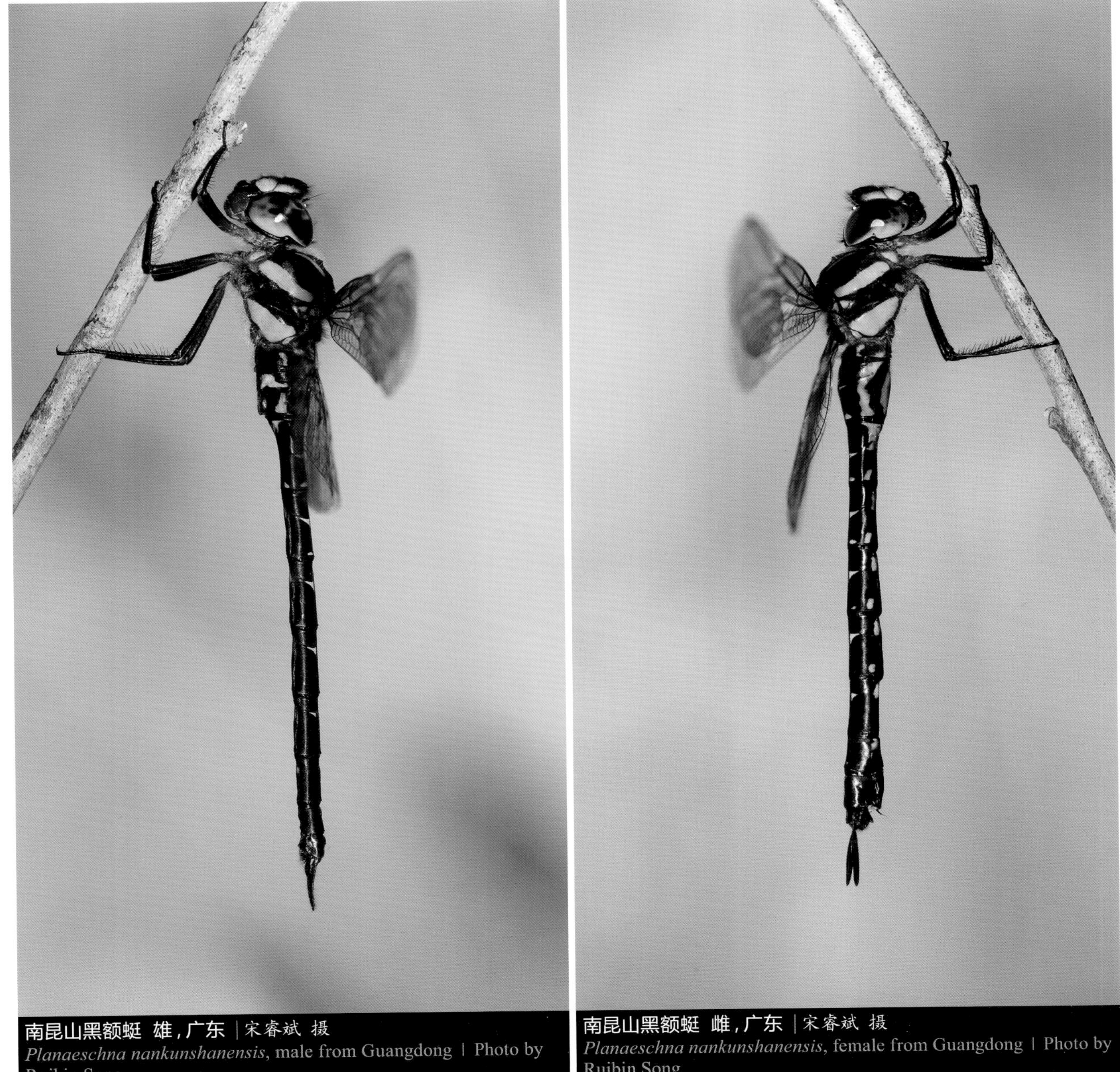

南昆山黑额蜓 雄，广东 | 宋睿斌 摄
Planaeschna nankunshanensis, male from Guangdong | Photo by Ruibin Song

南昆山黑额蜓 雌，广东 | 宋睿斌 摄
Planaeschna nankunshanensis, female from Guangdong | Photo by Ruibin Song

【形态特征】雄性复眼蓝绿色，面部黄绿色，前额具1个甚大的黑色斑，后唇基中央具褐色条纹；胸部黑色，肩前条纹甚阔，合胸侧面具2条宽阔的黄绿色条纹，后胸前侧板具1个三角形小黄斑，足黑色，翅透明；腹部黑色具黄绿色条纹。雌性复眼黄绿色，尾毛极长，长度约与第9~10节两节等长。【长度】体长 63~71 mm，腹长 49~55 mm，后翅 41~48 mm。【栖息环境】海拔 300~1000 m的山区小溪。【分布】中国广东特有。【飞行期】9—12月。

[Identification] Male eyes bluish green, face yellowish green, antefrons with a large black spot, postclypeus with a brown marking centrally. Thorax black with broad antehumeral stripes, sides with two broad yellowish green stripes, metepisternum with a small yellow triangular spot, legs black, wings hyaline. Abdomen brown with yellowish green stripes. Female eyes yellowish green, cerci very long, as long as S9+S10. **[Measurements]** Total length 63-71 mm, abdomen 49-55 mm, hind wing 41-48 mm. **[Habitat]** Exposed montane streams at 300-1000 m elevation. **[Distribution]** Endemic to Guangdong of China. **[Flight Season]** September to December.

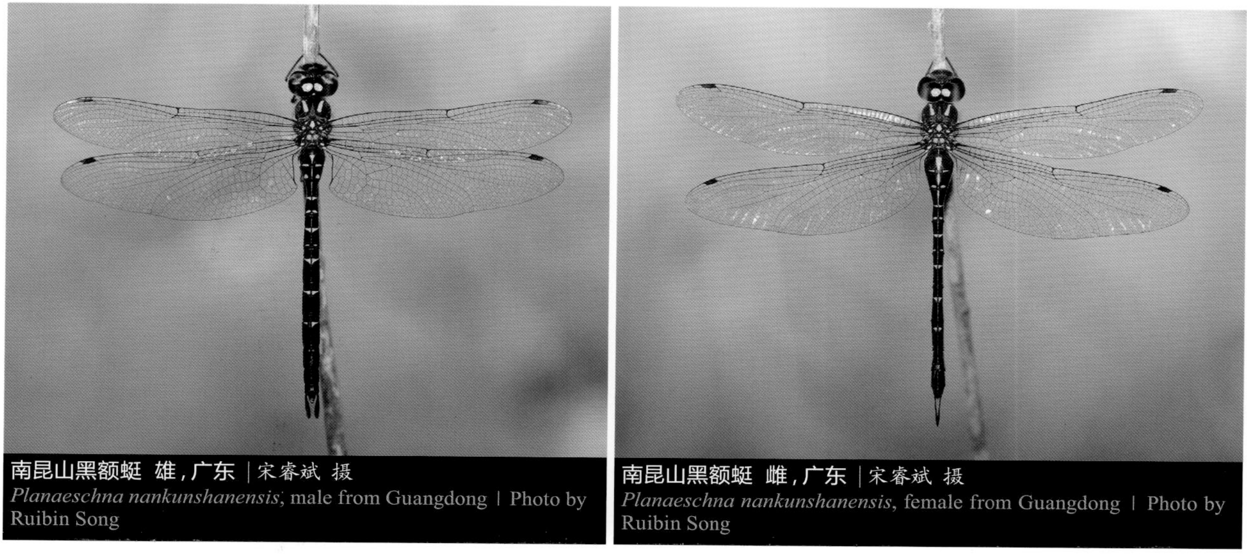

南昆山黑额蜓 雄，广东｜宋睿斌 摄
Planaeschna nankunshanensis, male from Guangdong | Photo by Ruibin Song

南昆山黑额蜓 雌，广东｜宋睿斌 摄
Planaeschna nankunshanensis, female from Guangdong | Photo by Ruibin Song

南岭黑额蜓 *Planaeschna nanlingensis* Wilson & Xu, 2008

南岭黑额蜓 雄，广东
Planaeschna nanlingensis, male from Guangdong

【形态特征】复眼深蓝色，面部黄色，前额具1个甚大的黑色斑，后唇基中央具深褐色条纹；胸部黑色，肩前条纹甚阔，合胸侧面具2条宽阔的黄绿色条纹，后胸前侧板具1个三角形小黄斑，足黑色，翅透明；腹部黑色具黄色斑点。雌性尾毛稍长于第10节。【长度】体长 64~67 mm，腹长 50~52 mm，后翅 44~46 mm。【栖息环境】海拔1000~1700 m的山区小溪。【分布】中国广东特有。【飞行期】7—9月。

[Identification] Eyes dark blue, face yellow, antefrons with a large black spot, postclypeus with a dark brown marking centrally. Thorax black with broad antehumeral stripes, sides with two broad yellowish green stripes, metepisternum with a small triangular yellow spot, legs black, wings hyaline. Abdomen black with yellow stripes. Female cerci slightly longer than S10. **[Measurements]** Total length 64-67 mm, abdomen 50-52 mm, hind wing 44-46 mm. **[Habitat]** Montane streams at 1000-1700 m elevation. **[Distribution]** Endemic to Guangdong of China. **[Flight Season]** July to September.

南岭黑额蜓 雄，广东
Planaeschna nanlingensis, male from Guangdong

南岭黑额蜓 雌，广东
Planaeschna nanlingensis, female from Guangdong

褐面黑额蜓 *Planaeschna owadai* Karube, 2002

【形态特征】雄性复眼蓝绿色，面部黄褐色；胸部褐色，肩前条纹甚阔，合胸侧面具2条宽阔的黄绿色条纹，足大面积黑色，翅透明；腹部黑褐色具黄绿色斑点。雌性体色稍淡，尾毛与第10节等长。【长度】体长 55～58 mm，腹长 42～45 mm，后翅 40～41 mm。【栖息环境】海拔 800～1500 m的山区狭窄沟渠和水流较小的小型瀑布。【分布】云南（西双版纳、普洱、临沧、德宏）；越南。【飞行期】9—12月。

[Identification] Male eyes bluish green, face yellowish brown. Thorax brown with broad antehumeral stripes, sides with two broad yellowish green stripes, legs largely black, wings hyaline. Abdomen blackish brown with yellowish green markings. Female paler, cerci as long as S10. **[Measurements]** Total length 55-58 mm, abdomen 42-45 mm, hind wing 40-41 mm. **[Habitat]** Narrow ditches and small waterfalls at 800-1500 m elevation. **[Distribution]** Yunnan (Xishuangbanna, Pu'er, Lincang, Dehong); Vietnam. **[Flight Season]** September to December.

褐面黑额蜓 雄，云南（德宏）
Planaeschna owadai, male from Yunnan (Dehong)

褐面黑额蜓 雄，云南（西双版纳）| 莫善濂 摄
Planaeschna owadai, male from Yunnan (Xishuangbanna) | Photo by Shanlian Mo

褐面黑额蜓 雌，云南（西双版纳）| 莫善濂 摄
Planaeschna owadai, female from Yunnan (Xishuangbanna) | Photo by Shanlian Mo

李氏黑额蜓 *Planaeschna risi* Asahina, 1964

【形态特征】雄性复眼绿色，面部黄色和黑色，上唇黑色具1个黄色斑，前唇基和后唇基下缘黑色，前额具1个甚大的黑色斑；胸部黑色，肩前条纹甚阔，合胸侧面具2条宽阔的黄条纹，后胸前侧板具2个甚小的黄斑，足大面积黑色，翅透明；腹部黑色具黄色斑点。雌性尾毛长度约为第10节的1.5倍。【长度】体长 63~72 mm，腹长 48~57 mm，后翅 42~48 mm。【栖息环境】海拔 500~2500 m的山区小溪。【分布】广西、台湾；日本。【飞行期】4—12月。

李氏黑额蜓 雄，台湾
Planaeschna risi, male from Taiwan

李氏黑额蜓 雌，台湾
Planaeschna risi, female from Taiwan

[Identification] Male eyes green, face yellow and black, labrum black with a yellow spot, anteclypeus and lower edge of postclypeus black, antefrons with a large black spot. Thorax black with broad antehumeral stripes, sides with two broad yellow stripes, metepisternum with two small yellow spots, legs largely black, wings hyaline. Abdomen black with yellow markings. Female cerci about one and a half times as long as S10. **[Measurements]** Total length 63-72 mm, abdomen 48-57 mm, hind wing 42-48 mm. **[Habitat]** Montane streams at 500-2500 m elevation. **[Distribution]** Guangxi, Taiwan; Japan. **[Flight Season]** April to December.

粗壮黑额蜓 *Planaeschna robusta* Zhang & Cai, 2013

【形态特征】雄性复眼蓝色，面部黄绿色，前额具1个黑色斑；胸部黑色，具肩前条纹和肩前下点，合胸侧面具2条宽阔的黄绿色条纹，后胸前侧板具3个大小不等的绿色斑点，足黑色，翅透明；腹部黑色具黄绿色斑点。雌性复眼绿色，合胸的黄绿色条纹更发达，尾毛较长，为第10节的2.5倍。【长度】体长 69~75 mm，腹长 54~58 mm，后翅 46~51 mm。【栖息环境】海拔 800~1200 m的山区小溪。【分布】中国四川特有。【飞行期】7—9月。

[Identification] Male eyes blue, face yellowish green, antefrons with a black spot. Thorax black with antehumeral stripes and spots, sides with two broad yellowish green stripes, metepisternum with three green spots, legs black, wings hyaline. Abdomen black with yellowish green markings. Female eyes green, thoracic markings broader, cerci long, about two and a half times as long as S10. **[Measurements]** Total length 69-75 mm, abdomen 54-58 mm, hind wing 46-51 mm. **[Habitat]** Montane streams at 800-1200 m elevation. **[Distribution]** Endemic to Sichuan of China. **[Flight Season]** July to September.

粗壮黑额蜓 雄，四川
Planaeschna robusta, male from Sichuan

粗壮黑额蜓 雌，四川
Planaeschna robusta, female from Sichuan

山西黑额蜓 *Planaeschna shanxiensis* Zhu & Zhang, 2001

【形态特征】雄性复眼蓝色，面部黄绿色，前额具1个黑色斑，上额具1个"T"形斑；胸部黑色，具肩前条纹和肩前下点，合胸侧面具2条宽阔的黄绿色条纹，后胸前侧板具2个大小不等的黄色斑点，足黑褐色，翅透明；腹部黑色具发达的黄绿色斑点。雌性翅稍染褐色，基部染有橙黄色，尾毛甚短，约与第10节等长。【长度】体长 68~70 mm，腹长 52~54 mm，后翅 46~50 mm。【栖息环境】海拔 1500 m以下的山区小溪。【分布】中国特有，分布于北京、山西、河南、湖北。【飞行期】7—11月。

[Identification] Male eyes blue, face yellowish green, antefrons with a black spot, top of frons with a T-mark. Thorax black with antehumeral stripes and spots, sides with two broad yellowish green stripes, metepisternum with two yellow spots, legs blackish brown, wings hyaline. Abdomen black with yellowish green markings. Female wings tinted with light brown, bases tinted with orange, cerci short, about same length as S10. **[Measurements]** Total length 68-70 mm, abdomen 52-54 mm, hind wing 46-50 mm. **[Habitat]** Montane streams below 1500 m elevation. **[Distribution]** Endemic to China, recorded from Beijing, Shanxi, Henan, Hubei. **[Flight Season]** July to November.

山西黑额蜓 雌，北京
Planaeschna shanxiensis, female from Beijing

山西黑额蜓 雄，北京
Planaeschna shanxiensis, male from Beijing

幽灵黑额蜓 *Planaeschna skiaperipola* Wilson & Xu, 2008

【形态特征】雄性复眼绿色，面部黄绿色，前额具1个甚大的黑色斑，后唇基中央具浅褐色条纹；胸部黑色，肩前条纹甚阔，合胸侧面具2条宽阔的黄绿色条纹，后胸前侧板具3～4个大小不等的黄斑，足大面积黑色，翅透明；腹部黑色具发达的黄色斑点。雌性后胸前侧板具较发达的黄色条纹，尾毛较长，长度约为第10节的2倍。【长度】体长 63～67 mm，腹长 49～52 mm，后翅 39～46 mm。【栖息环境】海拔 1000 m以下的山区小溪。【分布】中国特有，分布于福建、广东、香港。【飞行期】7—12月。

[Identification] Male eyes green, face yellowish green, antefrons with a large black spot, postclypeus with a light brown marking centrally. Thorax black with broad antehumeral stripes, sides with two broad yellowish green stripes, metepisternum with 3-4 yellow spots, legs largely black, wings hyaline. Abdomen black with numerous yellow markings. Female metepisternum with a linear stripe, cerci about twice as long as S10. **[Measurements]** Total length 63-67 mm, abdomen 49-52 mm, hind wing 39-46 mm. **[Habitat]** Montane streams below 1000 m elevation. **[Distribution]** Endemic to China, recorded from Fujian, Guangdong, Hong Kong. **[Flight Season]** July to December.

幽灵黑额蜓 雄，广东 | 宋睿斌 摄
Planaeschna skiaperipola, male from Guangdong | Photo by Ruibin Song

幽灵黑额蜓 雌，广东 | 宋睿斌 摄
Planaeschna skiaperipola, female from Guangdong | Photo by Ruibin Song

幽灵黑额蜓 雌，广东 | 宋睿斌 摄
Planaeschna skiaperipola, female from Guangdong | Photo by Ruibin Song

幽灵黑额蜓 雄，广东 | 宋睿斌 摄
Planaeschna skiaperipola, male from Guangdong | Photo by Ruibin Song

遂昌黑额蜓 *Planaeschna suichangensis* Zhou & Wei, 1980

遂昌黑额蜓 雌，浙江
Planaeschna suichangensis, female from Zhejiang

遂昌黑额蜓 雌，广东
Planaeschna suichangensis, female from Guangdong

遂昌黑额蜓 雄，广东
Planaeschna suichangensis, male from Guangdong

【形态特征】雄性复眼绿色，面部黄色，前额具1个甚大的黑色斑；胸部黑色，肩前条纹甚阔，合胸侧面具2条宽阔的黄条纹，后胸前侧板具1个甚小的黄斑，足大部分黑色，翅透明；腹部黑色具黄色斑点。雌性与雄性相似，尾毛长度约为第10节的1.5倍。【长度】体长 65~70 mm，腹长 50~54 mm，后翅 45~50 mm。【栖息环境】海拔500~1000 m的山区小溪。【分布】中国特有，分布于浙江、福建、广东、广西。【飞行期】6—11月。

[Identification] Male eyes green, face yellow, antefrons with a large black spot. Thorax black with broad antehumeral stripes, sides with two broad yellow stripes, metepisternum with a small yellow spot, legs largely black, wings hyaline. Abdomen black with yellow markings. Female similar to male, cerci about one and a half times as long as S10. **[Measurements]** Total length 65-70 mm, abdomen 50-54 mm, hind wing 45-50 mm. **[Habitat]** Montane streams at 500-1000 m elevation. **[Distribution]** Endemic to China, recorded from Zhejiang, Fujian, Guangdong, Guangxi. **[Flight Season]** June to November.

台湾黑额蜓 *Planaeschna taiwana* Asahina, 1951

【形态特征】雄性复眼绿色，面部黄色和黑色，上唇黑色具1个黄色斑，前唇基和后唇基下缘黑色，前额具1个甚大的黑色斑；胸部黑色，肩前条纹甚阔，肩前下点甚小，合胸侧面具2条宽阔的黄绿条纹，后胸前侧板具2个甚小的黄斑，足大面积黑色，翅透明；腹部黑色具黄色斑点。雌性与雄性相似，尾毛约与第10节等长。【长度】体长 64~71 mm，腹长 51~57 mm，后翅 44~51 mm。【栖息环境】海拔 1000 m以下的山区小溪。【分布】中国台湾特有。【飞行期】5—11月。

台湾黑额蜓 雄，台湾
Planaeschna taiwana, male from Taiwan

台湾黑额蜓 雌，台湾
Planaeschna taiwana, female from Taiwan

[Identification] Male eyes green, face yellow and black, labrum black with a yellow spot, anteclypeus and lower edge of postclypeus black, antefrons with a large black spot. Thorax black with broad antehumeral stripes and tiny antehumeral spot, sides with two broad yellowish green stripes, metepisternum with two small yellow spots, legs largely black, wings hyaline. Abdomen black with yellow markings. Female similar to male, cerci about same length as S10. **[Measurements]** Total length 64-71 mm, abdomen 51-57 mm, hind wing 44-51 mm. **[Habitat]** Montane streams below 1000 m elevation. **[Distribution]** Endemic to Taiwan of China. **[Flight Season]** May to November.

黑额蜓属待定种1 *Planaeschna* sp. 1

【形态特征】雄性复眼蓝绿色，面部黄色；胸部褐色，肩前条纹甚阔，合胸侧面具2条黄绿相间的宽阔条纹，足大面积黑色，中足的腿节具甚大的黄色斑，翅透明；腹部黑色具黄色斑点。雌性体色稍淡，尾毛长度约为第10节的1.5倍。【长度】体长 61~66 mm，腹长 47~51 mm，后翅 45~49 mm。【栖息环境】海拔 1800~2000 m的山区小溪。【分布】云南（昆明）。【飞行期】8—10月。

[Identification] Male eyes bluish green, face yellow. Thorax brown with broad antehumeral stripes, sides with two broad yellowish green stripes, legs largely black, femora of mid legs with yellow spots, wings hyaline. Abdomen black with yellow markings. Female paler, cerci about one and a half times as long as S10. **[Measurements]** Total length 61-66 mm, abdomen 47-51 mm, hind wing 45-49 mm. **[Habitat]** Montane streams at 1800-2000 m elevation. **[Distribution]** Yunnan (Kunming). **[Flight Season]** August to October.

黑额蜓属待定种1 雄，云南（昆明）
Planaeschna sp. 1, male from Yunnan (Kunming)

黑额蜓属待定种1 雌，云南（昆明）
Planaeschna sp. 1, female from Yunnan (Kunming)

黑额蜓属待定种2 *Planaeschna* sp. 2

【形态特征】雄性复眼蓝绿色，面部黄褐色，前额具褐色斑；胸部黑色，肩前条纹甚阔，绿色，合胸侧面具2条黄绿相间的宽阔条纹，基本覆盖整个合胸侧面，足大面积黑色，前足和中足的腿节具甚大的黄色斑，翅透明；腹部黑色具黄色斑点。雌性体色稍淡，尾毛长度约为第10节的1.5倍。【长度】体长 66~68 mm，腹长 51~53 mm，后翅 46~47 mm。【栖息环境】海拔 1500 m左右的山区开阔小溪。【分布】云南（德宏）。【飞行期】9—11月。

黑额蜓属待定种2 雄，云南（德宏）
Planaeschna sp. 2, male from Yunnan (Dehong)

黑额蜓属待定种2 雌，云南（德宏）
Planaeschna sp. 2, female from Yunnan (Dehong)

[Identification] Male eyes bluish green, face yellowish brown, antefrons with brown spots. Thorax black with broad green antehumeral stripes, sides largely covered by two broad green and yellow stripes, legs largely black, femora of fore and mid legs with large yellow spots, wings hyaline. Abdomen black with yellow markings. Female paler, cerci about one and a half times as long as S10. **[Measurements]** Total length 66-68 mm, abdomen 51-53 mm, hind wing 46-47 mm. **[Habitat]** Exposed montane streams at about 1500 m elevation. **[Distribution]** Yunnan (Dehong). **[Flight Season]** September to November.

黑额蜓属待定种2 雄，云南（德宏）
Planaeschna sp. 2, male from Yunnan (Dehong)

黑额蜓属待定种3 *Planaeschna* sp. 3

【形态特征】雄性复眼蓝绿色，面部黄色，上额具黑色斑；胸部褐色，具宽阔的肩前条纹，合胸侧面具2条宽阔的黄绿色条纹，足大面积黑色，前足和中足具黄斑，翅透明；腹部黑色，第1~8节具背中条纹。【长度】体长 62~63 mm，腹长 48~49 mm，后翅 41~42 mm。【栖息环境】海拔 1500 m左右的山区开阔小溪。【分布】云南（德宏）。【飞行期】9—11月。

[Identification] Male eyes bluish green, face yellow, antefrons with black spot. Thorax brown with broad antehumeral stripes, sides with two broad yellowish green stripes, legs largely black, fore and mid legs with yellow spots, wings hyaline. Abdomen black, S1-S8 with dorsal stripes along the carina. **[Measurements]** Total length 62-63 mm, abdomen 48-49 mm, hind wing 41-42 mm. **[Habitat]** Exposed montane streams at about 1500 m elevation. **[Distribution]** Yunnan (Dehong). **[Flight Season]** September to November.

黑额蜓属待定种3 雄，云南（德宏）
Planaeschna sp. 3, male from Yunnan (Dehong)

黑额蜓属待定种3 雌，云南（德宏）
Planaeschna sp. 3, female from Yunnan (Dehong)

黑额蜓属待定种4 *Planaeschna* sp. 4

【形态特征】雄性复眼蓝绿色，面部黄色，前额具黑色斑；胸部黑色，具肩前条纹和肩前下点，合胸侧面具2条宽阔的黄绿色条纹，后胸前侧板具1个短条纹，足大面积黑色，翅透明；腹部黑色具黄色斑点。雌性与雄性相似，尾毛稍长于第10节。【长度】体长 60~62 mm，腹长 47~48 mm，后翅 41~44 mm。【栖息环境】海拔 1500 m左右的山区开阔小溪。【分布】云南（德宏）。【飞行期】9—11月。

[Identification] Male eyes bluish green, face yellow, antefrons with black spot. Thorax black with antehumeral stripes and spots, sides with two broad yellowish green stripes, metepisternum with a short stripe, legs largely black, wings hyaline. Abdomen black with yellow markings. Female similar to male, cerci slightly longer than S10. **[Measurements]** Total length 60-62 mm, abdomen 47-48 mm, hind wing 41-44 mm. **[Habitat]** Exposed montane streams at about 1500 m elevation. **[Distribution]** Yunnan (Dehong). **[Flight Season]** September to November.

黑额蜓属待定种4 雄，云南（德宏）
Planaeschna sp. 4, male from Yunnan (Dehong)

黑额蜓属待定种4 雌，云南（德宏）
Planaeschna sp. 4, female from Yunnan (Dehong)

黑额蜓属待定种5 *Planaeschna* sp. 5

【形态特征】雄性复眼深蓝色，面部棕黄色；胸部褐色，肩前条纹甚阔，合胸侧面具2条宽阔的黄绿色条纹，后胸前侧板具1个较小的三角形斑，足大面积黑色，翅透明；腹部黑色具发达的黄色斑点。【长度】雄性体长 66～67 mm，腹长 52～53 mm，后翅 42～43 mm。【栖息环境】海拔 1500 m左右的山区开阔小溪。【分布】云南（德宏、临沧）。【飞行期】9—11月。

[Identification] Male eyes dark blue, face brownish yellow. Thorax brown with broad antehumeral stripes, sides with two broad yellowish green stripes, metepisternum with a small triangular spot, legs largely black, wings hyaline. Abdomen black with numerous yellow markings. **[Measurements]** Total length 66-67 mm, abdomen 52-53 mm, hind wing 42-43 mm. **[Habitat]** Exposed montane streams at about 1500 m elevation. **[Distribution]** Yunnan (Dehong, Lincang). **[Flight Season]** September to November.

黑额蜓属待定种5 雄，云南（德宏）
Planaeschna sp. 5, male from Yunnan (Dehong)

黑额蜓属待定种5 雌，云南（临沧）
Planaeschna sp. 5, female from Yunnan (Lincang)

黑额蜓属待定种6 *Planaeschna* sp. 6

【形态特征】雄性复眼蓝绿色，面部黄褐色，上额色彩较深；胸部黑色，具肩前条纹和肩前下点，合胸侧面具2条宽阔的黄绿色条纹，后胸前侧板具1个细条纹，足大面积黑色，翅透明；腹部黑色具黄色斑点。【长度】雄性体长 60 mm，腹长 48 mm，后翅 40 mm。【栖息环境】海拔 1500 m左右的山区开阔小溪。【分布】云南（德宏）。【飞行期】10—12月。

[Identification] Male eyes bluish green, face yellow, top of frons darkened. Thorax black with antehumeral stripes and spots, sides with two broad yellowish green stripes, metepisternum with a narrow stripe, legs largely black, wings hyaline. Abdomen black with yellow spots. **[Measurements]** Male total length 60 mm, abdomen 48 mm, hind wing 40 mm. **[Habitat]** Exposed montane streams at about 1500 m elevation. **[Distribution]** Yunnan (Dehong). **[Flight Season]** October to December.

黑额蜓属待定种6 雄，云南（德宏）
Planaeschna sp. 6, male from Yunnan (Dehong)

黑额蜓属待定种7 *Planaeschna* sp. 7

黑额蜓属待定种7 雌，海南
Planaeschna sp. 7, female from Hainan

【形态特征】雌性面部黄色和黑色，上唇完全黑色，前额具1个较大的黑斑；胸部黑色，具黄色肩前条纹，合胸侧面具2条黄色的条纹，足黑色，翅透明；腹部黑色，第3~7节具黄环，尾毛较长。【长度】雌性腹长 52 mm，后翅 43 mm。【栖息环境】海拔 1000 m处森林中的溪流。【分布】海南。【飞行期】6—8月。

[Identification] Female face black and yellow, labrum entirely black, antefrons with a large black spots. Thorax black with yellow antehumeral stripes, sides with two yellow stripes, legs black, wings hyaline. Abdomen black with yellow rings on S3-S7, cerci long. **[Measurements]** Female abdomen 52 mm, hind wing 43 mm. **[Habitat]** Streams in forest at 1000 m elevation. **[Distribution]** Hainan. **[Flight Season]** June to August.

多棘蜓属 Genus *Polycanthagyna* Fraser, 1933

本属全球已知的4种都分布于中国。赵氏多棘蜓在此作为红褐多棘蜓的异名处理。本属蜻蜓体大型；复眼绿色或者蓝色，身体黑褐色具苹果绿色或黄色条纹；翅透明，基室无横脉，IR3叉状，臀三角室3室。雌性的产卵器发达，具长刺，产卵管长。

本属蜻蜓栖息于静水环境，偏爱小型的深坑水潭。雄性会吊挂在水潭上方的树枝上等待雌性。雌性在陡坡的泥土和苔藓上产卵。

Four described species, all recorded from China, have been included in this genus. *Polycanthagyna chaoi* Yang & Li, 1994 is here regarded as a synonym of *P. erythromelas*. Species of the genus are large-sized. Eyes blue or green,

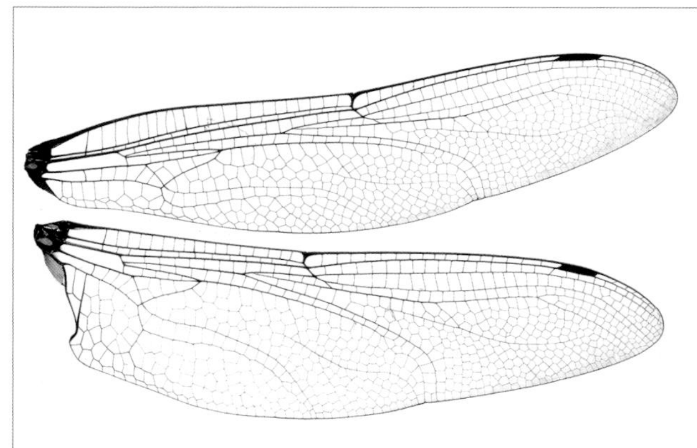
红褐多棘蜓 雄翅
Polycanthagyna erythromelas, male wings

红褐多棘蜓 雄
Polycanthagyna erythromelas, male

body blackish brown with apple green or yellow stripes. Wings hyaline, median space without crossveins, IR3 forked, anal triangle 3-celled. Female ovipositor long, vulvar lamina with stout apical spines.

Polycanthagyna species inhabit standing water habitats, frequenting small, deep pools in forest. Males hang on the trees above the pools and wait for females. Females lay eggs into mud and moss on steep slopes and embankments.

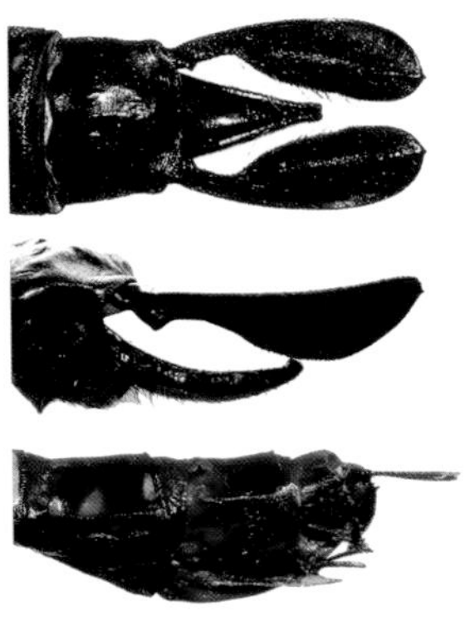

红褐多棘蜓
Polycanthagyna erythromelas

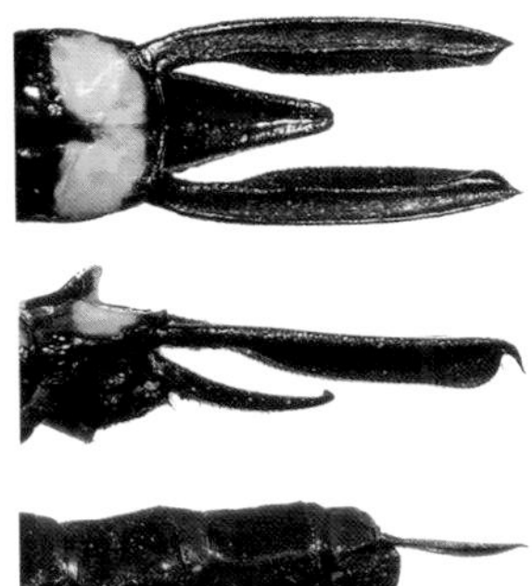

黄绿多棘蜓
Polycanthagyna melanictera

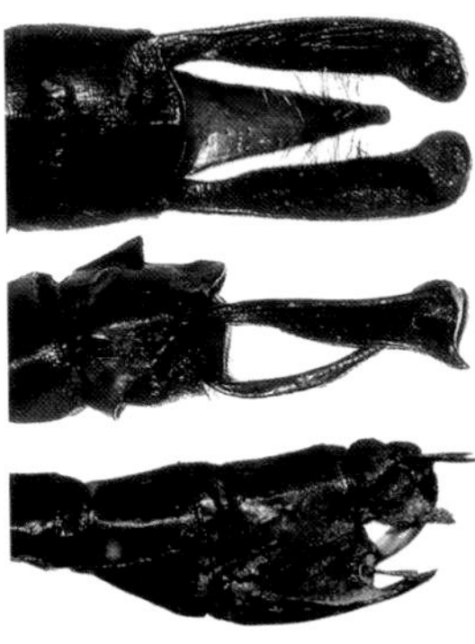

蓝黑多棘蜓
Polycanthagyna ornithocephala

多棘蜓属 雄性肛附器和雌性产卵器
Genus *Polycanthagyna*, male anal appendages and female ovipositor

红褐多棘蜓 *Polycanthagyna erythromelas* (McLachlan, 1896)

红褐多棘蜓 雄，贵州
Polycanthagyna erythromelas, male from Guizhou

红褐多棘蜓 雌，云南（普洱）
Polycanthagyna erythromelas, female from Yunnan (Pu'er)

【形态特征】雄性复眼绿色，面部黄白色，额具黑色斑；胸部肩前条纹甚阔，合胸侧面具2条宽阔的苹果绿色条纹，翅透明；腹部黑色具较发达的苹果绿色斑点，第2节侧方具1个小蓝斑；雌性复眼蓝绿色，面部黄色，年轻的雌性腹部第3~7节为红褐色，随年纪的增长色彩逐渐加深至黑褐色。【长度】体长 80~86 mm，腹长 61~66 mm，后翅 50~53 mm。【栖息环境】海拔 1500 m以下森林中水草匮乏的小至中型静水水潭。【分布】陕西、四川、云南、贵州、湖北、湖南、江西、福建、广东、广西、海南、香港、台湾；不丹、印度、尼泊尔、巴基斯坦、缅甸、老挝、泰国、越南。【飞行期】3—12月。

[Identification] Male eyes green, face yellowish white, frons black spot. Thorax black with broad antehumeral stripes, sides with two broad apple green stripes, wings hyaline. Abdomen black with apple green spots, S2 with a small blue spot laterally. Female eyes bluish green, face yellow, S3-S7 in young female reddish brown, with age the color darkens. **[Measurements]** Total length 80-86 mm, abdomen 61-66 mm, hind wing 50-53 mm. **[Habitat]** Small to medium sized pools lacking aquatic plants in forest below 1500 m elevation. **[Distribution]** Shaanxi, Sichuan, Yunnan, Guizhou, Hubei, Hunan, Jiangxi, Fujian, Guangdong, Guangxi, Hainan, Hong Kong, Taiwan; Bhutan, India, Nepal, Pakistan, Myanmar, Laos, Thailand, Vietnam. **[Flight Season]** March to December.

黄绿多棘蜓 *Polycanthagyna melanictera* (Selys, 1883)

【形态特征】雄性复眼深蓝色，面部淡蓝色，额具褐色斑；胸部黑褐色，肩前条纹甚阔，合胸侧面具2条宽阔的黄色条纹，足黑色，翅透明；腹部褐色具黄绿色条纹，第2节侧面具1个甚大的蓝斑，第10节具1个角锥形突起和1个甚大黄色背斑。雌性复眼黄绿色，随着年纪增长逐渐变为蓝色，身体黑褐色具发达的黄色斑点和条纹，其中腹部第7节的黄色斑点发达。【长度】体长 73~81 mm，腹长 54~62 mm，后翅 50~53 mm。【栖息环境】海拔 1500 m以下森林中水草匮乏的小至中型静水水潭。【分布】山东、河南、湖北、湖南、陕西、山西、四川、贵州、浙江、广东、台湾；朝鲜半岛、日本。【飞行期】5—9月。

[Identification] Male eyes dark blue, face pale blue, frons with brown spot. Thorax blackish brown with broad antehumeral stripes, sides with two broad yellow stripes, legs black, wings hyaline. Abdomen brown with yellowish green markings, S2 with a large blue spot laterally, dorsum of S10 with pyramidal prominence and a large yellow spot.

黄绿多棘蜓 雄，贵州
Polycanthagyna melanictera, male from Guizhou

Female eyes yellowish green, and darkening to blue with age, body blackish brown with numerous yellow markings, S7 with large yellow spots. **[Measurements]** Total length 73-81 mm, abdomen 54-62 mm, hind wing 50-53 mm. **[Habitat]** Small to medium sized pools lacking aquatic plants in forest below 1500 m elevation. **[Distribution]** Shandong, Henan, Hubei, Hunan, Shaanxi, Shanxi, Sichuan, Guizhou, Zhejiang, Guangdong, Taiwan; Korean peninsula, Japan. **[Flight Season]** May to September.

黄绿多棘蜓 雌，产卵，贵州 | 莫善濂 摄
Polycanthagyna melanictera, female laying eggs from Guizhou | Photo by Shanlian Mo

黄绿多棘蜓 雄，贵州
Polycanthagyna melanictera, male from Guizhou

蓝黑多棘蜓 *Polycanthagyna ornithocephala* (McLachlan, 1896)

蓝黑多棘蜓 雌，广东
Polycanthagyna ornithocephala, female from Guangdong

蓝黑多棘蜓 雄，重庆
Polycanthagyna ornithocephala, male from Chongqing

【形态特征】雄性复眼蓝褐色，面部白色，额具黑色斑；胸部黑褐色，肩前条纹甚阔，合胸侧面具2条宽阔的黄色条纹，足黑色，翅透明；腹部黑褐色并具黄色条纹。雌性胸部褐色具黄色条纹，翅稍染淡琥珀色；腹部具黄色条纹，第2~6节背面红褐色。【长度】体长 73~78 mm，腹长 55~58 mm，后翅 47~55 mm。【栖息环境】海拔 2000 m以下森林中水草匮乏的小型静水水潭（包括人工水塘）及林道上的暂时性积水潭。【分布】四川、重庆、贵州、江苏、湖北、湖南、福建、广东、广西、台湾；孟加拉国、泰国、印度。【飞行期】6—10月。

[Identification] Male eyes bluish brown, face white, frons with black spot. Thorax blackish brown with broad antehumeral stripes, sides with two broad yellow stripes, legs black, wings hyaline. Abdomen dark brown with yellow

蓝黑多棘蜓 雄，重庆
Polycanthagyna ornithocephala, male from Chongqing

蓝黑多棘蜓 雌，广东
Polycanthagyna ornithocephala, female from Guangdong

markings. Female thorax brown with yellow stripes, wings tinted with light amber. Abdomen with yellow stripes, dorsum of S2-S6 reddish brown. **[Measurements]** Total length 73-78 mm, abdomen 55-58 mm, hind wing 47-55 mm. **[Habitat]** Small bodies of standing water, including man made pools, and the temporary pools in the forested paths below 2000 m elevation. **[Distribution]** Sichuan, Chongqing, Guizhou, Jiangsu, Hubei, Hunan, Fujian, Guangdong, Guangxi, Taiwan; Bangladesh, Thailand, India. **[Flight Season]** June to October.

多棘蜓属待定种1 *Polycanthagyna* sp. 1

多棘蜓属待定种1 雄，云南（德宏）
Polycanthagyna sp. 1, male from Yunnan (Dehong)

【形态特征】雄性复眼深蓝色，面部蓝白色，额具黑色斑；胸部红褐色，肩前条纹甚阔，合胸侧面具2条宽阔的黄绿色条纹，足黑色，基部红褐色，翅透明；腹部红褐色并具黄绿色条纹。雌性较粗壮，复眼蓝色，面部黄绿色，仅上额具黑色斑；翅稍染淡琥珀色；腹部红褐色并具发达黄绿色条纹。**【长度】**体长 66~73 mm，腹长 47~55 mm，后翅 46~51 mm。**【栖息环境】**海拔 1500 m以下森林中的小至中型静水水潭。**【分布】**云南（西双版纳、德宏）。**【飞行期】**8—12月。

[Identification] Male eyes dark blue, face bluish white, frons with black spot. Thorax reddish brown with broad antehumeral stripes, sides with two broad yellowish green stripes, legs black with reddish brown bases, wings hyaline. Abdomen reddish brown with yellowish green markings. Female short and robust, eyes blue, face yellowish green. Wings tinted with light amber. Abdomen reddish brown with yellowish green markings. **[Measurements]** Total length 66-73 mm, abdomen 47-55 mm, hind wing 46-51 mm. **[Habitat]** Small to medium sized ponds in dense forest below 1500 m elevation. **[Distribution]** Yunnan (Xishuangbanna, Dehong). **[Flight Season]** August to December.

多棘蜓属待定种1 雄，云南（德宏）
Polycanthagyna sp. 1, male from Yunnan (Dehong)

多棘蜓属待定种1 雌，云南（德宏）
Polycanthagyna sp. 1, female from Yunnan (Dehong)

多棘蜓属待定种2 *Polycanthagyna* sp. 2

【形态特征】雌性复眼深蓝色，面部白色，额具1个深褐色斑；胸部黑褐色，肩前条纹甚阔，合胸侧面具2条宽阔的黄色条纹，足黑色，翅透明；腹部褐色具黄绿色条纹。【长度】雌性体长 73～81 mm，腹长 54～62 mm，后翅 50～53 mm。【栖息环境】海拔 1000～2000 m森林中水草匮乏的小至中型静水水潭。【分布】云南（红河）。【飞行期】4—6月。

多棘蜓属待定种2 雌，云南（红河）
Polycanthagyna sp. 2, female from Yunnan (Honghe)

[Identification] Female eyes dark blue, face white, frons with a dark brown spot. Thorax blackish brown with broad antehumeral stripes, sides with two broad yellow stripes, legs black, wings hyaline. Abdomen brown with yellowish green markings. **[Measurements]** Female total length 73-81 mm, abdomen 54-62 mm, hind wing 50-53 mm. **[Habitat]** Small to medium sized pools lacking aquatic plants in forest at 1000-2000 m elevation. **[Distribution]** Yunnan (Honghe). **[Flight Season]** April to June.

沼蜓属 Genus *Sarasaeschna* Karube & Yeh, 2001

本属全球已知18种，主要分布在亚洲东部和日本，中国已知9种。本属蜻蜓是一类体型较小的稀有物种。复眼绿色或蓝绿色，面部黄色和黑色，通常在前额具1个黑斑，上额具1个“T”形斑；胸部和腹部黑褐色具黄色斑纹，翅透明。不同种类雄性的肛附器差异显著，是有效的区分特征。

本属蜻蜓主要栖息于山区的沼泽地，通常是具有一定海拔高度的森林沼泽，有些种类也可以在流速缓慢的溪流中生活。雄性具领域行为，在沼泽上来回飞行，并时而悬停，雌性产卵于苔藓或者泥土中。

The genus contains 18 species, mainly confined to eastern Asia and Japan, nine species are recorded from China. Species of the genus are small for the family and rarely seen. Eyes green or bluish green, face yellow and black, antefrons with a black spots and top of frons with a T-mark. Thorax and abdomen blackish brown with yellow markings, wings hyaline. Each species possesses an unique structure of the male anal appendages, a key diagnostic character.

Sarasaeschna species inhabit montane marshes with moderate elevation, some species can also live in slow flowing streams. Males exhibit territorial behavior by flying above water, and sometimes hover. Females lay eggs into mud or moss.

刃尾沼蜓 雄
Sarasaeschna sabre, male

源垭沼蜓 雄翅
Sarasaeschna pyanan, male wings

源垭沼蜓
Sarasaeschna pyanan

沼蜓属 雄性肛附器和雌性产卵器
Genus *Sarasaeschna*, male anal appendages and female ovipositor

连氏沼蜓
Sarasaeschna

刃尾沼蜓
Sarasaeschna sabre

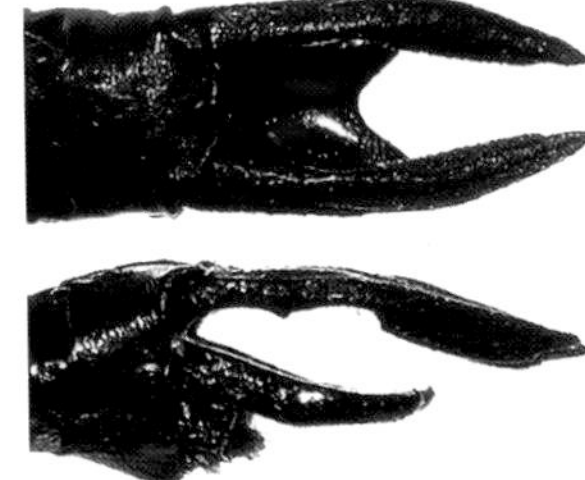

锋刀沼蜓
Sarasaeschna tsaopiensis

沼蜓属 雄性肛附器
Genus *Sarasaeschna*, male anal appenages

连氏沼蜓 *Sarasaeschna lieni* (Yeh & Chen, 2000)

连氏沼蜓 雄，台湾
Sarasaeschna lieni, male from Taiwan

【形态特征】雄性复眼深绿色，面部黄色，前额具1个甚大的黑斑，上额具1个"T"形斑；胸部黑褐色，肩前条纹甚阔，合胸侧面具2条黄色条纹，后胸前侧板具1个较小的三角形斑，足大面积黑色，翅透明；腹部黑色，第1～6节具黄色斑，第1～2节和第4～6节宽，第3节和第7～10节收缩显著。【长度】雄性体长 57 mm，腹长 43 mm，后翅 34 mm。【栖息环境】海拔 500～2000 m森林中的渗流地和沼泽地。【分布】中国台湾特有。【飞行期】4—7月。

[Identification] Male eyes dark green, face yellow, antefrons with a large black spot, top of frons with a T-mark. Thorax blackish brown with broad antehumeral stripes, sides with two yellow stripes, metepisternum with a small triangular spot, legs largely black, wings hyaline. Abdomen black, S1-S6 with yellow spots, S1-S2 and S4-S6 broad, S3 and S7-S10 stalked. **[Measurements]** Male total length 57 mm, abdomen 43 mm, hind wing 34 mm. **[Habitat]** Seepages and marshes in forest at 500-2000 m elevation. **[Distribution]** Endemic to Taiwan of China. **[Flight Season]** April to July.

高峰沼蜓 *Sarasaeschna gaofengensis* Yeh & Kiyoshi, 2015

高峰沼蜓 雄，越南 | Tom Kompier 摄
Sarasaeschna gaofengensis, male from Vietnam | Photo by Tom Kompier

【形态特征】雄性复眼深绿色，面部大面积黄色，前额具1个甚大的黑斑，上额具1个"T"形斑；胸部黑褐色，肩前条纹甚阔，合胸侧面具2条黄色条纹，后胸前侧板具黄斑，足大面积黑色，翅透明；腹部黑色，第2~7节具黄色斑。【长度】雄性腹长 43 mm，后翅 38 mm。【栖息环境】海拔 500~2000 m的山区。【分布】云南（楚雄）；越南。【飞行期】不详。

[Identification] Male eyes dark green, face largely yellow, antefrons with a large black spot, top of frons with a T-mark. Thorax blackish brown with broad antehumeral stripes, sides with two yellow stripes, metepisternum with small spots, legs largely black, wings hyaline. Abdomen black, S2-S7 with yellow spots. **[Measurements]** Male abdomen 43 mm, hind wing 38 mm. **[Habitat]** Montane area at 500-2000 m elevation. **[Distribution]** Yunnan (Chuxiong); Vietnam. **[Flight Season]** Unknown.

尼氏沼蜓 *Sarasaeschna niisatoi* (Karube, 1998)

【形态特征】雄性复眼深绿色，面部大面积黄色，前额具1个甚大的黑斑，上额具1个甚大"T"形斑；胸部黑褐色，肩前条纹甚阔，合胸侧面具2条黄色条纹，后胸前侧板具黄斑，足大面积黑色，翅透明；腹部黑色，第2~6节具黄色斑。【长度】不详。【栖息环境】海拔 1000 m以下森林中的渗流地和沼泽地。【分布】海南；越南。【飞行期】不详。

[Identification] Male eyes dark green, face largely yellow, antefrons with a large black spot, top of frons with a T-mark. Thorax blackish brown with broad antehumeral stripes, sides with two yellow stripes, metepisternum with small spots, legs largely black, wings hyaline. Abdomen black, S2-S6 with yellow spots. **[Measurements]** Unknown. **[Habitat]** Seepages and marshes in forest below 1000 m elevation. **[Distribution]** Hainan; Vietnam. **[Flight Season]** Unknown.

尼氏沼蜓 雄，越南 | Tom Kompier 摄
SSarasaeschna niisatoi, male from Vietnam | Photo by Tom Kompier

源埡沼蜓 *Sarasaeschna pyanan* (Asahina, 1951)

【形态特征】雄性复眼深绿色，面部大面积黄色，上唇黑色，前额具1个甚大的黑斑，上额具1个“T”形斑；胸部黑褐色，肩前条纹甚阔，合胸侧面具2条黄色条纹，后胸前侧板具1个较小的三角形斑，足大面积黑色，翅透明；腹部黑色，第1～6节具黄色斑，第1～2节和第4～6节宽，第3节和第7～10节收缩显著。雌性腹部基方2节膨大，体斑与雄性相似。【长度】雄性体长 57～68 mm，腹长 43～53 mm，后翅 40 mm。【栖息环境】海拔 500～2000 m森林中的渗流地和沼泽地。【分布】中国台湾特有。【飞行期】5—8月。

源埡沼蜓 雄，台湾 | 嘎嘎 摄
Sarasaeschna pyanan, male from Taiwan | Photo by Gaga

[Identification] Male eyes dark green, face largely yellow, labrum black, antefrons with a large black spot, top of frons with a T-mark. Thorax blackish brown with broad antehumeral stripes, sides with two yellow stripes, metepisternum with a small triangular spot, legs largely black, wings hyaline. Abdomen black, S1-S6 with yellow spots, S1-S2 and S4-S6 broad, S3 and S7-S10 stalked. Female similar to male, S1-S2 expanded. **[Measurements]** Male total length 57-68 mm, abdomen 43-53 mm, hind wing 40 mm. **[Habitat]** Seepages and marshes in forest at 500-2000 m elevation. **[Distribution]** Endemic to Taiwan of China. **[Flight Season]** May to August.

刃尾沼蜓 *Sarasaeschna sabre* (Wilson & Reels, 2001)

【形态特征】雄性复眼深绿色，面部黄色，前额具1个甚大的黑斑，上额具1个“T”形斑；胸部黑褐色，肩前条纹甚阔，合胸侧面具2条黄绿色条纹，后胸前侧板具1个较小的三角形斑，足大面积黑色，翅透明；腹部黑色，第1～6节具黄绿色斑，第1～2节和第4～6节宽，第3节和第7～10节收缩显著。【长度】雄性体长 55～57 mm，腹长 42～44 mm，后翅 35～36 mm。【栖息环境】海拔 1000 m以下森林中的渗流地和沼泽地。【分布】中国海南特有。【飞行期】4—6月。

[Identification] Male eyes dark green, face yellow, antefrons with a large black spot, top of frons with a T-mark. Thorax blackish brown with broad antehumeral stripes, sides with two yellowish green stripes, metepisternum with a small triangular spot, legs largely black, wings hyaline. Abdomen black, S1-S6 with yellowish green spots, S1-S2 and S4-S6 broad, S3 and S7-S10 stalked. **[Measurements]** Male total length 55-57 mm, abdomen 42-44 mm, hind wing 35-36 mm. **[Habitat]** Seepages and marshes in forest below 1000 m elevation. **[Distribution]** Endemic to Hainan of China. **[Flight Season]** April to June.

刃尾沼蜓 雄，海南
Sarasaeschna sabre, male from Hainan

锋刀沼蜓 *Sarasaeschna tsaopiensis* (Yeh & Chen, 2000)

【形态特征】雄性复眼深绿色，面部大面积黑色，后唇基黄色，上额具1个“T”形斑；胸部黑褐色，肩前条纹甚阔，合胸侧面具2条黄色条纹，后胸前侧板具1个较小的三角形斑，足大面积黑色，翅透明；腹部黑色，第1～6节具黄色斑，第1～2节和第4～6节宽，第3节和7～10节收缩显著。雌性腹部基方2节膨大，体斑与雄性相似。【长度】雄性体长 61～67 mm，腹长 45～50 mm，后翅 37～41 mm。【栖息环境】海拔 1000 m以下森林中的渗流地和沼泽地。【分布】中国台湾特有。【飞行期】5—8月。

[Identification] Male eyes dark green, face largely black, postclypeus yellow, top of frons with a T-mark. Thorax blackish brown with broad antehumeral stripes, sides with two yellow stripes, metepisternum with a small triangular spot, legs largely black, wings hyaline. Abdomen black, S1-S6 with yellow spots, S1-S2 and S4-S6 broad, S3 and S7-S10 stalked. Female similar to male, S1-S2 expanded. **[Measurements]** Male total length 61-67 mm, abdomen 45-50 mm, hind wing 37-41 mm. **[Habitat]** Seepages and marshes in forest below 1000 m elevation. **[Distribution]** Endemic to Taiwan of China. **[Flight Season]** May to August.

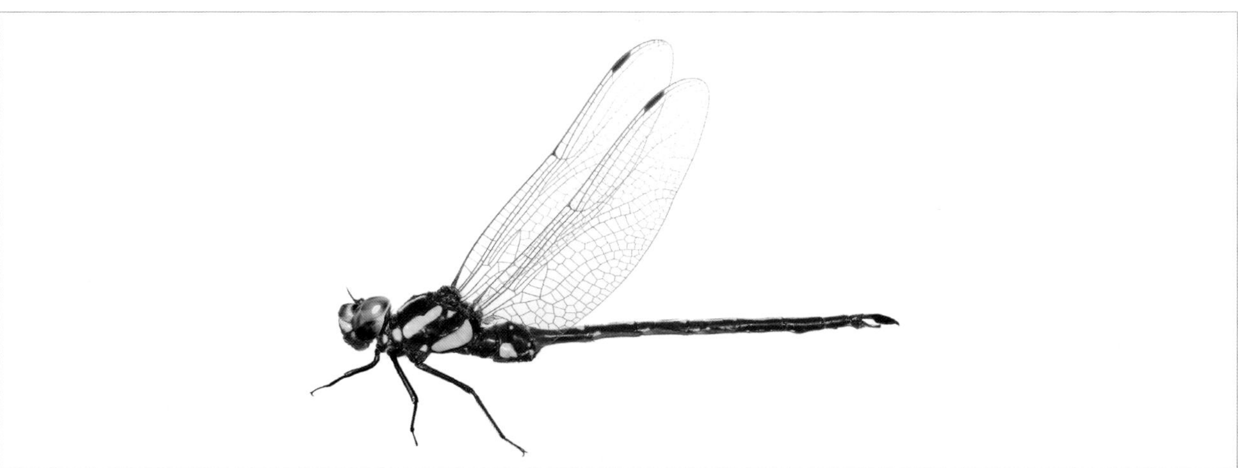

锋刀沼蜓 雄，台湾
Sarasaeschna tsaopiensis, male from Taiwan

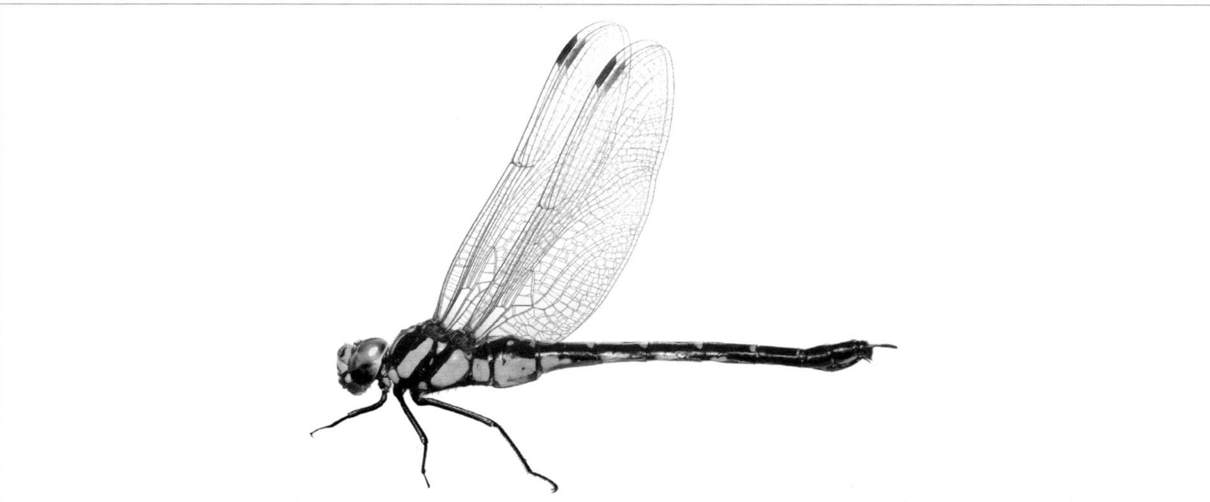

锋刀沼蜓 雌，台湾
Sarasaeschna tsaopiensis, female from Taiwan

短痣蜓属 Genus *Tetracanthagyna* Selys, 1883

本属全世界已知5种，主要分布在亚洲的热带区域。中国已知仅1种，分布于华南和西南地区。本属是一类体型粗壮且巨大的蜻蜓，体色通常很暗淡，无鲜艳的条纹和斑点；面部很窄，复眼发达；翅透明，翅痣短，基室无横脉，IR3叉状，Rspl和Mspl以上具3~4列翅室。

本属蜻蜓主要栖息于溪流和沟渠。成虫白天主要停落在茂密的丛林中，黄昏时较活跃。雄性无领域行为，很难遇见。雌性则较常见，午后和黄昏时分在小溪边缘的朽木和泥土中产卵。

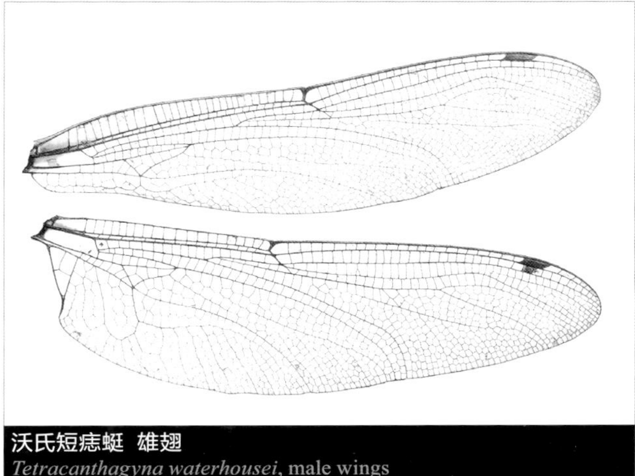

沃氏短痣蜓 雄翅
Tetracanthagyna waterhousei, male wings

This genus contains five species, mainly confined to tropical Asia. In China only one species is recorded, distributed in the South and Southwest regions. Species of the genus are robust and very large to enormous dragonflies, they usually have a dark ground color without bright colored stripes. Face narrow, eyes are large. Wings hyaline, pterostigma short, median space without corssveins, IR3 forked, 3-4 rows of cells above Rspl and Mspl.

Tetracanthagyna species inhabit streams and ditches. They perch in dense forest during the daytime, and become active at twilight. Territorial behavior has not been observed, so males are seldom seen. Females are more commonly found when they lay eggs into deadwood or soil at stream margins in the afternoon or at twilight.

沃氏短痣蜓 雌
Tetracanthagyna waterhousei, female

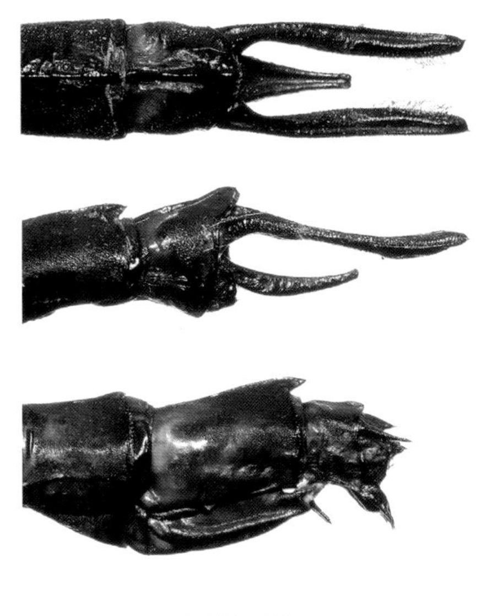

沃氏短痣蜓
Tetracanthagyna waterhousei,

短痣蜓属 雄性肛附器和雌性产卵器
Genus *Tetracanthagyna*, male anal appendages and female ovipositor

沃氏短痣蜓 *Tetracanthagyna waterhousei* McLachlan, 1898

沃氏短痣蜓 雄，云南（西双版纳）
Tetracanthagyna waterhousei, male from Yunnan (Xishuangbanna)

【形态特征】雄性复眼蓝褐色，面部黄褐色，额具黑斑；胸部深褐色，肩前条纹较短，合胸侧面具2条宽阔的黄条纹，足基部至腿节顶端红褐色，翅甚阔，透明，基方具黑色斑；腹部黑褐色具细小的黄斑，第8~10节背面末端具较短的锥形突起，肛附器长。雌性体色稍浅，复眼褐色，腹部粗壮。【长度】体长 75~87 mm，腹长 57~67 mm，后翅 57~62 mm。【栖息环境】海拔 1000 m以下的山区林荫小溪和宽阔河流河岸带植被茂盛的河段。【分布】云南、福建、广东、广西、海南、香港；孟加拉国、印度、马来西亚、印度尼西亚、缅甸、泰国、柬埔寨、老挝、越南。【飞行期】3—7月。

[Identification] Male eyes bluish brown, face yellowish brown, frons with black spots. Thorax dark brown with short anterhumeral stripes, sides with two broad yellow stripes, legs reddish brown from bases to the end of femora, wings hyaline and broad, bases with black spots. Abdomen blackish brown with small yellow spots, S8-S10 with short pyramidal prominences apically, anal appendages long. Female paler, eyes brown, abdomen more robust. **[Measurements]** Total length 75-87 mm, abdomen 57-67 mm, hind wing 57-62 mm. **[Habitat]** Shady streams and rivers with plenty of marginal vegetation below 1000 m elevation. **[Distribution]** Yunnan, Fujian, Guangdong, Guangxi, Hainan, Hong Kong; Bangladesh, India, Malaysia, Indonesia, Myanmar, Thailand, Cambodia, Laos, Vietnam. **[Flight Season]** March to July.

沃氏短痣蜓 雌，云南（西双版纳）
Tetracanthagyna waterhousei, female from Yunnan (Xishuangbanna)

沃氏短痣蜓 雄，云南（西双版纳）
Tetracanthagyna waterhousei, male from Yunnan (Xishuangbanna)

沃氏短痣蜓 雌，云南（西双版纳）
Tetracanthagyna waterhousei, female from Yunnan (Xishuangbanna)

2 春蜓科 Family Gomphidae

本科世界性分布，全球已知超过100属近1000种。中国已经发现37属200余种。本科蜻蜓既包括细小的“侏儒”，也包括粗壮的“巨兽”。它们最显著的特征是复眼较小，通常绿色，在头顶分离较远。身体通常黑色或褐色具黄色或绿色的条纹和斑点；绝大多数种类翅透明；腹部较长。雄性的肛附器、阳茎和钩片的构造以及雌性头部、下生殖板的形态是重要的辨识特征。

本科蜻蜓主要栖息于流水环境，包括河流和清澈的山区溪流；少数栖息于静水环境，如池塘、湖泊和沼泽地。雄性具有显著的领域行为，停落在水面附近占据领地，等待配偶并驱逐入侵的雄性。有些种类，如环尾春蜓，具有悬停飞行的本领。雌性较难遇见，仅在产卵时才会靠近水面。

The family is distributed worldwide, contains nearly 1000 species among over 100 genera. Over 200 species among 37 genera are recorded from China. Some species are small while others are "robust" and "large". Eyes are small and usually green, separated above by the frons and occiput, a remarkable characters for the family. Body usually black or brown marked with yellow or green stripes (thorax) and rings or spots (abdomen). Wings hyaline in most species. Abdomen long. Male anal appendages, penis and hamulus, female face, vulvar lamina are important characters in separating species.

Species of the family mostly occur along running water habitats, including rivers and clean montane streams; a few occur standing water habitats including ponds, lakes and marshes. Males exhibit territorial behavior by perching close to the water's edge darting out to meet and fend off an intruder before returning to its perch. Some species (e. g. *Lamelligomphus*) can hover for a long time. Females are seldom seen and only approach water for laying eggs.

安娜环尾春蜓 雄

Lamelligomphus annakarlorum, male

金黄显春蜓 雄

Phaenandrogomphus aureus, male

安春蜓属 Genus *Amphigomphus* Chao, 1954

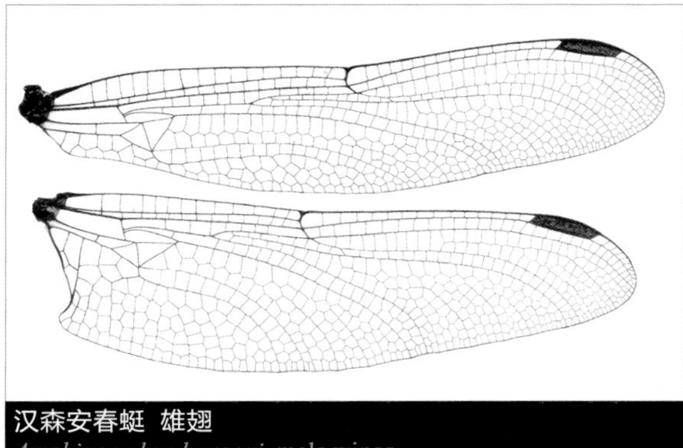
汉森安春蜓 雄翅
Amphigomphus hansoni, male wings

本属亚洲特有，全球已知3种都在中国分布。本属蜻蜓体中型；体色通常绿色；翅痣具支持脉，三角室、上三角室和下三角室无横脉，前翅盘区具2列翅室，臀圈不发达，具2～3室，前翅的臀区在臀脉与翅后缘之间仅有2列翅室，基臀区具1条横脉，后翅基缘呈锐角，臀三角室4室。雄性上肛附器的长度为下肛附器的2倍，末端向下弯曲，下肛附器的两枝分歧的角度很大。前钩片末端分两枝，后钩片锥状。阳茎具1对长鞭。

本属蜻蜓栖息于森林中的小溪。雄性具领域行为，停落在水边的枝头或者岩石上。

The genus contains three Asian species and all of which occur in China. Species of the genus are medium-sized. Body usually greenish. The pterostigma well braced, no crossvein in triangle, hypertriangle and subtriangle, discoidal field with 2 rows of cells in fore wings, anal loop poorly developed, usually 2-3 celled, 2 rows of cells between A and hind margin in fore wings, cubital space with 1 crossvein, tornus strongly angled, anal triangle 4-celled. Male superior appendages twice as long as the inferior appendage, with the tips curved downwards. The inferior appendage with two branches strongly divergent. Penis with a pair of long flagella.

Amphigomphus species inhabit streams in forest. Territorial males perch on branches or rocks close to water.

娜卡安春蜓 雄
Amphigomphus nakamurai, male

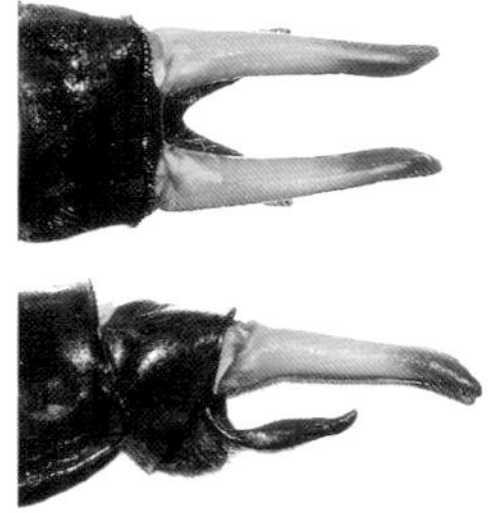

汉森安春蜓
Amphigomphus hansoni

娜卡安春蜓
Amphigomphus nakamurai

索氏安春蜓
Amphigomphus somnuki

安春蜓属 雄性肛附器
Genus *Amphigomphus*, male anal appenages

汉森安春蜓 *Amphigomphus hansoni* Chao, 1954

【形态特征】雄性复眼深绿色，面部大面积黑色，额黄绿色；胸部黑色，背条纹绿色，甚阔，胸部侧面具2条黄绿色宽条纹，后胸前侧板上缘具黄绿色条纹，足上具黄斑；腹部黑色具黄斑，上肛附器黄色，末端黑褐色，下肛附器黑色，长度为上肛附器的2/3。雌性与雄性相似，尾毛白色。本种肛附器与其他两种明显不同。【长度】体长 60～64 mm，腹长 44～50 mm，后翅 35～40 mm。【栖息环境】海拔 1000 m以下森林中的小溪和沟渠。【分布】中国特有，分布于江西、浙江、福建、广东、海南。【飞行期】4—8月。

[Identification] Male eyes dark green, face largely black, frons yellowish green. Thorax black with green and broad antehumeral stripes, sides with two broad yellowish green stripes, metepisternum with yellowish green stripes,

汉森安春蜓 雄，广东 | 宋睿斌 摄
Amphigomphus hansoni, male from Guangdong | Photo by Ruibin Song

legs with yellow spots. Abdomen black with yellow markings, superior appendages yellow with black tips, inferior appendage black, two thirds as long as superiors. Female similar to male with white cerci. The male appendages differ from other two species. **[Measurements]** Total length 60-64 mm, abdomen 44-50 mm, hind wing 35-40 mm. **[Habitat]** Montane streams and ditches below 1000 m elevation. **[Distribution]** Endemic to China, recorded from Jiangxi, Zhejiang, Fujian, Guangdong, Hainan. **[Flight Season]** April to August.

汉森安春蜓 雄，广东
Amphigomphus hansoni, male from Guangdong

汉森安春蜓 雌，广东
Amphigomphus hansoni, female from Guangdong

娜卡安春蜓 *Amphigomphus nakamurai* Karube, 2001

【形态特征】雄性复眼深绿色，面部黑色，额黄绿色；胸部黑色，背条纹绿色，甚阔，胸部侧面具2条绿色宽条纹，后胸前侧板上缘具1个较小的三角形斑；腹部黑色具黄斑，第7节的黄斑较大，上肛附器黄色，末端黑褐色，下肛附器黑色，长度约为上肛附器的1/2。雌性色彩较淡。【长度】体长 55~61 mm，腹长 42~45 mm，后翅 36~37 mm。【栖息环境】海拔 500~1500 m茂密森林中的狭窄溪流。【分布】云南（红河）；老挝、越南。【飞行期】4—6月。

[Identification] Male eyes dark green, face black, frons yellowish green. Thorax black with green and broad antehumeral stripes, sides with two broad green stripes, metepisternum with a small triangular spot. Abdomen black

娜卡安春蜓 雄，云南（红河）
Amphigomphus nakamurai, male from Yunnan (Honghe)

娜卡安春蜓 雌，云南（红河）
Amphigomphus nakamurai, male from Yunnan (Honghe)

with yellow spots, superior appendages yellow with blackish brown tips, inferior appendage black, about half length of superiors. Female paler. **[Measurements]** Total length 55-61 mm, abdomen 42-45 mm, hind wing 36-37 mm. **[Habitat]** Narrow streams in dense forest at 500-1500 m elevation. **[Distribution]** Yunnan (Honghe); Laos, Vietnam. **[Flight Season]** April to June.

娜卡安春蜓 雄，云南（红河）
Amphigomphus nakamurai, male from Yunnan (Honghe)

索氏安春蜓 *Amphigomphus somnuki* Hämäläinen, 1996

【形态特征】雄性复眼深绿色，面部大面积褐色，额黄绿色；胸部棕色，背条纹绿色，甚阔，胸部侧面具2条绿色宽条纹，后胸前侧板上缘具1个较小的三角形斑；腹部黑色具黄斑，上肛附器黄色，末端黑褐色，下肛附器黑色，长度为约上肛附器的1/2。本种和娜卡安春蜓的肛附器构造相似，但足基方红褐色，后者完全黑色。【长度】雄性体长55 mm，腹长 42 mm，后翅 34 mm。【栖息环境】海拔 500～1000 m茂密森林中的狭窄溪流。【分布】云南（西双版纳）；泰国、老挝。【飞行期】4—7月。

[Identification] Male eyes dark green, face largely brown, frons yellowish green. Thorax brown with green and broad antehumeral stripes, sides with two broad green stripes, metepisternum with a small triangular spot. Abdomen black with yellow spots, superior appendages yellow with blackish brown tips, inferior appendage black, about half length of superiors. The species is similar to *A. nakamurai*, but legs reddish brown at bases, legs of *A. nakamurai* entirely black. **[Measurements]** Male total length 55 mm, abdomen 42 mm, hind wing 34 mm. **[Habitat]** Narrow streams in dense forest at 500-1000 m elevation. **[Distribution]** Yunnan (Xishuangbanna); Thailand, Laos. **[Flight Season]** April to July.

索氏安春蜓 雄，云南（西双版纳）
Amphigomphus somnuki, male from Yunnan (Xishuangbanna)

安春蜓属待定种 *Amphigomphus* sp.

【形态特征】雄性面部大面积黑色，额黄绿色，上唇具1对甚小的黄斑；胸部黑色，背条纹绿色，甚阔，胸部侧面具2条黄绿色宽条纹；腹部黑色，第1～7节具黄斑。本种雄性的肛附器与汉森安春蜓相似，但足上无黄斑，腹部斑纹也不同。【长度】雄性体长 58～64 mm，腹长 45～47 mm，后翅 36～38 mm。【栖息环境】海拔 500 m以下森林中的小溪和沟渠。【分布】广西。【飞行期】4—7月。

[Identification] Male face largely black, frons yellowish green, labrum with a pair of small yellow spots. Thorax black with green and broad antehumeral stripes, sides with two broad yellowish green stripes. Abdomen black, S1-S7 with yellow markings. Male anal appendages of this species is similar to *A. hansoni*, but legs without yellow markings and abdominal markings are different. **[Measurements]** Male total length 58-64 mm, abdomen 45-47 mm, hind wing 36-38 mm. **[Habitat]** Montane streams and ditches below 500 m elevation. **[Distribution]** Guangxi. **[Flight Season]** April to July.

安春蜓属待定种 雄，广西
Amphigomphus sp., male from Guangxi

异春蜓属 Genus *Anisogomphus* Selys, 1858

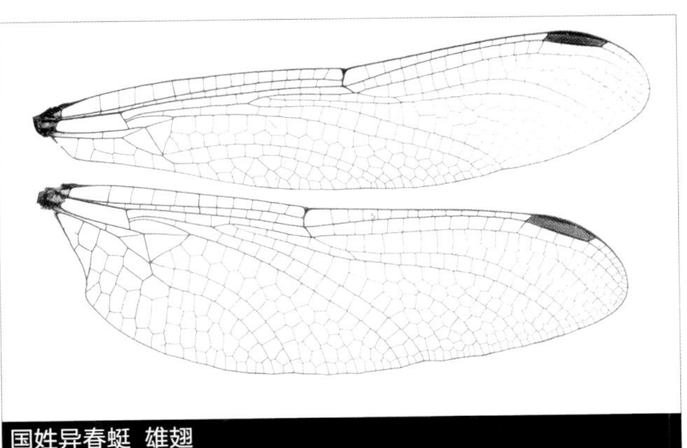
国姓异春蜓 雄翅
Anisogomphus koxingai, male wings

本属全球约有20种，主要分布于亚洲的亚热带地区。中国已知约15种，但一些种类已经被移入新建立的筒尾春蜓属*Euthygomphus* Kosterin, 2016。本属蜻蜓体中型；上颚外方白色，额横纹黄白色，甚阔；翅透明，翅痣具支持脉，三角室、上三角室和下三角室通常无横脉，前翅盘区具2列翅室，基臀区具1条横脉，后翅基缘未呈角状，臀三角室的基边甚斜，通常3～4室；足长，后足腿节伸达腹部第2节中央或更长；雄性第7～9节膨大，下肛附器中央凹陷甚阔，两枝分歧的角度较大；前钩片末端钩曲，后钩片叶片状，末端向前方突起，阳茎无鞭。本属雄性可以通过肛附器、后钩片和阳茎的构造区分。

本属蜻蜓栖息于山区的开阔溪流和大型河流。雄性常见于停落在水面的岩石上或者植物上。

The genus contains about 20 species mostly distributed in subtropical Asia. About 15 species are recorded from China, but some of our species have recently been placed to a new genus, *Euthygomphus* Kosterin, 2016. Species of the genus are medium-sized. Base of mandibles white, frons with a broad yellowish white stripe. Wings hyaline, pterostigma well braced, no crossvein in triangle, hypertriangle and subtriangle, discoidal field with 2 rows of cells in fore wings, cubital space with one crossvein, tornus not angled, anal triangle 3- or 4-celled with basal side oblique. Legs long, hind legs reaching mid S2 or even longer. Male S7-S9 expanded, inferior appendage with a broad median hollow and forming two divergent branches. Anterior hamulus curved distally, posteriors leaf-shaped with tips protruding, penis without flagella. Males of the genus can be separated by structure of anal appendages, posterior hamulus and penis.

Anisogomphus species inhabit exposed montane streams and rivers. Males usually perch on rocks or plants above water.

安氏异春蜓 雄
Anisogomphus anderi, male

安氏异春蜓
Anisogomphus anderi

镰尾异春蜓
Anisogomphus caudalis

高山异春蜓
Anisogomphus nitidus

异春蜓属 雄性肛附器
Genus *Anisogomphus*, male anal appenages

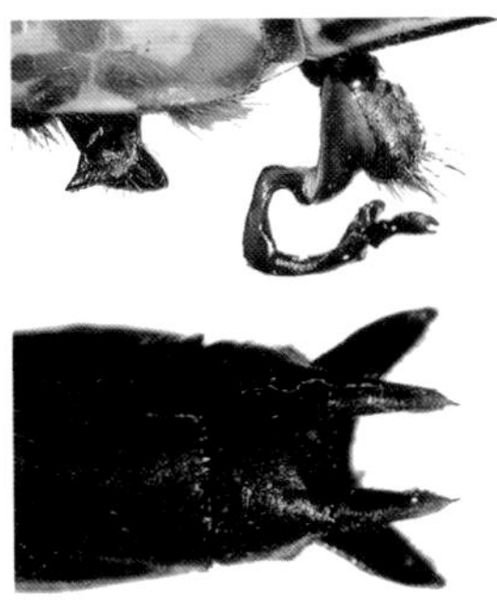

双条异春蜓
Anisogomphus bivittatus

福氏异春蜓
Anisogomphus forresti

马奇异春蜓
Anisogomphus maacki

异春蜓属 雄性肛附器和次生殖器
Genus *Anisogomphus*, male appendages and secondary genitalia

国姓异春蜓
Anisogomphus koxingai

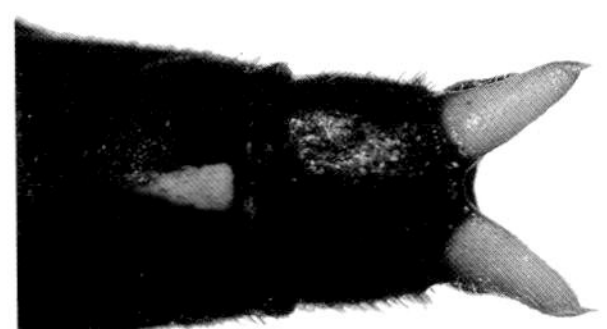

云南异春蜓
Anisogomphus yunnanensis

异春蜓属 雄性肛附器和阳茎
Genus *Anisogomphus*, male appendages and penis

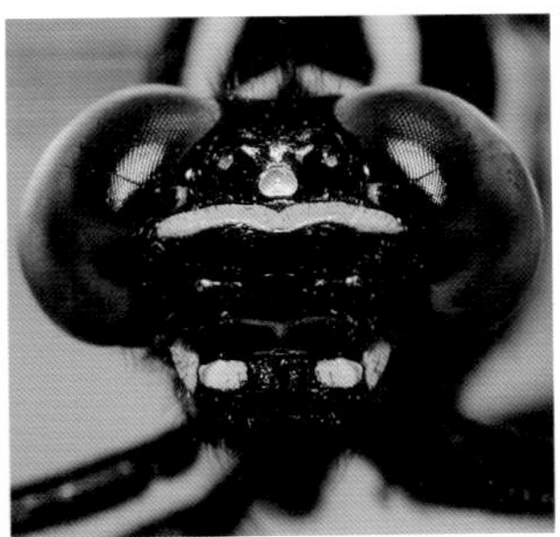
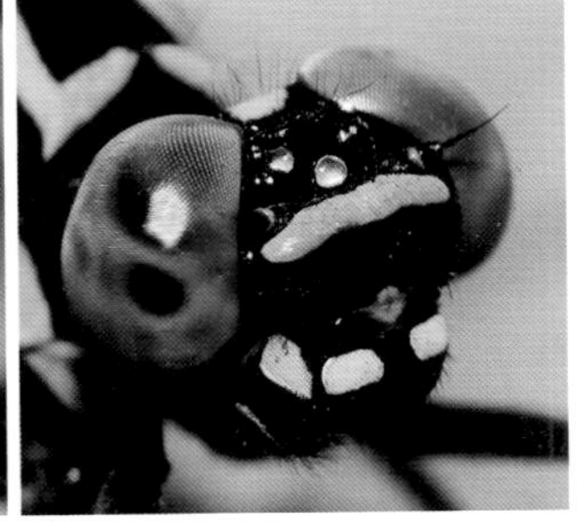

双条异春蜓
Anisogomphus bivittatus

镰尾异春蜓
Anisogomphus caudalis

国姓异春蜓
Anisogomphus koxingai

云南异春蜓
Anisogomphus yunnanensis

异春蜓属 雄性面部
Genus *Anisogomphus*, male face

安氏异春蜓 *Anisogomphus anderi* Lieftinck, 1948

【形态特征】雄性上唇具1对白色斑点；胸部黑色，背条纹与领条纹和肩前上点相连，形成“Z”形条纹，具细而短的肩前下条纹，合胸侧面具3条黄色条纹；腹部黑色具黄色斑纹，上肛附器上面白色，下端黑色，下肛附器黑色。雌性与雄性相似，尾毛白色。**【长度】**体长 52~55 mm，腹长 38~40 mm，后翅 33~38 mm。**【栖息环境】**海拔 1500 m以下森林中的开阔溪流。**【分布】**中国特有，分布于云南、贵州、四川、湖北、湖南、江西、浙江、福建、广西、广东。**【飞行期】**4—9月。

[Identification] Male labrum with a pair of white spots. Thorax black, collar stripes connecting with dorsal stipes and superior spots, forming Z-shaped markings, lower antehumeral stripes short and narrow, sides with three yellow stripes. Abdomen black with yellow markings, superior appendages white above and black below, inferior appendage black. Female similar to male with white cerci. **[Measurements]** Total length 52-55 mm, abdomen 38-40 mm, hind wing 33-38 mm. **[Habitat]** Exposed streams in forest below 1500 m elevation. **[Distribution]** Endemic to China, recorded from Yunnan, Guizhou, Sichuan,Hubei, Hunan, Jiangxi, Zhejiang, Fujian, Guangxi, Guangdong. **[Flight Season]** April to September.

安氏异春蜓 雄，贵州
Anisogomphus anderi, male from Guizhou

安氏异春蜓 雌，贵州
Anisogomphus anderi, female from Guizhou

双条异春蜓 *Anisogomphus bivittatus* (Selys, 1854)

双条异春蜓 雄，云南（普洱）
Anisogomphus bivittatus, male from Yunnan (Pu'er)

【形态特征】雄性上唇黄色；胸部黑色，背条纹与领条纹相连，形成"7"形条纹，肩前条纹细而长，合胸侧面具3条黄色条纹；腹部黑色具黄色斑纹，肛附器黑色。雌性与雄性相似。【长度】体长 49～53 mm，腹长 37～39 mm，后翅 30～35 mm。【栖息环境】海拔 500～1000 m森林中的开阔溪流。【分布】云南（普洱）；不丹、印度、尼泊尔。【飞行期】5—7月。

[Identification] Male labrum yellow. Thorax black, collar stripes connecting with dorsal stripes to form 7-shaped markings, antehumeral stripes long, sides with three yellow stripes. Abdomen black with yellow markings, anal appendages black. Female similar to male. [Measurements] Total length 49-53 mm, abdomen 37-39 mm, hind wing 30-35 mm. [Habitat] Exposed streams in forest at 500-1000 m elevation. [Distribution] Yunnan (Pu'er); Bhutan, India, Nepal. [Flight Season] May to July.

镰尾异春蜓 *Anisogomphus caudalis* Fraser, 1926

镰尾异春蜓 雌，云南（德宏）
Anisogomphus caudalis, female from Yunnan (Dehong)

【形态特征】雄性上唇黑色具1对白色斑点；胸部黑色，背条纹与领条纹相连，肩前上点三角形，肩前下条纹细而短，合胸侧面具3条黄色条纹；腹部黑色具黄斑，上肛附器白色，较短小，下肛附器黑色。雌性与雄性相似。【长度】雄性体长 49 mm，腹长 38 mm，后翅 30 mm。【栖息环境】海拔 500～1000 m森林中的狭窄小溪。【分布】云南（德宏）；印度、缅甸。【飞行期】5—7月。

[Identification] Male labrum black with a pair of white spots. Thorax black, collar stripes connecting with dorsal stipes, superior spots triangular, lower antehumeral stripes narrow and short, sides with three yellow stripes. Abdomen black with yellow markings, superior appendages white and short, inferior appendage black. Female similar to male. **[Measurements]** Male total length 49 mm, abdomen 38 mm, hind wing 30 mm. **[Habitat]** Narrow streams in forest at 500-1000 m elevation. **[Distribution]** Yunnan (Dehong); India, Bruma. **[Flight Season]** May to July.

镰尾异春蜓 雄，云南（德宏）
Anisogomphus caudalis, male from Yunnan (Dehong)

黄脸异春蜓 *Anisogomphus flavifacies* Klots, 1947

【形态特征】雄性复眼灰绿色；胸部黑色，背条纹与领条纹相连，形成“7”形条纹，合胸侧面大面积黄色；腹部黑色具黄色斑纹，肛附器黑色。【长度】腹长 37~42 mm，后翅 33~36 mm。【栖息环境】海拔 1500~2000 m的山区溪流。【分布】中国云南（昆明）特有。【飞行期】5—7月。

[Identification] Male eyes greyish green. Thorax black, collar stripes connecting with dorsal stripes to form 7-shaped markings, sides mostly yellow. Abdomen black with yellow markings, anal appendages black. **[Measurements]** Abdomen 37-42 mm, hind wing 33-36 mm. **[Habitat]** Montane streams at 1500-2000 m elevation. **[Distribution]** Endemic to Yunnan (Kunming) of China. **[Flight Season]** May to July.

黄脸异春蜓 雄，云南（昆明）| 陈尽 摄
Anisogomphus flavifacies, male from Yunnan (Kunming) | Photo by Jin Chen

福氏异春蜓 *Anisogomphus forresti* (Morton, 1928)

【形态特征】雄性上唇大面积黄色；胸部黑色，背条纹与领条纹相连，形成“7”形条纹，肩前条纹较长，合胸侧面具3条黄色条纹；腹部黑色具黄色斑纹，肛附器黑色。雌性与雄性相似，腹部的黄色条纹更发达。【长度】体长 51~56 mm，腹长 38~41 mm，后翅 31~38 mm。【栖息环境】海拔 1500~2500 m森林中的开阔溪流。【分布】中国云南（大理、德宏、红河、文山）特有。【飞行期】5—7月。

[Identification] Male labrum largely yellow. Thorax black, collar stripes connecting with dorsal stripes to form 7-shaped markings, antehumeral stripes long, sides with three yellow stripes. Abdomen black with yellow markings, anal appendages black. Female similar to male, abdomen with broader yellow markings. **[Measurements]** Total length 51-56 mm, abdomen 38-41 mm, hind wing 31-38 mm. **[Habitat]** Exposed streams in forest at 1500-2500 m elevation. **[Distribution]** Endemic to Yunnan (Dali, Dehong, Honghe, Wenshan) of China. **[Flight Season]** May to July.

福氏异春蜓 雄，云南（大理）
Anisogomphus forresti, male from Yunnan (Dali)

福氏异春蜓 雌，云南（大理）
Anisogomphus forresti, female from Yunnan (Dali)

国姓异春蜓 *Anisogomphus koxingai* Chao, 1954

国姓异春蜓 雄，广西
Anisogomphus koxingai, male from Guangxi

【形态特征】雄性上唇黑色具1对黄斑；胸部黑色，背条纹与领条纹不相连，肩前上点三角形，肩前下条纹细而短，合胸侧面具3条黄色条纹；腹部黑色具黄色斑纹，第7节基方具1个甚大黄斑，上肛附器白色，下肛附器黑色。雌性与雄性相似，后头缘两侧各具1个甚小的刺突；腹部的黄色条纹更发达，尾毛白色。【长度】体长 46~57 mm，腹长 36~43 mm，后翅 28~33 mm。【栖息环境】海拔 1000 m以下的开阔溪流和河流。【分布】云南、贵州、福建、广西、广东、海南、香港、台湾；越南。【飞行期】3—8月。

[Identification] Male labrum black with a pair of yellow spots. Thorax black, collar stripes not connecting with dorsal stripes, superior spots triangular, antehumeral stripes narrow and short, sides with three yellow stripes. Abdomen black with yellow markings, S7 with a large basal spot, superior appendages white, inferior appendage black. Female similar to male, occipital margin with a small horn. Abdomen with broader yellow markings, cerci white. **[Measurements]** Total length 46-57 mm, abdomen 36-43 mm, hind wing 28-33 mm. **[Habitat]** Exposed streams and rivers below 1000 m elevation. **[Distribution]** Yunnan, Guizhou, Fujian, Guangxi, Guangdong, Hainan, Hong Kong, Taiwan; Vietnam. **[Flight Season]** March to August.

国姓异春蜓 雌，广东 | 宋睿斌 摄
Anisogomphus koxingai, female from Guangdong | Photo by Ruibin Song

国姓异春蜓 雄，广西
Anisogomphus koxingai, male from Guangxi

马奇异春蜓 *Anisogomphus maacki* (Selys, 1872)

【形态特征】雄性上唇主要黄色；胸部黑色，背条纹与领条纹相连，形成"7"形条纹，肩前条纹细而长，合胸侧面第2条纹中央间断，第3条纹完整；腹部黑色具黄色斑纹，肛附器黑色。雌性与雄性相似，腹部的黄色条纹更发达。【长度】体长 49～54 mm，腹长 36～39 mm，后翅 30～34 mm。【栖息环境】海拔 1500 m以下的开阔溪流和河流。【分布】中国东北至华中和西南地区；朝鲜半岛、日本、俄罗斯远东、越南。【飞行期】6—9月。

[Identification] Male labrum largely yellow. Thorax black, collar stripes connecting with dorsal stripes to form 7-shaped markings, antehumeral stripes long, second lateral stripe interrupted medially and third lateral stripe complete.

马奇异春蜓 雄，北京
Anisogomphus maacki, male from Beijing

Abdomen black with yellow markings, anal appendages black. Female similar to male, abdomen with more extensive yellow markings. **[Measurements]** Total length 49-54 mm, abdomen 36-39 mm, hind wing 30-34 mm. **[Habitat]** Exposed streams and rivers below 1500 m elevation. **[Distribution]** From the Northeast to the Central and Southwest China; Korean peninsula, Japan, Russian Far East, Vietnam. **[Flight Season]** June to September.

马奇异春蜓 雄，北京
Anisogomphus maacki, male from Beijing

马奇异春蜓 雌，北京
Anisogomphus maacki, female from Beijing

高山异春蜓 *Anisogomphus nitidus* Yang & Davies, 1993

【形态特征】雄性上唇具1对黄色斑点；胸部黑色，背条纹与领条纹和肩前条纹相连，合胸侧面具3条黄色条纹；腹部黑色具黄色斑纹，上肛附器白色，下肛附器黑色。【长度】雄性体长 47 mm，腹长 35 mm，后翅 31 mm。【栖息环境】海拔 1500～2500 m森林中的开阔溪流。【分布】中国云南（大理）特有。【飞行期】7—9月。

[Identification] Male labrum with a pair of yellow spots. Thorax black, collar stripes connecting with dorsal stripes and antehumeral stripes, sides with three yellow stripes. Abdomen black with yellow markings, superior appendages white, inferior appendage black. **[Measurements]** Male total length 47 mm, abdomen 35 mm, hind wing 31 mm. **[Habitat]** Exposed streams in forest at 1500-2500 m elevation. **[Distribution]** Endemic to Yunnan (Dali) of China. **[Flight Season]** July to September.

高山异春蜓 雄，云南（大理）
Anisogomphus nitidus, male from Yunnan (Dali)

云南异春蜓 *Anisogomphus yunnanensis* Zhou & Wu, 1992

【形态特征】雄性上唇黑色具1对黄斑；胸部黑色，背条纹与领条纹相连，肩前上点较小，有时具较细的肩前条纹，合胸侧面具3条黄色条纹；腹部黑色具黄色斑纹，第7节基方具1个甚大黄斑，上肛附器白色，下肛附器黑色。雌性与雄性相似，腹部的黄色条纹更发达，尾毛白色。【长度】体长 45~46 mm，腹长 35~36 mm，后翅 26~29 mm。【栖息环境】海拔 1000 m以下的开阔溪流和河流。【分布】云南（普洱、西双版纳）；柬埔寨。【飞行期】4—6月。

[Identification] Male labrum black with a pair of yellow spots. Thorax black, collar stripes connecting with dorsal stripes, superior spots small, antehumeral stripes sometimes present, sides with three yellow stripes. Abdomen black with yellow markings, S7 with a large basal spot, superior appendages white, inferior appendage black. Female similar to male, abdomen with broader yellow markings, cerci white. **[Measurements]** Total length 45-46 mm, abdomen 35-36 mm, hind wing 26-29 mm. **[Habitat]** Exposed streams and rivers below 1000 m elevation. **[Distribution]** Yunnan (Pu'er, Xishuangbanna); Cambodia. **[Flight Season]** April to June.

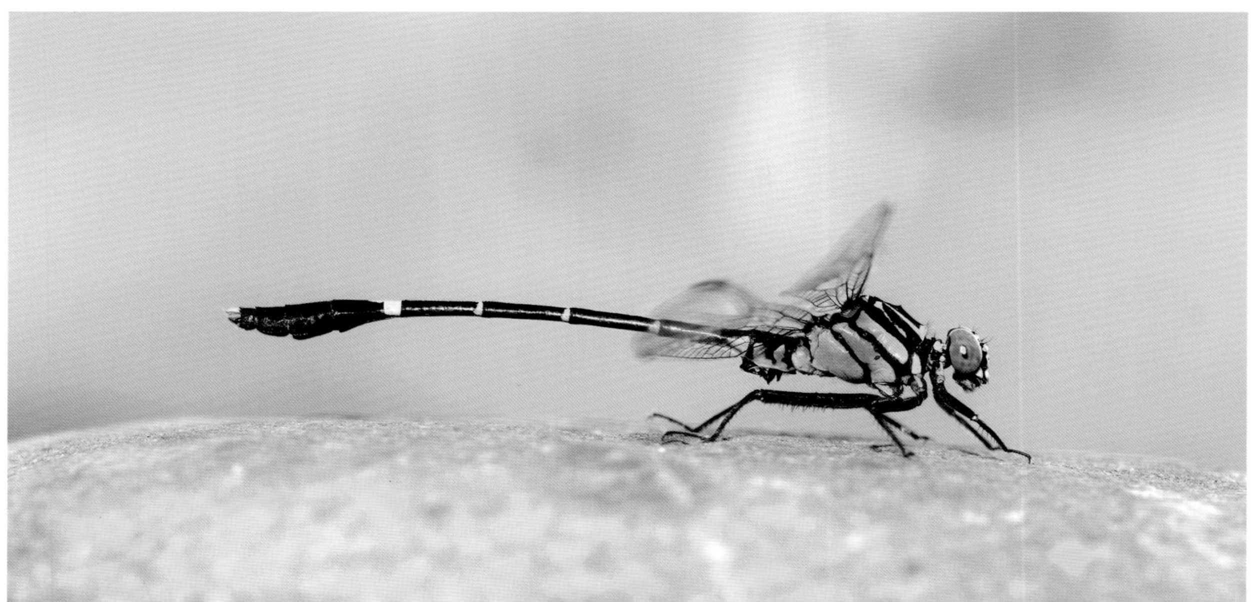

云南异春蜓 雄，云南（西双版纳）
Anisogomphus yunnanensis, male from Yunnan (Xishuangbanna)

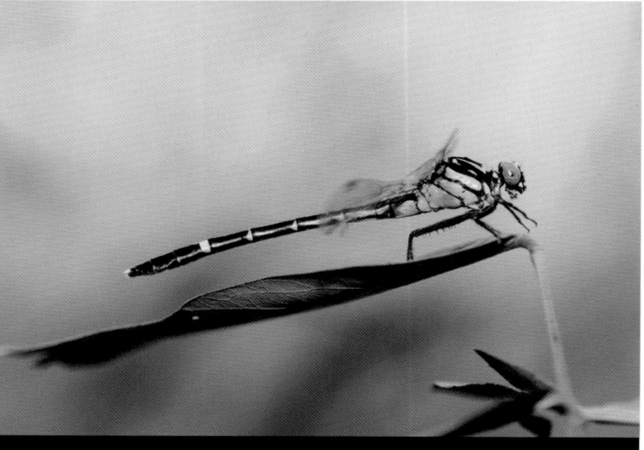

云南异春蜓 雌，云南（西双版纳）
Anisogomphus yunnanensis, female from Yunnan (Xishuangbanna)

异春蜓属待定种 *Anisogomphus* sp.

异春蜓属待定种 雄，云南（普洱）
Anisogomphus sp., male from Yunnan (Pu'er)

【形态特征】雄性上唇黑色具1对黄色斑点，后头黄色；胸部黑色，背条纹与领条纹几乎相连，肩前条纹甚长，合胸侧面具3条黄色条纹；腹部黑色具黄色斑纹，上肛附器白色，较短小，下肛附器黑色。雌性与雄性相似。【长度】体长 56～59 mm，腹长 41～44 mm，后翅 33～38 mm。【栖息环境】海拔 1000～1500 m森林中的开阔小溪。【分布】云南（普洱）。【飞行期】5—7月。

[Identification] Male labrum black with a pair of yellow spots, occiput yellow. Thorax black, collar stripes almost connecting with dorsal stripes, antehumeral stripes long, sides with three yellow stripes. Abdomen black with yellow markings, superior appendages white and short, inferior appendage black. Female similar to male. **[Measurements]** Total length 56-59 mm, abdomen 41-44 mm, hind wing 33-38 mm. **[Habitat]** Exposed streams in forest at 1000-1500 m elevation. **[Distribution]** Yunnan (Pu'er). **[Flight Season]** May to July.

亚春蜓属 Genus *Asiagomphus* Asahina, 1985

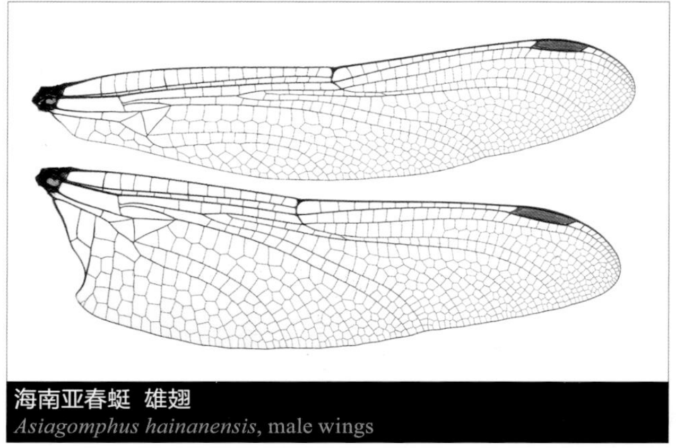
海南亚春蜓 雄翅
Asiagomphus hainanensis, male wings

本属全球已知约25种，分布于亚洲。中国已知10余种。本属蜻蜓体中型；翅透明，翅痣具支持脉，三角室、上三角室和下三角室通常无横脉，前翅盘区具2列翅室，基臀区具1条横脉，后翅基缘略呈角状，臀三角室的基边甚斜，通常3室。雄性上肛附器的两枝分歧，无齿，下肛附器的两枝与上肛附器近等长，分歧的角度也几乎一致。前钩片基半部的宽度为端半部宽度的2倍，后钩片末端钩曲，阳茎无鞭。目前本属的分类学尚不完善，许多种类变异较大。身体色彩、头部构造、后钩片及下生殖板形状是较重要的区分特征。

本属蜻蜓栖息于山区溪流，有些喜欢较阴暗的环境，有些喜欢阳光充足的开阔溪流。雄性领域行为显著，通常停在溪流边缘的地面上或者岩石上，时而悬停飞行，飞行时后足伸展。

The genus contains about 25 species distributed in Asia. Over ten are recorded from China. Species of the genus are medium-sized dragonflies. Wings hyaline, pterostigma braced, no crossvein in triangle, hypertriangle and subtriangle, discoidal field with 2 rows of cells in fore wings, cubital space with one crossvein, tornus slightly angled, anal triangle usually 3-celled with basal side oblique. Male superior appendages divergent without teeth, the inferior almost same length as superiors. Anterior hamulus with basal half broad, twice width of the distal half, posterior hamulus with the tip curved, penis without flagella. Taxonomic study for the genus is inadequate and many species are variable. Body markings and the structure of head, posterior hamulus and female vulvar lamina are useful diagnostic characters.

Asiagomphus species inhabit montane streams, some species prefer the shade, and others prefer to perch in the sun. Males exhibit territorial behavior by alighting on the rocks or ground, sometimes they hover with the hind legs extended.

海南亚春蜓 雄
Asiagomphus hainanensis, male

黄基亚春蜓
Asiagomphus acco

海南亚春蜓
Asiagomphus hainanensis

和平亚春蜓
Asiagomphus pacificus

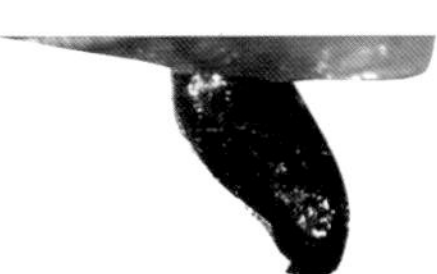
三角亚春蜓
Asiagomphus perlaetus

凹缘亚春蜓
Asiagomphus septimus

凸缘亚春蜓
Asiagomphus xanthenatus

亚春蜓属 雄性后钩片
Genus *Asiagomphus*, male posterior hamulus

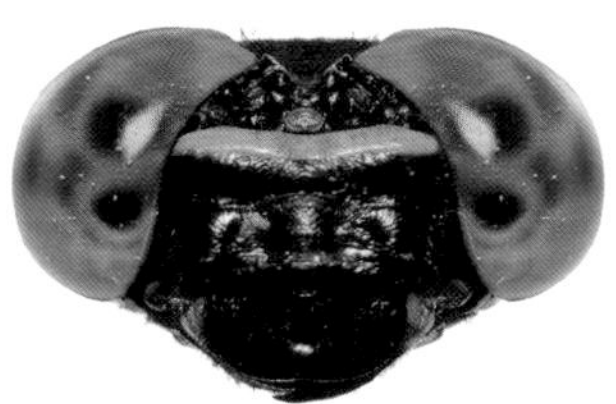
海南亚春蜓
Asiagomphus hainanensis

和平亚春蜓
Asiagomphus pacificus

凹缘亚春蜓
Asiagomphus septimus

凸缘亚春蜓
Asiagomphus xanthenatus

三角亚春蜓
Asiagomphus perlaetus

西南亚春蜓
Asiagomphus hesperius

异春蜓属 雌性头部正面观
Genus *Asiagomphus*, female head in frontal view

黄基亚春蜓 *Asiagomphus acco* Asahina, 1996

【形态特征】雄性面部大面积黑色，额横纹黄色；胸部黑色，背条纹与领条纹相连，无肩前条纹，合胸侧面第2条纹和第3条纹合并；腹部黑色，第1～2节大面积黄色，第9节背面后方具1个大黄斑，肛附器黑色。雌性与雄性相似，但第4～7节具黄斑。【长度】体长 60～62 mm，腹长 45～46 mm，后翅 39～40 mm。【栖息环境】海拔 500 m以下的开阔溪流。【分布】云南（红河）、广西；老挝、越南。【飞行期】4—6月。

[Identification] Male face largely black, frons with a yellow stripe. Thorax black, collar stripes connecting with dorsal stripes, second and third lateral stripes combined. Abdomen black, S1-S2 largely yellow, S9 with a large yellow dorsal spot posteriorly, anal appendages black. Female similar to male but S4-S7 with yellow spots. **[Measurements]** Total length 60-62 mm, abdomen 45-46 mm, hind wing 39-40 mm. **[Habitat]** Exposed montane streams below 500 m elevation. **[Distribution]** Yunnan (Honghe), Guangxi; Laos, Vietnam. **[Flight Season]** April to June.

黄基亚春蜓 雄，云南（红河）
Asiagomphus acco, male from Yunnan (Honghe)

黄基亚春蜓 雌，云南（红河）
Asiagomphus acco, female from Yunnan (Honghe)

黄基亚春蜓 雄，广西
Asiagomphus acco, male from Guangxi

金斑亚春蜓 *Asiagomphus auricolor* (Fraser, 1926)

【形态特征】雄性面部大面积黑色，额横纹黄色；胸部黑色，背条纹与领条纹相连，具甚小的肩前上点，合胸侧面第2条纹和第3条纹完整，有时合并；腹部黑色，第1~2节背面和侧面具黄斑，第3~7节基方具黄环，第9节背面后方具1个大黄斑，肛附器黑色。雌性后头缘中央具1个甚小突起，侧单眼后方具2对角状突起。【长度】体长 60~65 mm，腹长 45~47 mm，后翅 38~44 mm。【栖息环境】海拔 500 m以下的开阔溪流和河流。【分布】云南（红河）；越南。【飞行期】4—6月。

[Identification] Male face largely black, frons with a yellow stripe. Thorax black, collar stripes connecting with dorsal stripes, superior spots small, second and third lateral stripes complete or combined. Abdomen black, S1-S2 with dorsal and lateral yellow markings, S3-S7 with yellow rings basally, S9 with a large yellow dorsal spot posteriorly, anal appendages black. Female occipital margin with a small prominence in the middle, vertex with two pairs of horns behind lateral ocelli. **[Measurements]** Total length 60-65 mm, abdomen 45-47 mm, hind wing 38-44 mm. **[Habitat]** Exposed streams and rivers below 500 m elevation. **[Distribution]** Yunnan (Honghe); Vietnam. **[Flight Season]** April to June.

金斑亚春蜓 雄，云南（红河）
Asiagomphus auricolor, male from Yunnan (Honghe)

海南亚春蜓 *Asiagomphus hainanensis* (Chao, 1953)

海南亚春蜓 雌，广西
Asiagomphus hainanensi, female from Guangxi

海南亚春蜓 雄，广西
Asiagomphus hainanensi, male from Guangxi

【形态特征】雄性上唇黑色，有时具1对黄斑；胸部黑色，背条纹与领条纹相连，肩前条纹甚细，有时仅有肩前上点，合胸侧面第2条纹和第3条纹完整；腹部黑色，第1～8节背面和侧面具黄斑，第9节背面后方具1个大黄斑。雌性与雄性相似，但更粗壮。**【长度】**体长 61～71 mm，腹长 45～53 mm，后翅 39～46 mm。**【栖息环境】**海拔 1000 m以下的开阔溪流。**【分布】**湖南、江西、浙江、福建、海南、广东、香港、台湾；越南。**【飞行期】**3—7月。

[Identification] Male labrum black sometimes with a pair of yellow spots. Thorax black, collar stripes connecting with dorsal stripes, antehumeral stripes narrow, sometimes only a small superior spot present, second and third lateral stripes complete. Abdomen black, S1-S8 with dorsal and lateral yellow markings, S9 with a large yellow dorsal spot posteriorly. Female similar to male but stouter. **[Measurements]** Total length 61-71 mm, abdomen 45-53 mm, hind wing 39-46 mm. **[Habitat]** Exposed montane streams below 1000 m elevation. **[Distribution]** Hunan, Jiangxi, Zhejiang, Fujian, Hainan, Guangdong, Hong Kong, Taiwan; Vietnam. **[Flight Season]** March to July.

西南亚春蜓 *Asiagomphus hesperius* (Chao, 1953)

西南亚春蜓 雌，贵州
Asiagomphus hesperius, female from Guizhou

【形态特征】 雌性上唇具1对黄斑，额横纹黄色，后头缘具3个角状突起；胸部黑色，背条纹与领条纹相连，肩前条纹较长，合胸侧面第2条纹中央大面积间断，第3条纹完整；腹部黑色，第1~9节背面和侧面具黄斑，其中第7节前方黄斑甚大。雄性与雌性相似。**【长度】** 体长 63~69 mm，腹长 42~46 mm，后翅 35~36 mm。**【栖息环境】** 海拔1500 m以下的开阔溪流。**【分布】** 中国特有，分布于陕西、四川、贵州。**【飞行期】** 4—7月。

[Identification] Female labrum with a pair of yellow spots, frons with a yellow stripe, occipital margin with three horns. Thorax black, collar stripes connecting with dorsal stripes, antehumeral stripes long, second lateral stripe largely interrupted medially, third lateral stripes complete. Abdomen black, S1-S9 with lateral and dorsal stripes, S7 with a large yellow spot anteriorly. Male similar to female. **[Measurements]** Total length 63-69 mm, abdomen 42-46 mm, hind wing 35-36 mm. **[Habitat]** Exposed montane streams below 1500 m elevation. **[Distribution]** Endemic to China, recorded from Shaanxi, Sichuan, Guizhou. **[Flight Season]** April to July.

和平亚春蜓 *Asiagomphus pacificus* (Chao, 1953)

【形态特征】雄性面部大面积黑色，额横纹黄色；胸部黑色，背条纹与领条纹相连，肩前条纹细而长，合胸侧面第2条纹和第3条纹完整；腹部黑色，第1～8节背面和侧面具黄斑，第9节背面后方具1个大黄斑，肛附器黑色。雌性与雄性相似，侧单眼后方具1对角状突起。【长度】体长 63～65 mm，腹长 47～49 mm，后翅 40～42 mm。【栖息环境】海拔 1500 m以下的开阔溪流。【分布】中国特有，分布于贵州、河南、湖北、湖南、浙江、福建、广西、广东、台湾。【飞行期】4—8月。

[Identification] Male face largely black, frons with a yellow stripe. Thorax black, collar stripes connecting with dorsal stripes, antehumeral stripes long and narrow, second and third lateral stripes complete. Abdomen black, S1-S8 with dorsal and lateral yellow markings, S9 with a large yellow dorsal spot posteriorly, anal appendages black. Female similar to male, head with a pair of horns behind the lateral ocelli. **[Measurements]** Total length 63-65 mm, abdomen 47-49 mm, hind wing 40-42 mm. **[Habitat]** Exposed montane streams below 1500 m elevation. **[Distribution]** Endemic to China, recorded from Guizhou, Henan, Hubei, Hunan, Zhejiang, Fujian, Guangxi, Guangdong, Taiwan. **[Flight Season]** April to August.

和平亚春蜓 雌，浙江
Asiagomphus pacificus, female from Zhejiang

和平亚春蜓 雌，广东 | 宋睿斌 摄
Asiagomphus pacificus, female from Guangdong | Photo by Ruibin Song

和平亚春蜓 雄，广西
Asiagomphus pacificus, male from Guangxi

三角亚春蜓 *Asiagomphus perlaetus* (Chao, 1953)

【形态特征】雄性面部大面积黑色，额横纹黄色；胸部黑色，背条纹与领条纹相连，肩前条纹细长，合胸侧面第2条纹中央间断，有时完整，第3条纹完整；腹部黑色，第1~7节背面和侧面具黄斑，其中第7节基方黄斑甚大，第9节背面后方具1个大黄斑，肛附器黑色。雌性与雄性相似，后头缘中央具3个角状突起。【长度】体长 60~62 mm，腹长 44~45 mm，后翅 37~40 mm。【栖息环境】海拔 500 m以下的开阔溪流和河流。【分布】中国特有，分布于贵州、湖北、浙江、福建、广东、台湾。【飞行期】4—7月。

三角亚春蜓 雄，广东 | 吴宏道 摄
Asiagomphus perlaetus, male from Guangdong | Photo by Hongdao Wu

三角亚春蜓 雄，湖北
Asiagomphus perlaetus, male from Hubei

三角亚春蜓 雌，广东 | 宋睿斌 摄
Asiagomphus perlaetus, female from Guangdong | Photo by Ruibin Song

[Identification] Male face largely black, frons with a yellow stripe. Thorax black, collar stripes connecting with dorsal stripes, antehumeral stripes long and narrow, second lateral stripe interrupted medially or complete, third lateral stripes complete. Abdomen black, S1-S7 with dorsal and lateral yellow markings, S7 with a large yellow spot basally, S9 with a large yellow dorsal spot posteriorly, anal appendages black. Female similar to male, head with three horns in the middle of occipital margin. **[Measurements]** Total length 60-62 mm, abdomen 44-45 mm, hind wing 37-40 mm. **[Habitat]** Exposed montane streams and rivers below 500 m elevation. **[Distribution]** Endemic to China, recorded from Guizhou, Hubei, Zhejiang, Fujian, Guangdong, Taiwan. **[Flight Season]** April to July.

面具亚春蜓 *Asiagomphus personatus* (Selys, 1873)

【形态特征】雄性上唇具1对黄斑，额横纹甚阔，后头黄色；胸部黑色，背条纹与领条纹相连，肩前条纹细而长，合胸侧面第2条纹和第3条纹完整；腹部黑色，各节具黄斑，肛附器黑色。雌性与雄性相似，侧单眼后方具1对角状突起。**【长度】**体长 60～63 mm，腹长 42～47 mm，后翅 35～40 mm。**【栖息环境】**海拔 1500 m以下的开阔溪流。**【分布】**云南（西双版纳、普洱）；印度、缅甸。**【飞行期】**4—7月。

[Identification] Male labrum with a pair of yellow spots, frons with a broad yellow stripe, occiput yellow. Thorax black, collar stripes connecting with dorsal stripes, antehumeral stripes long and narrow, second and third lateral stripes

面具亚春蜓 雄，云南（普洱）
Asiagomphus personatus, male from Yunnan (Pu'er)

complete. Abdomen black with yellow markings on all segments, anal appendages black. Female similar to male, head with a pair of horns behind lateral ocelli. **[Measurements]** Total length 60-63 mm, abdomen 42-47 mm, hind wing 35-40 mm. **[Habitat]** Exposed montane streams below 1500 m elevation. **[Distribution]** Yunnan (Xishuangbanna, Pu'er); India, Myanmar. **[Flight Season]** April to July.

面具亚春蜓 雄，云南（普洱）
Asiagomphus personatus, male from Yunnan (Pu'er)

凹缘亚春蜓 *Asiagomphus septimus* (Needham, 1930)

【形态特征】雄性面部大面积黑色，额横纹黄色；胸部黑色，背条纹与领条纹相连，肩前条纹较短，合胸侧面第2条纹和第3条纹完整；腹部黑色，第1~7节背面和侧面具黄斑，第8节侧缘下方有时具黄斑，第9节后方具1个大黄斑，肛附器黑色。雌性与雄性相似，后头缘中央稍微凹陷；下生殖板向下伸出。【长度】体长 62~70 mm，腹长 47~52 mm，后翅 39~45 mm。【栖息环境】海拔 1000 m以下的林荫小溪和渗流地。【分布】中国特有，分布于江西、福建、广东、香港、台湾。【飞行期】4—7月。

[Identification] Male face largely black, frons with a yellow stripe. Thorax black, collar stripes connecting with dorsal stripes, antehumeral stripes short, second and third lateral stripes complete. Abdomen black, S1-S7 with dorsal and lateral yellow markings, S8 with lateral yellow stripes sometimes, S9 with a large yellow dorsal spot posteriorly, anal appendages black. Female similar to male, occipital margin slightly depressed medially. Vulvar lamina projecting. **[Measurements]** Total length 62-70 mm, abdomen 47-52 mm, hind wing 39-45 mm. **[Habitat]** Shady streams and seepages below 1000 m elevation. **[Distribution]** Endemic to China, recorded from Jiangxi, Fujian, Guangdong, Hong Kong, Taiwan. **[Flight Season]** April to July.

凹缘亚春蜓 雌，香港 | 梁嘉景 摄
Asiagomphus septimus, female from Hong Kong | Photo by Kenneth Leung

凹缘亚春蜓 雄，香港 | 祁麟峰 摄
Asiagomphus septimus, male from Hong Kong | Photo by Mahler Ka

凹缘亚春蜓 雄，香港 | 祁麟峰 摄
Asiagomphus septimus, male from Hong Kong | Photo by Mahler Ka

凸缘亚春蜓 *Asiagomphus xanthenatus* (Williamson, 1907)

【形态特征】雄性上唇具1对黄斑，额横纹黄色，后头大面积黄色；胸部黑色，背条纹与领条纹有时相连，有时相距较近，有时具较短而细的肩前条纹，合胸侧面第2条纹和第3条纹完整或合并；腹部黑色，第1～7节背面和侧面具黄斑，第8节侧后方有时具黄斑，第9节背面后方具1个大黄斑，肛附器黑色。雌性后头缘中央凸起，具黄斑，侧单眼后方具1对角状突起。【长度】体长 58～63 mm，腹长 44～47 mm，后翅 36～40 mm。【栖息环境】海拔 1000 m以下森林中的溪流。【分布】云南（普洱、临沧）；缅甸、泰国、老挝。【飞行期】4—6月。

[Identification] Male labrum with a pair of yellow spots, frons with a yellow stripe, occiput largely yellow. Thorax black, collar stripes connecting with dorsal stripes or slightly separated, antehumeral stripes short and narrow, second and third lateral stripes complete or combined. Abdomen black, S1-S7 with dorsal and lateral yellow markings, S8

凸缘亚春蜓 雄，云南（普洱）
Asiagomphus xanthenatus, male from Yunnan (Pu'er)

with lateral yellow stripes sometimes, S9 with a large yellow dorsal spot posteriorly, anal appendages black. Female occipital margin erect medially with yellow spot, vertex with a pair of horns behind lateral ocelli. **[Measurements]** Total length 58-63 mm, abdomen 44-47 mm, hind wing 36-40 mm. **[Habitat]** Streams in forest below 1000 m elevation. **[Distribution]** Yunnan (Pu'er, Lincang); Myanmar, Thailand, Laos. **[Flight Season]** April to June.

凸缘亚春蜓 雄，云南（普洱）
Asiagomphus xanthenatus, male from Yunnan (Pu'er)

凸缘亚春蜓 雌，云南（普洱）
Asiagomphus xanthenatus, female from Yunnan (Pu'er)

亚春蜓属待定种1 *Asiagomphus* sp. 1

【形态特征】雄性上唇具1对黄斑，额横纹黄色；胸部黑色，背条纹与领条纹相连，具肩前上点，合胸侧面第2条纹和第3条纹完整；腹部黑色，第1～8节背面和侧面具黄斑，第9节背面后方具1个大黄斑，肛附器黑色。雌性与雄性相似。【长度】体长 54～65 mm，腹长 40～48 mm，后翅 35～41 mm。【栖息环境】海拔 1000 m以下的开阔溪流、河流和沟渠。【分布】海南。【飞行期】4—6月。

[Identification] Male labrum with a pair of yellow spots, frons with a yellow stripe. Thorax black, collar stripes connecting with dorsal stripes, superior spots present, second and third lateral stripes complete. Abdomen black, S1-S8

亚春蜓属待定种1 雄，海南
Asiagomphus sp. 1, male from Hainan

with dorsal and lateral yellow markings, S9 with a large yellow dorsal spot posteriorly, anal appendages black. Female similar to male. **[Measurements]** Total length 54-65 mm, abdomen 40-48 mm, hind wing 35-41 mm. **[Habitat]** Exposed streams, rivers and ditches below 1000 m elevation. **[Distribution]** Hainan. **[Flight Season]** April to June.

亚春蜓属待定种1 雌，海南
Asiagomphus sp. 1, female from Hainan

亚春蜓属待定种2 *Asiagomphus* sp. 2

【形态特征】雄性面部大面积黑色，额横纹黄色，后头中央具1个矮脊，两侧稍凹陷；胸部黑色，背条纹与领条纹相连，具甚小的肩前上点，合胸侧面第2条纹和第3条纹合并；腹部黑色，第1～7节背面和侧面具黄斑，第9节背面后方具1个较窄的黄斑，肛附器黑色。**【长度】**体长 57～61 mm，腹长 41～44 mm，后翅 35～36 mm。**【栖息环境】**海拔500 m以下森林中的溪流。**【分布】**云南（红河）。**【飞行期】**4—6月。

[Identification] Male face largely black, frons with a yellow stripe, occiput with a median short carina, depressed laterally. Thorax black, collar stripes connecting with dorsal stripes, superior spots small, second and third lateral stripes combined. Abdomen black, S1-S7 with dorsal and lateral yellow markings, S9 with a narrow yellow dorsal spot posteriorly, anal appendages black. **[Measurements]** Total length 57-61 mm, abdomen 41-44 mm, hind wing 35-36 mm. **[Habitat]** Streams in forest below 500 m elevation. **[Distribution]** Yunnan (Honghe). **[Flight Season]** April to June.

亚春蜓属待定种2 雄，云南（红河）
Asiagomphus sp. 2, male from Yunnan (Honghe)

亚春蜓属待定种3 *Asiagomphus* sp. 3

亚春蜓属待定种3 雄，云南（西双版纳）
Asiagomphus sp. 3, male from Yunnan (Xishuangbanna)

亚春蜓属待定种3 雌，云南（西双版纳）
Asiagomphus sp. 3, female from Yunnan (Xishuangbanna)

【形态特征】雄性上唇具1对黄斑，额横纹黄色，甚阔，后头黑色；胸部黑色，背条纹与领条纹相连，合胸侧面第2条纹和第3条纹合并；腹部黑色，第1～7节背面和侧面具黄斑，第9节背面后方具1个大黄斑，肛附器黑色。雌性与雄性相似，但后头缘具黄斑，稍微隆起。【长度】体长 56～60 mm，腹长 41～45 mm，后翅 34～40 mm。【栖息环境】海拔 1000 m以下的开阔溪流。【分布】云南（西双版纳）。【飞行期】4—6月。

[Identification] Male labrum with a pair of yellow spots, frons with a broad yellow stripe, occiput black. Thorax black, collar stripes connecting with dorsal stripes, second and third lateral stripes combined. Abdomen black, S1-S7 with dorsal and lateral yellow markings, S9 with a large yellow dorsal spot posteriorly, anal appendages black. Female similar to male, occiput with yellow spots and slightly raised. **[Measurements]** Total length 56-60 mm, abdomen 41-45 mm, hind wing 34-40 mm. **[Habitat]** Exposed streams below 1000 m elevation. **[Distribution]** Yunnan (Xishuangbanna). **[Flight Season]** April to June.

缅春蜓属 Genus *Burmagomphus* Williamson, 1907

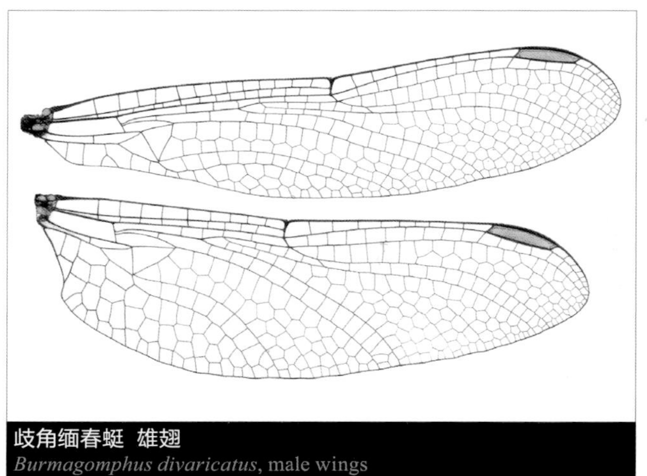
歧角缅春蜓 雄翅
Burmagomphus divaricatus, male wings

本属全球已知30种，主要分布于亚洲的亚热带和热带区域。中国已知15种，分布广泛。本属蜻蜓是一类体型较小的春蜓；翅透明，翅痣具支持脉，三角室、上三角室和下三角室通常无横脉，前翅盘区具2列翅室，基臀区具1条横脉，后翅基缘未呈角状，臀三角室的基边甚斜，通常3室。雄性腹部第7～9节扩大，上肛附器黑色，两枝平行或稍分歧，基方相距甚远，腹方具突起，下肛附器和上肛附器等长，中央凹陷甚阔，两枝分歧，末端向上弯曲。前钩片较小，指状，后钩片较大，扁平状，前缘末端具1个小钩，阳茎具1对长鞭。

本属蜻蜓栖息于溪流和河流。雄性停落在水面附近的岩石、沙滩或植物上占据领地。

The genus contains 30 subtropical and tropical Asian species. 15 species are recorded from China and are widespread in our country. Species of the genus are relative small for the family. Wings hyaline, pterostigma braced, no crossvein in triangle, hypertriangle and subtriangle, discoidal field with 2 rows of cells in fore wings, cubital space with one crossvein, tornus not angled, anal triangle usually 3-celled and its basal side oblique. Male abdomen expanded at S7-S9, two branches of superior appendages parallel or slightly divergent, each widely separated at base, ventral teeth present, the inferiors almost same length as superiors with a deep and broad median hollow, forming two divergent branches whose tips are curved upwards. Anterior hamulus small and finger shaped, posterior hamulus flattened and broad, the anterior margin with a small hook, penis with a pair of long flagella.

Burmagomphus species inhabit streams and rivers. Males exhibit territorial behavior by perching on the rocks, sandy beach or plants near water.

歧角缅春蜓 雄
Burmagomphus divaricatus, male wings

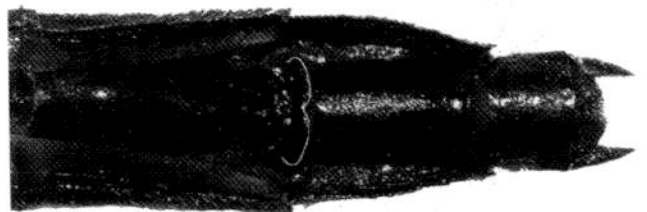

联纹缅春蜓
Burmagomphus vermicularis

威廉缅春蜓
Burmagomphus williamsoni

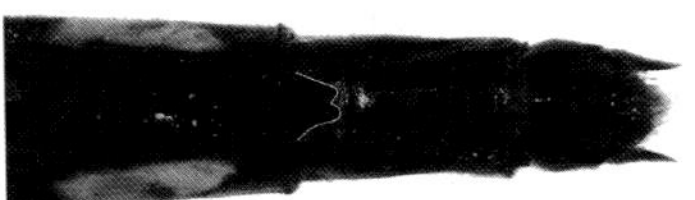

索氏缅春蜓
Burmagomphus sowerbyi

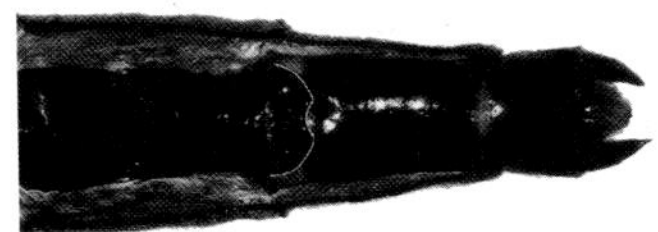

朝比奈缅春蜓
Burmagomphus asahinai

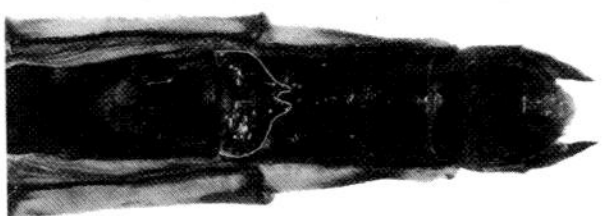

领纹缅春蜓
Burmagomphus collaris

歧角缅春蜓
Burmagomphus divaricatus

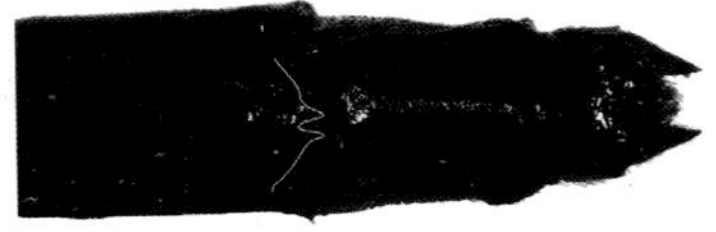

溪居缅春蜓
Burmagomphus intinctus

巨缅春蜓
Burmagomphus magnus

缅春蜓属 雌性下生殖板
Genus *Burmagomphus*, female vulvar lamina

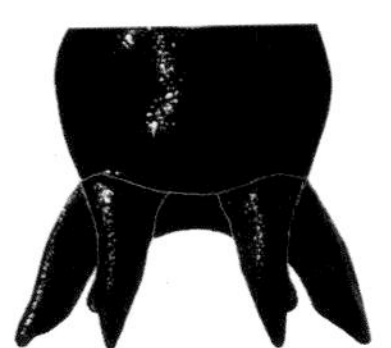

太阳缅春蜓
Burmagomphus apricus

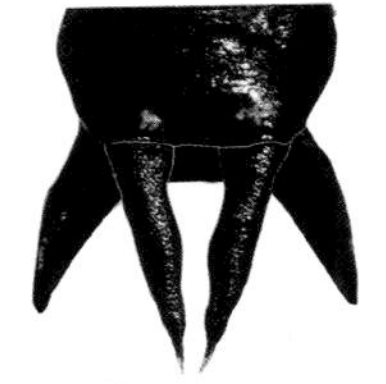

齿尾缅春蜓
Burmagomphus dentatus

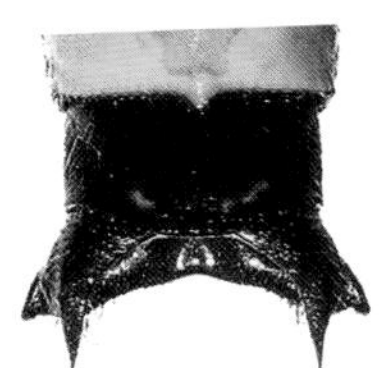

领纹缅春蜓
Burmagomphus collaris

朝比奈缅春蜓
Burmagomphus asahinai

歧角缅春蜓
Burmagomphus divaricatus

欢乐缅春蜓
Burmagomphus gratiosus

缅春蜓属 雄性肛附器
Genus *Burmagomphus*, male anal appendages

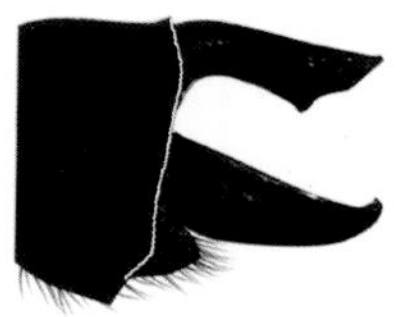
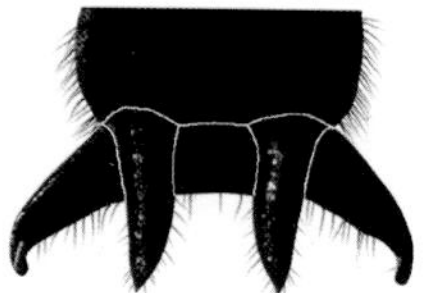
溪居缅春蜓
Burmagomphus intinctus

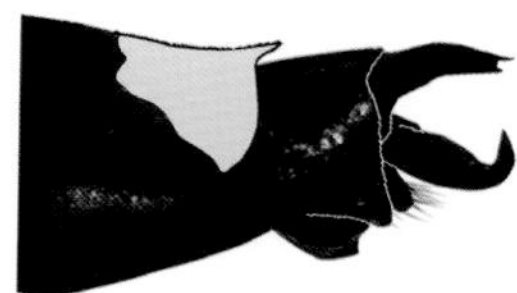

林神缅春蜓
Burmagomphus latescens

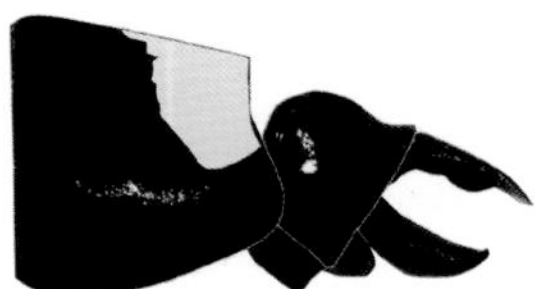
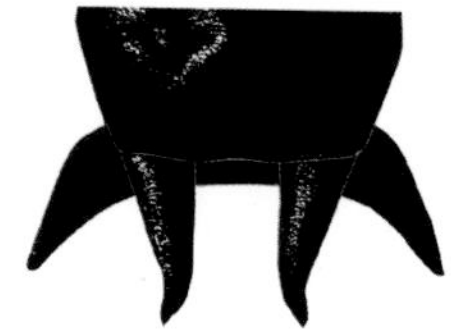
巨缅春蜓
Burmagomphus magnus

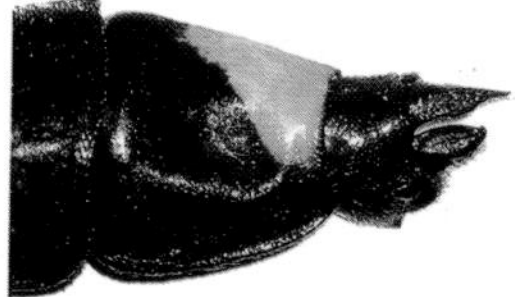
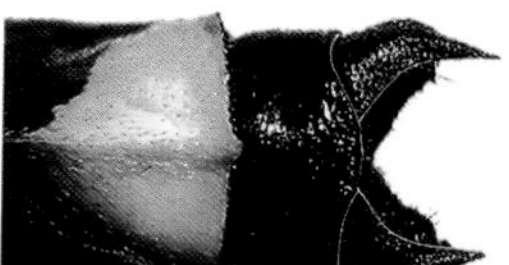
索氏缅春蜓
Burmagomphus sowerbyi

联纹缅春蜓
Burmagomphus vermicularis

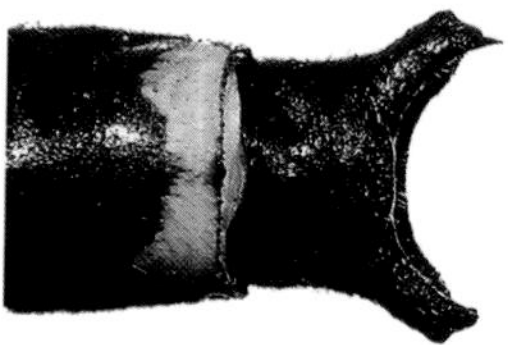
威廉缅春蜓
Burmagomphus williamsoni

缅春蜓属 雄性肛附器
Genus *Burmagomphus*, male anal appendages

太阳缅春蜓 雄
Burmagomphus apricus, male

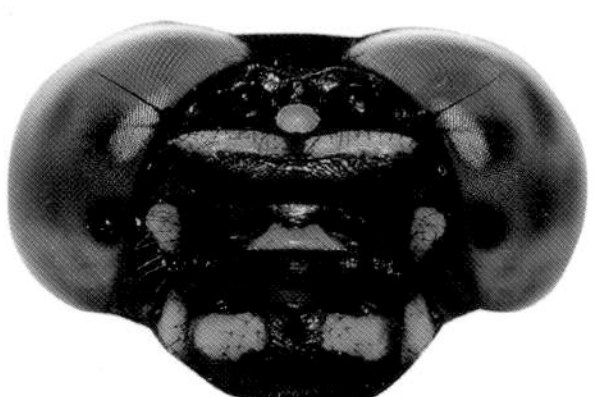
朝比奈缅春蜓 雄
Burmagomphus asahinai, male

歧角缅春蜓 雄
Burmagomphus divaricatus, male

领纹缅春蜓 雄
Burmagomphus collaris, male

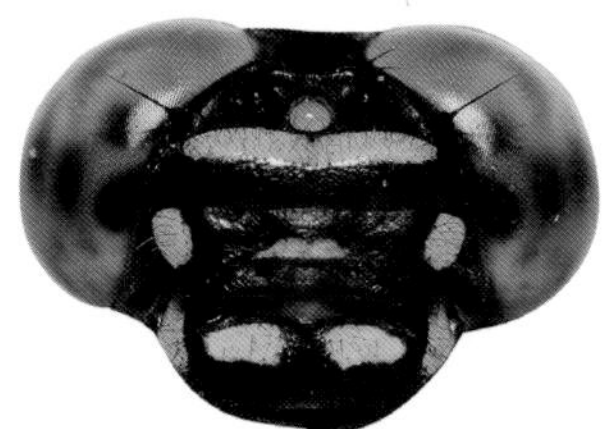
林神缅春蜓 雄
Burmagomphus latescens, male

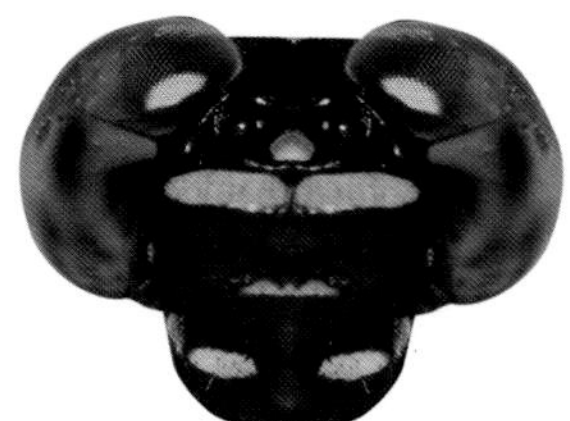
巨缅春蜓 雄
Burmagomphus magnus, male

缅春蜓属 面部正面观
Genus *Burmagomphus*, head in frontal view

联纹缅春蜓 雄
Burmagomphus vermicularis, male

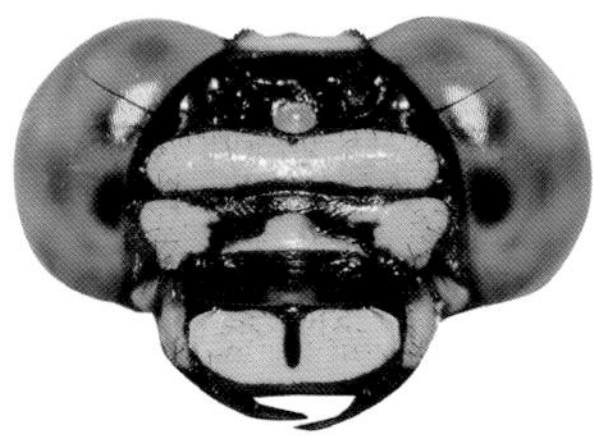
领纹缅春蜓 雌
Burmagomphus collaris, female

朝比奈缅春蜓 雌
Burmagomphus asahinai, female

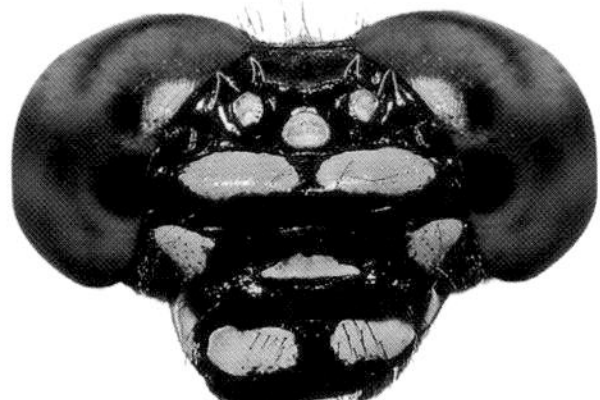
歧角缅春蜓 雌
Burmagomphus divaricatus, female

欢乐缅春蜓 雌
Burmagomphus gratiosus, female

联纹缅春蜓 雌
Burmagomphus vermicularis, female

索氏缅春蜓 雌
Burmagomphus sowerbyi, female

威廉缅春蜓 雌
Burmagomphus williamsoni, female

缅春蜓属 面部正面观
Genus *Burmagomphus*, head in frontal view

太阳缅春蜓 *Burmagomphus apricus* Zhang, Kosterin & Cai, 2015

【形态特征】雄性上唇具1对黄斑，后唇基侧缘和下缘中央具黄斑，额横纹黄色，甚阔；胸部黑色，背条纹与领条纹相连，肩前条纹长而阔，合胸侧面第2条纹和第3条纹完整；腹部黑色，第1~7节背面和侧面具黄斑，第9节具1个甚大黄斑。【长度】雄性体长 48 mm，腹长 35 mm，后翅 28 mm。【栖息环境】海拔 1000 m以下的开阔溪流。【分布】中国云南（西双版纳）特有。【飞行期】5—7月。

[Identification] Male labrum with a pair of yellow spots, lower edge of postclypeus with yellow spots, frons with a broad yellow stripe. Thorax black, collar stripes connecting with dorsal stripes, antehumeral stripes broad and long, second and third lateral stripe complete. Abdomen black, S1-S7 with dorsal and lateral yellow markings, S9 with a large yellow spot. **[Measurements]**

太阳缅春蜓 雄，云南（西双版纳）
Burmagomphus apricus, male from Yunnan (Xishuangbanna)

Male total length 48 mm, abdomen 35 mm, hind wing 28 mm. **[Habitat]** Exposed streams below 1000 m elevation. **[Distribution]** Endemic to Yunnan (Xishuangbanna) of China. **[Flight Season]** May to July.

太阳缅春蜓 雄，云南（西双版纳）
Burmagomphus apricus, male from Yunnan (Xishuangbanna)

朝比奈缅春蜓 *Burmagomphus asahinai* Kosterin, Makbun & Dawwrueng, 2012

【形态特征】雄性上唇具1对黄斑，后唇基下缘中央和侧面具黄斑，额横纹中央间断；胸部黑色，背条纹与领条纹相距较远，具甚小的肩前上点，肩前下条纹紧靠下缘，合胸侧面第2条纹仅有下半段，第3条纹完整，后胸前侧板上缘具小黄斑；腹部黑色，第1～9节具黄斑。雌性与雄性相似，但后头缘中央隆起，两侧各具较小的刺突。【长度】体长 43～45 mm，腹长 31～33 mm，后翅 25～27 mm。【栖息环境】海拔 500 m以下的开阔小溪和河流。【分布】云南（西双版纳）；泰国、柬埔寨、越南。【飞行期】5—7月。

[Identification] Male labrum with a pair of yellow spots, lower edge and sides of postclypeus with yellow spots, frons with a yellow stripe interrupted medially. Thorax black, collar stripes separated from dorsal stipes for a distance, superior spots small, antehumeral stripes close to lower margin, upper half of second lateral stripe absent, third lateral stripe complete, metepisternum with small yellow spots. Abdomen black, S1-S9 with yellow markings. Female similar to male, occipital margin with a median pyramidal prominence and small spines laterally. [Measurements] Total length 43-45 mm, abdomen 31-33 mm, hind wing 25-27 mm. [Habitat] Exposed streams and rivers below 500 m elevation. [Distribution] Yunnan (Xishuangbanna); Thailand, Cambodia, Vietnam. [Flight Season] May to July.

朝比奈缅春蜓 雌，云南（西双版纳）
Burmagomphus asahinai, female from Yunnan (Xishuangbanna)

朝比奈缅春蜓 雄，云南（西双版纳）
Burmagomphus asahinai, male from Yunnan (Xishuangbanna)

朝比奈缅春蜓 雄，云南（西双版纳）
Burmagomphus asahinai, male from Yunnan (Xishuangbanna)

领纹缅春蜓 *Burmagomphus collaris* (Needham, 1930)

领纹缅春蜓 雄，北京
Burmagomphus collaris, male from Beijing

领纹缅春蜓 雌，北京
Burmagomphus collaris, female from Beijing

领纹缅春蜓 雄，北京
Burmagomphus collaris, male from Beijing

【形态特征】雄性上唇具1对大黄斑，后唇基下缘中央和侧面具黄斑，额横纹甚阔，后头黄色；胸部黑色，背条纹与领条纹不相连，肩前条纹甚阔，合胸侧面第2条纹上方间断，第3条纹完整，甚细；腹部黑色，各节具黄斑，第9节具1个甚大的三角形黄斑。雌性与雄性相似，但后头具刺突。**【长度】**体长 42~46 mm，腹长 32~35 mm， 后翅 22~29 mm。**【栖息环境】**海拔 500 m以下的开阔小溪和河流。**【分布】**河北、江苏、浙江、北京；韩国。**【飞行期】**6—9月。

[Identification] Male labrum with a pair of yellow spots, lower edge and sides of postclypeus with yellow spots, frons with a broad yellow stripe, occiput yellow. Thorax black, collar stripes not connecting with dorsal stripes, antehumeral stripes very broad, second lateral stripe largely interrupted above, third lateral stripe complete but narrow. Abdomen black with yellow markings, S9 with a large triangular yellow spot. Female similar to male, occiput with spines. **[Measurements]** Total length 42-46 mm, abdomen 32-35 mm, hind wing 22-29 mm. **[Habitat]** Exposed streams and rivers below 500 m elevation. **[Distribution]** Hebei, Jiangsu, Zhejiang, Beijing; South Korea. **[Flight Season]** June to September.

齿尾缅春蜓 *Burmagomphus dentatus* Zhang, Kosterin & Cai, 2015

【形态特征】雄性上唇具1对黄斑，后唇基下缘中央具细黄纹，额横纹黄色，甚阔；胸部黑色，背条纹与领条纹不相连，肩前条纹长而阔，合胸侧面第2条纹和第3条纹完整；腹部黑色，第1~7节背面和侧面具黄斑，第9节具1个甚大黄斑。雌性侧单眼后方具角状突起。【长度】体长 49~53 mm，腹长 37~40 mm，后翅 28~33 mm。【栖息环境】海拔 500 m以下的宽阔河流。【分布】中国贵州特有。【飞行期】6—8月。

[Identification] Male labrum with a pair of yellow spots, lower edge of postclypeus with fine yellow stripes, frons with a broad yellow stripe. Thorax black, collar stripes not connecting with dorsal stripes, antehumeral stripes broad and long, second and third lateral stripes complete. Abdomen black, S1-S7 with dorsal and lateral yellow markings, S9 with a large yellow spot. Female vertex with horns behind lateral ocelli. **[Measurements]** Total length 49-53 mm, abdomen 37-40 mm, hind wing 28-33 mm. **[Habitat]** Rivers below 500 m elevation. **[Distribution]** Endemic to Guizhou of China. **[Flight Season]** June to August.

齿尾缅春蜓 雄，贵州 | 苏毅雄 摄
Burmagomphus dentatus, male from Guizhou | Photo by Samson So

齿尾缅春蜓 雄，贵州
Burmagomphus dentatus, male from Guizhou

歧角缅春蜓 *Burmagomphus divaricatus* Lieftinck, 1964

【形态特征】雄性上唇具1对黄斑，后唇基下缘中央和侧面具黄斑，额横纹甚阔；胸部黑色，背条纹与领条纹不相连，较倾斜，与肩前下条纹相连，肩前上点甚小，合胸侧面第2条纹仅有下方的2/3，第3条纹“Y”形；腹部黑色，第1～9节具黄斑。雌性与雄性相似，头顶具2对角状突起，后头黄色，后头缘稍微隆起。【长度】体长 39～43 mm，腹长 29～32 mm，后翅 22～27 mm。【栖息环境】海拔 1000 m以下森林中的小溪。【分布】云南（西双版纳、德宏、普洱）；泰国、老挝、柬埔寨、越南、马来半岛、新加坡。【飞行期】5—12月。

[Identification] Male labrum with a pair of yellow spots, lower edge and sides of postclypeus with yellow spots, frons with a broad yellow stripe. Thorax black, dorsal stripes not connecting with collar stripes but connecting with the lower stripes, superior spots small, lower two thirds of second lateral stripe present, third lateral stripe Y-shaped.

歧角缅春蜓 雄，云南（西双版纳）
Burmagomphus divaricatus, male from Yunnan (Xishuangbanna)

Abdomen black, S1-S9 with yellow markings. Female similar to male, vertex with two pairs of horns, occiput yellow and slightly erect medially. **[Measurements]** Total length 39-43 mm, abdomen 29-32 mm, hind wing 22-27 mm. **[Habitat]** Montane streams below 1000 m elevation. **[Distribution]** Yunnan (Xishuangbanna, Dehong, Pu'er); Thailand, Laos, Cambodia, Vietnam, Peninsular Malaysia, Singapore. **[Flight Season]** May to December.

歧角缅春蜓 雌，云南（西双版纳）
Burmagomphus divaricatus, female from Yunnan (Xishuangbanna)

歧角缅春蜓 雄，云南（西双版纳）
Burmagomphus divaricatus, male from Yunnan (Xishuangbanna)

欢乐缅春蜓 *Burmagomphus gratiosus* Chao, 1954

【形态特征】雄性上唇具1对黄斑，额横纹中央间断；胸部黑色，背条纹较倾斜，与领条纹不相连，肩前条纹仅为下方的一小段，无肩前上点，合胸侧面第2条纹仅有下方的2/3，第3条纹“Y”形，翅透明；腹部黑色，第1~7节具黄斑，第9节背面后方具黄斑。雌性与雄性相似，侧单眼后方具1对刺状突起。【长度】体长 46~48 mm，腹长 34~36 mm，后翅 27~29 mm。【栖息环境】海拔 500 m以下的河流和开阔溪流。【分布】中国特有，分布于贵州、福建。【飞行期】6—8月。

欢乐缅春蜓 雄，贵州 | 苏毅雄 摄
Burmagomphus gratiosus, male from Guizhou | Photo by Samson So

欢乐缅春蜓 雌，贵州 | 莫善濂 摄
Burmagomphus gratiosus, female from Guizhou | Photo by Shanlian Mo

[Identification] Male labrum with a pair of yellow spots, frons with a broad yellow stripe interrupted medially. Thorax black, dorsal stripes oblique, not connecting with collar stripes, antehumeral stripes absent, lower two thirds of second lateral stripe present, third lateral stripe Y-shaped, wings hyaline. Abdomen black, S1- S7 with yellow markings, S9 with a yellow spot posteriorly. Female similar to male, vertex with a pair of spines behind lateral ocelli. **[Measurements]** Total length 46-48 mm, abdomen 34-36 mm, hind wing 27-29 mm. **[Habitat]** Exposed streams and rivers below 500 m elevation. **[Distribution]** Endemic to China, recorded from Guizhou, Fujian. **[Flight Season]** June to August.

欢乐缅春蜓 雄，贵州 | 苏毅雄 摄
Burmagomphus gratiosus, male from Guizhou | Photo by Samson So

溪居缅春蜓 *Burmagomphus intinctus* (Needham, 1930)

【形态特征】雄性上唇具1对黄斑，后唇基下缘中央具黄斑，额横纹黄色，甚阔；胸部黑色，背条纹与领条纹不相连，肩前条纹长而阔，合胸侧面第2条纹和第3条纹完整；腹部黑色，第1～7节具黄斑，第9节黄斑甚大。雌性与雄性相似。【长度】体长 49～50 mm，腹长 35～36 mm，后翅 30～32 mm。【栖息环境】海拔 1000 m以下的河流和开阔溪流。【分布】中国特有，分布于浙江、福建。【飞行期】5—8月。

[Identification] Male labrum with a pair of yellow spots, lower edge of postclypeus with yellow spots, frons with a broad yellow stripe. Thorax black, collar stripes not connecting with dorsal stripes, antehumeral stripes broad and long, second and third lateral stripes complete. Abdomen black, S1-S7 with dorsal and lateral yellow makrings, S9 with a large yellow spot posteriorly. Female similar to male. **[Measurements]** Total length 49-50 mm, abdomen 35-36 mm, hind wing 30-32 mm. **[Habitat]** Exposed streams and rivers below 1000 m elevation. **[Distribution]** Endemic to China, recorded from Zhejiang, Fujian. **[Flight Season]** May to August.

溪居缅春蜓 雄，浙江
Burmagomphus intinctus, male from Zhejiang

溪居缅春蜓 雌，浙江
Burmagomphus intinctus, female from Zhejiang

溪居缅春蜓 雄，浙江
Burmagomphus intinctus, male from Zhejiang

林神缅春蜓 *Burmagomphus latescens* Zhang, Kosterin & Cai, 2015

【形态特征】雄性上唇具1对黄斑，后唇基下缘中央和侧面具黄斑，额横纹黄色，甚阔；胸部黑色，背条纹较倾斜，与领条纹不相连，但与肩前下条纹相连，肩前上点甚小，合胸侧面第2条纹仅有下方的2/3，第3条纹“Y”形；腹部黑色，各节具黄斑。【长度】雄性体长 46~47 mm，腹长 33~34 mm，后翅 27 mm。【栖息环境】海拔 500 m处森林中的溪流。【分布】中国云南（临沧）特有。【飞行期】8—10月。

[Identification] Male labrum with a pair of yellow spots, lower edge and sides of postclypeus with yellow spots, frons with a broad yellow stripe. Thorax black, dorsal stripes oblique, not connecting with collar stripes but connecting with the lower stripes, superior spots very small, lower two thirds of second lateral stripe present, third lateral stripe Y-shaped. Abdomen black with yellow makrings. **[Measurements]** Male total length 46-47 mm, abdomen 33-34 mm, hind wing 27 mm. **[Habitat]** Streams in forest at 500 m elevation. **[Distribution]** Endemic to Yunnan (Lincang) of China. **[Flight Season]** August to October.

林神缅春蜓 雄，云南（德宏）
Burmagomphus latescens, male from Yunnan (Dehong)

巨缅春蜓 *Burmagomphus magnus* Zhang, Kosterin & Cai, 2015

巨缅春蜓 雄，广西
Burmagomphus magnus, male from Guangxi

巨缅春蜓 雌，广西
Burmagomphus magnus, male from Guangxi

【形态特征】雄性上唇具1对黄斑，后唇基侧缘和下缘中央具黄斑，额横纹甚阔；胸部黑色，背条纹与领条纹不相连，肩前条纹长而阔，合胸侧面第2条纹和第3条纹完整；腹部黑色，第1~7节背面和侧面具黄斑，第9节具1个甚大黄斑。雌性侧单眼后方具角状突起。【长度】体长 50~53 mm，腹长 37~40 mm，后翅 30~34 mm。【栖息环境】海拔 500 m以下的开阔溪流和河流。【分布】中国特有，分布于云南（红河）、广西、广东。【飞行期】4—8月。

[Identification] Male labrum with a pair of yellow spots, lower edge and sides of postclypeus with yellow spots, frons with a broad yellow stripe. Thorax black, collar stripes not connecting with dorsal stripes, antehumeral stripes broad and long, second and third lateral stripes complete. Abdomen black, S1-S7 with dorsal and lateral yellow markings, S9 with a large yellow spot. Female vertex with horns behind lateral ocelli. **[Measurements]** Total length 50-53 mm, abdomen 37-40 mm, hind wing 30-34 mm. **[Habitat]** Exposed streams and rivers below 500 m elevation. **[Distribution]** Endemic to China, recorded from Yunnan (Honghe), Guangxi, Guangdong. **[Flight Season]** April to August.

索氏缅春蜓 *Burmagomphus sowerbyi* (Needham, 1930)

索氏缅春蜓 雄，湖北
Burmagomphus sowerbyi, male from Hubei

【形态特征】雄性上唇具1对黄斑，后唇基下缘中央具黄斑，额横纹甚阔，后头黄色；胸部黑色，背条纹与领条纹不相连，肩前条纹长而阔，合胸侧面第2条纹中央间断，上半部倾斜，第3条纹完整；腹部黑色，第1～7节侧面和背面具黄斑，第9节黄斑甚大。雌性与雄性相似，黄斑更发达。【长度】体长 47～48 mm，腹长 35～36 mm，后翅 27～29 mm。【栖息环境】海拔 500 m以下的开阔溪流和河流。【分布】中国特有，分布于河南、江苏、贵州、湖北、湖南、福建、广西。【飞行期】5—8月。

索氏缅春蜓 雌，湖北
Burmagomphus sowerbyi, female from Hubei

[Identification] Male labrum with a pair of yellow spots, lower edge of postclypeus with a median yellow spot, frons with a broad yellow stripe, occiput yellow. Thorax black, collar stripes not connecting with dorsal stripes, antehumeral stripes long and broad, second lateral stripe interrupted medially with the upper part oblique, third lateral stripe complete. Abdomen black, S1-S7 with dorsal and lateral yellow markings, S9 with a large yellow spot. Female similar to male with more extensive yellow markings. **[Measurements]** Total length 47-48 mm, abdomen 35-36 mm, hind wing 27-29 mm. **[Habitat]** Exposed streams and rivers below 500 m elevation. **[Distribution]** Endemic to China, recorded from Henan, Jiangsu, Guizhou, Hubei, Hunan, Fujian, Guangxi. **[Flight Season]** May to August.

联纹缅春蜓 *Burmagomphus vermicularis* (Martin, 1904)

【形态特征】雄性上唇具1对黄斑，后唇基下缘中央和侧面具黄斑，额横纹甚阔；胸部黑色，背条纹与领条纹不相连，与肩前下条纹相连，肩前上点甚小，合胸侧面第2条纹仅有下方的2/3，第3条纹“Y”形；腹部黑色，第1～9节具黄斑。雌性与雄性相似。【长度】体长 37～45 mm，腹长 28～34 mm，后翅 22～28 mm。【栖息环境】海拔 1000 m以下的开阔溪流和河流。【分布】云南、福建、广西、广东、海南、香港、台湾；老挝、越南。【飞行期】5—10月。

[Identification] Male labrum with a pair of yellow spots, lower edge and sides of postclypeus with yellow spots, frons with a broad stripe. Thorax black, dorsal stripes not connecting with collar stripes but connecting with the lower stripes, superior spots small, lower two thirds of second lateral stripe present, third lateral stripe Y-shaped. Abdomen black, S1-S9 with yellow markings. Female similar to male. **[Measurements]** Total length 37-45 mm, abdomen 28-34 mm, hind wing 22-28 mm. **[Habitat]** Exposed streams and rivers below 1000 m elevation. **[Distribution]** Yunnan, Fujian, Guangxi, Guangdong, Hainan, Hong Kong, Taiwan; Laos, Vietnam. **[Flight Season]** May to October.

联纹缅春蜓 雄，海南
Burmagomphus vermicularis, male from Hainan

联纹缅春蜓 雌，广西
Burmagomphus vermicularis, female from Guangxi

联纹缅春蜓 雄，广西
Burmagomphus vermicularis, male from Guangxi

威廉缅春蜓 *Burmagomphus williamsoni* Förster, 1914

【形态特征】雄性上唇具1对黄斑，后唇基下缘和侧面具黄斑，额横纹甚阔；胸部黑色，背条纹与领条纹不相连，较倾斜，与肩前下条纹相连，肩前上点甚小，合胸侧面第2条纹仅有下方的2/3，第3条纹"Y"形；腹部黑色具黄斑。雌性与雄性相似，但后头黄色，后头缘锥形隆起甚高。【长度】体长 39～40 mm，腹长 28～29 mm，后翅 23～25 mm。【栖息环境】海拔 1000 m以下的开阔溪流和河流。【分布】云南（西双版纳）；缅甸、泰国、马来半岛、印度尼西亚。【飞行期】5—7月。

[Identification] Male labrum with a pair of yellow spots, lower edge and sides of postclypeus with yellow spots, frons with a broad stripe. Thorax black, dorsal stripes not connecting with collar stripes but connecting with the lower stripes, superior spots small, lower two thirds of second lateral stripe present, third lateral stripe Y-shaped. Abdomen black with yellow markings. Female similar to male, occiput yellow and erected in the middle. **[Measurements]** Total length 39-40 mm, abdomen 28-29 mm, hind wing 23-25 mm. **[Habitat]** Exposed streams and rivers below 1000 m elevation. **[Distribution]** Yunnan (Xishuangbanna); Myanmar, Thailand, Peninsular Malaysia, Indonesia. **[Flight Season]** May to July.

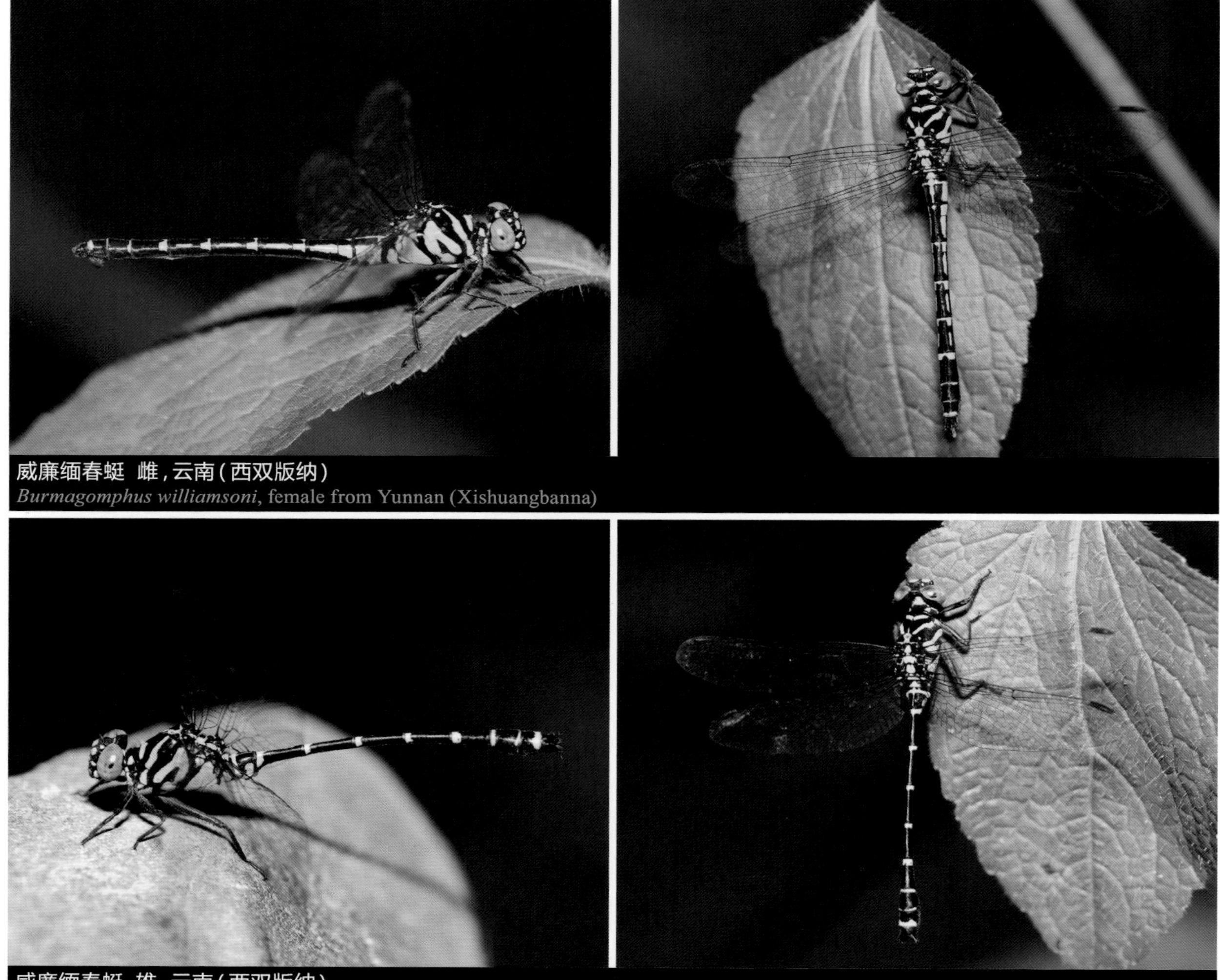

威廉缅春蜓 雌，云南（西双版纳）
Burmagomphus williamsoni, female from Yunnan (Xishuangbanna)

威廉缅春蜓 雄，云南（西双版纳）
Burmagomphus williamsoni, male from Yunnan (Xishuangbanna)

戴春蜓属 Genus *Davidius* Selys, 1878

本属全球已知20余种，主要分布于亚洲的亚热带区域。中国已知10余种。本属蜻蜓是一类小型春蜓；翅透明，翅痣具支持脉，后翅三角室长，约为前翅三角室的2倍，前翅的臀脉与翅后缘之间仅有1列翅室，基臀区具1条横脉，后翅基缘呈锐角，臀三角室的基边甚斜，后翅三角室常具1条横脉。雄性上肛附器锥形，强烈分歧，基方内部具突起，下肛附器中裂较深，向上弯曲。前钩片末端分两枝。阳茎无鞭。

本属蜻蜓栖息于山区溪流。雄性仅在阳光充足时出现在水面的岩石或植物上占据领地。

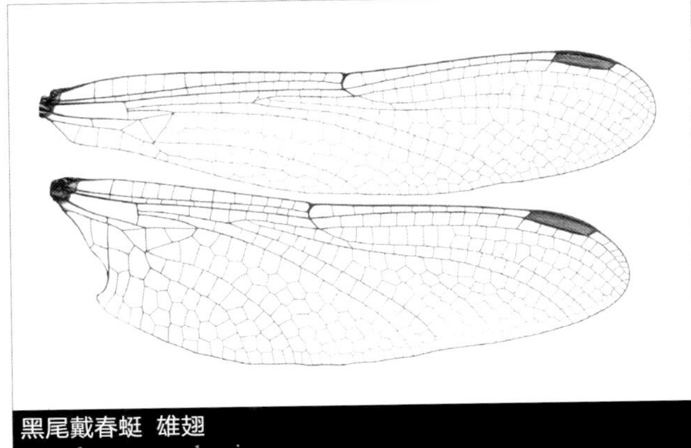

黑尾戴春蜓 雄翅
Davidius trox, male wings

The genus contains over 20 species, mainly distributed in subtropical Asia. Over ten species are recored from China. Species of the genus are relative small-sized in this family. Wings hyaline, pterostigma braced, triangle in hind wings long, about twice as long as the triangle in fore wings, a single row of cells between A and hind margin in fore wings, cubital space with one crossvein, tornus strongly angled, anal triangle with basal side oblique, triangle in hind wing usually with one crossvein. Male superior appendages strongly divergent, pyramidal in shape, with inner teeth basally, inferior appendage with a deep median hollow, the tips curved upwards. Anterior hamulus divided into two branches distally. Penis without flagella.

Davidius species inhabit montane streams. Males exhibit territorial behavior by perching on the rocks or plants near water in sunny days.

方钩戴春蜓 雄
Davidius squarrosus, male

弗鲁戴春蜓
Davidius fruhstorferi

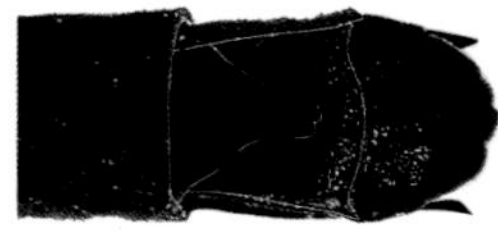

方钩戴春蜓
Davidius squarrosus

黑尾戴春蜓
Davidius trox

扎洛克戴春蜓
Davidius zallorensis

周氏戴春蜓
Davidius zhoui

戴春蜓属 雌性下生殖板
Genus *Davidius*, female vulvar lamina

细纹戴春蜓
Davidius delineatus

弗鲁戴春蜓
Davidius fruhstorferi

方钩戴春蜓
Davidius squarrosus

双角戴春蜓
Davidius bicornutus

扎洛克戴春蜓
Davidius zallorensis

周氏戴春蜓
Davidius zhoui

黑尾戴春蜓
Davidius trox

戴春蜓属 雄性肛附器
Genus *Davidius*, male anal appendages

双角戴春蜓 *Davidius bicornutus* Selys, 1878

【形态特征】雄性面部大面积黄色；胸部黑色，背条纹与领条纹不相连，肩前条纹不稳定，合胸侧面第2条纹和第3条纹大面积合并；腹部黑色，第1～9节具黄斑，肛附器黑色，上肛附器较长。雌性侧单眼后方具1对角状突起。【长度】体长 56～62 mm，腹长 40～46 mm，后翅 37～38 mm。【栖息环境】海拔 500～1000 m森林中的溪流。【分布】中国特有，分布于北京、河北、陕西、四川。【飞行期】6—9月。

[Identification] Male face largely yellow. Thorax black, dorsal stripes not connecting with collar stripes, antehumeral stripes variable, second and third lateral stripes largely combined. Abdomen black, S1-S9 with yellow markings, anal appendages black, superior appendages long. Female vertex with a pair of horns behind lateral ocelli. **[Measurements]** Total length 56-62 mm, abdomen 40-46 mm, hind wing 37-38 mm. **[Habitat]** Streams in forest at 500-1000 m elevation. **[Distribution]** Endemic to China, recorded from Beijing, Hebei, Shaanxi, Sichuan. **[Flight Season]** June to September.

双角戴春蜓 雄，北京
Davidius bicornutus, male from Beijing

细纹戴春蜓 *Davidius delineatus* Fraser, 1926

【形态特征】雄性面部大面积黑色，额横纹黄色，甚阔；胸部黑色，背条纹与领条纹相连，合胸侧面第2条纹中央间断较长，第3条纹完整，中央甚细；腹部黑色，第1～6节具黄斑，肛附器白色。【长度】雄性体长 34 mm，腹长 26 mm，后翅 20～21 mm。【栖息环境】海拔 1000～1500 m森林中的溪流。【分布】云南（德宏）；印度、尼泊尔。【飞行期】4—7月。

[Identification] Male face largely black, frons with a broad stripe. Thorax black, dorsal stripes connecting with collar stripes, second lateral stripe largely interrupted medially, third lateral stripe complete, narrow medially. Abdomen black, S1-S6 with yellow markings, anal appendages white. **[Measurements]** Male total length 34 mm, abdomen 26 mm, hind wing 20-21 mm. **[Habitat]** Streams in forest below 1000-1500 m elevation. **[Distribution]** Yunnan (Dehong); India, Nepal. **[Flight Season]** April to July.

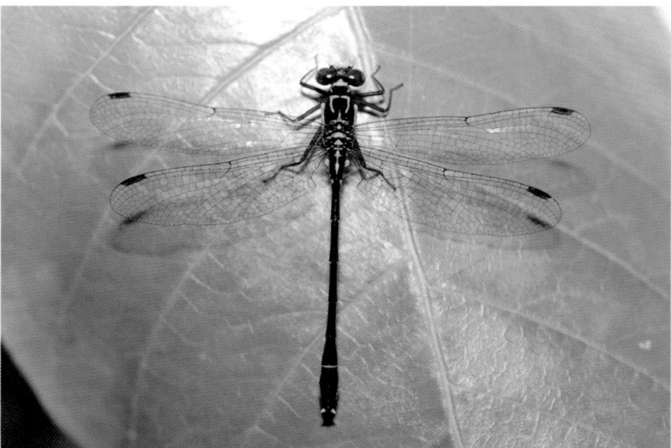

细纹戴春蜓 雄，云南（德宏）
Davidius delineatus, male from Yunnan (Dehong)

弗鲁戴春蜓 *Davidius fruhstorferi* Martin, 1904

【形态特征】雄性面部大面积黑色，额横纹黄色，甚阔；胸部黑色，背条纹与领条纹相连，合胸侧面第2条纹中央间断较长，第3条纹完整；腹部黑色，第1~5节具较小的黄斑，肛附器白色。雌性与雄性相似，腹部侧面具更多黄斑。【长度】体长 37~41 mm，腹长 28~31 mm，后翅 20~24 mm。【栖息环境】海拔 2000 m以下森林中的溪流。【分布】贵州、广西、广东、福建；泰国、老挝、越南。【飞行期】4—8月。

弗鲁戴春蜓 雄，广西
Davidius fruhstorferi, male from Guangxi

[Identification] Male face largely black, frons with a broad yellow stripe. Thorax black, dorsal stripes connecting with collar stripes, second lateral stripe largely interrupted medially, third lateral stripe complete. Abdomen black, S1-S5 with small yellow markings, anal appendages white. Female similar to male, sides of abdomen with more yellow markings. **[Measurements]** Total length 37-41 mm, abdomen 28-31 mm, hind wing 20-24 mm. **[Habitat]** Streams in forest below 2000 m elevation. **[Distribution]** Guizhou, Guangxi, Guangdong, Fujian; Thailand, Laos, Vietnam. **[Flight Season]** April to August.

弗鲁戴春蜓 雄，广西
Davidius fruhstorferi, male from Guangxi

方钩戴春蜓 *Davidius squarrosus* Zhu, 1991

方钩戴春蜓 雄，湖北
Davidius squarrosus, male from Hubei

方钩戴春蜓 雄，湖北
Davidius squarrosus, male from Hubei

方钩戴春蜓 交尾，湖北
Davidius squarrosus, mating pair from Hubei

【形态特征】雄性面部大面积黑色，额横纹黄色，较阔；胸部黑色，背条纹与领条纹相连，合胸侧面第2条纹中央间断较长，第3条纹完整；腹部黑色，第1～6节具小黄斑，肛附器白色。雌性与雄性相似，腹部侧面具黄斑。【长度】体长 38～42 mm，腹长 28～32 mm，后翅 23～25 mm。【栖息环境】海拔 500～1500 m森林中的溪流。【分布】中国特有，分布于湖北、陕西。【飞行期】6—8月。

[Identification] Male face largely black, frons with a broad yellow stripe. Thorax black, dorsal stripes connecting with collar stripes, second lateral stripe largely interrupted medially, third lateral stripe complete. Abdomen black, S1-S6 with small yellow markings, anal appendages white. Female similar to male, sides of abdomen with yellow markings. **[Measurements]** Total length 38-42 mm, abdomen 28-32 mm, hind wing 23-25 mm. **[Habitat]** Streams in forest at 500-1500 m elevation. **[Distribution]** Endemic to China, recorded from Hubei, Shaanxi. **[Flight Season]** June to August.

黑尾戴春蜓 *Davidius trox* Needham, 1931

黑尾戴春蜓 雌，贵州 | 宋黎明 摄
Davidius trox, female from Guizhou | Photo by Liming Song

黑尾戴春蜓 雌，贵州 | 莫善濂 摄
Davidius trox, female from Guizhou | Photo by Shanlian Mo

黑尾戴春蜓 雄，贵州 | 宋黎明 摄
Davidius trox, male from Guizhou | Photo by Liming Song

【形态特征】雄性面部大面积黑色，额横纹黄色，较阔；胸部黑色，背条纹与领条纹相连，具甚小的肩前上点，合胸侧面第2条纹中央间断较长，第3条纹完整；腹部黑色，第1~9节具较小的黄斑，肛附器黑色。雌性与雄性相似，腹部侧面具较多黄斑。【长度】体长 45~46 mm，腹长 34~35 mm，后翅 28~30 mm。【栖息环境】海拔1000~1500 m森林中的溪流。【分布】中国特有，分布于四川、贵州。【飞行期】5—8月。

[Identification] Male face largely black, frons with a broad yellow stripe. Thorax black, dorsal stripes connecting with collar stripes, superior spots small, second lateral stripe largely interrupted medially, third lateral stripe complete. Abdomen black, S1-S9 with small yellow markings, anal appendages black. Female similar to male, sides of abdomen with more yellow markings. **[Measurements]** Total length 45-46 mm, abdomen 34-35 mm, hind wing 28-30 mm. **[Habitat]** Streams in forest at 1000-1500 m elevation. **[Distribution]** Endemic to China, recorded from Sichuan, Guizhou. **[Flight Season]** May to August.

扎洛克戴春蜓 *Davidius zallorensis* Hagen, 1878

【形态特征】雄性面部大面积黑色，上唇下缘黄色，额黄色；胸部黑色，背条纹与领条纹相连，具甚小的肩前上点，合胸侧面第2条纹中央间断，第3条纹完整；腹部黑色，第1~9节具甚小黄斑，上肛附器白色，下肛附器黑色。雌性与雄性相似，腹部侧面具更多黄斑。【长度】体长 45~47 mm，腹长 33~36 mm，后翅30~33 mm。【栖息环境】海拔 2000~2500 m森林中的开阔溪流。【分布】云南（大理）；印度、尼泊尔。【飞行期】5—7月。

[Identification] Male face largely black, lower edge of labrum yellow, frons yellow. Thorax black, dorsal stripes connecting with collar stripes, superior spots small, second lateral stripe interrupted medially, third lateral stripe complete. Abdomen black, S1-S9 with small yellow markings, superior appendages white, inferior appendage black. Female similar to male, sides of abdomen with more yellow markings. **[Measurements]** Total length 45-47 mm, abdomen 33-36 mm, hind wing 30-33 mm. **[Habitat]** Exposed streams in forest at 2000-2500 m elevation. **[Distribution]** Yunnan (Dali); India, Nepal. **[Flight Season]** May to July.

扎洛克戴春蜓 雌，云南（大理）
Davidius zallorensis, female from Yunnan (Dali)

扎洛克戴春蜓 雄，云南（大理）
Davidius zallorensis, male from Yunnan (Dali)

扎洛克戴春蜓 雄，云南（大理）
Davidius zallorensis, male from Yunnan (Dali)

周氏戴春蜓 *Davidius zhoui* Chao, 1995

周氏戴春蜓 雄，云南（大理）
Davidius zhoui, male from Yunnan (Dali)

【形态特征】雄性面部大面积黑色，额横纹黄色，较阔；胸部黑色，背条纹与领条纹相连，具甚小的肩前上点，合胸侧面第2条纹中央间断，第3条纹完整；腹部黑色，第1~6节具较小的黄斑，肛附器白色。雌性与雄性相似，腹部侧面具较多的黄斑。【长度】体长 42~43 mm，腹长 32~33 mm，后翅 28~29 mm。【栖息环境】海拔 1500~2500 m森林中的开阔狭窄溪流。【分布】中国云南特有。【飞行期】5—7月。

[Identification] Male face largely black, frons with a broad yellow stripe. Thorax black, dorsal stripes connecting with collar stripes, superior spots small, second lateral stripe interrupted medially, third lateral stripe complete.

Abdomen black, S1-S6 with small yellow markings, anal appendages white. Female similar to male, sides of abdomen with more yellow markings. **[Measurements]** Total length 42-43 mm, abdomen 32-33 mm, hind wing 28-29 mm. **[Habitat]** Exposed and narrow streams in forest at 1500-2500 m elevation. **[Distribution]** Endemic to Yunnan of China. **[Flight Season]** May to July.

周氏戴春蜓 雌，云南（大理）
Davidius zhoui, female from Yunnan (Dali)

周氏戴春蜓 雄，云南（大理）
Davidius zhoui, male from Yunnan (Dali)

戴春蜓属待定种1 *Davidius* sp. 1

【形态特征】雄性上唇黑色具1对小黄斑，前唇基黑色，后唇基和额黄色；胸部黑色，背条纹与领条纹不相连，具甚小的肩前上点，合胸侧面第2条纹和第3条纹大面积合并；腹部黑色，第1～9节具黄斑，肛附器黑色，上肛附器较长。【长度】雄性体长 56 mm，腹长 42 mm，后翅 35 mm。【栖息环境】海拔 1500 m森林中的开阔溪流。【分布】云南（德宏）。【飞行期】9—11月。

[Identification] Male labrum black with a pair of yellow spots, anteclypeus black, postclypeus and frons yellow. Thorax black, dorsal stripes not connecting with collar stripes, superior spots small, second lateral stripe and third lateral stripe largely combined. Abdomen black, S1-S9 with yellow markings, anal appendages black, superior appendages long. **[Measurements]** Male total length 56 mm, abdomen 42 mm, hind wing 35 mm. **[Habitat]** Exposed streams in forest at 1500 m elevation. **[Distribution]** Yunnan (Dehong). **[Flight Season]** September to November.

戴春蜓属待定种1 雄，云南（德宏）
Davidius sp. 1, male from Yunnan (Dehong)

戴春蜓属待定种1 雄，云南（德宏）
Davidius sp. 1, male from Yunnan (Dehong)

戴春蜓属待定种2 *Davidius* sp. 2

【形态特征】雄性面部大面积黄色，上唇具黑斑，后头黑色；胸部黑色，背条纹甚短，合胸脊黄色，与领条纹在中央相连，具甚小的肩前上点，合胸侧面第2条纹和第3条纹缺如；腹部黑色，各节具较小的黄斑，肛附器黑色。雌性与雄性相似。【长度】体长 42~44 mm，腹长 31~32 mm，后翅 24~25 mm。【栖息环境】海拔 2000~2500 m森林中的开阔溪流。【分布】云南（大理）。【飞行期】5—7月。

[Identification] Male face largely yellow, labrum with black spots, occiput black. Thorax black, dorsal stripes very short, dorsal carina yellow and connecting with collar stripes in the middle, superior spots small, second and third lateral stripes absent. Abdomen black with small yellow markings on each segment, anal appendages black. Female similar to male. **[Measurements]** Total length 42-44 mm, abdomen 31-32 mm, hind wing 24-25 mm. **[Habitat]** Exposed streams in forest at 2000-2500 m elevation. **[Distribution]** Yunnan (Dali). **[Flight Season]** May to July.

戴春蜓属待定种2 雄，云南（大理）
Davidius sp. 2, male from Yunnan (Dali)

戴春蜓属待定种2 雌，云南（丽江）
Davidius sp. 2, female from Yunnan (Lijiang)

戴春蜓属待定种2 雄，云南（大理）
Davidius sp. 2, male from Yunnan (Dali)

戴春蜓属待定种3 *Davidius* sp. 3

【形态特征】雄性面部大面积黑色，额横纹黄色，甚阔；胸部黑色，背条纹与领条纹相连，合胸侧面第2条纹中央间断较长，第3条纹完整，翅基方1/2染有琥珀色；腹部黑色，第1~3节具较小的黄斑，肛附器白色。雌性与雄性相似，翅上的琥珀色面积更大。【长度】体长 37~40 mm，腹长 28~31 mm，后翅 24~25 mm。【栖息环境】海拔500~1000 m森林中的溪流。【分布】广西。【飞行期】5—7月。

[Identification] Male face largely black, frons with a broad yellow stripe. Thorax black, dorsal stripes connecting with collar srtipes, second lateral stripe largely interrupted medially, third lateral stripe complete, wings with amber at basal half. Abdomen black, S1-S3 with small yellow markings, anal appendages white. Female similar to male, wings with larger amber color. **[Measurements]** Total length 37-40 mm, abdomen 28-31 mm, hind wing 24-25 mm. **[Habitat]** Streams in forest at 500-1000 m elevation. **[Distribution]** Guangxi. **[Flight Season]** May to July.

戴春蜓属待定种3 雄，广西
Davidius sp. 3, male from Guangxi

戴春蜓属待定种3 雌，广西
Davidius sp. 3, female from Guangxi

戴春蜓属待定种3 交尾，广西
Davidius sp. 3, mating pair from Guangxi

闽春蜓属 Genus *Fukienogomphus* Chao, 1954

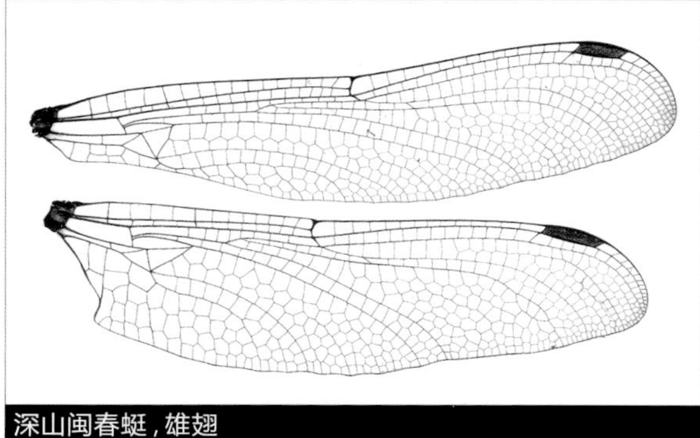

深山闽春蜓，雄翅
Fukienogomphus prometheus, male wings

本属分布于中国、老挝和越南，全球已知的3种都在中国分布。本属蜻蜓体中型；翅透明，翅痣具支持脉，三角室通常无横脉，外边稍微弯曲，后翅三角室约为前翅三角室长度的2倍，前翅的臀脉与翅后缘之间仅有1列翅室，基臀区具1条横脉，后翅基缘略呈角度，臀三角室3～4室。雄性腹部第7～9节膨大，雌性腹部在基方最阔。雄性上肛附器白色，扁而阔，末端尖锐，内缘具齿，下肛附器强烈分歧。前钩片细长，后钩片粗，末端具1个短钩。阳茎无鞭。

本属蜻蜓栖息于茂盛森林中的池塘、小型水潭和流速较缓慢的溪流。雄性停落在水边的植物上占据领地。

The genus is confined to China, Laos and Vietnam with three described species and all are known from China. Species of the genus are medium-sized. Wings hyaline, pterostigma braced, triangle in hind wings long, about twice as long as the triangle in fore wings, a single row of cells between A and hind margin in fore wings, cubital space with one crossvein, tornus slightly angled, anal triangle 3-or 4-celled. Male abdomen expanded at S7-S9, female abdomen broadest at base. Male superior appendages white, flatterned and broad, the tips pointed, inner teeth present, the inferior appendage with two branches strongly divergent. Anterior hamulus narrow and long, posteriors broad with a distal hook. Penis without flagella.

Fukienogomphus species inhabit small pools and ponds in dense forest, also found in slow flowing streams. Males exhibit territorial behavior by perching on plants close to water.

深山闽春蜓 雄 | 宋黎明 摄
Fukienogomphus prometheus, male | Photo by Liming Song

深山闽春蜓，浙江
Fukienogomphus prometheus from Zhejiang

深山闽春蜓，广东
Fukienogomphus prometheus from Guangdong

显著闽春蜓
Fukienogomphus promineus

闽春蜓属 雄性肛附器
Genus *Fukienogomphus*, male anal appenages

赛芳闽春蜓 *Fukienogomphus choifongae* Wilson & Tam, 2006

【形态特征】雄性面部大面积黑色，上颚外方黄色，额横纹黄色，甚阔；胸部黑色，背条纹与领条纹不相连，肩前上点甚小，合胸侧面具3条黄色条纹；腹部黑色，第1~7节具黄斑，上肛附器白色，下肛附器黑色。雌性与雄性相似，尾毛白色。【长度】腹长 37~39 mm，后翅 32~33 mm。【栖息环境】溪流或泥潭。【分布】中国香港特有。【飞行期】4—6月。

[Identification] Male face largely black, base of mandible yellow, frons with a broad yellow stripe. Thorax black, dorsal stripes not connecting with collar stripes, superior spots small, second lateral stripe and third lateral stripe complete. Abdomen black, S1-S7 with yellow markings, superior appendages white, inferior appendage black. Female similar to male with white cerci. **[Measurements]** Abdomen 37-39 mm, hind wing 32-33 mm. **[Habitat]** Streams and muddy puddles. **[Distribution]** Endemic to Hong Kong of China. **[Flight Season]** April to June.

赛芳闽春蜓 雄，香港 | 张智民 摄
Fukienogomphus choifongae, male from Hong Kong | Photo by Cheung Chi Man

赛芳闽春蜓 雌，香港 | 张智民 摄
Fukienogomphus choifongae, female from Hong Kong | Photo by Cheung Chi Man

深山闽春蜓 *Fukienogomphus prometheus* (Lieftinck, 1939)

【形态特征】雄性面部大面积黑色，上颚外方黄色，额横纹黄色，甚阔；胸部黑色，背条纹与领条纹不相连，肩前上点甚小，合胸侧面第2条纹中央间断或完整，第3条纹完整；腹部黑色，第1~7节具黄斑，上肛附器白色，下肛附器黑色。雌性与雄性相似，尾毛白色。赛芳闽春蜓比本种体型小，下肛附器分歧的角度较小。【长度】体长 56~68 mm，腹长 44~51 mm，后翅 37~44 mm。【栖息环境】海拔 1000 m以下森林中的积水潭、沟渠和小型湿地。【分布】浙江、福建、广东、海南、台湾；越南。【飞行期】4—7月。

[Identification] Male face largely black, base of mandible yellow, frons with a broad yellow stripe. Thorax black, dorsal stripes not connecting with collar stripes, superior spots small, second lateral stripe interrupted medially, third lateral stripe complete. Abdomen black, S1-S7 with yellow markings, superior appendages white, inferior appendage black. Female similar to male with white cerci. *F. choifongae* is similar but smaller, the inferior appendage less divaricated. **[Measurements]** Total length 56-68 mm, abdomen 44-51 mm, hind wing 37-44 mm. **[Habitat]** Shallow pools, ditches and small marshes in forest below 1000 m elevation. **[Distribution]** Zhejiang, Fujian, Guangdong, Hainan, Taiwan; Vietnam. **[Flight Season]** April to July.

深山闽春蜓 雄，广东 | 吴宏道 摄
Fukienogomphus prometheus, male from Guangdong | Photo by Hongdao Wu

深山闽春蜓 雌，广东｜宋黎明 摄
Fukienogomphus prometheus, female from Guangdong | Photo by Liming Song

深山闽春蜓 雄，浙江
Fukienogomphus prometheus, male from Zhejiang

显著闽春蜓 *Fukienogomphus promineus* Chao, 1954

显著闽春蜓 雄，广东
Fukienogomphus promineus, male from Guangdong

显著闽春蜓 雄，广东 | 莫善濂 摄
Fukienogomphus promineus, male from Guangdong | Photo by Shanlian Mo

显著闽春蜓 雌，广东 | 莫善濂 摄
Fukienogomphus promineus, female from Guangdong | Photo by Shanlian Mo

【形态特征】雄性面部大面积黑色，上颚外方黄色，额横纹黄色，甚阔；胸部黑色，背条纹与领条纹不相连，肩前上点甚小，合胸侧面第2条纹和第3条纹完整并几乎合并；腹部黑色，第1~7节具黄斑，上肛附器白色，下肛附器黑色。雌性与雄性相似，尾毛白色。【长度】体长 70~79 mm，腹长 53~60 mm，后翅 47~52 mm。【栖息环境】海拔1000 m以下森林中的积水潭、沟渠和渗流地。【分布】福建、广东；越南。【飞行期】4—6月。

[Identification] Male face largely black, base of mandible yellow, frons with a broad yellow stripe. Thorax black, dorsal stripes not connecting with collar stripes, superior spots small, second and third lateral stripes complete and almost combined. Abdomen black, S1-S7 with yellow markings, superior appendages white, inferior appendage black. Female similar to male with white cerci. **[Measurements]** Total length 70-79 mm, abdomen 53-60 mm, hind wing 47-52 mm. **[Habitat]** Shallow pools, ditches and seepages in forest below 1000 m elevation. **[Distribution]** Fujian, Guangdong; Vietnam. **[Flight Season]** April to June.

长腹春蜓属 Genus *Gastrogomphus* Needham, 1941

本属全球仅1种，中国特有。本属春蜓体型较大，身体大面积黄色具黑色条纹，腹部甚长；翅透明，翅痣具支持脉，三角室、上三角室和下三角室无横脉，前翅盘区具2列翅室，前翅的臀脉与翅后缘之间仅有1列翅室，基臀区具1条横脉，臀三角室3室。雄性肛附器短，上肛附器与下肛附器等长，分歧的角度也一致。阳茎无鞭。

本属蜻蜓栖息于池塘和流速缓慢的开阔溪流。雄性经常停落在水边的地面或植物上。

The genus contains a single species endemic to China. Species of the genus is moderately large-sized, body largely yellow with black stripes, abdomen very long. Wings hyaline, pterostigma braced, no crossvein in triangle, hypertriangle and subtriangle, discoidal field with 2 rows of cells in fore wings, a single row of cells between A and hind margin in fore wings, cubital space with one crossvein, anal triangle 3-celled. Male anal appendages short, the superiors and inferior equal in length and divergent in same angle. Penis without flagella.

Gastrogomphus species inhabits ponds and slow flowing streams. Males usually perch on the ground or plants near water.

长腹春蜓 交尾 | 金洪光 摄
Gastrogomphus abdominalis, mating pair | Photo by Hongguang Jin

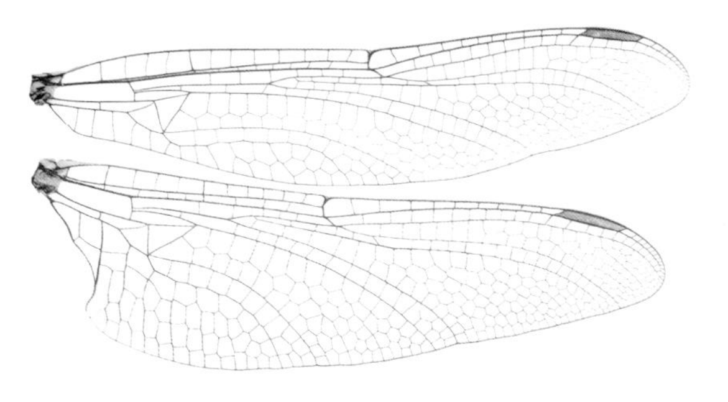

长腹春蜓 雄翅
Gastrogomphus abdominalis, male wings

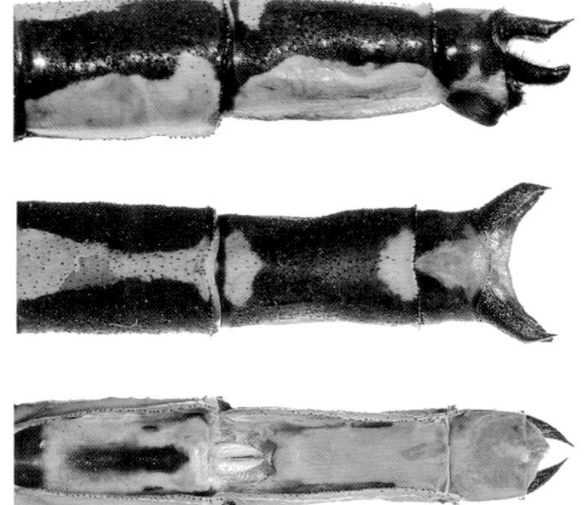

长腹春蜓
Gastrogomphus abdominalis

长腹春蜓属 雄性肛附器和雌性下生殖板
Genus *Gastrogomphus*, male anal appenages appendages and female vulvar lamina

长腹春蜓 *Gastrogomphus abdominalis* (McLachlan, 1884)

【形态特征】雄性复眼绿色，面部黄色；胸部黄色，合胸背面具黑色条纹，足黄色具黑色条纹；腹部黄色，侧面具黑色条纹，肛附器不发达，黑色。雌性与雄性相似，尾毛黑色。【长度】体长 62~66 mm，腹长 47~51 mm，后翅 35~42 mm。【栖息环境】海拔 500 m以下的池塘和流速缓慢的溪流。【分布】中国特有，分布于吉林、北京、河北、河南、江苏、安徽、湖北、湖南、浙江、福建。【飞行期】4—7月。

[Identification] Male eyes green, face yellow. Thorax yellow, dorsal part with black stripes, legs yellow with black markings. Abdomen yellow with lateral black stripes, anal appendages short and black. Female similar to male with black cerci. **[Measurements]** Total length 62-66 mm, abdomen 47-51 mm, hind wing 35-42 mm. **[Habitat]** Ponds and slow flowing streams below 500 m elevation. **[Distribution]** Endemic to China, recorded from Jilin, Beijing, Hebei, Henan, Jiangsu, Anhui, Hubei, Hunan, Zhejiang, Fujian. **[Flight Season]** April to July.

长腹春蜓 雄，湖北
Gastrogomphus abdominalis, male from Hubei

长腹春蜓 雄，湖北
Gastrogomphus abdominalis, male from Hubei

长腹春蜓 雌，安徽
Gastrogomphus abdominalis, female from Anhui

小叶春蜓属 Genus *Gomphidia* Selys, 1854

本属全球已知20余种，分布于亚洲和非洲。中国已知5种，分布于华中、华南和西南地区。本属蜻蜓是一类体型较大且粗壮的春蜓；翅透明，翅痣甚长，具支持脉，三角室3～4室，上三角室1～3室，前翅下三角室2～3室，后翅下三角室通常1室，具较小的臀圈，前翅臀脉与翅后缘之间具2列翅室，基臀区具2条横脉，后翅基缘呈钝角。雄性的下肛附器甚短。阳茎具1对鞭。

本属蜻蜓主要栖息于山区溪流，少数生活在静水环境。雄性经常停立在枝头上占据领地，并时而飞行巡逻，雄性间经常展开争斗。交尾时停落在水附近的树丛中。雌性产卵时先排出巨大的卵块，然后将其投入水中。

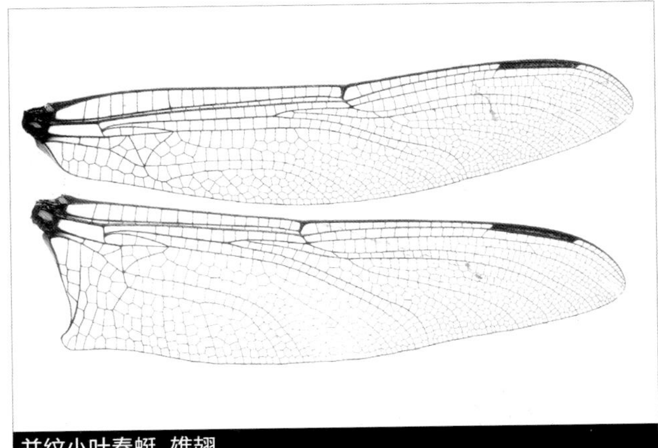

并纹小叶春蜓 雄翅
Gomphidia kruegeri, male wings

The genus contains over 20 species, distributed in Asia and Africa. Five species are recorded from China, found in the Central, South and Southwest. Species of the genus are large-sized and robust species. Wings hyaline, pterostigma long and braced, triangle 3-4 celled, hypertriangle 1-3 celled, subtriangle in fore wing 2-3 celled and usually 1 in hind wing, anal loop small, 2 rows of cells between A and hind margin in fore wings, cubital space with two crossveins, tornus obtuse angled. Male inferior appendage very short. Penis with a pair of flagella.

并纹小叶春蜓 雄
Gomphidia kruegeri, male

Gomphidia species mainly inhabit montane streams, a few species found also at standing water. Males usually perch on top of branches within their territory, sometimes patrol, males fight frequently. Mating can be seen in trees nearby water. Females exude a huge egg mass first and then deposit it into water.

福建小叶春蜓
Gomphidia fukienensis

克氏小叶春蜓
Gomphidia kelloggi

并纹小叶春蜓
Gomphidia kruegeri

黄纹小叶春蜓
Gomphidia abbotti

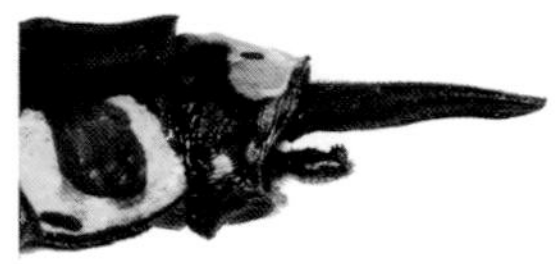

联纹小叶春蜓
Gomphidia confluens

小叶春蜓属 雄性肛附器
Genus *Gomphidia*, male anal appendages

福建小叶春蜓
Gomphidia fukienensis

并纹小叶春蜓
Gomphidia kruegeri

小叶春蜓属 雄性后钩片
Genus *Gomphidia*, Male posterior hamulus

福建小叶春蜓
Gomphidia fukienensis

克氏小叶春蜓
Gomphidia kelloggi

并纹小叶春蜓
Gomphidia kruegeri

黄纹小叶春蜓
Gomphidia abbotti

联纹小叶春蜓
Gomphidia confluens

小叶春蜓属 雌性下生殖板
Genus *Gomphidia*, female vulvar lamina

黄纹小叶春蜓 *Gomphidia abbotti* Williamson, 1907

【形态特征】雄性面部大面积黑褐色，上唇具1对甚大的黄斑，后唇基侧下缘黄色，额横纹黄色，甚阔，侧单眼后方具1对锥形凸起，后头缘稍微隆起；胸部黑褐色，背条纹与领条纹不相连，合胸侧面具2条黄色条纹；腹部黑色，第2~7节基方具甚大的黄斑，第8~10节黄斑较小。雌性与雄性相似，但更粗壮。【长度】体长 68~70 mm，腹长 50~52 mm，后翅 39~40 mm。【栖息环境】海拔 500 m以下的河流、开阔溪流和沟渠。【分布】云南、广西、海南；缅甸、泰国、柬埔寨、老挝、越南、马来半岛、苏门答腊。【飞行期】3—6月。

[Identification] Male face largely blackish brown, labrum with a pair of large spots, lower lateral edge of postclypeus yellow, frons with a broad yellow stripe, vertex with a pair of pyramidal prominences behind lateral ocelli, occipital margin slightly raised. Thorax blackish brown, dorsal stripes and collar stripes not connecting, sides with two broad yellow stripes. Abdomen black, S2-S7 with large basal spots, S8-S10 with smaller spots. Female similar to male but stouter. **[Measurements]** Total length 68-70 mm, abdomen 50-52 mm, hind wing 39-40 mm. **[Habitat]** Rivers, exposed streams and ditches below 500 m elevation. **[Distribution]** Yunnan, Guangxi, Hainan; Myanmar, Thailand, Cambodia, Laos, Vietnam, Peninsular Malaysia, Sumatra. **[Flight Season]** March to June.

黄纹小叶春蜓 雄，云南（西双版纳）
Gomphidia abbotti, male from Yunnan (Xishuangbanna)

黄纹小叶春蜓 雌，海南 | 莫善濂 摄
Gomphidia abbotti, female from Hainan | Photo by Shanlian Mo

黄纹小叶春蜓 雄，云南（西双版纳）
Gomphidia abbotti, male from Yunnan (Xishuangbanna)

联纹小叶春蜓 *Gomphidia confluens* Selys, 1878

联纹小叶春蜓 雄，安徽 | 秦彧 摄
Gomphidia confluens, male from Anhui | Photo by Yu Qin

联纹小叶春蜓 雄，浙江
Gomphidia confluens, male from Zhejiang

【形态特征】雄性面部大面积黄色，侧单眼后方具1对锥形凸起，后头黑色，后头缘稍微隆起；胸部黑褐色，背条纹与领条纹相连，具甚细小的肩前条纹和肩前上点，合胸侧面大面积黄色，后胸侧缝线黑色；腹部黑色，各节具大小和形状不同的黄斑。雌性与雄性相似，但更粗壮。**【长度】**体长 73~75 mm，腹长 53~54 mm，后翅 46~48 mm。**【栖息环境】**海拔 1000 m以下的池塘、河流、开阔溪流和沟渠。**【分布】**黑龙江、吉林，辽宁、北京、河北、安徽、江苏、湖北、浙江、福建、广东；朝鲜半岛、俄罗斯远东、越南。**【飞行期】**4—8月。

[Identification] Male face largely yellow, vertex with a pair of pyramidal prominences behind lateral ocelli, occipital margin slightly raised and black. Thorax blackish brown, dorsal stripes and collar stripes connecting, superior spots small, antehumeral stripes narrow, sides largely yellow, metapleural suture black. Abdomen black with yellow markings on each segments. Female similar to male but stouter. **[Measurements]** Total length 73-75 mm, abdomen 53-54 mm, hind wing 46-48 mm. **[Habitat]** Ponds, rivers, exposed streams and ditches below 1000 m elevation. **[Distribution]** Heilongjiang, Jilin, Liaoning, Beijing, Hebei, Anhui, Jiangsu, Hubei, Zhejiang, Fujian, Guangdong; Korean peninsula, Russian Far East, Vietnam. **[Flight Season]** April to August.

福建小叶春蜓 *Gomphidia fukienensis* Chao, 1955

【形态特征】与并纹小叶春蜓相似，但后唇基侧下缘具黄斑；雄性肛附器、后钩片以及雌性下生殖板的构造不同。曾被认为是并纹小叶春蜓的亚种，此处提升至种。【长度】体长 78～82 mm，腹长 57～61 mm，后翅 48～50 mm。【栖息环境】海拔 1000 m以下的河流、溪流和沟渠。【分布】中国特有，分布于贵州、湖北、浙江、福建、广东、台湾。【飞行期】4—8月。

[Identification] The specie is similar to *G. kruegeri*, but lateral lower edge of postclypeus with yellow spots, male anal appendages, posterior hamulus and female vulvar lamina are different. The species was previously regarded as a subspecies of *G. kruegeri*, which is raised to species level here. **[Measurements]** Total length 78-82 mm, abdomen 57-61 mm, hind wing 48-50 mm. **[Habitat]** Rivers, streams and ditches below 1000 m elevation. **[Distribution]** Endemic to China, recorded from Guizhou, Hubei, Zhejiang, Fujian, Guangdong, Taiwan. **[Flight Season]** April to August.

福建小叶春蜓 雄，浙江
Gomphidia fukienensis, male from Zhejiang

克氏小叶春蜓 *Gomphidia kelloggi* Needham, 1930

【形态特征】雄性面部黑色具黄斑，上唇中央具1个甚大黄斑，侧单眼后方具1对锥形凸起，后头黑色，后头缘稍微隆起；胸部黑色，背条纹与领条纹不相连，具甚细小的肩前上点，合胸侧面第2条纹和第3条纹完整，甚阔；腹部黑色，除第9节外各节具黄斑。雌性与雄性相似，但更粗壮。【长度】体长 68~74 mm，腹长 51~55 mm，后翅 42~46 mm。【栖息环境】海拔 500 m以下的开阔溪流和沟渠。【分布】中国特有，分布于福建、广东、香港。【飞行期】4—6月。

[Identification] Male face black with yellow markings, labrum with a large yellow spot, vertex with a pair of pyramidal prominences behind lateral ocelli, occiput black, occipital margin slightly raised. Thorax black, dorsal stripes and collar stripes not connecting, superior spots very small, second and third lateral stripes complete and broad. Abdomen black with yellow markings except S9. Female similar to male but stouter. **[Measurements]** Total length 68-74 mm, abdomen 51-55 mm, hind wing 42-46 mm. **[Habitat]** Exposed streams and ditches below 500 m elevation. **[Distribution]** Endemic to China, recorded from Fujian, Guangdong, Hong Kong. **[Flight Season]** April to June.

克氏小叶春蜓 雄，广东 | 宋黎明 摄
Gomphidia kelloggi, male from Guangdong | Photo by Liming Song

克氏小叶春蜓 雄，广东 | 吴宏道 摄
Gomphidia kelloggi, male from Guangdong | Photo by Hongdao Wu

克氏小叶春蜓 雌，广东 | 吴宏道 摄
Gomphidia kelloggi, female from Guangdong | Photo by Hongdao Wu

并纹小叶春蜓 *Gomphidia kruegeri* Martin, 1904

【形态特征】雄性面部黑色具黄斑，上唇中央具1个甚大黄斑，额横纹甚阔，侧单眼后方具1对锥形凸起，后头黑色，后头缘稍微隆起；胸部黑色，背条纹与领条纹不相连，具甚细小的肩前上点，合胸侧面第2条纹和第3条纹完整，甚阔；腹部黑色，除第9节外各节具黄斑，下肛附器几乎退化。雌性与雄性相似，但更粗壮，后头后方中央具1个瘤状隆起。【长度】体长 78～84 mm，腹长 60～62 mm，后翅 45～53 mm。【栖息环境】1000 m以下的河流、溪流和沟渠。【分布】贵州、云南、福建、广西、广东、海南；泰国、老挝、越南。【飞行期】3—9月。

[Identification] Male face black with yellow markings, labrum with a large yellow spot, frons with a broad yellow stripe, vertex with a pair of pyramidal prominences behind lateral ocelli, occiput black, occipital margin slightly raised. Thorax black, dorsal stripes and collar stripes not connecting, superior spots very small, second and third lateral stripes complete and broad. Abdomen black with yellow markings except S9, inferior appendage reduced. Female similar to male but stouter, occipital margin with a median tubercle. **[Measurements]** Total length 78-84 mm, abdomen 60-62 mm, hind wing 45-53 mm. **[Habitat]** Rivers, streams and ditches below 1000 m elevation. **[Distribution]** Guizhou, Yunnan, Fujian, Guangxi, Guangdong, Hainan; Thailand, Laos, Vietnam. **[Flight Season]** March to September.

并纹小叶春蜓 雄，云南（西双版纳）
Gomphidia kruegeri, male from Yunnan (Xishuangbanna)

并纹小叶春蜓 雄，广东
Gomphidia kruegeri, male from Guangdong

并纹小叶春蜓 雌，海南
Gomphidia kruegeri, female from Hainan

小叶春蜓属待定种1 *Gomphidia* sp. 1

【形态特征】雄性面部黑色具黄斑，上唇中央具1个甚大黄斑，额横纹甚阔，侧单眼后方具1对锥形凸起，后头黑色，后头缘稍微隆起；胸部黑色，背条纹与领条纹不相连，具甚细小的肩前上点，合胸侧面第2条纹和第3条纹几乎合并；腹部黑色具黄斑，下肛附器几乎退化。雌性与雄性相似，后头后方中央具1个瘤状隆起。【长度】体长 71~73 mm，腹长 53~54 mm，后翅 42~45 mm。【栖息环境】海拔 1000 m以下的溪流和沟渠。【分布】云南（西双版纳）。【飞行期】4—6月。

[Identification] Male face black with yellow markings, labrum with a large yellow spot, frons with a broad yellow stripe, vertex with a pair of pyramidal prominences behind lateral ocelli, occiput black, occipital margin slightly raised. Thorax black, dorsal stripes and collar stripes not connecting, superior spots very small, second and third lateral stripes almost combined. Abdomen black with yellow markings, inferior appendage reduced. Female similar to male, occipital margin with a median tubercle. **[Measurements]** Total length 71-73 mm, abdomen 53-54 mm, hind wing 42-45 mm. **[Habitat]** Streams and ditches below 1000 m elevation. **[Distribution]** Yunnan (Xishuangbanna). **[Flight Season]** April to June.

小叶春蜓属待定种1 雄，云南（西双版纳）
Gomphidia sp. 1, male from Yunnan (Xishuangbanna)

小叶春蜓属待定种1 雌，云南（西双版纳）
Gomphidia sp. 1, female from Yunnan (Xishuangbanna)

小叶春蜓属待定种1 雄，云南（西双版纳）
Gomphidia sp. 1, male from Yunnan (Xishuangbanna)

小叶春蜓属待定种2 *Gomphidia* sp. 2

小叶春蜓属待定种2 雌，云南（西双版纳）
Gomphidia sp. 2, female from Yunnan (Xishuangbanna)

小叶春蜓属待定种2 雄，云南（西双版纳）
Gomphidia sp. 2, male from Yunnan (Xishuangbanna)

【形态特征】雄性面部几乎完全黄色，侧单眼后方具1对锥形凸起，后头黑色，后头缘稍微隆起；胸部黑色，背条纹与领条纹不相连，具甚细小的肩前上点，合胸侧面第2条纹和第3条纹几乎合并；腹部黑色，第2～10节具黄斑，下肛附器几乎退化。雌性与雄性相似，后头中央呈锥状隆起。【长度】体长 68～73 mm，腹长 51～54 mm，后翅 42～45 mm。【栖息环境】海拔 500 m以下的开阔溪流和河流。【分布】云南（西双版纳）。【飞行期】3—5月。

[Identification] Male face mostly yellow, vertex with a pair of pyramidal prominences behind lateral ocelli, occiput black, occipital margin slightly raised. Thorax black, dorsal stripes and collar stripes not connecting, superior spots very small, second and third lateral stripes almost combined. Abdomen black, S2-S10 with yellow markings, inferior appendage reduced. Female similar to male, occipital margin pyramidally erected. **[Measurements]** Total length 68-73 mm, abdomen 51-54 mm, hind wing 42-45 mm. **[Habitat]** Exposed streams and rivers below 500 m elevation. **[Distribution]** Yunnan (Xishuangbanna). **[Flight Season]** March to May.

类春蜓属 Genus *Gomphidictinus* Fraser, 1942

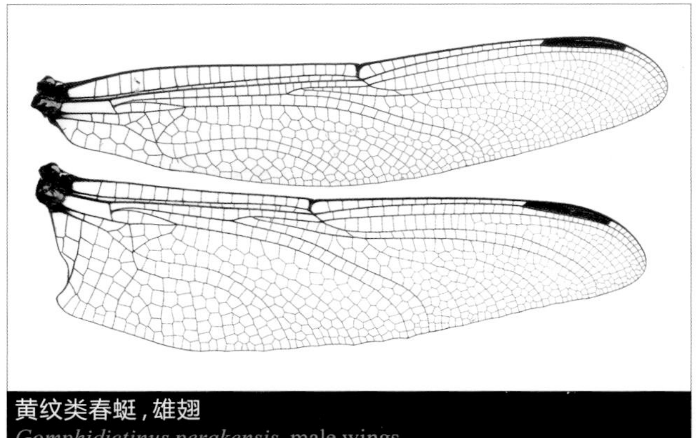

黄纹类春蜓，雄翅
Gomphidictinus perakensis, male wings

本属全球仅知3种，主要分布于亚洲的热带区域。中国已知2种，分布于云南、广西和海南。本属蜻蜓是一类体型较大且粗壮的春蜓，外观与小叶春蜓相似；翅透明，翅痣甚长，具支持脉，三角室4室，上三角室3～4室，下三角室3室，臀圈开放，前翅臀脉与翅后缘之间仅具2列翅室，基臀区具2条横脉，后翅基缘呈钝角。雄性下肛附器较短。

本属蜻蜓栖息于山区的清澈溪流。雄性经常停在枝头上占据领地，并时而飞行巡逻。雄性间经常展开争斗。

The genus contains three species distributed in tropical Asia. Two species are recorded from China, found in Yunnan, Guangxi and Hainan. Species of the genus are large-sized and robust, the habitus similar to species of *Gomphidia*. Wings hyaline, pterostigma long and braced, triangle 4-celled, hypertriangle 3-4 celled, subtriangle 3-celled, anal loop open, 2 rows of cells between A and hind margin in fore wings, cubital space with two crossveins, tornus obtuse angled. Male inferior appendage short.

Gomphidictinus species inhabit montane streams. Males usually perch on top of branches, and sometimes patrol. Males fight frequently.

黄纹类春蜓
Gomphidictinus perakensis

童氏类春蜓
Gomphidictinus tongi

类春蜓属 雌性下生殖板
Genus *Gomphidictinus*, female vulvar lamina

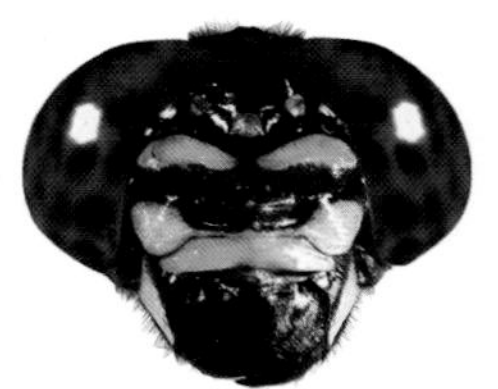

黄纹类春蜓
Gomphidictinus perakensis

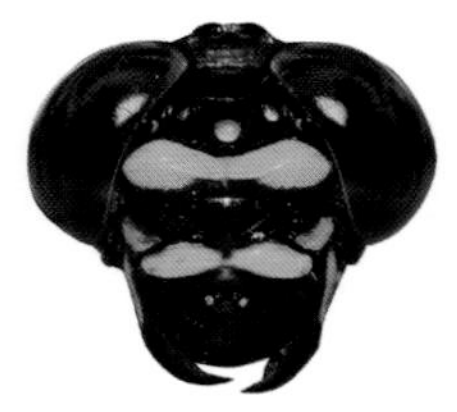

童氏类春蜓
Gomphidictinus tongi

类春蜓属 雌性头部
Genus *Gomphidictinus*, female head in frontal view

童氏类春蜓
Gomphidictinus tongi

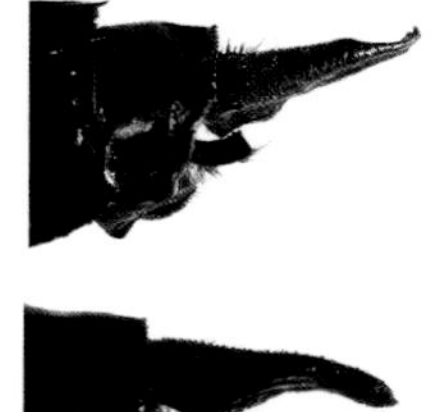

黄纹类春蜓
Gomphidictinus perakensis

类春蜓属 雄性肛附器
Genus *Gomphidictinus*, male anal appendages

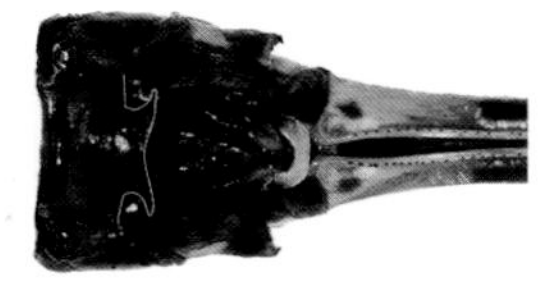

黄纹类春蜓
Gomphidictinus perakensis

类春蜓属 雄性腹部基方，示第1节突起
Genus *Gomphidictinus*, male abdominal base, showing the ventral tubercle

黄纹类春蜓 雄
Gomphidictinus perakensis, male

黄纹类春蜓 *Gomphidictinus perakensis* (Laidlaw, 1902)

【形态特征】雄性面部黑色具黄斑，额横纹中央间断，侧单眼后方具1对锥形凸起，后头缘中央稍微隆起；胸部黑色，背条纹与领条纹不相连，具甚小的肩前上点，合胸侧面第2条纹和第3条纹合并；腹部黑色，第1节腹面具鲸尾状突起，第2～6节侧面和背面基方具黄条纹，第7节具甚大黄斑。雌性与雄性相似，但更粗壮，未熟时两性腹部基方黄色。【长度】体长 73～77 mm，腹长 53～57 mm，后翅 49～53 mm。【栖息环境】海拔 1000 m以下森林中的小溪。【分布】云南（临沧、普洱、西双版纳）；泰国、老挝、柬埔寨、越南、马来半岛。【飞行期】5—9月。

[Identification] Male face black with yellow markings, frons with a broad stripe interrupted medially, vertex with a pair of pyramidal prominences behind lateral ocelli, occipital margin slightly raised. Thorax black, dorsal stripes and collar stripes not connecting, superior spots very small, second and third lateral stripes almost combined. Abdomen black, S1 with a whale tail shaped tubercle ventrally, S2-S6 with yellow spots laterally and dorsally, S7 with a large

黄纹类春蜓 雄，云南（西双版纳）
Gomphidictinus perakensis, male from Yunnan (Xishuangbanna)

yellow spot. Female similar but stouter. Abdominal base yellow in immature. **[Measurements]** Total length 73-77 mm, abdomen 53-57 mm, hind wing 49-53 mm. **[Habitat]** Montane streams below 1000 m elevation. **[Distribution]** Yunnan (Lincang, Pu'er, Xishuangbanna); Thailand, Laos, Cambodia, Vietnam, Peninsular Malaysia. **[Flight Season]** May to September.

黄纹类春蜓 雄，云南（西双版纳）
Gomphidictinus perakensis, male from Yunnan (Xishuangbanna)

黄纹类春蜓 雌，云南（普洱）
Gomphidictinus perakensis, female from Yunnan (Pu'er)

童氏类春蜓 *Gomphidictinus tongi* Zhang, Guan & Wang, 2017

【形态特征】雄性面部黑色具黄斑，额横纹甚阔，侧单眼后方具1对锥形凸起，后头缘稍微隆起；胸部黑色，背条纹与领条纹不相连，具甚细小的肩前上点，合胸侧面第2条纹和第3条纹几乎合并；腹部黑色，第1～7节侧面和背面基方具黄斑，其中第7节黄斑甚大，第10节背面具1对小黄斑。雌性与雄性相似，但更粗壮，后头缘具1对片状突起。【长度】体长 78～85 mm，腹长 58～63 mm，后翅 48～54 mm。【栖息环境】海拔 1000 m以下的林荫小溪。【分布】广西、海南；越南。【飞行期】5—7月。

童氏类春蜓 雄，海南
Gomphidictinus tongi, male from Hainan

[Identification] Male face black with yellow markings, frons with a broad stripe, vertex with a pair of pyramidal prominences behind lateral ocelli, occipital margin slightly raised. Thorax black, dorsal stripes and collar stripes not connecting, superior spots very small, second and third lateral stripes almost combined. Abdomen black, S1-S7 with yellow spots laterally and dorsally, S7 with a large yellow spot, S10 with a pair of small yellow spots. Female similar to male but stouter, occipital margin with a pair of lobed projections. **[Measurements]** Total length 78-85 mm, abdomen 58-63 mm, hind wing 48-54 mm. **[Habitat]** Shady streams in forest below 1000 m elevation. **[Distribution]** Guangxi, Hainan; Vietnam. **[Flight Season]** May to July.

童氏类春蜓 雄，海南
Gomphidictinus tongi, male from Hainan

童氏类春蜓 雌，海南
Gomphidictinus tongi, female from Hainan

曦春蜓属 Genus *Heliogomphus* Laidlaw, 1922

本属全球已知20余种，分布于亚洲的热带和亚热带区域。中国已知3种，分布于华南和西南地区。本属蜻蜓是一类体中型的春蜓；翅透明，三角室、下三角室和上三角室通常没有横脉，前翅臀脉与翅后缘之间仅有1列翅室，基臀区具1~2条横脉，后翅基缘呈钝角。雄性上肛附器末端弯曲如牛角。

本属蜻蜓栖息于山区的清澈溪流。雄性经常停在水面的岩石上占据领地。有些种类的雌性产卵前先排出1个巨大的卵块，然后将其投入水中。

The genus contains over 20 species, distributed in tropical and subtropical Asia. Three species are recorded from China, found in the South and Southwest. Species of the genus are medium-sized. Wings hyaline, triangle, hypertriangle and subtriangle without crossvein, a single row of cells between A and hind margin in fore wings, cubital space with 1-2 crossveins, tornus obtuse angled. Male superior appendages strongly curved like bull horns.

Heliogomphus species inhabit montane streams. Males usually perch on rocks above water for territory. Some females exude an egg mass first and then deposit it into water.

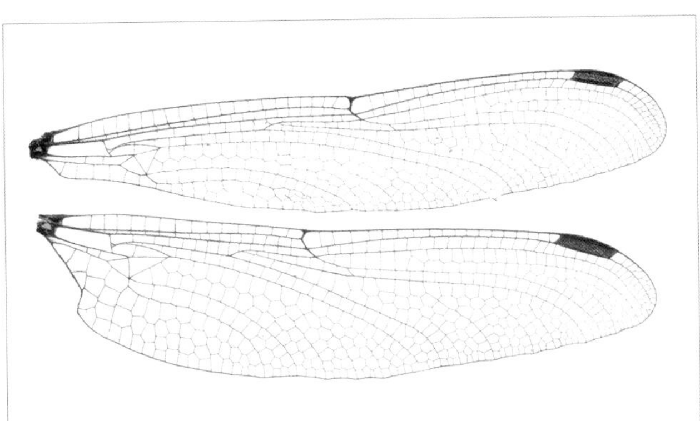

独角曦春蜓 雄翅
Heliogomphus scorpio, male wings

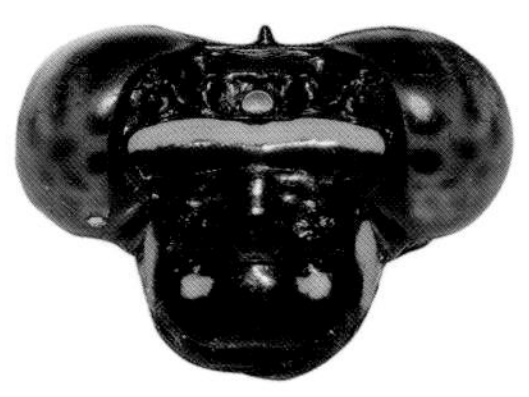

独角曦春蜓
Heliogomphus scorpio

扭尾曦春蜓
Heliogomphus retroflexus

独角曦春蜓
Heliogomphus scorpio

赛丽曦春蜓
Heliogomphus selysi

曦春蜓属 雌性头部
Genus *Heliogomphus*, female head in frontal view

曦春蜓属 雄性肛附器
Genus *Heliogomphus*, male appendages

独角曦春蜓 雄
Heliogomphus scorpio, male

扭尾曦春蜓 *Heliogomphus retroflexus* (Ris, 1912)

【形态特征】雄性面部黑色具黄斑，额横纹甚阔；胸部黑色，背条纹与领条纹不相连，合胸侧面第2条纹和第3条纹完整；腹部黑色，第1～3节侧面具黄斑，第2～7节背面基方具黄环纹，上肛附器黄白色，末端扭曲，下肛附器黑色。雌性与雄性相似，尾毛白色。【长度】体长 50～52 mm，腹长 37～39 mm，后翅 31～34 mm。【栖息环境】海拔 1000 m以下森林中的小溪。【分布】福建、浙江、广东、海南、台湾；老挝、越南。【飞行期】3—10月。

[Identification] Male face black with yellow markings, frons with a broad stripe. Thorax black, dorsal stripes and collar stripes not connecting, second and third lateral stripes complete. Abdomen black, S1-S3 with lateral yellow spots, S2-S7 with basal yellow rings, superior appendages yellowish white and curved apically, inferior appendage black. Female similar to male, cerci white. **[Measurements]** Total length 50-52 mm, abdomen 37-39 mm, hind wing 31-34 mm. **[Habitat]** Montane streams below 1000 m elevation. **[Distribution]** Fujian, Zhejiang, Guangdong, Hainan, Taiwan; Laos, Vietnam. **[Flight Season]** March to October.

扭尾曦春蜓 雄，广东
Heliogomphus retroflexus, male from Guangdong

扭尾曦春蜓 雌，广东
Heliogomphus retroflexus, female from Guangdong

独角曦春蜓 *Heliogomphus scorpio* (Ris, 1912)

【形态特征】雄性面部黑色具黄斑，额横纹中央间断；胸部黑色，背条纹与领条纹相连，肩前条纹仅有上方的一段，合胸侧面第2条纹和第3条纹完整；腹部黑色，第1～3节侧面具黄斑，第2～6节具甚细的背中条纹，第7节基方具甚大的黄白色斑。肛附器黑色，上肛附器叉状，甚阔。雌性与雄性相似，后头缘中央具1个角状突起。【长度】体长 57～60 mm，腹长 43～45 mm，后翅 38～40 mm。【栖息环境】海拔 1000 m以下森林中的小溪。【分布】云南（红河）、福建、浙江、广东、广西、海南、香港；老挝、越南。【飞行期】4—9月。

[Identification] Male face black with yellow markings, frons with a broad stripe interrupted medially. Thorax black, dorsal stripes and collar stripes connecting, antehumeral stripes with only upper half present, second and third lateral stripes complete. Abdomen black, S1-S3 with lateral yellow spots, S2-S6 with narrow stripes along the carina, S7 with a large yellowish white spot, superior appendages black, fork-shaped and expanded laterally. Female similar to male, occiput with a median horn. **[Measurements]** Total length 57-60 mm, abdomen 43-45 mm, hind wing 38-40 mm. **[Habitat]** Montane streams below 1000 m elevation. **[Distribution]** Yunnan (Honghe), Fujian, Zhejiang, Guangdong, Guangxi, Hainan, Hong Kong; Laos, Vietnam. **[Flight Season]** April to September.

独角曦春蜓 雄，云南（红河）
Heliogomphus scorpio, male from Yunnan (Honghe)

独角曦春蜓 雌，广东
Heliogomphus scorpio, female from Guangdong

独角曦春蜓 雄，广西
Heliogomphus scorpio, male from Guangxi

赛丽曦春蜓 *Heliogomphus selysi* Fraser, 1925

【形态特征】雄性面部黑色具黄斑，上唇具1对黄斑，额横纹甚阔；胸部黑色，背条纹与领条纹不相连，具甚小的肩前上点，合胸侧面第2条纹和第3条纹完整；腹部黑色，第1~3节侧面具黄斑，第1~7节具黄色的背中条纹，上肛附器黄白色，末端扭曲，下肛附器黑色。【长度】雄性体长 48 mm，腹长 36 mm，后翅 30 mm。【栖息环境】海拔 1000 m 以下森林中的狭窄小溪。【分布】云南（西双版纳）；印度、缅甸、泰国、老挝。【飞行期】5—7月。

[Identification] Male face black with yellow markings, labrum with a pair of yellow spots, frons with a broad stripe. Thorax black, dorsal stripes and collar stripes not connecting, superior spots very small, second and third lateral stripes complete. Abdomen black, S1-S3 with lateral yellow spots, S1-S7 with yellow stripes along the carina, superior appendages yellowish white and curved at tip, inferior appendage black. **[Measurements]** Male total length 48 mm, abdomen 36 mm, hind wing 30 mm. **[Habitat]** Narrow montane streams below 1000 m elevation. **[Distribution]** Yunnan (Xishuangbanna); India, Myanmar, Thailand, Laos. **[Flight Season]** May to July.

赛丽曦春蜓 雄，云南（西双版纳）
Heliogomphus selysi, male from Yunnan (Xishuangbanna)

赛丽曦春蜓 雄，云南（西双版纳）
Heliogomphus selysi, male from Yunnan (Xishuangbanna)

曦春蜓属待定种 *Heliogomphus* sp.

【形态特征】雌性面部黑色具黄斑，额横纹甚阔，后头缘中央具1个角状突起；胸部褐色，背条纹与领条纹相连，肩前上点三角形，合胸侧面第2条纹和第3条纹完整；腹部黑色，第1～6节侧面具黄斑，第2～6节具甚细的背中条纹，第7节基方具甚大的黄斑。【长度】雌性体长 53 mm，腹长 42 mm，后翅 38 mm。【栖息环境】海拔 800 m处茂密森林中的狭窄小溪。【分布】云南（西双版纳）。【飞行期】4—5月。

[Identification] Female face black with yellow markings, frons with a broad stripe, occiput with a median horn. Thorax brown, dorsal stripes and collar stripes connecting, superior spots triangular shaped, second and third lateral stripes complete. Abdomen black, S1-S6 with lateral yellow spots, S2-S6 with narrow stripes along carina, S7 with a large yellow spot. **[Measurements]** Female total length 53 mm, abdomen 42 mm, hind wing 38 mm. **[Habitat]** Narrow streams in dense forest at 800 m elevation. **[Distribution]** Yunnan (Xishuangbanna). **[Flight Season]** April to May.

曦春蜓属待定种 雌，云南（西双版纳）
Heliogomphus sp., female from Yunnan (Xishuangbanna)

叶春蜓属 Genus *Ictinogomphus* Cowley, 1934

本属全球已知约15种，分布于亚洲、非洲和大洋洲。中国已知2种，分布于华中、华南和西南等地。本属蜻蜓是一类体型较大的春蜓；翅透明，翅痣甚长，具支持脉，三角室2~3室，上三角室2~3室，前翅下三角室2室，后翅1室，臀圈4室，前翅臀脉与翅后缘之间仅具2列翅室，基臀区具2条横脉，后翅基缘呈钝角。腹部第8节侧缘具叶片状突起。雄性下肛附器较短。

本属蜻蜓栖息于各类静水水域和流速缓慢的溪流。雄性经常停在枝头上占据领地，并时而飞行巡逻。雄性间经常展开争斗。交尾时间较短，在空中进行。雄性护卫产卵。

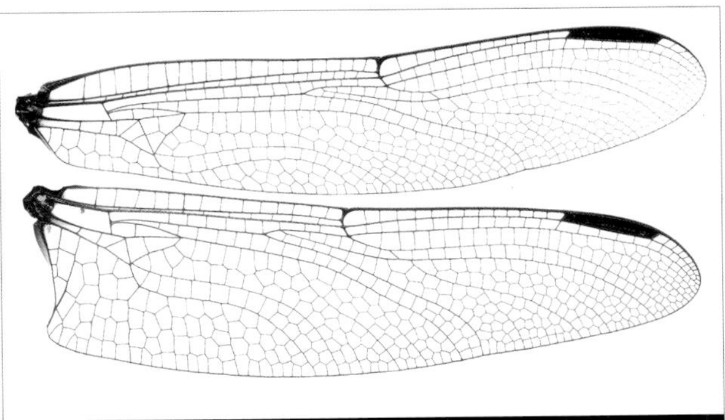
华饰叶春蜓 雄翅
Ictinogomphus decoratus, male wings

The genus contains 15 species, widespread in Asia, Africa and Australia. Two are recorded from China, found in the Central, South and Southwest regions. Species of the genus are large-sized gomphids. Wings hyaline, pterostigma long and braced, triangle 2-3 celled, hypertriangle 2-3 celled, subtriangle in fore wing 2-celled and only 1-celled in hind wing, anal loop 4-celled, 2 rows of cells between A and hind margin in fore wings, cubital space with two crossveins, tornus obtuse angled. S8 with lateral leaf-like projection. Male inferior appendage short.

Ictinogomphus species inhabit both standing water and slow flowing streams. Males usually perch on top of branches, sometimes patrol. Males fight frequently. Mating duration is short and in flight. Females oviposit with male guarding.

华饰叶春蜓
Ictinogomphus decoratus

霸王叶春蜓
Ictinogomphus pertinax

叶春蜓属 雄性腹部末端
Genus *Ictinogomphus*, male distal abdomen

华饰叶春蜓
Ictinogomphus decoratus

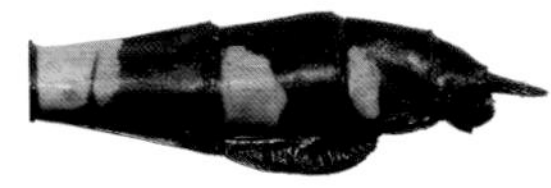
霸王叶春蜓
Ictinogomphus pertinax

叶春蜓属 雌性腹部末端
Genus *Ictinogomphus*, female distal abdomen

华饰叶春蜓 雄
Ictinogomphus decoratus, male

华饰叶春蜓 *Ictinogomphus decoratus* (Selys, 1854)

【形态特征】雄性面部黑色具黄斑，上唇具1对黄斑，额横纹甚阔，侧单眼后方具角状突起，后头黄色；胸部黑色，背条纹与领条纹不相连，肩前上点甚小，合胸侧面第2条纹和第3条纹合并；腹部黑色，第2～9节具黄斑。雌性与雄性相似。**【长度】**体长 65～67 mm，腹长 47～49 mm，后翅 37～40 mm。**【栖息环境】**海拔 1000 m以下的池塘。**【分布】**云南（德宏、临沧、普洱、西双版纳）；东南亚广布。**【飞行期】**3—12月。

[Identification] Male face black with yellow markings, labrum with a pair of yellow spots, frons with a broad stripe, vertex with pyramidal prominences behind lateral ocelli, occiput yellow. Thorax black, dorsal stripes and collar stripes not connecting, superior spots very small, second and third lateral stripes combined. Abdomen black, S2-S9 with lateral yellow markings. Female similar to male. **[Measurements]** Total length 65-67 mm, abdomen 47-49 mm, hind wing 37-40 mm. **[Habitat]** Ponds below 1000 m elevation. **[Distribution]** Yunnan (Dehong, Lincang, Pu'er, Xishuangbanna); Widespread in Southeast Asia. **[Flight Season]** March to December.

华饰叶春蜓 雄，云南（西双版纳）
Ictinogomphus decoratus, male from Yunnan (Xishuangbanna)

华饰叶春蜓 雄，云南（西双版纳）
Ictinogomphus decoratus, male from Yunnan (Xishuangbanna)

霸王叶春蜓 *Ictinogomphus pertinax* (Hagen, 1854)

【形态特征】雄性面部黑色具黄斑，上唇具1对黄斑，额横纹甚阔，侧单眼后方具角状突起，后头黄色；胸部黑色，背条纹与领条纹不相连，具肩前上点和肩前下条纹，合胸侧面第2条纹和第3条纹完整；腹部黑色，第2～9节具黄斑。雌性与雄性相似。【长度】体长 68～72 mm，腹长 49～54 mm，后翅 40～45 mm。【栖息环境】海拔 1500 m以下的池塘、河流和溪流。【分布】中国南方广布；印度、缅甸、老挝、越南、日本。【飞行期】3—12月。

[Identification] Male face black with yellow markings, labrum with a pair of yellow spots, frons with a broad stripe, vertex with pyramidal prominences behind lateral ocelli, occiput yellow. Thorax black, dorsal stripes and collar stripes not connecting, superior spots and antehumeral stripes present, second and third lateral stripes complete.

霸王叶春蜓 雌，广东 | 吴宏道 摄
Ictinogomphus pertinax, female from Guangdong | Photo by Hongdao Wu

Abdomen black, S2-S9 with lateral yellow markings. Female similar to male. **[Measurements]** Total length 68-72 mm, abdomen 49-54 mm, hind wing 40-45 mm. **[Habitat]** Ponds, rivers and streams below 1500 m elevation. **[Distribution]** Widespread in the south of China; India, Myanmar, Laos, Vietnam, Japan. **[Flight Season]** March to December.

霸王叶春蜓 雄，广东
Ictinogomphus pertinax, male from Guangdong

霸王叶春蜓 雄，广东 | 宋黎明 摄
Ictinogomphus pertinax, male from Guangdong | Photo by Liming Song

猛春蜓属 Genus *Labrogomphus* Needham, 1931

本属全球仅1种，分布于中国、老挝和越南。本属蜻蜓是体型较大的春蜓；翅透明，翅痣甚长，具支持脉，三角室、上三角室和下三角室无横脉，臀圈3室，基臀区具1条横脉，臀三角室3室；腹部第9节甚长，超过第8节长度的2倍，长度约为第10节的5倍。雄性上肛附器的两枝平行，末端尖锐，内缘腹面具齿，下肛附器中央凹陷较深，两枝分歧。

本属蜻蜓栖息于山区溪流和河流。雄性经常停在水面的大岩石或者河岸的沙滩上占据领地，时而飞行巡逻。

The genus contains a single species from China, Laos and Vietnam. The species is large-sized. Wings hyaline, pterostigma long and braced, no crossvein in triangle, hypertriangle and subtriangle, anal loop 3-celled, cubital space with one crossvein. S9 very long, exceeding two times of the length of S8, about five times as long as S10. Male superior appendages with two branches parallel, tips pointed, ventral teeth present, inferior appendage with a deep median hollow, forming two divergent branches.

Labrogomphus species inhabits montane streams and rivers. Males usually perch on big rocks or sandy beach for territory and sometimes patrol.

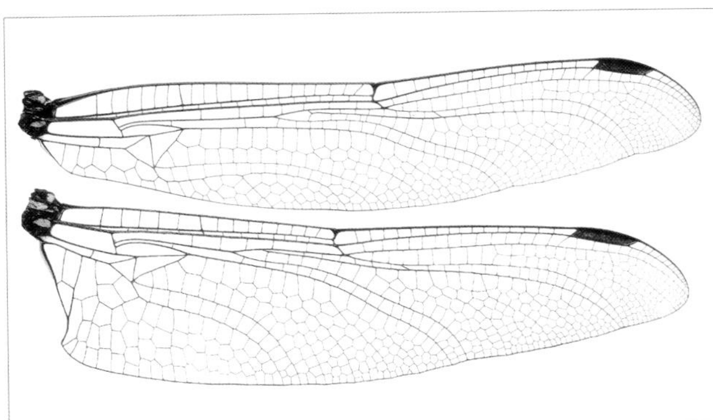

凶猛春蜓 雄翅
Labrogomphus torvus, male wings

凶猛春蜓
Labrogomphus torvus

猛春蜓属 雄性腹部末端
Genus *Labrogomphus*, male distal abdomen

凶猛春蜓 雌
Labrogomphus torvus, female

凶猛春蜓 *Labrogomphus torvus* Needham, 1931

凶猛春蜓 雄
Labrogomphus torvus, male

【形态特征】雄性面部黑色具黄斑，上唇具1对黄斑，额横纹甚阔，后头缘中央突起较高；胸部黑色，背条纹与领条纹不相连，肩前条纹细长，合胸侧面第2条纹和第3条纹完整；腹部黑色，第1～8节具黄斑，其中第7节黄斑甚大，第8节侧下方具较短的叶片状突起，第9节甚长。雌性与雄性相似。【长度】体长77～83 mm，腹长 56～62 mm，后翅 46～50 mm。【栖息环境】海拔 1000 m以下的开阔溪流和河流。【分布】安徽、湖北、贵州、福建、广东、广西、海南、香港；老挝、越南。【飞行期】4—10月。

[Identification] Male face black with yellow markings, labrum with a pair of yellow spots, frons with a broad stripe, occipital margin highly raised. Thorax black, dorsal stripes and collar stripes not connecting, antehumeral stripes long, second and third lateral stripes complete. Abdomen black, S1-S8 with yellow markings, S7 with a big yellow spot, lateral lower eage of S8 with short leaf-like flaps, S9 elongated. Female similar to male. **[Measurements]** Total length 77-83 mm, abdomen 56-62 mm, hind wing 46-50 mm. **[Habitat]** Exposed streams and rivers below 1000 m elevation. **[Distribution]** Anhui, Hubei, Guizhou, Fujian, Guangdong, Guangxi, Hainan, Hong Kong; Laos, Vietnam. **[Flight Season]** April to October.

凶猛春蜓 雄，广东
Labrogomphus torvus, male from Guangdong

环尾春蜓属 Genus *Lamelligomphus* Fraser, 1922

本属全球已知10余种，但目前放在钩尾春蜓属的某些种类可能会被移入该属。中国已知10余种，分布广泛。本属蜻蜓体中型；翅透明，翅痣甚长，具支持脉，三角室、上三角室和下三角室无横脉，臀圈2室，基臀区具1条横脉。雄性上肛附器的两枝平行，末端向下钩曲，弯曲处的腹方具许多细齿，下肛附器长于上肛附器，末端包在上肛附器外方，因此形成1个圆腔。阳茎末端具鞭。雌性后头缘具角状突起。

本属蜻蜓栖息于水面开阔、具碎石的溪流。雄性停落在大岩石或者水边的植物上占据领地，时而飞行巡逻，有时长时间近水面定点悬停飞行。雌性空投式产卵。

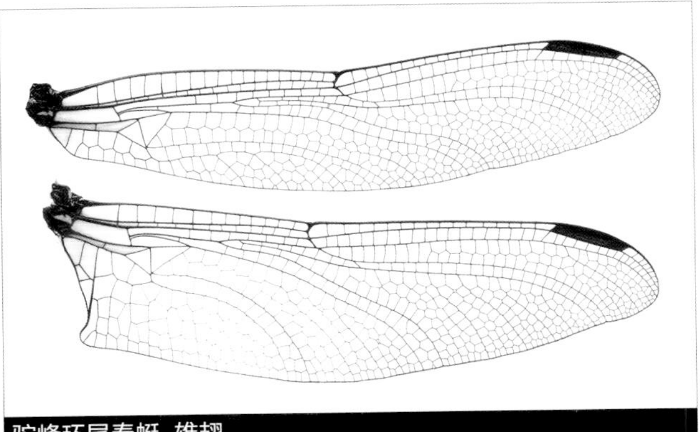

驼峰环尾春蜓 雄翅
Lamelligomphus camelus, male wings

The genus contains over ten species, but more species may be transferred from the genus *Onychogomphus*. Over ten species are recorded from China. Species of the genus are medium-sized. Wings hyaline, pterostigma long and braced, no crossvein in triangle, hypertriangle and subtriangle, anal loop 2-celled, cubital space with one crossvein. Male superior appendages with two branches parallel, the tips strongly curved downwards, and a row of small teeth present at the curved point, inferior appendage longer, curved upwards thus forming a circular cavity. Penis with a pair of flagella. Female head with occipital horns.

驼峰环尾春蜓 雄
Lamelligomphus camelus, male

Lamelligomphus species inhabit exposed montane streams with big rocks. Terrtorial males usually perch on the rocks or plants near water, sometimes patrol a short distance, they may hover for long time close to the water's surface. Females oviposit by dropping the egg mass into water when hovering.

黄尾环尾春蜓
Lamelligomphus biforceps

驼峰环尾春蜓
Lamelligomphus camelus

周氏环尾春蜓指名亚种
Lamelligomphus choui choui

台湾环尾春蜓
Lamelligomphus formosanus

海南环尾春蜓
Lamelligomphus hainanensis

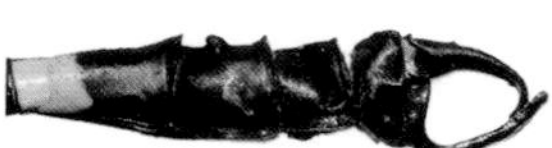

安娜环尾春蜓
Lamelligomphus annakarlorum

环纹环尾春蜓
Lamelligomphus ringens

李氏环尾春蜓
Lamelligomphus risi

双髻环尾春蜓
Lamelligomphus tutulus

环尾春蜓属 雄性腹部末端
Genus *Lamelligomphus*, male distal abdomen

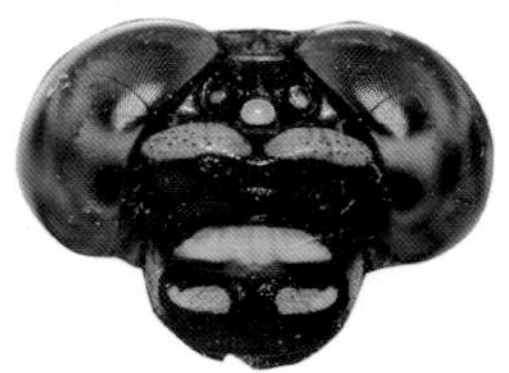

安娜环尾春蜓 雄
Lamelligomphus annakarlorum, male

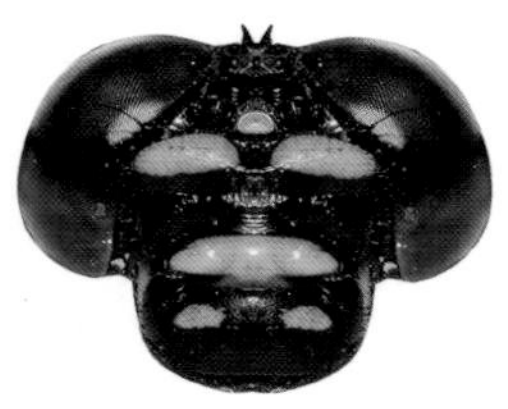

安娜环尾春蜓 雌
Lamelligomphus annakarlorum, female

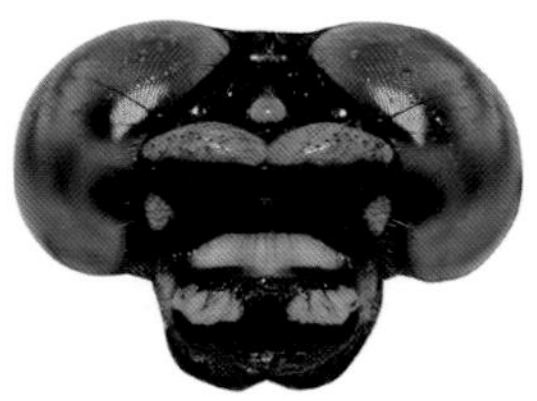

黄尾环尾春蜓 雄
Lamelligomphus biforceps, male

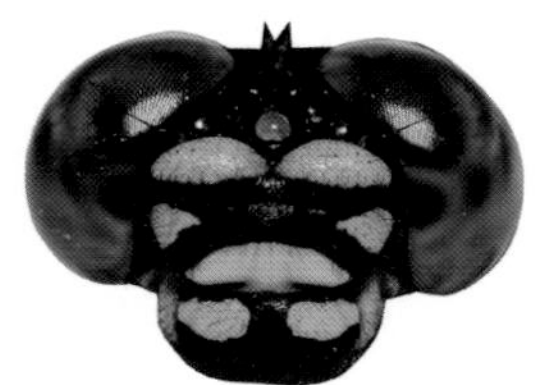

黄尾环尾春蜓 雌
Lamelligomphus biforceps, female

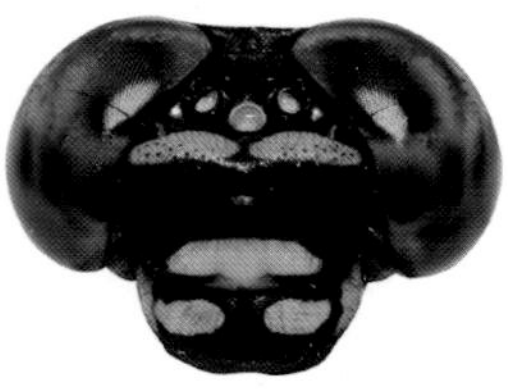

驼峰环尾春蜓 雄
Lamelligomphus camelus, male

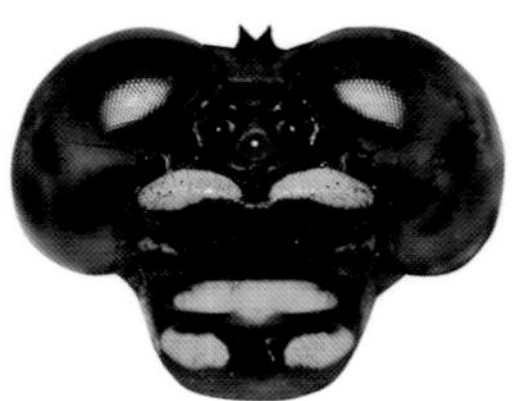

驼峰环尾春蜓 雌
Lamelligomphus camelus, female

环尾春蜓属 头部正面观
Genus *Lamelligomphus*, head in frontal view

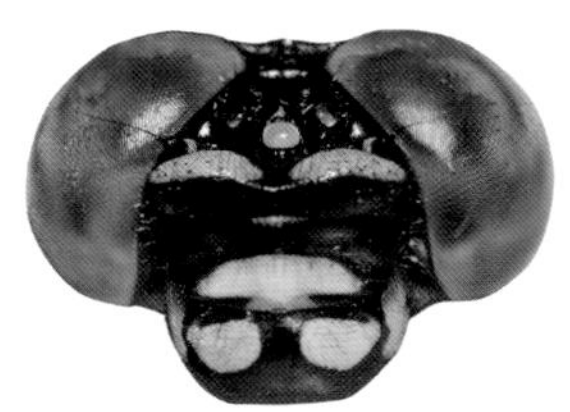
台湾环尾春蜓 雄
Lamelligomphus formosanus, male

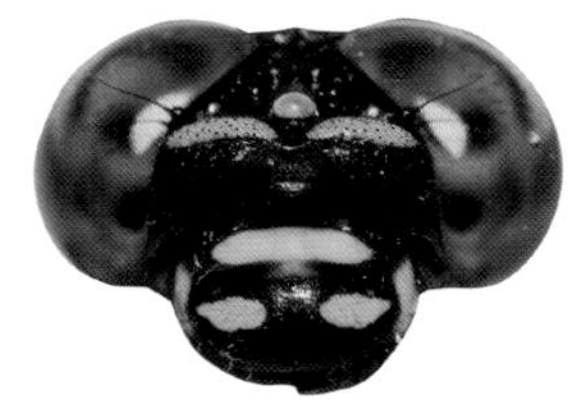
海南环尾春蜓 雄
Lamelligomphus hainanensis, male

环纹环尾春蜓 雄
Lamelligomphus ringens, male

环纹环尾春蜓 雌
Lamelligomphus ringens, female

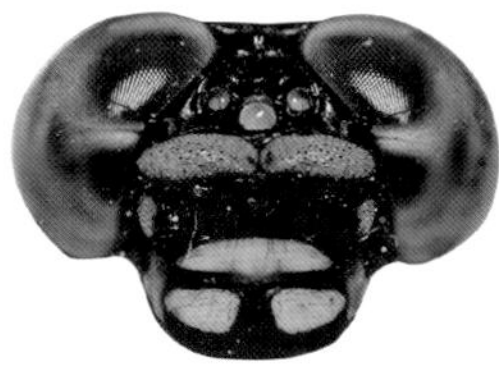
李氏环尾春蜓 雄
Lamelligomphus risi, male

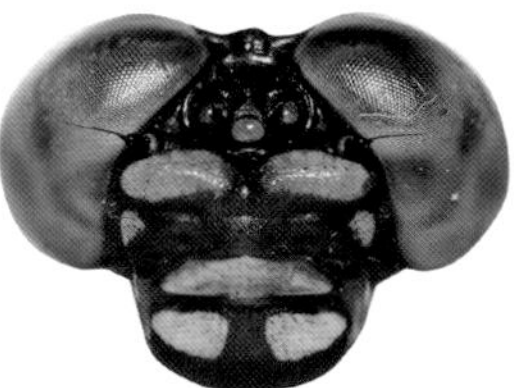
李氏环尾春蜓 雌
Lamelligomphus risi, female

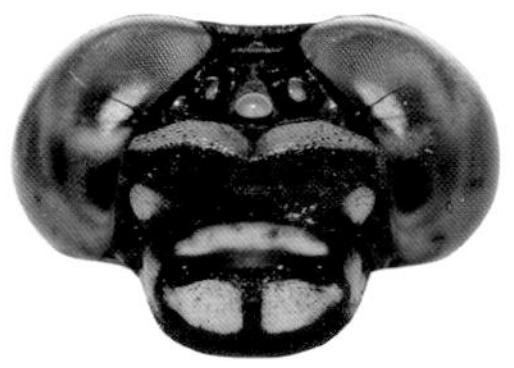
环尾春蜓属待定种1 雄
Lamelligomphus sp. 1, male

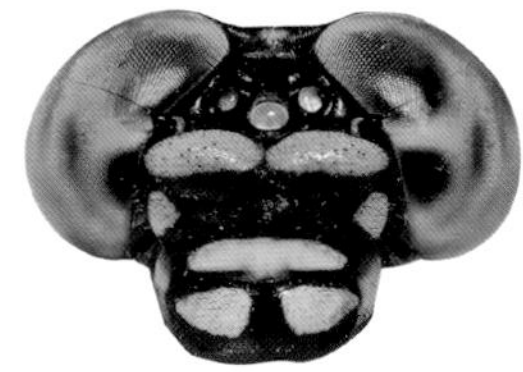
环尾春蜓属待定种2 雄
Lamelligomphus sp. 2, male

环尾春蜓属待定种4 雄
Lamelligomphus sp. 4, male

环尾春蜓属 头部正面观
Genus *Lamelligomphus*, head in frontal view

安娜环尾春蜓 *Lamelligomphus annakarlorum* Zhang, Yang & Cai, 2016

【形态特征】雄性面部黑色具黄斑，上唇具1对黄斑，额横纹甚阔；胸部大面积黑色，背条纹与领条纹相连，合胸侧面具2条黄色条纹，其中后胸后侧板的黄条纹甚短；腹部黑色，第1～7节具黄斑，第8节背面具1对驼峰状突起，肛附器黑色。雌性与雄性相似。【长度】体长 67～69 mm，腹长 49～51 mm，后翅 42～44 mm。【栖息环境】海拔 1000 m以下森林中的溪流。【分布】中国云南（西双版纳）特有。【飞行期】5—10月。

安娜环尾春蜓 雌，云南（西双版纳）
Lamelligomphus annakarlorum, female from Yunnan (Xishuangbanna)

[Identification] Male face black with yellow markings, labrum with a pair of yellow spots, frons with a broad stripe. Thorax largely black, dorsal stripes and collar stripes connecting, sides with two yellow stripes, the yellow stripe on metepimeron short. Abdomen black, S1-S7 with yellow markings, S8 with a pair of hump shaped prominences, anal appendages black. Female similar to male. **[Measurements]** Total length 67-69 mm, abdomen 49-51 mm, hind wing 42-44 mm. **[Habitat]** Montane streams below 1000 m elevation. **[Distribution]** Endemic to Yunnan (Xishuangbanna) of China. **[Flight Season]** May to October.

安娜环尾春蜓 雄，云南（西双版纳）
Lamelligomphus annakarlorum, male from Yunnan (Xishuangbanna)

安娜环尾春蜓 雌，云南（西双版纳）
Lamelligomphus annakarlorum, female from Yunnan (Xishuangbanna)

黄尾环尾春蜓 *Lamelligomphus biforceps* (Selys, 1878)

【形态特征】雄性面部黑色具黄斑，上唇具1对黄斑，额横纹中央间断；胸部黑色，背条纹与领条纹不相连，肩前条纹细长，合胸侧面第2条纹和第3条纹完整；腹部黑色，第1~8节具黄斑，上肛附器外方黄色。雌性与雄性相似。云南环尾春蜓此处作为本种的异名。【长度】体长 55~57 mm，腹长 41~43 mm，后翅 32~35 mm。【栖息环境】海拔 1500~2000 m的开阔溪流。【分布】云南（大理）；印度、不丹、尼泊尔。【飞行期】6—10月。

[Identification] Male face black with yellow markings, labrum with a pair of yellow spots, frons with a stripe interrupted medially. Thorax black, dorsal stripes and collar stripes not connecting, antehumeral stripes long and narrow, second and third lateral stripes complete. Abdomen black, S1-S8 with yellow markings, superior appendages with yellow on outer margin. Female similar to male. *L. laetus* Yang & Davies, 1993 is considered as a junior synonym of this species here. **[Measurements]** Total length 55-57 mm, abdomen 41-43 mm, hind wing 32-35 mm. **[Habitat]** Exposed streams at 1500-2000 m elevation. **[Distribution]** Yunnan (Dali); India, Bhutan, Nepal. **[Flight Season]** June to October.

黄尾环尾春蜓 雄，云南（大理）
Lamelligomphus biforceps, male from Yunnan (Dali)

黄尾环尾春蜓 雌，云南（大理）
Lamelligomphus biforceps, female from Yunnan (Dali)

黄尾环尾春蜓 雄，云南（大理）
Lamelligomphus biforceps, male from Yunnan (Dali)

驼峰环尾春蜓 *Lamelligomphus camelus* (Martin, 1904)

驼峰环尾春蜓 雄，云南（红河）
Lamelligomphus camelus, male from Yunnan (Honghe)

【形态特征】雄性面部黑色具黄斑，上唇具1对黄斑，额横纹中央间断；胸部黑色，背条纹与领条纹不相连，具甚小的肩前上点，合胸侧面第2条纹和第3条纹有时完整，有时完全合并；腹部黑色，第1~7节具黄斑，第8节背面中央具驼峰状突起，肛附器黑色。雌性与雄性相似。本种与安娜环尾春蜓相似，但雄性后钩片及雌性下生殖板的构造不同。【长度】体长 67~70 mm，腹长 51~52 mm，后翅 39~42 mm。【栖息环境】海拔 1000 m以下的山区开阔溪流。【分布】云南、贵州、浙江、福建、广东、广西、海南；老挝、越南。【飞行期】4—9月。

[Identification] Male face black with yellow markings, labrum with a pair of yellow spots, frons with a stripe interrupted medially. Thorax black, dorsal stripes and collar stripes not connecting, superior spots very small, second and third lateral stripes complete or combined. Abdomen black, S1-S7 with yellow markings, S8 with a pair of hump shaped prominences, anal appendages black. Female similar to male. Similar to *L. annakarlorum*, but male anal appendages and female vulvar lamina are different. **[Measurements]** Total length 67-70 mm, abdomen 51-52 mm, hind wing 39-42 mm. **[Habitat]** Exposed montane streams below 1000 m elevation. **[Distribution]** Yunnan, Guizhou, Zhejiang, Fujian, Guangdong, Guangxi, Hainan; Laos, Vietnam. **[Flight Season]** April to September.

驼峰环尾春蜓 雌，广东 | 宋睿斌 摄
Lamelligomphus camelus, female from Guangdong | Photo by Ruibin Song

驼峰环尾春蜓 雄，广东
Lamelligomphus camelus, male from Guangdong

周氏环尾春蜓指名亚种 *Lamelligomphus choui choui* Chao & Liu, 1989

【形态特征】雄性面部黑色具黄斑，上唇具1对黄斑，额横纹较宽阔；胸部黑色，背条纹与领条纹不相连，具较细的肩前条纹，合胸侧面第2条纹和第3条纹完整；腹部黑色，各节具黄斑，肛附器黑色。雌性与雄性相似，后头缘具1对角状突起。【长度】体长 60～65 mm，腹长 45～48 mm，后翅 36～38 mm。【栖息环境】海拔 1000 m以下的山区开阔溪流。【分布】中国特有，分布于安徽、四川、广东。【飞行期】6—9月。

[Identification] Male face black with yellow markings, labrum with a pair of yellow spots, frons with a broad stripe. Thorax black, dorsal stripes and collar stripes not connecting, antehumeral stripes narrow, second and third

周氏环尾春蜓指名亚种 雌，广东 | 莫善濂 摄
Lamelligomphus choui choui, female from Guangdong | Photo by Shanlian Mo

周氏环尾春蜓指名亚种 雄，广东 | 莫善濂 摄
Lamelligomphus choui choui, male from Guangdong | Photo by Shanlian Mo

周氏环尾春蜓指名亚种 雄，广东 | 莫善濂 摄
Lamelligomphus choui choui, male from Guangdong | Photo by Shanlian Mo

lateral stripes complete. Abdomen black with yellow markings on all segments, anal appendages black. Female similar to male, occipital margin with a pair of occipital horns. **[Measurements]** Total length 60-65 mm, abdomen 45-48 mm, hind wing 36-38 mm. **[Habitat]** Exposed montane streams below 1000 m elevation. **[Distribution]** Endemic to China, recorded from Anhui, Sichuan, Guangdong. **[Flight Season]** June to September.

台湾环尾春蜓 *Lamelligomphus formosanus* (Matsumura, 1926)

【形态特征】雄性面部黑色具黄斑，上唇具1对黄斑，额横纹黄色，中央间断；胸部黑色，背条纹与领条纹不相连，具较细的肩前条纹，合胸侧面第2条纹和第3条纹完整，下方大面积合并；腹部黑色，第1～7节具黄斑，肛附器黑色。雌性与雄性相似。**【长度】**体长 64～69 mm，腹长 47～52 mm，后翅 38～39 mm。**【栖息环境】**海拔 1000 m以下的山区开阔溪流。**【分布】**湖北、贵州、福建、广东、广西、台湾；越南。**【飞行期】**4—9月。

[Identification] Male face black with yellow markings, labrum with a pair of yellow spots, frons with a stripe interrupted medially. Thorax black, dorsal stripes and collar stripes not connecting, antehumeral stripes narrow, second and third lateral stripes complete and combined ventrally. Abdomen black, S1-S7 with yellow markings, anal appendages black. Female similar to male. **[Measurements]** Total length 64-69 mm, abdomen 47-52 mm, hind wing 38-39 mm. **[Habitat]** Exposed montane streams below 1000 m elevation. **[Distribution]** Hubei, Guizhou, Fujian, Guangdong, Guangxi, Taiwan; Vietnam. **[Flight Season]** April to September.

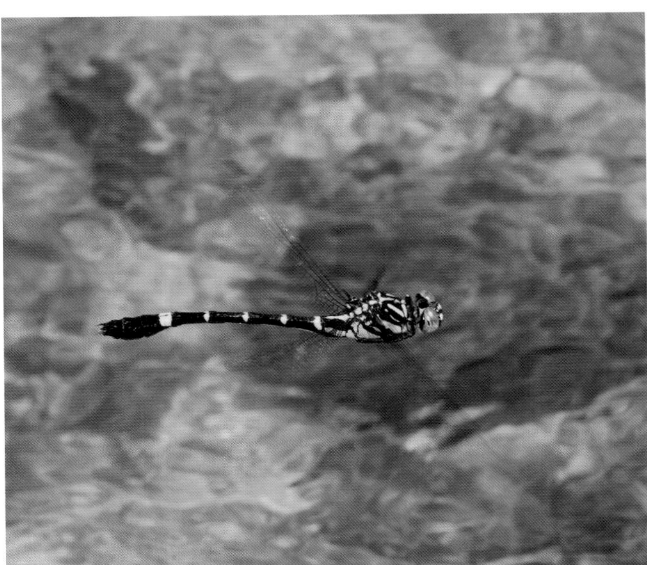
台湾环尾春蜓 雌，广东
Lamelligomphus formosanus, female from Guangdong

台湾环尾春蜓 雄，广东
Lamelligomphus formosanus, male from Guangdong

台湾环尾春蜓 雄，贵州
Lamelligomphus formosanus, male from Guizhou

海南环尾春蜓 *Lamelligomphus hainanensis* (Chao, 1954)

【形态特征】雄性面部黑色具黄斑，上唇具1对黄斑，额横纹黄色，中央间断；胸部黑色，背条纹与领条纹不相连，无肩前条纹，合胸侧面第2条纹和第3条纹合并；腹部黑色，第1~7节具黄斑，肛附器黑色。雌性与雄性相似。【长度】体长 60~64 mm，腹长 46~48 mm，后翅 35~36 mm。【栖息环境】海拔 1000 m以下的溪流和河流。【分布】中国特有，分布于广东、海南、香港。【飞行期】4—8月。

[Identification] Male face black with yellow markings, labrum with a pair of yellow spots, frons with a stripe interrupted medially. Thorax black, dorsal stripes and collar stripes not connecting, antehumeral stripes absent, second and third lateral stripes combined. Abdomen black, S1- S7 with yellow markings, anal appendages black. Female similar to male. **[Measurements]** Total length 60-64 mm, abdomen 46-48 mm, hind wing 35-36 mm. **[Habitat]** Streams and rivers below 1000 m elevation. **[Distribution]** Endemic to China, recorded from Guangdong, Hainan, Hong Kong. **[Flight Season]** April to August.

海南环尾春蜓 雄，海南
Lamelligomphus hainanensis, male from Hainan

海南环尾春蜓 交尾，广东 | 宋睿斌 摄
Lamelligomphus hainanensis, mating pair from Guangdong | Photo by Ruibin Song

墨脱环尾春蜓 *Lamelligomphus motuoensis* Chao, 1983

墨脱环尾春蜓 雄，西藏｜吴超 摄
Lamelligomphus motuoensis, male from Tibet | Photo by Chao Wu

【形态特征】雄性面部黑色具黄斑，上唇具1对黄斑，额横纹黄色，中央间断；胸部黑色，背条纹与领条纹不相连，无肩前条纹，合胸侧面第2条纹和第3条纹合并；腹部黑色，第1～7节基方具黄斑，肛附器黑色。【长度】体长60～62 mm，腹长 44～47 mm，后翅 36～42 mm。【栖息环境】海拔 1500 m以下的山区溪流。【分布】中国西藏特有。【飞行期】不详。

[Identification] Male face black with yellow markings, labrum with a pair of yellow spots, frons with a yellow stripe interrupted medially. Thorax black, dorsal stripes and collar stripes not connecting, antehumeral stripes absent, second and third lateral stripes combined. Abdomen black, S1-S7 with yellow spots, anal appendages black. [Measurements] Total length 60-62 mm, abdomen 44-47 mm, hind wing 36-42 mm. [Habitat] Montane streams below 1500 m elevation. [Distribution] Endemic to Tibet of China. [Flight Season] Unknown.

环纹环尾春蜓 *Lamelligomphus ringens* (Needham, 1930)

环纹环尾春蜓 雄，北京
Lamelligomphus ringens, male from Beijing

环纹环尾春蜓 雌，安徽 | 秦彧 摄
Lamelligomphus ringens, female from Anhui | Photo by Yu Qin

环纹环尾春蜓 雄，北京
Lamelligomphus ringens, male from Beijing

【形态特征】雄性面部黑色具黄斑，上唇具1对黄斑，额横纹黄色，甚阔；胸部黑色，背条纹与领条纹不相连，无肩前条纹，合胸侧面第2条纹和第3条纹合并；腹部黑色，各节具黄斑，上肛附器大面积黑色，末端具黄色。雌性与雄性相似。【长度】体长 61～63 mm，腹长 45～47 mm，后翅 37～39 mm。【栖息环境】海拔 1000 m以下的山区溪流。【分布】黑龙江、吉林、辽宁、北京、河北、山西、安徽、湖北、重庆、四川；朝鲜半岛。【飞行期】6—9月。

[Identification] Male face black with yellow markings, labrum with a pair of yellow spots, frons with a broad yellow stripe. Thorax black, dorsal stripes and collar stripes not connecting, antehumeral stripes absent, second and third lateral stripes combined. Abdomen black with yellow markings in all segments, superior appendages largely black with tips yellow. Female similar to male. **[Measurements]** Total length 61-63 mm, abdomen 45-47 mm, hind wing 37-39 mm. **[Habitat]** Montane streams below 1000 m elevation. **[Distribution]** Heilongjiang, Jilin, Liaoning, Beijing, Hebei, Shanxi, Anhui, Hubei, Chongqing, Sichuan; Korean peninsula. **[Flight Season]** June to September.

李氏环尾春蜓 *Lamelligomphus risi* (Fraser, 1922)

李氏环尾春蜓 雌，云南（德宏）
Lamelligomphus risi, female from Yunnan (Dehong)

李氏环尾春蜓 雄，云南（德宏）
Lamelligomphus risi, male from Yunnan (Dehong)

李氏环尾春蜓 雄，云南（德宏）
Lamelligomphus risi, male from Yunnan (Dehong)

【形态特征】雄性面部黑色具黄斑，上唇具1对黄斑，额横纹黄色，中央间断；胸部黑色，背条纹与领条纹相连，合胸侧面第2条纹和第3条纹合并；腹部黑色，第1~7节具黄斑，肛附器大面积黑色，上肛附器外缘黄色。雌性与雄性相似。本种雌性后头无角状突起。【长度】体长 55~63 mm，腹长 41~47 mm，后翅 33~41 mm。【栖息环境】海拔500 m以下森林中的溪流。【分布】云南（德宏）；印度、尼泊尔。【飞行期】8—11月。

[Identification] Male face black with yellow markings, labrum with a pair of yellow spots, frons with a yellow stripe interrupted medially. Thorax black, dorsal stripes and collar stripes connecting, second and third lateral stripes combined. Abdomen black, S1-S7 with yellow markings, anal appendages largely black with outer margin yellow. Female similar to male, occipital horn absent. **[Measurements]** Total length 55-63 mm, abdomen 41-47 mm, hind wing 33-41 mm. **[Habitat]** Streams in forest below 500 m elevation. **[Distribution]** Yunnan (Dehong); India, Nepal. **[Flight Season]** August to November.

双髻环尾春蜓 *Lamelligomphus tutulus* Liu & Chao, 1990

【形态特征】雄性面部黑色具黄斑，上唇具1对黄斑，额横纹黄色，中央间断；胸部黑色，背条纹与领条纹不相连，肩前条纹甚细，合胸侧面第2条纹和第3条纹几乎合并；腹部黑色，各节具黄斑，肛附器黑色。雌性与雄性相似。【长度】体长 63～65 mm，腹长 45～47 mm，后翅 37～40 mm。【栖息环境】海拔 500～1500 m的开阔溪流。【分布】中国特有，分布于贵州、重庆、广西。【飞行期】6—9月。

[Identification] Male face black with yellow markings, labrum with a pair of yellow spots, frons with a yellow stripe interrupted medially. Thorax black, dorsal stripes and collar stripes not connecting, antehumeral stripes narrow, second and third lateral stripes almost combined. Abdomen black with yellow markings in all segments, anal appendages black. Female similar to male. **[Measurements]** Total length 63-65 mm, abdomen 45-47 mm, hind wing 37-40 mm. **[Habitat]** Exposed streams at 500-1500 m elevation. **[Distribution]** Endemic to China, recorded from Guizhou, Chongqing, Guangxi. **[Flight Season]** June to September.

双髻环尾春蜓 雄，贵州
Lamelligomphus tutulus, male from Guizhou

双髻环尾春蜓 雌，贵州
Lamelligomphus tutulus, female from Guizhou

双髻环尾春蜓 雄，贵州
Lamelligomphus tutulus, male from Guizhou

环尾春蜓属待定种1 *Lamelligomphus* sp. 1

【形态特征】雄性面部黑色具黄斑，上唇具1对黄斑，额横纹黄色；胸部黑色，背条纹与领条纹不相连，无肩前条纹，合胸侧面第2条纹和第3条纹合并；腹部黑色，第1～8节具黄斑，上肛附器基方1/2黄色。雌性与雄性相似。【长度】体长 61～63 mm，腹长 46～47 mm，后翅 37～42 mm。【栖息环境】海拔 1000 m以下的河流和溪流。【分布】云南（西双版纳）。【飞行期】5—11月。

[Identification] Male face black with yellow markings, labrum with a pair of yellow spots, frons with a yellow stripe. Thorax black, dorsal stripes and collar stripes not connecting, antehumeral stripes absent, second and third lateral stripes combined. Abdomen black, S1-S8 with yellow markings, superior appendages with basal half yellow. Female similar to male. **[Measurements]** Total length 61-63 mm, abdomen 46-47 mm, hind wing 37-42 mm. **[Habitat]** Rivers and streams below 1000 m elevation. **[Distribution]** Yunnan (Xishuangbanna). **[Flight Season]** May to November.

环尾春蜓属待定种1 雄，云南（西双版纳）
Lamelligomphus sp. 1, male from Yunnan (Xishuangbanna)

环尾春蜓属待定种1 雌，云南（西双版纳）
Lamelligomphus sp. 1, female from Yunnan (Xishuangbanna)

环尾春蜓属待定种2 *Lamelligomphus* sp. 2

环尾春蜓属待定种2 雄，云南（临沧）
Lamelligomphus sp. 2, male from Yunnan (Lincang)

【形态特征】雄性面部黑色具黄斑，上唇具1对黄斑，额横纹黄色，中央间断；胸部黑色，背条纹与领条纹不相连，肩前条纹甚细，合胸侧面第2条纹和第3条纹几乎合并；腹部黑色，第1～9节具黄斑，肛附器黄褐色。【长度】雄性体长 59 mm，腹长 44 mm，后翅 35 mm。【栖息环境】海拔 1000 m以下的开阔溪流。【分布】云南（临沧）。【飞行期】8—10月。

[Identification] Male face black with yellow markings, labrum with a pair of yellow spots, frons with a yellow stripe interrupted medially. Thorax black, dorsal stripes and collar stripes not connecting, antehumeral stripes narrow, second and third lateral stripes almost combined. Abdomen black, S1-S9 with yellow markings, anal appendages yellowish brown. **[Measurements]** Male total length 59 mm, abdomen 44 mm, hind wing 35 mm. **[Habitat]** Exposed streams below 1000 m elevation. **[Distribution]** Yunnan (Lincang). **[Flight Season]** August to October.

环尾春蜓属待定种3 *Lamelligomphus* sp. 3

环尾春蜓属待定种3 雄，云南（普洱）
Lamelligomphus sp. 3, male from Yunnan (Pu'er)

【形态特征】雄性面部黑色具黄斑，上唇具1对黄斑，额横纹黄色；胸部黑色，背条纹与领条纹相连，无肩前条纹，合胸侧面第2条纹和第3条纹几乎合并；腹部黑色，第1～8节具黄斑，第3节背面中央具1个五角星状黄斑，肛附器黑色。【长度】雄性体长 52 mm，腹长 39 mm，后翅 30 mm。【栖息环境】海拔 1000 m以下森林中的溪流。【分布】云南（西双版纳、普洱）。【飞行期】5—10月。

[Identification] Male face black with yellow markings, labrum with a pair of yellow spots, frons with a yellow stripe. Thorax black, dorsal stripes and collar stripes connecting, antehumeral stripes absent, second and third lateral stripes almost combined. Abdomen black, S1-S8 with yellow markings, S3 with a dorsal pentagram-shaped yellow spot, anal appendages black. **[Measurements]** Male total length 52 mm, abdomen 39 mm, hind wing 30 mm. **[Habitat]** Montane streams below 1000 m elevation. **[Distribution]** Yunnan (Xishuangbanna, Pu'er). **[Flight Season]** May to October.

环尾春蜓属待定种4 *Lamelligomphus* sp. 4

【形态特征】雄性面部主要黑色，前唇基黄色，额横纹黄色；胸部黑色，背条纹与领条纹不相连，无肩前条纹，合胸侧面第2条纹和第3条纹合并；腹部黑色，第1~7节具黄斑，肛附器黑色。【长度】雄性体长 58 mm，腹长 43 mm，后翅 35 mm。【栖息环境】海拔 1000 m处森林中的溪流。【分布】云南（红河）。【飞行期】5—7月。

[Identification] Male face largely black, anteclypeus yellow, frons with a yellow stripe. Thorax black, dorsal stripes and collar stripes not connecting, antehumeral stripes absent, second and third lateral stripes combined. Abdomen black, S1-S7 with yellow markings, anal appendages black. [Measurements] Male total length 58 mm, abdomen 43 mm, hind wing 35 mm. [Habitat] Montane streams at 1000 m elevation. [Distribution] Yunnan (Honghe). [Flight Season] May to July.

环尾春蜓属待定种4 雄，云南（红河）
Lamelligomphus sp. 4, male from Yunnan (Honghe)

纤春蜓属 Genus *Leptogomphus* Selys, 1878

本属全球已知约25种，分布于亚洲的亚热带和热带地区。中国已知10余种，主要分布于华南和西南等地。本属蜻蜓体中型；翅透明，翅痣较长，无显著的支持脉，三角室、上三角室和下三角室无横脉，基臀区具1条横脉，臀三角室的基边甚斜；雄性第8~10节向侧面膨大，上肛附器扁平，背面黄白色，腹面黑色，下肛附器约与上肛附器等长，黑色；阳茎通常无鞭，有时具1对短鞭；雌性头部后头缘通常具角状突起。

本属蜻蜓栖息于山区溪流环境，喜欢具有树荫遮盖的溪流和沟渠。雄性会停落在水边的植物上占据领地。雌性插秧式产卵。

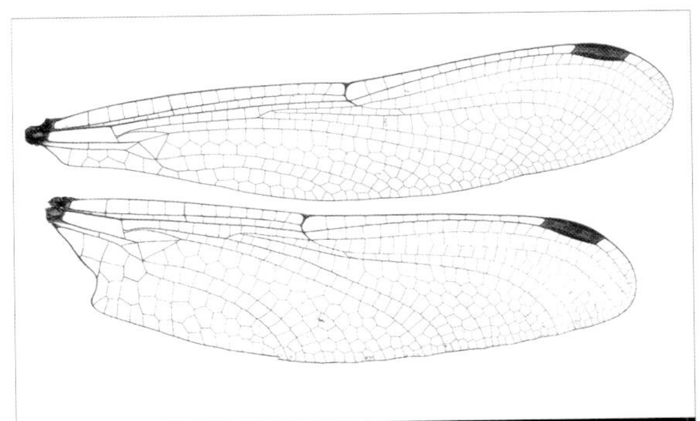

欢庆纤春蜓 雄翅
Leptogomphus celebratus, male wings

The genus contains about 25 species distributed throughout subtropical and tropical Asia. Over ten are recorded from China, mainly found in the South and Southwest. Species of the genus are medium-sized. Wings hyaline, pterostigma long but not braced, no crossvein in triangle, hypertriangle and subtriangle, cubital space with one crossvein, anal triangle with the basal side strongly oblique. Male S8-S10 expanded laterally, superior appendages flattened, yellowish

圆腔纤春蜓 雄
Leptogomphus perforatus, male

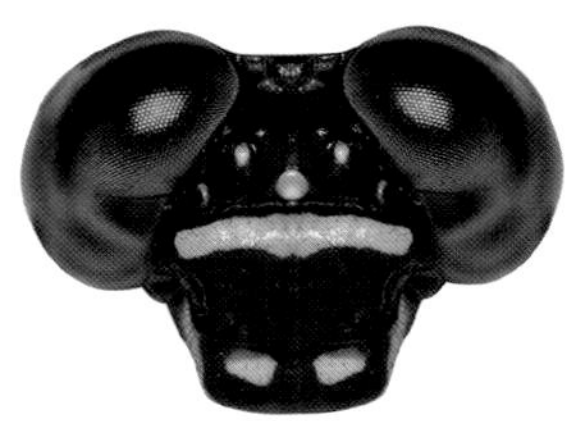

欢庆纤春蜓 雄
Leptogomphus celebrates, male

欢庆纤春蜓 雌
Leptogomphus celebratus, female

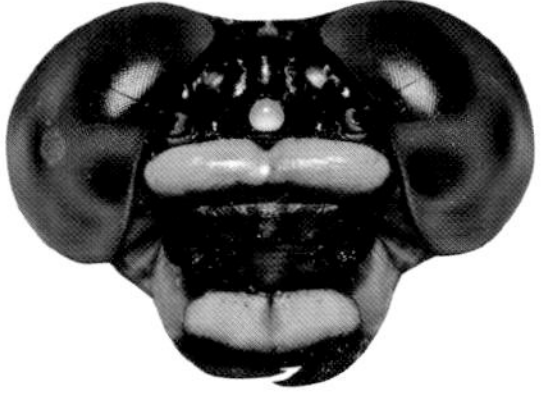

尖尾纤春蜓 雄
Leptogomphus gestroi, male

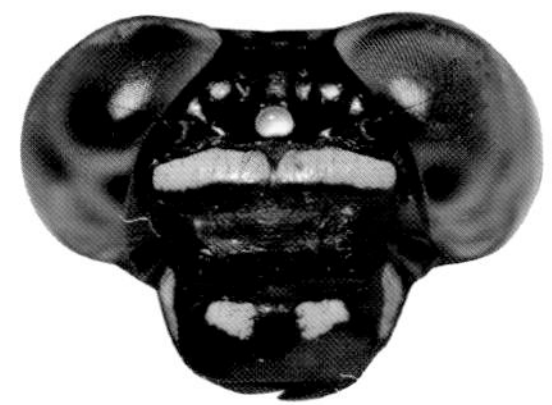

居间纤春蜓 雄
Leptogomphus intermedius, male

纤春蜓属 头部正面观
Genus *Leptogomphus*, head in frontal view

white dorsally and black ventrally, the inferior appendage equal in length, black. Penis usually without flagella but some species possess short a pair of flagella. Female occipital horns usually present.

Leptogomphus species inhabit montane streams, prefer shady and narrow streamlets. Males usually perch on plants near water for territory. Females lay eggs when the body erected.

居间纤春蜓 雌
Leptogomphus intermedius, female

圆腔纤春蜓 雄
Leptogomphus perforatus, male

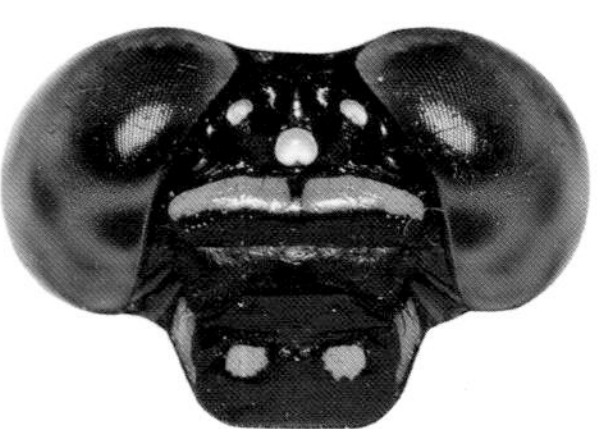

圆腔纤春蜓 雌
Leptogomphus perforatus, female

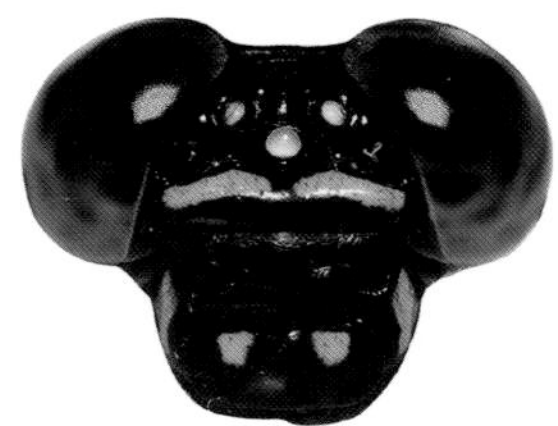

三道纤春蜓 雄
Leptogomphus tamdaoensis, male

羚角纤春蜓 雌
Leptogomphus uenoi, female

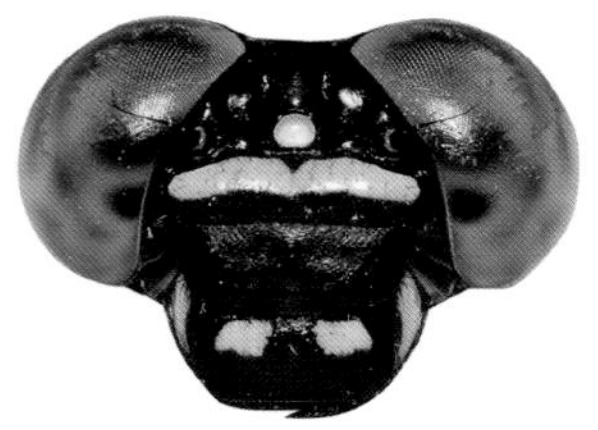

纤春蜓属待定种 雄
Leptogomphus sp. , male

纤春蜓属 头部正面观
Genus *Leptogomphus*, head in frontal view

欢庆纤春蜓
Leptogomphus celebrates

歧角纤春蜓
Leptogomphus divaricatus

优美纤春蜓
Leptogomphus elegans

尖尾纤春蜓
Leptogomphus gestroi

香港纤春蜓
Leptogomphus hongkongensis

居间纤春蜓
Leptogomphus intermedius

圆腔纤春蜓
Leptogomphus perforatus

苏氏纤春蜓指名亚种
Leptogomphus sauteri sauteri

三道纤春蜓
Leptogomphus tamdaoensis

纤春蜓属 雄性肛附器
Genus *Leptogomphus*, male appendages

纤春蜓属待定种
Leptogomphus sp.

歧角纤春蜓
Leptogomphus divaricatus

居间纤春蜓
Leptogomphus intermedius

圆腔纤春蜓
Leptogomphus perforatus

纤春蜓属 雄性肛附器
Genus *Leptogomphus*, male appendages

纤春蜓属 雄性次生殖器
Genus *Leptogomphus*, male secondary genitalia

欢庆纤春蜓 *Leptogomphus celebratus* Chao, 1982

【形态特征】雄性上唇具1对黄斑，额横纹黄色；胸部背条纹与领条纹不相连，具甚小的肩前上点和甚细的肩前下条纹，合胸侧面第2条纹和第3条纹完整；腹部黑色，第1～7节具黄斑。雌性与雄性相似，但后头中央具驼峰状隆起，后头后缘侧面具角状突起。【长度】体长 49～53 mm，腹长 39～41 mm，后翅 33～36 mm。【栖息环境】海拔1000 m以下茂密森林中的狭窄小溪和渗流地。【分布】中国海南特有。【飞行期】4—6月。

[Identification] Male labrum with a pair of yellow spots, frons with a yellow stripe. Thorax with dorsal stripes and collar stripes not connecting, superior spots small and lower stripes narrow, second and third lateral stripes complete. Abdomen black, S1-S7 with yellow spots. Female similar to male, occipital margin with a median hump-shaped prominence and small horns laterally. **[Measurements]** Total length 49-53 mm, abdomen 39-41 mm, hind wing 33-36 mm. **[Habitat]** Narrow streams and seepages in dense forest below 1000 m elevation. **[Distribution]** Endemic to Hainan of China. **[Flight Season]** April to June.

欢庆纤春蜓 雄，海南
Leptogomphus celebratus, male from Hainan

欢庆纤春蜓 雄，海南
Leptogomphus celebratus, male from Hainan

欢庆纤春蜓 雌，海南 | 莫善濂 摄
Leptogomphus celebratus, female from Hainan | Photo by Shanlian Mo

歧角纤春蜓 *Leptogomphus divaricatus* Chao, 1984

【形态特征】雄性上唇具1对黄斑，额横纹黄色；胸部背条纹与领条纹不相连，具甚细的肩前条纹，合胸侧面第2条纹和第3条纹完整；腹部黑色，第1~7节具黄斑。雌性与雄性相似，后头中央具1对角状突起。【长度】体长 59~62 mm，腹长 45~48 mm，后翅 37~42 mm。【栖息环境】海拔 500 m以下茂密森林中的溪流。【分布】广西、广东、福建；越南。【飞行期】4—8月。

[Identification] Male labrum with a pair of yellow spots, frons with a yellow stripe. Thorax with dorsal stripes and collar stripes not connecting, antehumeral stripes narrow, second and third lateral stripes complete. Abdomen black, S1-S7 with yellow markings. Female similar to male, occipital margin with a pair of median horns. **[Measurements]** Total length 59-62 mm, abdomen 45-48 mm, hind wing 37-42 mm. **[Habitat]** Streams in dense forest below 500 m elevation. **[Distribution]** Guangxi, Guangdong, Fujian; Vietnam. **[Flight Season]** April to August.

歧角纤春蜓 雄，广西
Leptogomphus divaricatus, male from Guangxi

优美纤春蜓 *Leptogomphus elegans* Lieftinck, 1948

【形态特征】雄性上唇具1对黄斑，额横纹黄色，中央间断；胸部背条纹与领条纹不相连，具甚细的肩前条纹，合胸侧面第2条纹和第3条纹完整；腹部黑色，第1~7节具黄斑。雌性与雄性相似，后头中央具1对角状突起。【长度】体长 57~65 mm，腹长 43~51 mm，后翅 38~45 mm。【栖息环境】海拔 1000 m以下茂密森林中的溪流。【分布】广西、广东、福建；越南。【飞行期】5—9月。

[Identification] Male labrum with a pair of yellow spots, frons with a yellow stripe interrupted medially. Thorax with dorsal stripes and collar stripes not connecting, antehumeral stripes narrow, second and third lateral stripes complete. Abdomen black, S1-S7 with yellow markings. Female similar to male, occipital margin with a pair of median horns. [Measurements] Total length 57-65 mm, abdomen 43-51 mm, hind wing 38-45 mm. [Habitat] Streams in dense forest below 1000 m elevation. [Distribution] Guangxi, Guangdong, Fujian; Vietnam. [Flight Season] May to September.

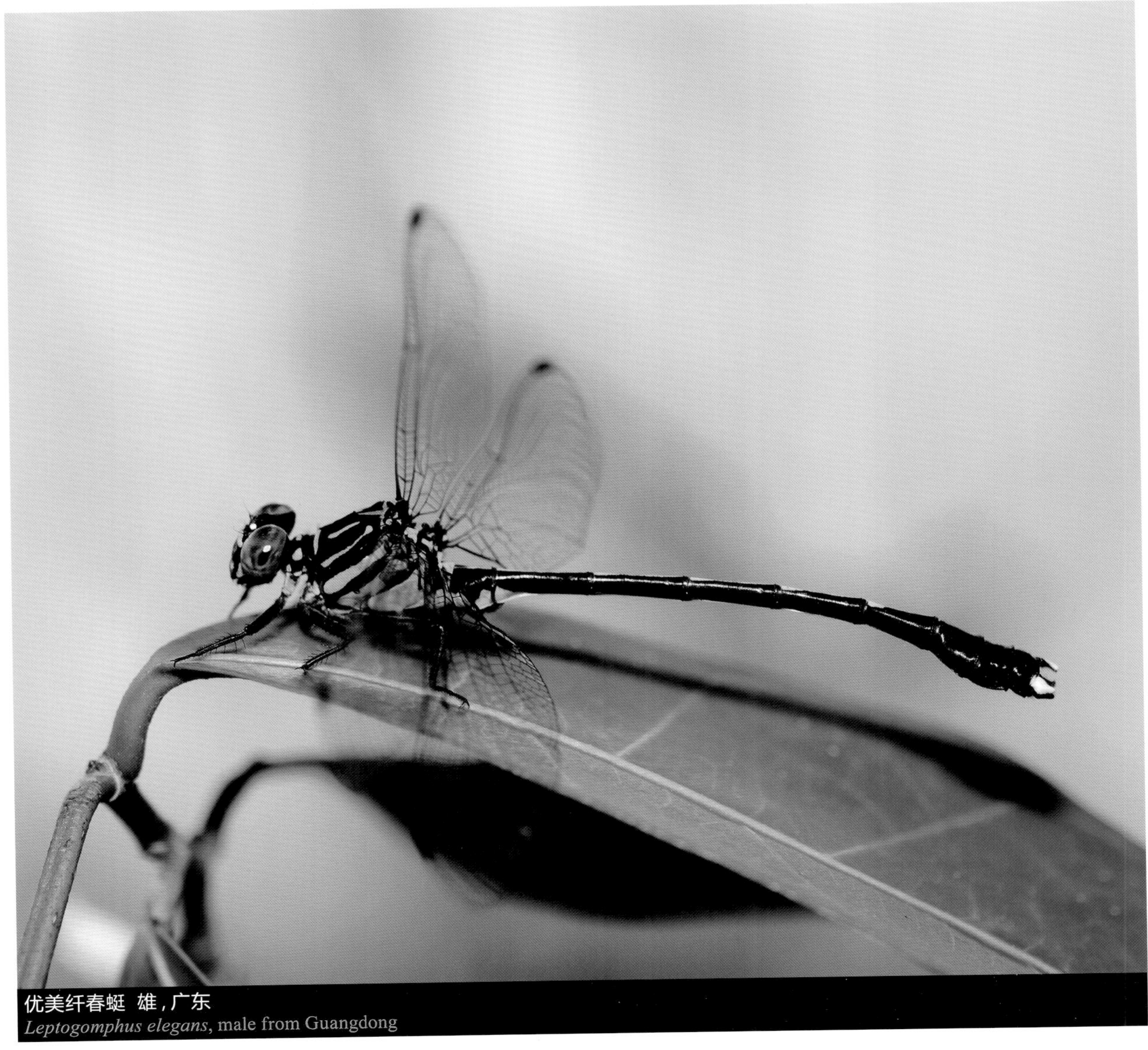

优美纤春蜓 雄，广东
Leptogomphus elegans, male from Guangdong

尖尾纤春蜓 *Leptogomphus gestroi* Selys, 1891

【形态特征】雄性上唇黄白色，额横纹黄色；胸部背条纹与领条纹不相连，肩前条纹细长，合胸侧面第2条纹和第3条纹完整，第3条纹甚细；腹部黑色，第1～8节具黄斑，上肛附器尖齿状。雌性与雄性相似，后头黄色。【长度】体长 47～48 mm，腹长 37～38 mm，后翅 30～32 mm。【栖息环境】海拔 500～1000 m茂密森林中的狭窄小溪。【分布】云南（西双版纳、德宏、普洱）；缅甸、泰国、越南。【飞行期】4—6月。

[Identification] Male labrum yellowish white, frons with a yellow stripe. Thorax with dorsal stripes and collar stripes not connecting, antehumeral stripes long, second and third lateral stripes complete, the third one very narrow. Abdomen black, S1-S8 with yellow markings, superior appendages tine-shaped. Female similar to male, occiput yellow. **[Measurements]** Total length 47-48 mm, abdomen 37-38 mm, hind wing 30-32 mm. **[Habitat]** Narrow streams in dense forest at 500-1000 m elevation. **[Distribution]** Yunnan (Xishuangbanna, Dehong, Pu'er); Myanmar, Thailand, Vietnam. **[Flight Season]** April to June.

尖尾纤春蜓 雄，云南（西双版纳）
Leptogomphus gestroi, male from Yunnan (Xishuangbanna)

尖尾纤春蜓 雌，云南（普洱）
Leptogomphus gestroi, female from Yunnan (Pu'er)

尖尾纤春蜓 雄，云南（西双版纳）
Leptogomphus gestroi, male from Yunnan (Xishuangbanna)

香港纤春蜓 *Leptogomphus hongkongensis* Asahina, 1988

【形态特征】雄性上唇具1对黄斑，额横纹黄色，中央间断；胸部背条纹与领条纹不相连，具细长的肩前条纹，合胸侧面第2条纹和第3条纹完整；腹部黑色，第1~7节具黄斑。雌性与雄性相似，后头中央具1对角状突起。【长度】体长 63~65 mm，腹长 46~49 mm，后翅 39~44 mm。【栖息环境】海拔 1000 m以下茂密森林中的狭窄小溪和渗流地。【分布】中国香港特有。【飞行期】4—8月。

[Identification] Male labrum with a pair of yellow spots, frons with a yellow stripe interrupted medially. Thorax with dorsal stripes and collar stripes not connecting, antehumeral stripes narrow, second and third lateral stripes complete. Abdomen black, S1-S7 with yellow markings. Female similar to male, occipital margin with a pair of median horns. **[Measurements]** Total length 63-65 mm, abdomen 46-49 mm, hind wing 39-44 mm. **[Habitat]** Narrow streams and seepages in dense forest below 1000 m elevation. **[Distribution]** Endemic to Hong Kong of China. **[Flight Season]** April to August.

香港纤春蜓 雄，香港 | 祁麟峰 摄
Leptogomphus hongkongensis, male from Hong Kong | Photo by Mahler Ka

香港纤春蜓 雌，香港 | 梁嘉景 摄
Leptogomphus hongkongensis, female from Hong Kong | Photo by Kenneth Leung

香港纤春蜓 雄，香港 | 祁麟峰 摄
Leptogomphus hongkongensis, male from Hong Kong | Photo by Mahler Ka

居间纤春蜓 *Leptogomphus intermedius* Chao, 1982

居间纤春蜓 雄，广东 | 宋春斌 摄
Leptogomphus intermedius, male from Guangdong | Photo by Ruibin Song

【形态特征】雄性上唇具1对黄斑，额横纹黄色，中央间断；胸部背条纹与领条纹不相连，具甚细的肩前条纹，合胸侧面第2条纹和第3条纹完整；腹部黑色，第1～7节具黄斑。雌性与雄性相似，后头中央具1对较长的角状突起。【长度】体长 65～68 mm，腹长 49～51 mm，后翅 40～43 mm。【栖息环境】海拔 1000 m以下茂密森林中的狭窄小溪和渗流地。【分布】中国特有，分布于福建、广东。【飞行期】4—8月。

[Identification] Male labrum with a pair of yellow spots, frons with a yellow stripe interrupted medially. Thorax with dorsal stripes and collar stripes not connecting, antehumeral stripes narrow, second and third lateral stripes complete. Abdomen black, S1-S7 with yellow markings. Female similar to male, occipital margin with a pair of long horns. **[Measurements]** Total length 65-68 mm, abdomen 49-51 mm, hind wing 40-43 mm. **[Habitat]** Narrow streams and seepages in dense forest below 1000 m elevation. **[Distribution]** Endemic to China, recorded from Fujian, Guangdong. **[Flight Season]** April to August.

居间纤春蜓 雌，广东 | 莫善濂 摄
Leptogomphus intermedius, female from Guangdong | Photo by Shanlian Mo

居间纤春蜓 雄，广东 | 宋睿斌 摄
Leptogomphus intermedius, male from Guangdong | Photo by Ruibin Song

圆腔纤春蜓 *Leptogomphus perforatus* Ris, 1912

【形态特征】雄性上唇具1对黄斑，额横纹黄色，中央间断；胸部背条纹与领条纹不相连，具细长的肩前条纹，合胸侧面第2条纹和第3条纹完整；腹部黑色，第1～7节具黄斑。雌性与雄性相似，后头后缘两侧具小瘤状突起。【长度】体长 60～62 mm，腹长 46～47 mm，后翅 37～42 mm。【栖息环境】海拔 1000 m以下森林中的小溪。【分布】云南（红河）、广西、广东；越南。【飞行期】4—8月。

[Identification] Male labrum with a pair of yellow spots, frons with a yellow stripe interrupted medially. Thorax with dorsal stripes and collar stripes not connecting, antehumeral stripes narrow, second and third lateral stripes complete. Abdomen black, S1-S7 with yellow markings. Female similar to male, occipital margin with a pair of small tuberculate prominences. **[Measurements]** Total length 60-62 mm, abdomen 46-47 mm, hind wing 37-42 mm. **[Habitat]** Streams in dense forest below 1000 m elevation. **[Distribution]** Yunnan (Honghe), Guangxi, Guangdong; Vietnam. **[Flight Season]** April to August.

圆腔纤春蜓 雄，云南（红河）
Leptogomphus perforatus, male from Yunnan (Honghe)

圆腔纤春蜓 雌，云南（红河）
Leptogomphus perforates, female from Yunnan (Honghe)

苏氏纤春蜓台湾亚种 *Leptogomphus sauteri formosanus* Matsumura, 1926

【形态特征】雄性上唇具1对黄斑，额横纹黄色；胸部背条纹与领条纹不相连，具甚小的肩前上点，合胸侧面第2条纹和第3条纹完整；腹部黑色，第1～7节具黄斑。雌性与雄性相似，后头后缘具较短的瘤状突起。【长度】体长52～57 mm，腹长 40～44 mm，后翅 35～39 mm。【栖息环境】海拔 1000 m以下森林中的溪流。【分布】中国台湾特有。【飞行期】4—10月。

苏氏纤春蜓台湾亚种 雌，台湾 | 嘎嘎 摄
Leptogomphus sauteri formosanus, female from Taiwan | Photo by Gaga

苏氏纤春蜓台湾亚种 雄，台湾 | 嘎嘎 摄
Leptogomphus sauteri formosanus, male from Taiwan | Photo by Gaga

[Identification] Male labrum with a pair of yellow spots, frons with a yellow stripe. Thorax with dorsal stripes and collar stripes not connecting, superior spots very small, second and third lateral stripes complete. Abdomen black, S1-S7 with yellow markings. Female similar to male, occipital margin with short tuberculate prominences. **[Measurements]** Total length 52-57 mm, abdomen 40-44 mm, hind wing 35-39 mm. **[Habitat]** Streams in dense forest below 1000 m elevation. **[Distribution]** Endemic to Taiwan of China. **[Flight Season]** April to October.

苏氏纤春蜓指名亚种 *Leptogomphus sauteri sauteri* Ris, 1912

【形态特征】本亚种与苏氏纤春蜓台湾亚种近似，雄性的区别在于本亚种第6腹节基方具1个更大的黄斑，而台湾亚种第6节的黄斑甚小。两个亚种根据在台湾的分布划定，台湾南部为指名亚种，中部至北部为台湾亚种。【长度】体长 49～50 mm，腹长 36～37 mm，后翅 31～37 mm。【栖息环境】海拔 1000 m以下森林中的溪流。【分布】中国台湾特有。【飞行期】4—9月。

[Identification] Similar to *L. sauteri formosanus*, male distinguished by a larger yellow spot on S6. The two subspecies separated by distribution, the nominate subspecies confined to the south of Taiwan and *L. sauteri formosanus* occurs from the central to the north of Taiwan. **[Measurements]** Total length 49-50 mm, abdomen 36-37 mm, hind wing 31-37 mm. **[Habitat]** Streams in dense forest below 1000 m elevation. **[Distribution]** Endemic to Taiwan of China. **[Flight Season]** April to September.

三道纤春蜓 *Leptogomphus tamdaoensis* Karube, 2014

三道纤春蜓 雌，广西
Leptogomphus tamdaoensis, female from Guangxi

三道纤春蜓 雄，广西
Leptogomphus tamdaoensis, male from Guangxi

【形态特征】雄性上唇具1对黄斑，额横纹黄色，中央间断；胸部背条纹与领条纹不相连，具细长的肩前条纹，合胸侧面第2条纹和第3条纹完整；腹部黑色，第1~7节具黄斑。雌性与雄性相似。【长度】体长 60~63 mm，腹长 46~49 mm，后翅 38~42 mm。【栖息环境】海拔 1000 m以下森林中的小溪。【分布】广西；越南。【飞行期】5—7月。

[Identification] Male labrum with a pair of yellow spots, frons with a yellow stripe interrupted medially. Thorax with dorsal stripes and collar stripes not connecting, antehumeral stripes narrow, second and third lateral stripes complete. Abdomen black, S1-S7 with yellow markings. Female similar to male. **[Measurements]** Total length 60-63 mm, abdomen 46-49 mm, hind wing 38-42 mm. **[Habitat]** Streams in forest below 1000 m elevation. **[Distribution]** Guangxi; Vietnam. **[Flight Season]** May to July.

羚角纤春蜓 *Leptogomphus uenoi* Asahina, 1996

羚角纤春蜓 雌，广西
Leptogomphus uenoi, female from Guangxi

【形态特征】雌性上唇具1对黄斑，额横纹黄色，中央间断，后头缘中央具1对甚长的角状突起；胸部背条纹与领条纹不相连，具细长的肩前条纹，合胸侧面第2条纹和第3条纹完整；腹部黑色，第1～8节具黄斑。【长度】雌性体长54 mm，腹长 42 mm，后翅 37 mm。【栖息环境】海拔 1000 m以下森林中的小溪。【分布】广西；越南。【飞行期】5—7月。

[Identification] Female labrum with a pair of yellow spots, frons with a yellow stripe interrupted medially, occipital margin with a pair of long horns. Thorax with dorsal stripes and collar stripes not connecting, antehumeral stripes narrow, second and third lateral stripes complete. Abdomen black, S1-S8 with yellow markings. **[Measurements]** Female total length 54 mm, abdomen 42 mm, hind wing 37 mm. **[Habitat]** Streams in forest below 1000 m elevation. **[Distribution]** Guangxi; Vietnam. **[Flight Season]** May to July.

纤春蜓属待定种 *Leptogomphus* sp.

【形态特征】雄性上唇具1对黄斑，额横纹黄色；胸部背条纹与领条纹不相连，具细长的肩前条纹，合胸侧面第2条纹和第3条纹完整；腹部黑色，第1~7节具黄斑。【长度】雄性体长 54~60 mm，腹长 43~45 mm，后翅 36~37 mm。【栖息环境】海拔 1000 m以下茂密森林中的狭窄小溪和渗流地。【分布】云南（红河）。【飞行期】4—6月。

[Identification] Male labrum with a pair of yellow spots, frons with a yellow stripe. Thorax with dorsal stripes and collar stripes not connecting, antehumeral stripes narrow, second and third lateral stripes complete. Abdomen black, S1-S7 with yellow markings. **[Measurements]** Male total length 54-60 mm, abdomen 43-45 mm, hind wing 36-37 mm. **[Habitat]** Narrow streams and seepages in dense forest below 1000 m elevation. **[Distribution]** Yunnan (Honghe). **[Flight Season]** April to June.

纤春蜓属待定种 雄，云南（红河）
Leptogomphus sp., male from Yunnan (Honghe)

纤春蜓属待定种 雌，云南（红河）
Leptogomphus sp., female from Yunnan (Honghe)

大春蜓属 Genus *Macrogomphus* Selys, 1858

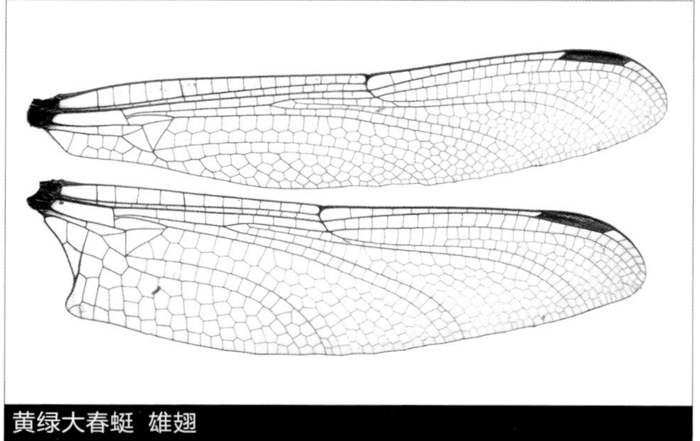
黄绿大春蜓 雄翅
Macrogomphus matsukii, male wings

本属全球已知接近20种，主要分布于亚洲的热带区域。中国已知3种，分布于华南和西南地区。本属春蜓体型较大；翅透明，翅痣甚长，三角室、上三角室和下三角室无横脉，基臀区具1条横脉，臀三角室3室；腹部第9节甚长，超过第8节长度的2倍，第10节甚短；雄性上肛附器内缘中央具1个分枝，末端尖锐，下肛附器中央凹陷较深，2枝分歧。

本属蜻蜓栖息于山区溪流。成虫偶见停在水边的树丛中。

The genus contains nearly 20 species, mainly distributed in tropical Asia. Three are recorded from China, found in the South and Southwest. Species of the genus are fairly large-sized dragonflies. Wings hyaline, pterostigma long, no crossvein in triangle, hypertriangle and subtriangle, cubital space with one crossvein, anal triangle 3-celled. S9 very long, exceeding two times as long as S8, S10 short. Male superior appendages with a median branch at the inner margin, tips pointed, the inferior appendage with a median hollow, forming two divergent branches.

Macrogomphus species inhabit montane streams. Adults occasionally perch in the thickets near water.

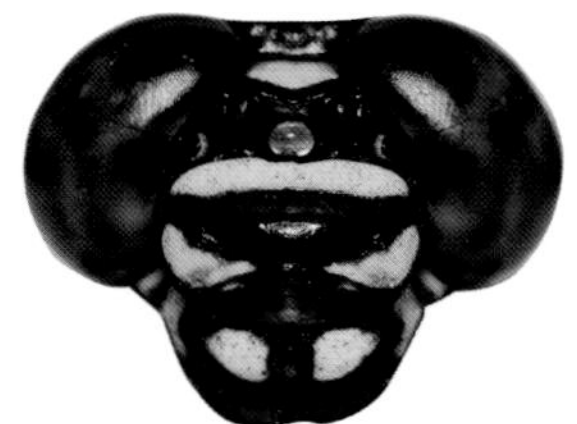
黄绿大春蜓 雄
Macrogomphus matsukii, male

黄绿大春蜓 雌
Macrogomphus matsukii, female

大春蜓属 头部正面观
Genus *Macrogomphus*, head in frontal view

黄绿大春蜓
Macrogomphus matsukii

大春蜓属 雌性下生殖板
Genus *Macrogomphus*, female vulvar lamina

黄绿大春蜓 雌
Macrogomphus matsukii, female

黄绿大春蜓
Macrogomphus matsukii

大春蜓属 雄性肛附器
Genus *Macrogomphus*, male appendages

黄绿大春蜓 *Macrogomphus matsukii* Asahina, 1986

【形态特征】雄性面部黑色具白斑，上唇具1对黄斑，额横纹黄色，头顶侧单眼后方黄色；胸部背条纹与领条纹相连，合胸侧面第2条纹和第3条纹完整，在下方大面积合并；腹部黑色，第1～9节具黄斑。雌性与雄性相似。【长度】体长 71～72 mm，腹长 54～55 mm，后翅 42～45 mm。【栖息环境】海拔 1000 m以下的泥沙溪流和沟渠。【分布】云南（西双版纳）；泰国、柬埔寨、老挝。【飞行期】4—7月。

[Identification] Male face black with white markings, labrum with a pair of yellow spots, frons with a yellow stripe, vertex yellow behind lateral ocelli. Thorax with dorsal stripes and collar stripes connecting, second and third lateral stripes complete and connected ventrally. Abdomen black, S1-S9 with yellow markings. Female similar to male. **[Measurements]** Total length 71-72 mm, abdomen 54-55 mm, hind wing 42-45 mm. **[Habitat]** Muddy streams and ditches below 1000 m elevation. **[Distribution]** Yunnan (Xishuangbanna); Thailand, Cambodia, Laos. **[Flight Season]** April to July.

黄绿大春蜓 雄，云南（西双版纳）
Macrogomphus matsukii, male from Yunnan (Xishuangbanna)

黄绿大春蜓 雌，云南（西双版纳）
Macrogomphus matsukii, female from Yunnan (Xishuangbanna)

大春蜓属待定种 *Macrogomphus* sp.

大春蜓属待定种 雄，海南
Macrogomphus sp., male from Hainan

大春蜓属待定种 雄，海南
Macrogomphus sp., male from Hainan

【形态特征】本种与黄绿大春蜓相似，但雄性的后钩片形状不同，身体的黄斑不如黄绿大春蜓发达。【长度】雄性体长 70 mm，腹长 53 mm，后翅 42 mm。【栖息环境】海拔 500 m以下的溪流。【分布】海南。【飞行期】4—5月。

[Identification] The species is similar to *M. matsukii* but the shape of posterior hamulus is different, yellow markings less developed. **[Measurements]** Male total length 70 mm, abdomen 53 mm, hind wing 42 mm. **[Habitat]** Streams below 500 m elevation. **[Distribution]** Hainan. **[Flight Season]** April to May.

硕春蜓属 Genus *Megalogomphus* Campion, 1923

本属全球已知10余种，主要分布在亚洲的热带地区。中国已知仅1种，分布于华南地区。本属春蜓体型较大；翅透明，翅痣甚长，三角室、上三角室和下三角室无横脉，臀圈2室，基臀区具1条横脉，臀三角室4室；雄性上肛附器较长，长度至少为第10节的2倍，末端稍微向下弯曲，下肛附器两枝分歧，末端叉状。

本属蜻蜓栖息于山区的溪流和河流。雄性经常停在水边的植物枝头上占据领地，并时而飞行巡逻。雄性之间经常展开争斗。

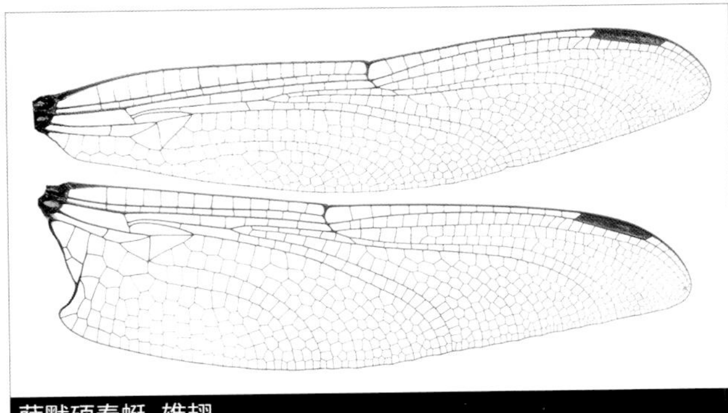

萨默硕春蜓 雄翅
Megalogomphus sommeri, male wings

The genus contains over ten species, mainly distributed in tropical Asia. Only one species is known from China, found in the South. Species of the genus are large-sized dragonflies. Wings hyaline, pterostigma long, no crossvein in triangle, hypertriangle and subtriangle, anal loop 2-celled, cubital space with one crossvein, anal triangle 4-celled. Male superior appendages long, about two times as long as S10, the tips curved slightly downwards, inferior appendage with two branches divergent, the tips fork-shaped.

Megalogomphus species inhabit montane streams and rivers. Males usually perch on top of branches for territory and patrol for short time. Males usually fight fiercely.

萨默硕春蜓
Megalogomphus sommeri

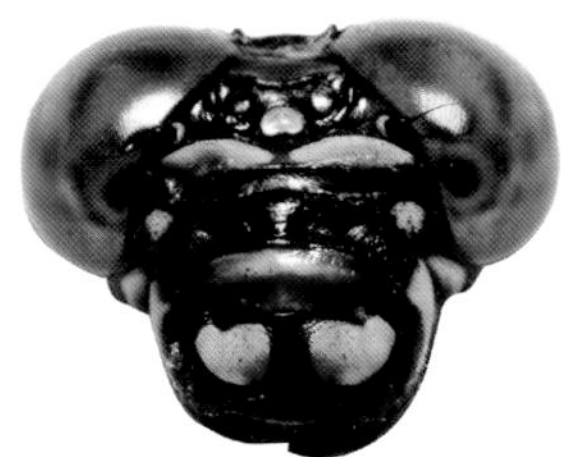

萨默硕春蜓，雌
Megalogomphus sommeri, female

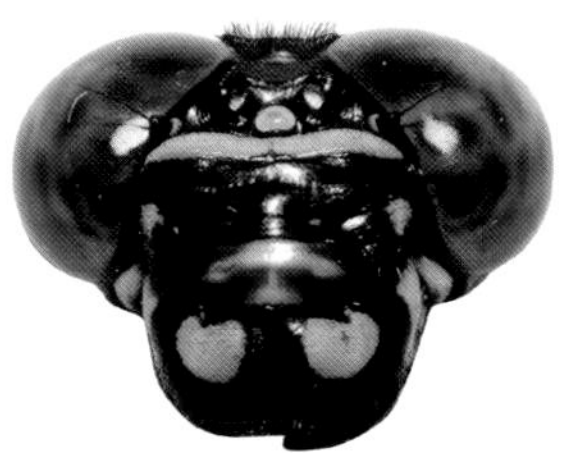

萨默硕春蜓，雄
Megalogomphus sommeri, male

硕春蜓属 雄性肛附器
Genus *Megalogomphus*, male appendages

硕春蜓属 头部正面观
Genus *Megalogomphus*, head in frontal view

萨默硕春蜓 雄
Megalogomphus sommeri, male

萨默硕春蜓 *Megalogomphus sommeri* (Selys, 1854)

【形态特征】雄性上唇具1对黄斑，额横纹黄色，中央间断；胸部黑色，背条纹与领条纹不相连，具甚小的肩前上点，合胸侧面第2条纹和第3条纹完整，在下方大面积合并；腹部黑色，第1～9节具黄斑，其中第7节黄斑甚大。雌性与雄性相似，但腹部较短。【长度】体长 78～80 mm，腹长 57～60 mm，后翅 50～55 mm。【栖息环境】海拔 1000 m以下的溪流和河流。【分布】广西、广东、海南、福建、香港；老挝、越南。【飞行期】4—10月。

[Identification] Male labrum with a pair of yellow spots, frons with a yellow stripe interrupted medially. Thorax black, dorsal stripes and collar stripes not connecting, superior spots small, second and third lateral stripes complete and connected ventrally. Abdomen black, S1-S9 with yellow markings, S7 with a large yellow spot. Female similar to male but abdomen shorter. **[Measurements]** Total length 78-80 mm, abdomen 57-60 mm, hind wing 50-55 mm. **[Habitat]** Rivers and streams below 1000 m elevation. **[Distribution]** Guangxi, Guangdong, Hainan, Fujian, Hong Kong; Laos, Vietnam. **[Flight Season]** April to October.

萨默硕春蜓 雌，广东
Megalogomphus sommeri, female from Guangdong

萨默硕春蜓 雄，海南
Megalogomphus sommeri, male from Hainan

萨默硕春蜓 雄，海南
Megalogomphus sommeri, male from Hainan

弯尾春蜓属 Genus *Melligomphus* Chao, 1990

本属全球已知6种，主要分布于亚洲的亚热带区域。中国已知5种，分布于华中、华南和西南地区。本属春蜓体中型；翅透明，翅痣甚长，具支持脉，三角室、上三角室和下三角室无横脉，基臀区具1条横脉。本属与环尾春蜓属非常相似，但雄性上肛附器末端仅稍微弯曲，而环尾春蜓属则强烈钩曲呈半圆弧形。

本属蜻蜓栖息于山区溪流和河流，喜欢开阔而多岩石的河段。雄性经常停在水面的大岩石上占据领地，有时低空悬停飞行。

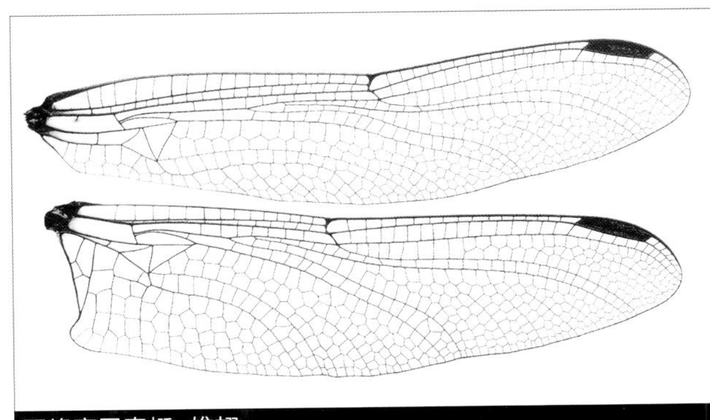

双峰弯尾春蜓，雄翅
Melligomphus ardens, male wings

The genus contains six species, mainly distributed in subtropical Asia. Five are recorded from China, found in the Central, South and Southwest. Species of the genus are medium-sized. Wings hyaline, pterostigma long and well braced, no crossvein in triangle, hypertriangle and subtriangle, cubital space with one crossvein. *Melligomphus* species are similar to *Lamelligomhus* species, but male superior appendages of *Melligomhus* only slightly curved, whereas in *Lamelligomhus* they are strongly curved and circular shape.

Melligomphus species inhabit montane streams, preferring exposed and rocky sections. Territorial males usually perch on the big rocks and sometimes hover above water.

双峰弯尾春蜓
Melligomphus ardens

无锋弯尾春蜓
Melligomphus ludens

弯尾春蜓属 雄性腹部末端
Genus *Melligomphus*, male distal abdomen

双峰弯尾春蜓 雄
Melligomphus ardens, male

双峰弯尾春蜓 *Melligomphus ardens* (Needham, 1930)

【形态特征】雄性上唇具1对黄斑，额横纹黄色；胸部背条纹与领条纹相连，具甚小的肩前上点，合胸侧面第2条纹和第3条纹完整，大面积合并；腹部黑色，第1～7节具黄斑，其中第7节黄斑甚大，第8节背面中央具较矮的驼峰状突起。雌性与雄性相似，后头缘具1对角状突起。【长度】体长 61～64 mm，腹长 46～48 mm，后翅 35～40 mm。【栖息环境】海拔 1000 m以下的开阔溪流。【分布】中国特有，分布于四川、贵州、安徽、湖南、浙江、福建、广东、广西、海南。【飞行期】4—8月。

双峰弯尾春蜓 雌，安徽 | 秦彧 摄
Melligomphus ardens, female from Anhui | Photo by Yu Qin

[Identification] Male labrum with a pair of yellow spots, frons with a yellow stripe. Thorax with dorsal stripes and collar stripes connecting, superior spots small, second and third lateral stripes complete and largely combined. Abdomen black, S1-S7 with yellow markings, S7 with a large yellow spot, S8 with a pair of hump-shaped prominences. Female similar to male, occipital margin with a pair of occipital horns. **[Measurements]** Total length 61-64 mm, abdomen 46-48 mm, hind wing 35-40 mm. **[Habitat]** Exposed streams below 1000 m elevation. **[Distribution]** Endemic to China, recorded from Sichuan, Guizhou, Anhui, Hunan, Zhejiang, Fujian, Guangdong, Guangxi, Hainan. **[Flight Season]** April to August.

双峰弯尾春蜓 雄，海南
Melligomphus ardens, male from Hainan

广东弯尾春蜓 *Melligomphus guangdongensis* (Chao, 1994)

【形态特征】雄性上唇具1对黄斑，额横纹黄色，中央间断；胸部背条纹与领条纹不相连，合胸侧面第2条纹和第3条纹完整并大面积合并；腹部黑色，第1～7节具黄斑，其中第7节黄斑甚大。雌性与雄性相似，尾毛白色。【长度】腹长 35～37 mm，后翅 29～30 mm。【栖息环境】海拔 1000 m以下森林中的溪流。【分布】中国特有，分布于广东、香港。【飞行期】4—7月。

广东弯尾春蜓 雄，香港 | 祁麟峰 摄
Melligomphus guangdongensis, male from Hong Kong | Photo by Mahler Ka

[Identification] Male labrum with a pair of yellow spots, frons with a yellow stripe interrupted medially. Thorax with dorsal stripes and collar stripes not connecting, second and third lateral stripes complete and largely combined. Abdomen black, S1-S7 with yellow markings, S7 with a large yellow spot. Female similar to male, cerci white. **[Measurements]** Abdomen 35-37 mm, hind wing 29-30 mm. **[Habitat]** Streams in forest below 1000 m elevation. **[Distribution]** Endemic to China, recorded from Guangdong, Hong Kong. **[Flight Season]** April to July.

广东弯尾春蜓 雌，香港 | 祁麟峰 摄
Melligomphus guangdongensis, female from Hong Kong | Photo by Mahler Ka

无峰弯尾春蜓 *Melligomphus ludens* (Needham, 1930)

【形态特征】雄性上唇具1对黄斑，额横纹黄色，中央间断；胸部背条纹与领条纹不相连，合胸侧面第2条纹和第3条纹合并；腹部黑色，第1~7节具黄斑，其中第7节黄斑甚大。【长度】雄性体长 62 mm，腹长 47 mm，后翅 34 mm。【栖息环境】海拔 500 m以下的溪流和河流。【分布】中国特有，分布于浙江、福建。【飞行期】6—8月。

[Identification] Male labrum with a pair of yellow spots, frons with a yellow stripe interrupted medially. Thorax with dorsal stripes and collar stripes not connecting, second and third lateral stripes combined. Abdomen black, S1-S7 with yellow markings, S7 with a large yellow spot. **[Measurements]** Male total length 62 mm, abdomen 47 mm, hind wing 34 mm. **[Habitat]** Streams and rivers below 500 m elevation. **[Distribution]** Endemic to China, recorded from Zhejiang, Fujian. **[Flight Season]** June to August.

无峰弯尾春蜓 雄，浙江
Melligomphus ludens, male from Zhejiang

长足春蜓属 Genus *Merogomphus* Martin, 1904

本属全球已知10余种，分布于亚洲的亚热带和热带地区。中国已知8种，分布于华中、华南和西南等地。本属春蜓体中型；翅透明，翅痣甚长，具支持脉，三角室、上三角室和下三角室无横脉，基臀区具1~2条横脉，臀三角3室。本属蜻蜓与异春蜓属相似，但雄性上肛附器腹方无突起。

本属蜻蜓栖息于山区溪流和沟渠。雄性经常停在水面附近的叶片上，并时而飞行巡逻，有些种类飞行时腹部弯曲。

The genus contains over ten species, distributed in subtropical and tropical Asia. Eight are recorded from China, found in the Central, South and Southwest. Species of the genus are medium-sized. Wings hyaline, pterostigma well braced, no crossvein in triangle, hypertriangle and subtriangle, cubital space with 1-2 crossveins, anal triangle 3-celled. Species of *Merogomphus* are similar to those of *Anisogomphus*, but the male superior appendages without ventral teeth.

Merogomphus species inhabit montane streams and ditches. Territorial males usually perch on leaves near water and sometimes patrol. Some species fly with the abdomen curved.

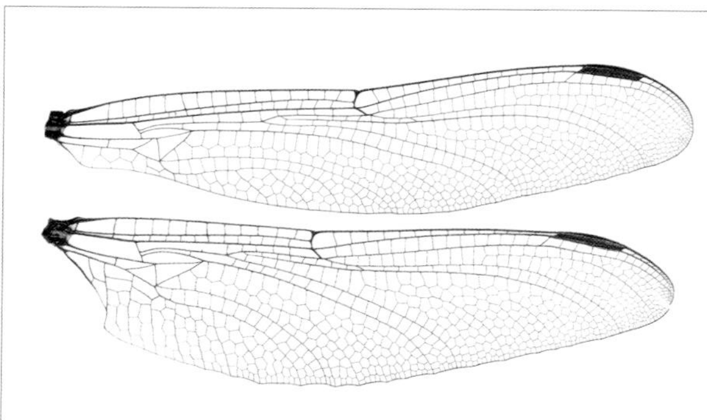

帕维长足春蜓 雄翅
Merogomphus pavici, male wings

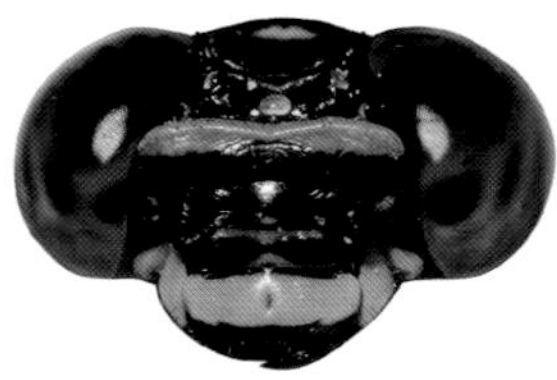

帕维长足春蜓 雄
Merogomphus pavici, male

帕维长足春蜓 雌
Merogomphus pavici, female

长足春蜓属 头部正面观
Genus *Merogomphus*, head in frontal view

帕维长足春蜓
Merogomphus pavici

四川长足春蜓
Merogomphus vespertinus

泰国长足春蜓
Merogomphus pinratani

越南长足春蜓
Merogomphus tamdaoensis

长足春蜓属 雄性肛附器
Genus *Merogomphus*, male appendages

帕维长足春蜓 雄
Merogomphus pavici, male

帕维长足春蜓 *Merogomphus pavici* Martin, 1904

【形态特征】雄性上唇黄色，额横纹甚阔；胸部背条纹与领条纹不相连，肩前条纹较长，合胸侧面第2条纹和第3条纹完整；腹部黑色，第1～7节具黄斑，其中第7节黄斑甚大，上肛附器白色，下肛附器黑色。雌性与雄性相似，侧单眼后方具角状突起。【长度】体长 67～72 mm，腹长 51～55 mm，后翅 40～47 mm。【栖息环境】海拔 1000 m以下的开阔溪流、河流和沟渠。【分布】湖北、贵州、浙江、福建、海南、广西、广东、台湾；泰国、老挝、越南。【飞行期】4—9月。

帕维长足春蜓 雌，广东
Merogomphus pavici, female from Guangdong

[Identification] Male labrum yellow, frons with a broad stripe. Thorax with dorsal stripes and collar stripes not connecting, antehumeral stripes long, second and third lateral stripes complete. Abdomen black, S1-S7 with yellow markings, S7 with a large yellow spot, superior appendages white, inferior appendage black. Female similar to male, vertex with horns behind lateral ocelli. **[Measurements]** Total length 67-72 mm, abdomen 51-55 mm, hind wing 40-47 mm. **[Habitat]** Exposed streams, rivers and ditches below 1000 m elevation. **[Distribution]** Hubei, Guizhou, Zhejiang, Fujian, Hainan, Guangxi, Guangdong, Taiwan; Thailand, Laos, Vietnam. **[Flight Season]** April to September.

帕维长足春蜓 雄，广东
Merogomphus pavici, male from Guangdong

泰国长足春蜓 *Merogomphus pinratani* (Hämäläinen, 1991)

【形态特征】雄性面部黑色，额横纹甚阔；胸部背条纹与领条纹相连，肩前条纹较长，合胸侧面第2条纹和第3条纹完整；腹部黑色，第1～8节具黄斑，肛附器褐色。雌性与雄性相似，后头缘中央凹陷较深。本种之前放在异春蜓属，此处移入长足春蜓属。【长度】体长 51～55 mm，腹长 39～42 mm，后翅 36～40 mm。【栖息环境】海拔500～1500 m森林中的狭窄小溪。【分布】云南（西双版纳、普洱）；泰国。【飞行期】6—10月。

[Identification] Male face black, frons with a broad stripe. Thorax with dorsal stripes and collar stripes connecting, antehumeral stripes long, second and third lateral stripes complete. Abdomen black, S1-S8 with yellow markings, anal appendages brown. Female similar to male, occipital margin concaved medially. The species was previously placed in genus *Anisogomphus* but it is moved to genus *Merogomphus* here. [Measurements] Total length 51-55 mm, abdomen 39-42 mm, hind wing 36-40 mm. [Habitat] Narrow streams in forest at 500-1500 m elevation. [Distribution] Yunnan (Xishuangbanna, Pu'er); Thailand. [Flight Season] June to October.

泰国长足春蜓 雄，云南（普洱）
Merogomphus pinratani, male from Yunnan (Pu'er)

泰国长足春蜓 雌，云南（西双版纳）
Merogomphus pinratani, female from Yunnan (Xishuangbanna)

泰国长足春蜓 雄，云南（普洱）
Merogomphus pinratani, male from Yunnan (Pu'er)

越南长足春蜓 *Merogomphus tamdaoensis* Karube, 2001

越南长足春蜓 雄，云南（红河）
Merogomphus tamdaoensis, male from Yunnan (Honghe)

越南长足春蜓 雌，云南（红河）
Merogomphus tamdaoensis, female from Yunnan (Honghe)

越南长足春蜓 雄，云南（红河）
Merogomphus tamdaoensis, male from Yunnan (Honghe)

【形态特征】雄性面部黑色，额具1对黄斑；胸部背条纹与领条纹相连，肩前条纹较细长，合胸侧面第2条纹和第3条纹完整；腹部黑色，第1～8节具黄斑，肛附器黑色。雌性与雄性相似，腹部的黄色条纹更发达。【长度】体长61～69 mm，腹长 46～52 mm，后翅 40～48 mm。【栖息环境】海拔 500～1000 m森林中的狭窄小溪。【分布】云南（红河）；越南。【飞行期】5—7月。

[Identification] Male face black, frons with a pair of yellow spots. Thorax with dorsal stripes and collar stripes connecting, antehumeral stripes narrow and long, second and third lateral stripes complete. Abdomen black, S1-S8 with yellow markings, anal appendages black. Female similar to male, abdominal yellow markings more developed. **[Measurements]** Total length 61-69 mm, abdomen 46-52 mm, hind wing 40-48 mm. **[Habitat]** Narrow streams in forest at 500-1000 m elevation. **[Distribution]** Yunnan (Honghe); Vietnam. **[Flight Season]** May to July.

江浙长足春蜓 *Merogomphus vandykei* Needham, 1930

【形态特征】雄性面部黄色，头顶黑色，后头黄色；胸部背条纹与领条纹不相连，肩前条纹较长，合胸侧面第2条纹缺如，第3条纹甚细；腹部黑色，第1~7节具黄斑，其中第7节黄斑甚大，上肛附器白色，下肛附器黑色。【长度】雄性腹长 53 mm，后翅 43 mm。【栖息环境】海拔1000 m以下的开阔溪流。【分布】中国特有，分布于河南、湖北、浙江、江苏。【飞行期】5—8月。

江浙长足春蜓 雄，湖北
Merogomphus vandykei, male from Hubei

[Identification] Male face yellow, vertex black, occiput yellow. Thorax with dorsal stripes and collar stripes not connecting, antehumeral stripes long, second lateral stripe absent, third lateral stripes very narrow. Abdomen black, S1-S7 with yellow markings, S7 with a large yellow spot, superior appendages white, inferior appendage black. **[Measurements]** Male abdomen 53 mm, hind wing 43 mm. **[Habitat]** Exposed streams below 1000 m elevation. **[Distribution]** Endemic to China, recorded from Henan, Hubei, Zhejiang, Jiangsu. **[Flight Season]** May to August.

江浙长足春蜓 雄，河南 | 计云 摄
Merogomphus vandykei, male from Henan | Photo by Yun Ji

四川长足春蜓 *Merogomphus vespertinus* Chao, 1999

【形态特征】雄性面部黑色具黄斑，上唇黄色，额横纹甚阔；胸部背条纹与领条纹不相连，肩前条纹较长，合胸侧面第2条纹和第3条纹完整或中央间断；腹部黑色，第1~7节具黄斑，上肛附器白色，下肛附器黑色。雌性与雄性相似，侧单眼后方具1对细长的角。【长度】体长 58~68 mm，腹长 45~53 mm，后翅 35~44 mm。【栖息环境】海拔1000 m以下的开阔溪流。【分布】中国特有，分布于四川、重庆。【飞行期】6—8月。

[Identification] Male face black with yellow markings, labrum yellow, frons with a broad stripe. Thorax with dorsal stripes and collar stripes not connecting, antehumeral stripes long, second and third lateral stripes complete or interrupted medially. Abdomen black, S1-S7 with yellow markings, superior appendages white, inferior appendage black. Female similar to male, vertex with a pair of long and narrow horns behind lateral ocelli. **[Measurements]** Total length 58-68 mm, abdomen 45-53 mm, hind wing 35-44 mm. **[Habitat]** Exposed streams below 1000 m elevation. **[Distribution]** Endemic to China, recorded from Sichuan, Chongqing. **[Flight Season]** June to August.

四川长足春蜓 雄，重庆
Merogomphus vespertinus, male from Chongqing

小春蜓属 Genus *Microgomphus* Selys, 1858

本属全球已知约15种，分布在亚洲和非洲的热带地区。中国已知2种，仅分布于云南。本属春蜓体型较小；翅透明，翅痣具支持脉，三角室、上三角室和下三角室无横脉，基臀区具1条横脉，盘区基方具2列翅室，前翅的臀脉与翅后缘之间仅具1列翅室；雄性上肛附器白色，内缘中央具1个分枝伸向下方，下肛附器向上弯曲，末端具1个短而窄的凹陷。

本属蜻蜓栖息于山区溪流，喜欢具有林荫和浅滩的溪流。雄性经常停落在沙滩上占据领地。雌性产卵时先停落在水边的树丛或者地面上排出卵块，然后将其投入水中。

The genus contains about 15 species, distributed in tropical Asia and Africa. Two are recorded from China but confined to Yunnan. Species of the genus are fairly small for the family. Wings hyaline, pterostigma braced, no crossvein in triangle, hypertriangle and subtriangle, discoidal field with 2 rows of cells, a single row of cells between A and hind margin in fore wings, cubital space with one crossvein. Male superior appendages white, inner margin with a branch pointed downwards, inferior appendage curved upwards, with a short and narrow cleft distally.

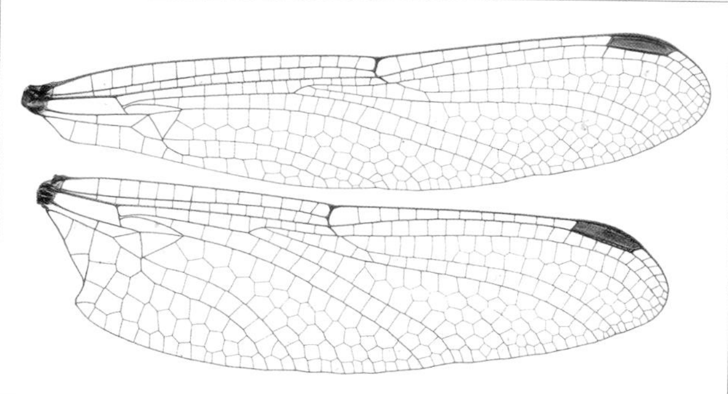

越南小春蜓 雄翅
Microgomphus jurzitzai, male wings

越南小春蜓 雄
Microgomphus jurzitzai, male

小春蜓属 头部正面观
Genus *Microgomphus*, head in frontal view

越南小春蜓 雄
Microgomphus jurzitzai, male

Microgomphus species inhabit montane streams, prefer streamlets with shady trees and shallow beaches. Males usually perch on the ground for territory. Females perch on ground or trees first and exude an egg mass, then deposit it into water.

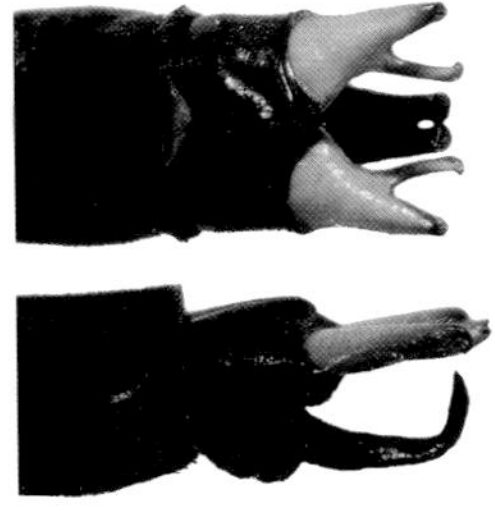

越南小春蜓
Microgomphus jurzitzai

小春蜓属待定种
Microgomphus sp.

小春蜓属 雄性肛附器
Genus *Microgomphus*, male appendages

越南小春蜓
Microgomphus jurzitzai

小春蜓属 雌性下生殖板
Genus *Microgomphus*, female vulvar lamina

越南小春蜓 *Microgomphus jurzitzai* Karube, 2000

【形态特征】雄性上唇具1对黄斑，额横纹甚阔；胸部背条纹与领条纹相连，具较小的肩前上点，合胸侧面第2条纹和第3条纹完整；腹部黑色，第1～6节背面和侧面具黄斑，第7节基方具较大的黄斑。雌性与雄性相似，侧单眼后方具1对角锥状突起，后头缘中央具1对刺状突起。【长度】体长 43～45 mm，腹长 32～34 mm，后翅 28～29 mm。【栖息环境】海拔 1500 m以下森林中的溪流。【分布】云南（红河、西双版纳）；越南。【飞行期】4—6月。

越南小春蜓 雄，云南（红河）
Microgomphus jurzitzai, male from Yunnan (Honghe)

[Identification] Male labrum with a pair of yellow spots, frons with a broad stripe. Thorax with dorsal stripes and collar stripes connecting, superior spots small, second and third lateral stripes complete. Abdomen black, S1-S6 with dorsal and lateral spots, S7 with a large yellow spot. Female similar to male, vertex with a pair of horns behind lateral ocelli, occipital margin with a pair of spines centrally. **[Measurements]** Total length 43-45 mm, abdomen 32-34 mm, hind wing 28-29 mm. **[Habitat]** Streams in forest below 1500 m elevation. **[Distribution]** Yunnan (Honghe, Xishuangbanna); Vietnam. **[Flight Season]** April to June.

越南小春蜓 雌，云南（红河）
Microgomphus jurzitzai, female from Yunnan (Honghe)

小春蜓属待定种 *Microgomphus* sp.

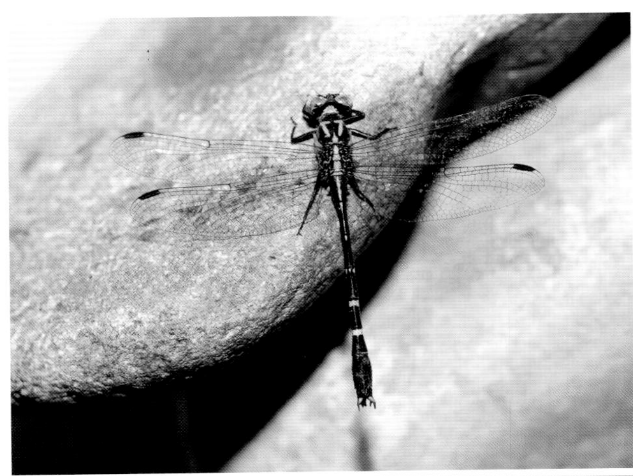

小春蜓属待定种 雄，云南（西双版纳）
Microgomphus sp., male from Yunnan (Xishuangbanna)

【形态特征】雄性面部黑色具白斑，额横纹甚阔；胸部背条纹与领条纹相连，合胸侧面第2条纹和第3条纹完整；腹部黑色，第1~7节具黄白色斑。【长度】雄性体长 41 mm，腹长 31 mm，后翅 25 mm。【栖息环境】海拔 500 m森林中的开阔溪流。【分布】云南（西双版纳）。【飞行期】4—6月。

[Identification] Male face black with white markings, frons with a broad stripe. Thorax with dorsal stripes and collar stripes connecting, second and third lateral stripes complete. Abdomen black, S1-S7 with yellowish white spots. **[Measurements]** Male total length 41 mm, abdomen 31 mm, hind wing 25 mm. **[Habitat]** Exposed stream in forest at 500 m elevation. **[Distribution]** Yunnan (Xishuangbanna). **[Flight Season]** April to June.

内春蜓属 Genus *Nepogomphus* Fraser, 1934

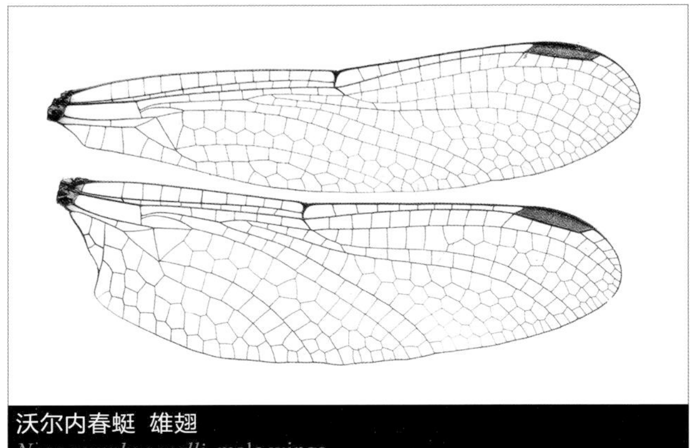

沃尔内春蜓 雄翅
Nepogomphus walli, male wings

本属全球已知3种，主要分布于亚洲的热带地区。中国已知2种，仅分布于西南地区。本属春蜓体小型；翅透明，翅痣较长，具支持脉，三角室、上三角室和下三角室无横脉，基臀区具1条横脉，盘区基方具2列翅室，前翅的臀脉与翅后缘之间仅具1列翅室，臀圈2室，臀三角室4室；雄性肛附器发达，上肛附器的两枝平行，末端向下弯曲，下肛附器的两枝平行，向上弯曲；阳茎具鞭。

本属蜻蜓栖息于山区溪流。它们经常穿梭于溪流边缘的树丛上，有时停落在靠近水面的叶片上。雄性会在水面上短暂悬停巡飞。

The genus contains three species distributed in tropical Asia. Two are recorded from China but only confined to the Southwest. Species of the genus are small for the family; wings hyaline, pterostigma long and braced, no crossvein in triangle, hypertriangle and subtriangle, cubital space with one crossvein, discoidal field with 2 rows of cells in fore wings, a single row of cells between A and hind margin in fore wings, anal loop 2-celled, anal triangle 4-celled. Male anal appendages large, two branches of superiors parallel, tips curved downwards, two branches of inferior appendage parallel and curved upwards. Penis with a pair of flagella.

Nepogomphus species inhabit montane streams. Individuals are usually seen among the short vegetation near water and sometimes perch on the leaves above water. Males patrol by hovering above water for a short time.

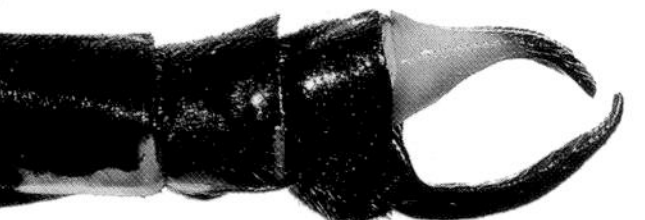

沃尔内春蜓
Nepogomphus walli

内春蜓属待定种
Nepogomphus sp.

内春蜓属 雄肛附器和雌下生殖板
Genus *Nepogomphus*, male anal appendages and female vulvar lamina

沃尔内春蜓 雄
Nepogomphus walli, male

沃尔内春蜓 *Nepogomphus walli* (Fraser, 1924)

【形态特征】雄性上唇具1对黄斑，额横纹黄色，中央间断；胸部背条纹与领条纹相连，合胸侧面第2条纹和第3条纹完整；腹部黑色，第1~7节具黄斑，第7节黄斑较大，上肛附器基方1/2黄色，端方黑色，下肛附器黑色。雌性后头缘具1簇甚小的棘状突起。【长度】体长 36~38 mm，腹长 26~28 mm，后翅 21~22 mm。【栖息环境】海拔 1500 m以下的山区溪流。【分布】云南广布；缅甸、泰国、柬埔寨、老挝、越南、马来半岛。【飞行期】6—12月。

[Identification] Male labrum with a pair of yellow spots, frons with a yellow stripe interrupted medially. Thorax with dorsal stripes and collar stripes connecting, second and third lateral stripes complete. Abdomen black, S1-S7 with yellow markings, S7 with a large yellow spot, superior appendages with basal half yellow, apical half black, inferior appendage black. Female occipital margin with a tuft of small spines. [Measurements] Total length 36-38 mm, abdomen 26-28 mm, hind wing 21-22 mm. [Habitat] Montane streams below 1500 m elevation. [Distribution] Widespread in Yunnan; Myanmar, Thailand, Cambodia, Laos, Vietnam, Peninsular Malaysia. [Flight Season] June to December.

沃尔内春蜓 雌，云南（西双版纳）
Nepogomphus walli, female from Yunnan (Xishuangbanna)

沃尔内春蜓 雄，云南（西双版纳）
Nepogomphus walli, male from Yunnan (Xishuangbanna)

内春蜓属待定种 *Nepogomphus* sp.

内春蜓属待定种 雄，云南（德宏）
Nepogomphus sp., male from Yunnan (Dehong)

【形态特征】本种与沃尔内春蜓相似，但体型稍大，雄性肛附器约与第10节等长，而沃尔内春蜓的肛附器长度约为第10节的1.5倍。【长度】体长 38~39 mm，腹长 28~29 mm，后翅 22~25 mm。【栖息环境】海拔 1000 m以下的山区溪流。【分布】云南（德宏）。【飞行期】6—11月。

[Identification] Similar to *N. walli* but larger, male anal appendages the same length as S10, 1.5 times length of S10 in *N. walli*. **[Measurements]** Total length 38-39 mm, abdomen 28-29 mm, hind wing 22-25 mm. **[Habitat]** Montane streams below 1000 m elevation. **[Distribution]** Yunnan (Dehong). **[Flight Season]** June to November.

日春蜓属 Genus *Nihonogomphus* Oguma, 1926

本属全球已知约20种，主要分布于亚洲的亚热带和热带区域。中国已知10余种，分布于华中、华南和西南地区。本属春蜓体中型，很多种类身体具鲜艳的绿色斑纹；翅透明，三角室、上三角室和下三角室无横脉，基臀区具1条横脉，臀圈1～2室，臀三角室4室；雄性肛附器发达，上肛附器的两枝在基方2/3平行，末端向内弯曲几乎成直角，下肛附器较短，向上弯曲，中央呈“U”形凹陷。阳茎具1对鞭。

本属蜻蜓栖息于山区溪流和河流。雄性停落在水面的岩石上或水边的沙滩上，时而巡飞。雌性空投式产卵。

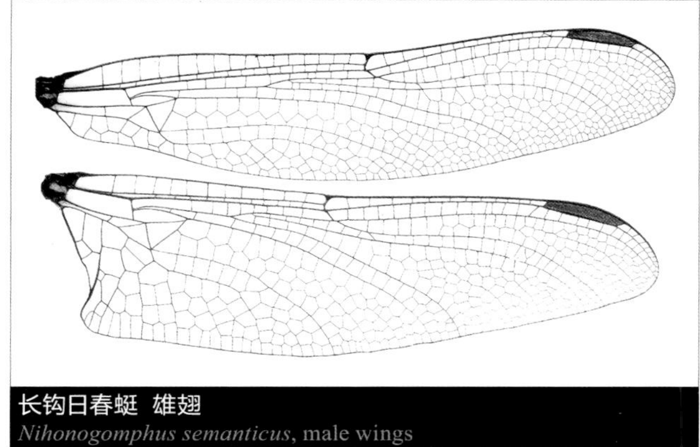

长钩日春蜓 雄翅
Nihonogomphus semanticus, male wings

The genus contains about 20 species, mainly distributed in subtropical and tropical Asia. Over ten are recorded from China, found in the Central, South and Southwest. Species of the genus are medium-sized, many species possess green markings. Wings hyaline, no crossvein in triangle, hypertriangle and subtriangle, cubital space with one crossvein, anal loop 1-2 celled, anal triangle 4-celled. Male superior appendages parallel at basal two thirds, tips strongly curved forming a right angle, inferior appendage short and curved upwards, with a median U shaped hollow. Penis with a pair of flagella.

黄侧日春蜓 雄 | 宋黎明 摄
Nihonogomphus luteolatus, male | Photo by Liming Song

Nihonogomphus species inhabit montane streams and rivers. Males perch on the big rocks or beach for territory, sometimes patrol along the stream. Females oviposit by dropping the eggs into water when hovering.

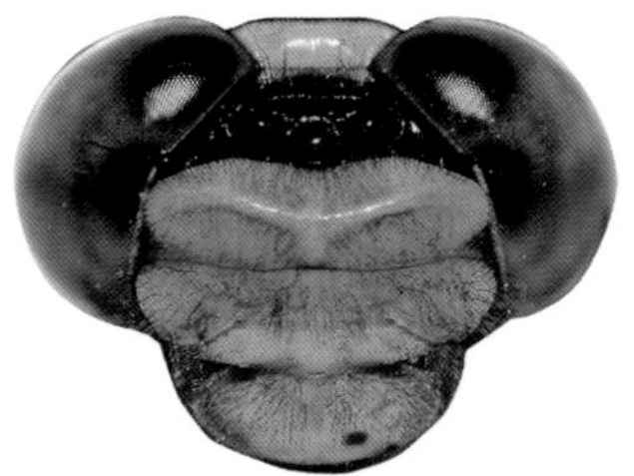

刀日春蜓 雄
Nihonogomphus cultratus, male

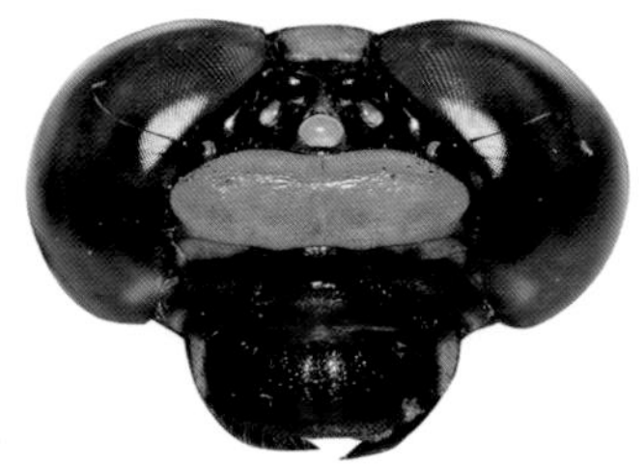

汤氏日春蜓 雄
Nihonogomphus thomassoni, male

黄侧日春蜓 雄
Nihonogomphus luteolatus, male

日春蜓属 头部正面观
Genus *Nihonogomphus*, head in frontal view

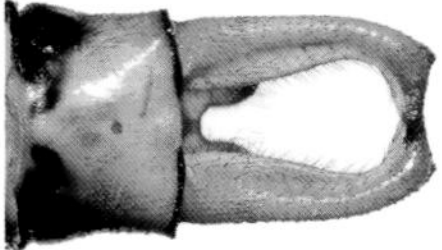

刀日春蜓
Nihonogomphus cultratus

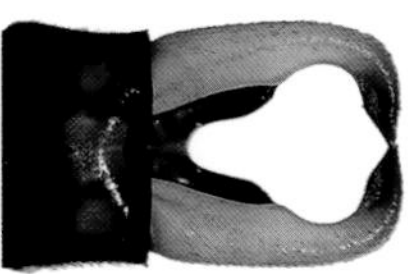

黄沙日春蜓
Nihonogomphus huangshaensis

黄侧日春蜓
Nihonogomphus luteolatus

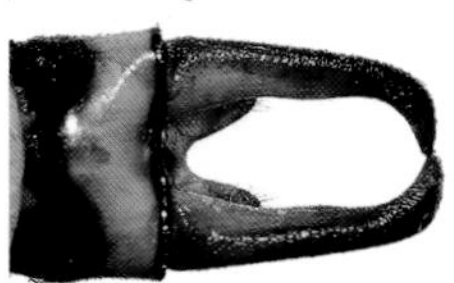

臼齿日春蜓
Nihonogomphus ruptus

长钩日春蜓
Nihonogomphus semanticus

汤氏日春蜓
Nihonogomphus thomassoni

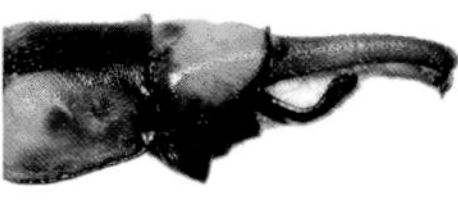

贝氏日春蜓
Nihonogomphus bequaerti

短翅日春蜓
Nihonogomphus brevipennis

日春蜓属 雄性肛附器
Genus *Nihonogomphus*, male anal appendages

贝氏日春蜓 *Nihonogomphus bequaerti* Chao, 1954

【形态特征】雄性面部大部分黄色，上唇黑色，中央具1个黄斑，头顶黑色，后头黄色；胸部大面积黄绿色，背条纹与领条纹相连，合胸侧面第2条纹缺如，第3条纹甚细；腹部黑色具黄斑。【长度】雄性体长 60 mm，腹长 44 mm，后翅 35 mm。【栖息环境】海拔 500 m以下的开阔溪流。【分布】中国特有，分布于安徽、湖北、浙江、福建。【飞行期】4—6月。

贝氏日春蜓 雄，安徽 | 陈尽 摄
Nihonogomphus bequaerti, male from Anhui | Photo by Jin Chen

[Identification] Male face largely yellow, labrum black with a centrally yellow spot, vertex black, occiput yellow. Thorax largely yellowish green, dorsal stripes and collar stripes connecting, second lateral stripe absent, third lateral stripe narrow. Abdomen black with yellow markings. **[Measurements]** Male total length 60 mm, abdomen 44 mm, hind wing 35 mm. **[Habitat]** Exposed streams below 500 m elevation. **[Distribution]** Endemic to China, recorded from Anhui, Hubei, Zhejiang, Fujian. **[Flight Season]** April to June.

短翅日春蜓 *Nihonogomphus brevipennis* (Needham, 1930)

【形态特征】雄性面部大部分黄色，头顶黑色；胸部大面积黄色，背条纹与领条纹相连，肩前条纹甚细，合胸侧面第2条纹上方缺失，第3条纹甚细；腹部黑色具黄斑。雌性与雄性相似，但黄斑更发达。【长度】体长 47 mm，腹长 34 mm，后翅 25~26 mm。【栖息环境】海拔 500 m以下的开阔溪流和河流。【分布】中国特有，分布于江苏、四川、湖北。【飞行期】4—6月。

[Identification] Male face largely yellow, vertex black. Thorax largely yellow, dorsal stripes and collar stripes connecting, antehumeral stripe narrow, second lateral stripe absent above, third lateral stripes narrow. Abdomen black with yellow markings. Female similar to male, yellow markings broader. **[Measurements]** Total length 47 mm, abdomen 34 mm, hind wing 25-26 mm. **[Habitat]** Exposed streams and rivers below 500 m elevation. **[Distribution]** Endemic to China, recorded from Jiangsu, Sichuan, Hubei. **[Flight Season]** April to June.

短翅日春蜓 雌，湖北
Nihonogomphus brevipennis, female from Hubei

短翅日春蜓 雄，湖北
Nihonogomphus brevipennis, male from Hubei

短翅日春蜓 雌，湖北
Nihonogomphus brevipennis, female from Hubei

刀日春蜓 *Nihonogomphus cultratus* Chao & Wang, 1990

刀日春蜓 雄，湖北
Nihonogomphus cultratus, male from Hubei

【形态特征】雄性面部大部分黄色，头顶黑色；胸部大面积黄色，背条纹甚阔，肩前条纹甚细，合胸侧面第2条纹上方缺失，第3条纹完整；腹部黑色具黄斑。雌性与雄性相似。【长度】体长 49～52 mm，腹长 35～38 mm，后翅 23～28 mm。【栖息环境】海拔 500 m以下的开阔溪流。【分布】中国特有，分布于河南、湖北。【飞行期】3—5月。

[Identification] Male face largely yellow, vertex black. Thorax largely yellow, dorsal stripes broad, antehumeral stripe narrow, second lateral stripe absent above, third lateral stripes complete. Abdomen black with yellow markings. Female similar to male. **[Measurements]** Total length 49 -52 mm, abdomen 35-38 mm, hind wing 23-28 mm. **[Habitat]** Exposed streams below 500 m elevation. **[Distribution]** Endemic to China, recorded from Henan, Hubei. **[Flight Season]** March to May.

黄沙日春蜓 *Nihonogomphus huangshaensis* Chao & Zhu, 1999

【形态特征】雄性面部大部分黑色，额黄绿色；胸部大面积黄绿色，背条纹与领条纹相连，合胸侧面第2条纹大面积缺失，第3条纹完整且宽阔；腹部黑色具黄斑。【长度】雄性体长 58～60 mm，腹长 43～45 mm，后翅 33～35 mm。【栖息环境】海拔 500 m以下森林中的溪流。【分布】中国广西特有。【飞行期】4—6月。

[Identification] Male face largely black, frons yellowish green. Thorax largely yellowish green, dorsal stripes and collar stripes connecting, second lateral stripe largely absent, third lateral stripes complete and wide. Abdomen black with yellow markings. **[Measurements]** Male total length 58-60 mm, abdomen 43-45 mm, hind wing 33-35 mm. **[Habitat]** Streams in forest below 500 m elevation. **[Distribution]** Endemic to Guangxi of China. **[Flight Season]** April to June.

黄沙日春蜓 雄，广西
Nihonogomphus huangshaensis, male from Guangxi

黄侧日春蜓 *Nihonogomphus luteolatus* Chao & Liu, 1990

【形态特征】雄性面部大部分黄色，上唇黑色，中央具1个甚大黄斑，头顶黑色，后头黄色；胸部大面积黄绿色，背条纹与领条纹相连，合胸侧面第2条纹缺失，第3条纹甚细；腹部黑色具黄斑。雌性与雄性相似，腹部黄斑较少。【长度】雄性体长 58 mm，腹长 43 mm，后翅 34 mm。【栖息环境】海拔 500 m以下的开阔溪流。【分布】中国特有，分布于广东、福建。【飞行期】3—6月。

黄侧日春蜓 雄，广东 | 吴宏道 摄
Nihonogomphus luteolatus, male from Guangdong | Photo by Hongdao Wu

[Identification] Male face largely yellow, labrum black with a centrally yellow spot, vertex black, occiput yellow. Thorax largely yellowish green, dorsal stripes and collar stripes connecting, second lateral stripe absent, third lateral stripe narrow. Abdomen black with yellow markings. Female similar to male, abdomen with less yellow markings. **[Measurements]** Male total length 58 mm, abdomen 43 mm, hind wing 34 mm. **[Habitat]** Exposed streams below 500 m elevation. **[Distribution]** Endemic to China, recorded from Guangdong, Fujian. **[Flight Season]** March to June.

黄侧日春蜓 雄，广东 | 吴宏道 摄
Nihonogomphus luteolatus, male from Guangdong | Photo by Hongdao Wu

黄侧日春蜓 雌，广东 | 宋黎明 摄
Nihonogomphus luteolatus, female from Guangdong | Photo by Liming Song

臼齿日春蜓 *Nihonogomphus ruptus* (Selys, 1858)

臼齿日春蜓 雄，黑龙江 | 莫善濂 摄
Nihonogomphus ruptus, male from Heilongjiang | Photo by Shanlian Mo

臼齿日春蜓 雌，黑龙江 | 莫善濂 摄
Nihonogomphus ruptus, female from Heilongjiang | Photo by Shanlian Mo

【形态特征】雄性面部大部分黄白色，头顶黑色，后头黄色；胸部大面积灰色，背条纹与领条纹相连，肩前条纹细长，在上方与背条纹相连，合胸侧面第2条纹上方缺失，第3条纹甚细；腹部黑色具黄斑。雌性与雄性相似，但腹部较短。【长度】体长 46～55 mm，腹长 31～39 mm，后翅 29～30 mm。【栖息环境】海拔 500 m以下的开阔溪流和河流。【分布】黑龙江；西伯利亚、俄罗斯远东、朝鲜半岛。【飞行期】5—7月。

[Identification] Male face largely yellowish white, vertex black, occiput yellow. Thorax largely grey, dorsal stripes and collar stripes connecting, antehumeral stripes narrow and connecting with dorsal stripes, second lateral stripe absent above, third lateral stripe narrow. Abdomen black with yellow markings. Female similar to male but abdomen shorter. **[Measurements]** Total length 46-55 mm, abdomen 31-39 mm, hind wing 29-30 mm. **[Habitat]** Exposed streams and rivers below 500 m elevation. **[Distribution]** Heilongjiang; Siberia, Russian Far East, Korean peninsula. **[Flight Season]** May to July.

长钩日春蜓 *Nihonogomphus semanticus* Chao, 1954

长钩日春蜓 雌，广东
Nihonogomphus semanticus, female from Guangdong

长钩日春蜓 雄，广东
Nihonogomphus semanticus, male from Guangdong

长钩日春蜓 雄，广东
Nihonogomphus semanticus, male from Guangdong

【形态特征】雄性面部大部分黑色，额和后头黄绿色；胸部大面积黄绿色，背条纹与领条纹相连，合胸侧面第2条纹上方大面积缺失，第3条纹完整；腹部黑色具黄斑。雌性与雄性相似。【长度】体长 59～60 mm，腹长 43～45 mm，后翅 35～37 mm。【栖息环境】海拔 1000 m以下森林中的开阔溪流。【分布】中国特有，分布于浙江、福建、广东。【飞行期】3—7月。

[Identification] Male face largely black, frons and occiput yellowish green. Thorax largely yellowish green, dorsal stripes and collar stripes connecting, second lateral stripe largely absent above, third lateral stripes complete. Abdomen black with yellow markings. Female similar to male. **[Measurements]** Total length 59-60 mm, abdomen 43-45 mm, hind wing 35-37 mm. **[Habitat]** Exposed streams in forest below 1000 m elevation. **[Distribution]** Endemic to China, recorded from Zhejiang, Fujian, Guangdong. **[Flight Season]** March to July.

汤氏日春蜓 *Nihonogomphus thomassoni* (Kirby, 1900)

【形态特征】雄性面部大部分黑色，额和后头黄绿色；胸部大面积黄绿色，背条纹与领条纹相连，合胸侧面第2条纹中央间断，第3条纹完整，有时第2条纹和第3条纹合并形成“Y”形；腹部黑色具黄斑。雌性与雄性相似。此处将黎氏日春蜓作为本种的异名。【长度】体长 60~63 mm，腹长 44~47 mm，后翅 34~37 mm。【栖息环境】海拔 1000 m以下的开阔溪流和河流。【分布】云南、贵州、广西、广东、海南、福建；越南。【飞行期】3—6月。

汤氏日春蜓 雄，云南（红河）
Nihonogomphus thomassoni, male from Yunnan (Honghe)

汤氏日春蜓 雌，海南
Nihonogomphus thomassoni, female from Hainan

[Identification] Male face largely black, frons and occiput yellowish green. Thorax largely yellowish green, dorsal stripes and collar stripes connecting, second lateral stripe interrupted medially, third lateral stripe complete, the second and third lateral stripes sometimes combined to form a Y-shaped stripe. Abdomen black with yellow markings. Female similar to male. *N. lieftincki* (Chao, 1954) is considered as a synonym of this species here. **[Measurements]** Total length 60-63 mm, abdomen 44-47 mm, hind wing 34-37 mm. **[Habitat]** Exposed streams and rivers below 1000 m elevation. **[Distribution]** Yunnan, Guizhou, Guangxi , Guangdong, Hainan, Fujian; Vietnam. **[Flight Season]** March to June.

汤氏日春蜓 雌，海南 | 莫善濂 摄
Nihonogomphus thomassoni, female from Hainan | Photo by Shanlian Mo

汤氏日春蜓 雄，广东 | 宋黎明 摄
Nihonogomphus thomassoni, male from Guangdong | Photo by Liming Song

汤氏日春蜓 雄，广西
Nihonogomphus thomassoni, male from Guangxi

日春蜓属待定种 *Nihonogomphus* sp.

日春蜓属待定种 雄，云南（西双版纳）
Nihonogomphus sp., male from Yunnan (Xishuangbanna)

【形态特征】雄性面部大部分黄色，头顶黑褐色，后头黄色；胸部大面积黄绿色，背条纹与领条纹相连，合胸侧面第2条纹大面积缺失，第3条纹完整，甚阔；腹部褐色，第1~7节具黄斑。雌性与雄性相似。【长度】体长 54~56 mm，腹长 39~42 mm，后翅 33~35 mm。【栖息环境】海拔 1000 m以下的沟渠和开阔溪流。【分布】云南（西双版纳）。【飞行期】3—5月。

[Identification] Male face largely yellow, vertex blackish brown, occiput yellow. Thorax largely yellowish green, dorsal stripes and collar stripes connecting, second lateral stripe largely absent, third lateral stripe complete and broad. Abdomen brown, S1-S7 with with yellow markings. Female similar to male. **[Measurements]** Total length 54-56 mm, abdomen 39-42 mm, hind wing 33-35 mm. **[Habitat]** Exposed streams and ditches below 1000 m elevation. **[Distribution]** Yunnan (Xishuangbanna). **[Flight Season]** March to May.

奈春蜓属 Genus *Nychogomphus* Carle, 1986

本属全球已知7种，分布于亚洲的亚热带区域。中国已知6种，仅分布于西南地区。本属蜻蜓是一类体中型的春蜓；面部黑色具黄斑，上唇具1对黄斑，额横纹较宽阔，中央间断；翅大面积透明，有时基方染有琥珀色，翅痣较长，具支持脉，三角室、上三角室和下三角室无横脉，基臀区具1条横脉，臀圈2室，臀三角室4室；雄性肛附器发达，上肛附器的两枝平行，末端向下弯曲，下肛附器向上弯曲，基部背方具1个矮突起。阳茎具1对鞭。

本属蜻蜓栖息于较宽阔的溪流和河流。成虫经常停落在溪流边缘茂盛的树荫中。

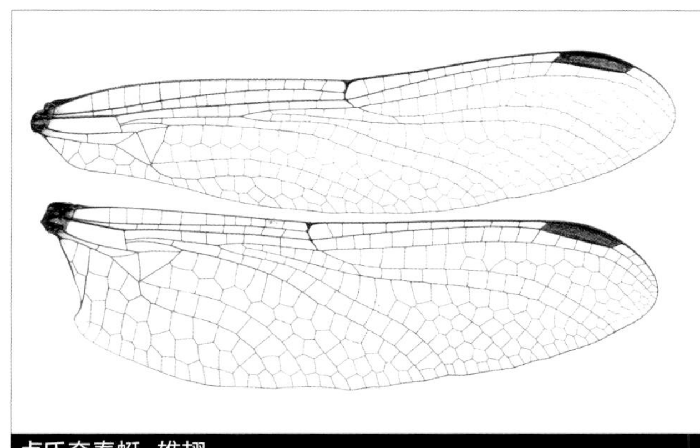
卢氏奈春蜓，雄翅
Nychogomphus lui, male wings

The genus contains seven species, distributed in subtropical Asia. Six species are recorded from China, only confined to the Southwest. Species of the genus are medium-sized dragonflies. Face black with yellow markings, labrum with a pair of yellow spots, frons with a broad stripe interrupted medially. Wings largely hyaline, in some species the wing bases tinted with amber, no crossvein in triangle, hypertriangle and subtriangle, cubital space with one crossvein, anal loop 2-celled, anal triangle 4-celled. Male superior appendages with two branches parallel, the tips curved downwards, the inferior appendage curved upwards, with a short dorsal prominence. Penis with a pair of flagella.

Nychogomphus species inhabit montane streams and rivers. Individuals usually perch in the shady forest near water.

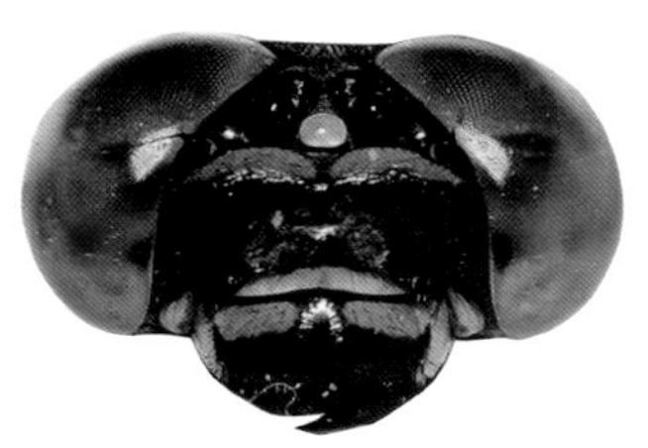
基齿奈春蜓 雄
Nychogomphus duaricus, male

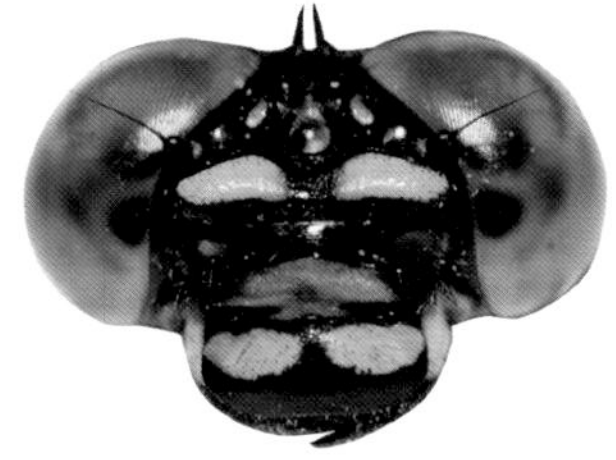
基齿奈春蜓 雌
Nychogomphus duaricus, female

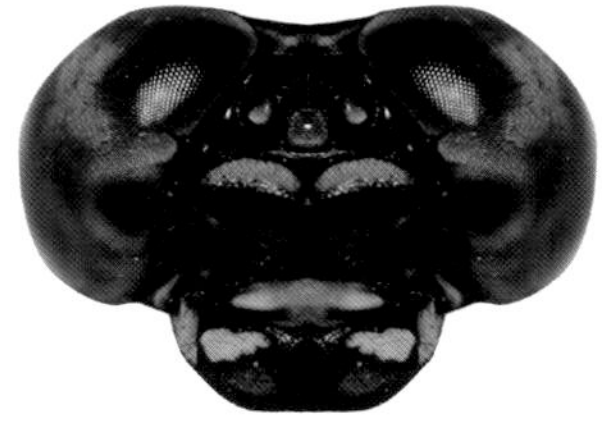
黄尾奈春蜓 雄
Nychogomphus flavicaudus, male

卢氏奈春蜓 雄
Nychogomphus lui, male

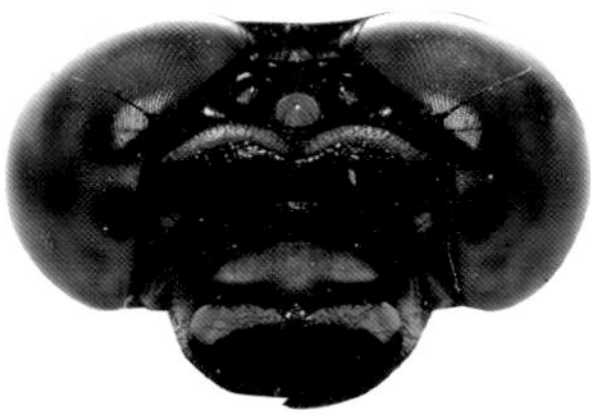
杨氏奈春蜓 雄
Nychogomphus yangi, male

杨氏奈春蜓 雌
Nychogomphus yangi, female

奈春蜓属 头部正面观
Genus *Nychogomphus*, head in frontal view

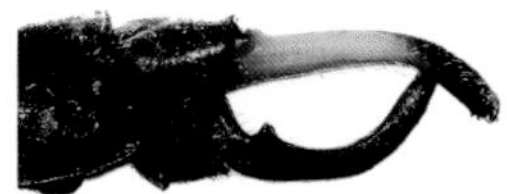

基齿奈春蜓
Nychogomphus duaricus

黄尾奈春蜓
Nychogomphus flavicaudus

卢氏奈春蜓
Nychogomphus lui

杨氏奈春蜓
Nychogomphus yangi

奈春蜓属 雄性肛附器
Genus *Nychogomphus*, male anal appendages

杨氏奈春蜓 雄
Nychogomphus yangi, male

基齿奈春蜓 *Nychogomphus duaricus* (Fraser, 1924)

【形态特征】雄性胸部黑色，背条纹与领条纹相连，肩前条纹甚细，合胸侧面第2条纹和第3条纹完整；腹部黑色，第1～9节具黄斑，上肛附器基方2/3黄白色，端方黑色，下肛附器黑色。雌性与雄性相似，后头中央具1对角状突起。【长度】体长 47～50 mm，腹长 35～37 mm，后翅 28～29 mm。【栖息环境】海拔 1000 m以下的溪流和河流。【分布】云南（西双版纳、普洱）；孟加拉国、印度、尼泊尔、泰国、柬埔寨、老挝、越南、马来半岛。【飞行期】4—11月。

[Identification] Male thorax black, dorsal stripes and collar stripes connecting, antehumeral stripes narrow, second and third lateral stripes complete. Abdomen black, S1-S9 with yellow markings, superior appendages with basal two thirds yellowish white and black apically, inferior appendage black. Female similar to male, occipital margin with a pair of median horns. [Measurements] Total length 47-50 mm, abdomen 35-37 mm, hind wing 28-29 mm. [Habitat] Streams and rivers below 1000 m elevation. [Distribution] Yunnan (Xishuangbanna, Pu'er); Bangladesh, India, Nepal, Thailand, Cambodia, Laos, Vietnam, Peninsular Malaysia. [Flight Season] April to November.

基齿奈春蜓 雄，云南（西双版纳）
Nychogomphus duaricus, male from Yunnan (Xishuangbanna)

基齿奈春蜓 雄，云南（西双版纳）
Nychogomphus duaricus, male from Yunnan (Xishuangbanna)

基齿奈春蜓 雌，云南（西双版纳）
Nychogomphus duaricus, female from Yunnan (Xishuangbanna)

黄尾奈春蜓 *Nychogomphus flavicaudus* (Chao, 1982)

【形态特征】雄性胸部黑色，背条纹与领条纹相连，合胸侧面第2条纹和第3条纹完整；腹部黑色，第1~7节具黄斑，上肛附器基方2/3黄白色，端方黑色，下肛附器黑色。雌性与雄性相似，后头中央具1对角状突起。与基齿奈春蜓相似，但体型稍大，无肩前条纹，雄性腹部第8~9节侧缘无黄斑。【长度】体长 55~57 mm，腹长 40~42 mm，后翅 31~35 mm。【栖息环境】海拔 1000 m以下的山区溪流。【分布】云南（红河）、海南；越南。【飞行期】4—6月。

[Identification] Male thorax black, dorsal stripes and collar stripes connecting, second and third lateral stripes complete. Abdomen black, S1-S7 with yellow markings, superior appendages with basal two thirds yellowish white and black apically, inferior appendage black. Female similar to male, occipital margin with a pair of median horns. Similar to *N. duaricus* but larger, antehumeral stripes absent, male S8-S9 without lateral yellow spots. **[Measurements]** Total length 55-57 mm, abdomen 40-42 mm, hind wing 31-35 mm. **[Habitat]** Montane streams below 1000 m elevation. **[Distribution]** Yunnan (Honghe), Hainan; Vietnam. **[Flight Season]** April to June.

黄尾奈春蜓 雄，云南（红河）
Nychogomphus flavicaudus, male from Yunnan (Honghe)

黄尾奈春蜓 雄，云南（红河）| 莫善濂 摄
Nychogomphus flavicaudus, male from Yunnan (Honghe) | Photo by Shanlian Mo

黄尾奈春蜓 交尾，云南（红河）| 莫善濂 摄
Nychogomphus flavicaudus, mating pair from Yunnan (Honghe) | Photo by Shanlian Mo

卢氏奈春蜓 *Nychogomphus lui* Zhou, Zhou & Lu, 2005

卢氏奈春蜓 雄，云南（红河）
Nychogomphus lui, male from Yunnan (Honghe)

【形态特征】雄性胸部黑色，背条纹与领条纹相连，肩前上点甚小，合胸侧面第2条纹和第3条纹完整；腹部黑色，第1~7节具黄斑，上肛附器基方2/3黄白色，端方黑色，下肛附器黑色，波状。雌性与雄性相似。【长度】体长52~54 mm，腹长 39~41 mm，后翅 30~34 mm。【栖息环境】海拔 1000 m以下的溪流和河流。【分布】云南（红河、文山）；越南。【飞行期】5—11月。

[Identification] Male thorax black, dorsal stripes and collar stripes connecting, superior spots small, second and third lateral stripes complete. Abdomen black, S1-S7 with yellow markings, superior appendages with basal two thirds yellowish white and black apically, inferior appendage undulate and black. Female similar to male. **[Measurements]** Total length 52-54 mm, abdomen 39-41 mm, hind wing 30-34 mm. **[Habitat]** Streams and rivers below 1000 m elevation. **[Distribution]** Yunnan (Honghe, Wenshan); Vietnam. **[Flight Season]** May to November.

杨氏奈春蜓 *Nychogomphus yangi* Zhang, 2014

【形态特征】雄性胸部黑色，背条纹与领条纹相连，肩前上点甚小，合胸侧面第2条纹和第3条纹完整；腹部黑色，第1~7节具黄斑，上肛附器基方1/2黄色，端方黑色，下肛附器黄色，波状。雌性与雄性相似。与卢氏奈春蜓相似，但雄性肛附器色彩不同。【长度】体长 52~54 mm，腹长 39~40 mm，后翅 29~32 mm。【栖息环境】海拔1000 m以下的溪流和河流。【分布】中国云南（西双版纳）特有。【飞行期】4—11月。

[Identification] Male thorax black, dorsal stripes and collar stripes connecting, superior spots small, second and third lateral stripes complete. Abdomen black, S1-S7 with yellow markings, superior appendages with basal half yellow and black apically, inferior appendage undulate and yellow. Female similar to male. Similar to *N. lui*, but the color of male anal appendages different. **[Measurements]** Total length 52-54 mm, abdomen 39-40 mm, hind wing 29-32 mm. **[Habitat]** Streams and rivers below 1000 m elevation. **[Distribution]** Endemic to Yunnan (Xishuangbanna) of China. **[Flight Season]** April to November.

杨氏奈春蜓 雌，云南（西双版纳）
Nychogomphus yangi, female from Yunnan (Xishuangbanna)

杨氏奈春蜓 雄，云南（西双版纳）
Nychogomphus yangi, male from Yunnan (Xishuangbanna)

钩尾春蜓属 Genus *Onychogomphus* Selys, 1854

钩尾春蜓属是包含肛附器十分发达、形状多变的一类中型春蜓，是分类上较困难的属。中国分布的环尾春蜓属、奈春蜓属和弯尾春蜓属等都和本属关系密切。本属中很多种类已经被移入这些近似的属中，在中国目前仅有1种，分布于新疆。

Onychogomphus contains a large group of medium-sized species with large and variably shaped anal appendages, the taxonomy of this genus is difficult. The genera *Lamelligomphus*, *Nychogomphus* and *Melligomphus* are similar and several species of *Onychogomphus* have been transferred to these related genera. Only one species of *Onychogomphus* is recorded from Xinjiang of China.

豹纹钩尾春蜓 *Onychogomphus forcipatus* (Linnaeus, 1758)

【形态特征】雄性复眼深绿色，面部大面积黄色；胸部黑色，背条纹与领条纹相连，肩前条纹较长，合胸侧面第2条纹中央稍微间断，第3条纹完整；腹部黑色具发达的黄色斑纹，肛附器褐色。雌性与雄性相似。【长度】体长46～50 mm，腹长 31～37 mm，后翅 25～30 mm。【栖息环境】开阔溪流和河流。【分布】新疆；非洲北部、欧洲、亚洲西南部。【飞行期】5—9月。

[Identification] Male eyes dark green, face largely yellow. Thorax largely black, dorsal stripes and collar stripes connecting, antehumeral stripes very long, second lateral stripes slightly interrupted medially, third lateral

豹纹钩尾春蜓 雄，芬兰 | Sami Karjalainen 摄
Onychogomphus forcipatus, male from Finland | Photo by Sami Karjalainen

stripes complete. Abdomen black with abundant yellow markings, anal appendages brown. Female similar to male. **[Measurements]** Total length 46-50 mm, abdomen 31-37 mm, hind wing 25-30 mm. **[Habitat]** Exposed streams and rivers. **[Distribution]** Xinjiang; North Africa, Europe and southwestern Asia. **[Flight Season]** May to September.

豹纹钩尾春蜓 雌，芬兰 | Sami Karjalainen 摄
Onychogomphus forcipatus, female from Finland | Photo by Sami Karjalainen

蛇纹春蜓属 Genus *Ophiogomphus* Selys, 1854

本属全球已知约30种，主要分布在北美洲，欧亚大陆种类较少。中国已知4种。本属春蜓体中型；翅透明，三角室、上三角室和下三角室无横脉，基臀区具1条横脉，臀圈2~3室，臀三角室4室。本属下分成3个亚属，指名亚属、长钩亚属和蛇斑亚属。指名亚属种类身体大面积黄绿色，侧面观时下肛附器与上肛附器相距较近，下肛附器基方无齿突。长钩亚属种类身体黑色具黄色条纹，雄性肛附器发达，钳状，上肛附器与下肛附器之间有1个较大空腔。

本属蜻蜓栖息于河流和山区溪流，雄性经常停落在溪流边缘的树丛或水面的大岩石上。

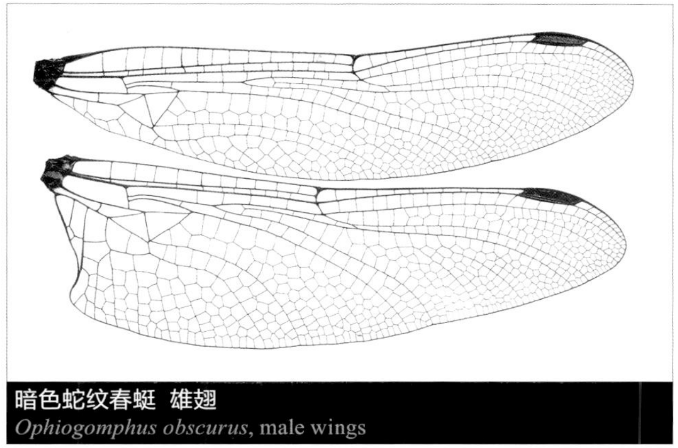

暗色蛇纹春蜓 雄翅
Ophiogomphus obscurus, male wings

The genus contains about 30 species, mainly distributed in North America with only a few species are known from Eurasia. Four species are recorded from China. Species of the genus are medium-sized dragonflies. Wings hyaline, wings without crossvein in triangle, hypertriangle and subtriangle, cubital space with one crossvein, anal loop 2-to 3-celled, anal triangle 4-celled. The genus has been divided into three subgenera, *Ophiogomphus, Ophionurus* and

Ophionuroides. Species of *Ophiogomphus* are largely yellowish green, in lateral view the inferior appendage lies close to the superior appendages, and no teeth are found on the base of the inferior appendage. Species of *Ophionurus* are fundamentally black with yellow markings, male anal appendages developed, forcepe-shaped, a big cavity present between the superiors and the inferior appendage.

Ophiogomphus species inhabit rivers and montane streams. Males usually perch on marginal trees or big rocks above water.

越南长钩春蜓
Ophiogomphus longihamulus

中华长钩春蜓
Ophiogomphus sinicus

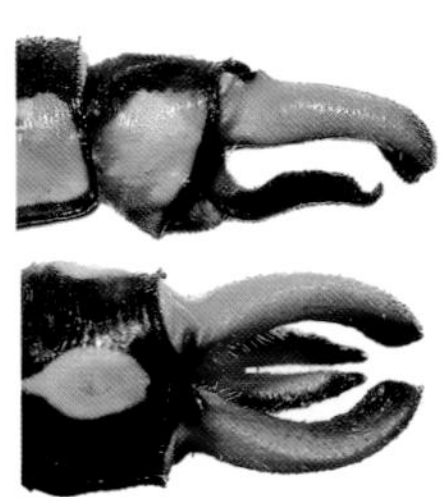

棘角蛇纹春蜓
Ophiogomphus spinicornis

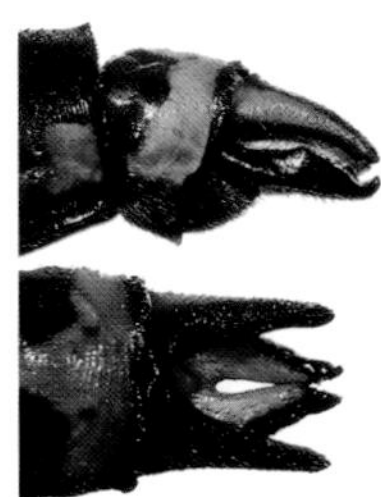

暗色蛇纹春蜓
Ophiogomphus obscurus

蛇纹春蜓属 雄性肛附器
Genus *Ophiogomphus*, male anal appendages

中华长钩春蜓
Ophiogomphus sinicus

暗色蛇纹春蜓
Ophiogomphus obscures

蛇纹春蜓属 雌性下生殖板
Genus *Ophiogomphus*, female vulvar lamina

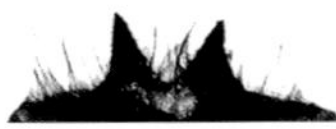

中华长钩春蜓
Ophiogomphus sinicus

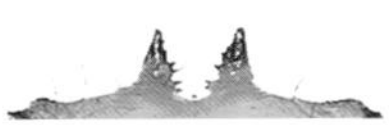

暗色蛇纹春蜓
Ophiogomphus obscures

蛇纹春蜓属 雌性后头缘
Genus *Ophiogomphus*, female occipital margin

棘角蛇纹春蜓 雄
Ophiogomphus spinicornis, male

中华长钩春蜓 雄 | 宋睿斌 摄
Ophiogomphus sinicus, male | Photo by Ruibin Song

越南长钩春蜓 *Ophiogomphus longihamulus* Karube, 2014

【形态特征】雄性上唇具1对黄斑，额横纹甚阔，中央间断；胸部黑色，背条纹与领条纹不相连，合胸侧面第2条纹和第3条纹大面积合并；腹部黑色，第1～7节具黄斑，肛附器黑色。雌性与雄性相似，后头具1对角状突起。【长度】体长 57～61 mm，腹长 42～46 mm，后翅 35～39 mm。【栖息环境】海拔 1000 m以下的山区溪流。【分布】云南（红河）；越南。【飞行期】5—7月。

[Identification] Male labrum with a pair of yellow spots, frons with a broad stripe interrupted medially. Thorax black, dorsal stripes and collar stripes not connecting, second and third lateral stripes largely combined. Abdomen black, S1-S7 with yellow markings, anal appendages black. Female similar to male, occiput with a pair of horns. [Measurements] Total length 57-61 mm, abdomen 42-46 mm, hind wing 35-39 mm. [Habitat] Montane streams below 1000 m elevation. [Distribution] Yunnan (Honghe); Vietnam. [Flight Season] May to July.

越南长钩春蜓 雄，云南（红河）
Ophiogomphus longihamulus, male from Yunnan (Honghe)

暗色蛇纹春蜓 *Ophiogomphus obscurus* Bartenev, 1909

【形态特征】雄性面部大面积黄绿色，头顶黑色；胸部绿色，背条纹甚阔，肩前条纹甚细，合胸侧面第2条纹大面积缺失，第3条纹完整，甚细；腹部黑色具丰富的黄绿色斑纹，上肛附器外方黄色，下肛附器黑色。雌性体色稍淡，后头缘具1对角状突起。【长度】体长 56～60 mm，腹长 41～42 mm，后翅 33～37 mm。【栖息环境】海拔 500 m以下的山区溪流和宽阔的河流。【分布】黑龙江、吉林、河北；朝鲜半岛、西伯利亚。【飞行期】6—9月。

[Identification] Male face largely yellowish green, vertex black. Thorax green, dorsal stripes broad, antehumeral stripes narrow, second lateral stripes largely absent, third lateral stripes complete but narrow. Abdomen black with abundant yellowish green markings, superior appendages with outer margin yellow, inferior appendage black. Female paler, occipical margin with a pair of horns. **[Measurements]** Total length 56-60 mm, abdomen 41-42 mm, hind wing 33-37 mm. **[Habitat]** Montane streams and rivers below 500 m elevation. **[Distribution]** Heilongjiang, Jilin, Hebei; Korean peninsula, Siberia. **[Flight Season]** June to September.

暗色蛇纹春蜓 雄，黑龙江
Ophiogomphus obscurus, male from Heilongjiang

暗色蛇纹春蜓 雌，黑龙江
Ophiogomphus obscurus, female from Heilongjiang

暗色蛇纹春蜓 雄，黑龙江
Ophiogomphus obscurus, male from Heilongjiang

中华长钩春蜓 *Ophiogomphus sinicus* (Chao, 1954)

【形态特征】雄性上唇具1对黄斑，额横纹甚阔，中央间断；胸部黑色，背条纹与领条纹不相连，合胸侧面第2条纹和第3条纹合并；腹部黑色，第1~7节具黄斑，肛附器主要黑色，上肛附器后缘黄色。雌性与雄性近似，后头缘中央具1对角状突起，尾毛白色。【长度】体长 59~62 mm，腹长 38~46 mm，后翅 31~38 mm。【栖息环境】海拔1000 m以下的山区溪流。【分布】江西、福建、广东、广西、香港；越南。【飞行期】5—9月。

[Identification] Male labrum with a pair of yellow spots, frons with a broad stripe interrupted medially. Thorax black, dorsal stripes and collar stripes not connecting, second and third lateral stripes combined. Abdomen black, S1-S7 with yellow markings, anal appendages largely black, superiors with apical yellow stripe. Female similar to male, occipical margin with a pair of horns, cerci white. [Measurements] Total length 59-62 mm, abdomen 38-46 mm, hind wing 31-38 mm. [Habitat] Montane streams below 1000 m elevation. [Distribution] Jiangxi, Fujian, Guangdong, Guangxi, Hong Kong; Vietnam. [Flight Season] May to September.

中华长钩春蜓 雄，广西
Ophiogomphus sinicus, male from Guangxi

中华长钩春蜓 雌，广西
Ophiogomphus sinicus, female from Guangxi

中华长钩春蜓 雄，广西
Ophiogomphus sinicus, male from Guangxi

棘角蛇纹春蜓 *Ophiogomphus spinicornis* Selys, 1878

棘角蛇纹春蜓 雄，北京
Ophiogomphus spinicornis, male from Beijing

棘角蛇纹春蜓 雌，北京
Ophiogomphus spinicornis, female from Beijing

【形态特征】雄性面部大面积黄绿色，头顶黑色；胸部黄绿色，背条纹甚阔，与肩前条纹在上方相连，合胸侧面第2条纹大面积缺失，第3条纹完整，甚细；腹部黑色，各节具丰富的黄绿色斑纹，上肛附器黄色，下肛附器黑色。雌性体色稍淡，后头缘两侧具突起。【长度】体长 57~63 mm，腹长 40~47 mm，后翅 32~40 mm。【栖息环境】海拔1000 m以下的山区溪流。【分布】北京、河北、山西、甘肃、青海、内蒙古；西伯利亚。【飞行期】7—9月。

[Identification] Male face largely yellowish green, vertex black. Thorax yellowish green, dorsal stripes broad, connecting with antehumeral above, second lateral stripes largely absent, third lateral stripes complete but narrow. Abdomen black with abundant yellowish green markings, superior appendages yellow, inferior appendage black. Female paler, occipical margin with prominences laterally. **[Measurements]** Total length 57-63 mm, abdomen 40-47 mm, hind wing 32-40 mm. **[Habitat]** Montane streams below 1000 m elevation. **[Distribution]** Beijing, Hebei, Shanxi, Gansu, Qinghai, Inner Mongolia; Siberia. **[Flight Season]** July to September.

东方春蜓属 Genus *Orientogomphus* Chao & Xu, 1987

本属全球已知7种，分布于亚洲的亚热带和热带区域。中国已知仅1种，分布于华南地区。本属春蜓体中型；翅透明，三角室、上三角室和下三角室无横脉，基臀区具1条横脉，臀圈2室，臀三角室4室；雄性肛附器发达，上肛附器向内和向下稍微弯曲，下肛附器长度为上肛附器的1/2，两枝显著分歧。

本属蜻蜓栖息于山区溪流，但很少遇见。雄性具领域行为，在黄昏时飞近水面悬停寻找配偶。白天多停落在溪流附近的茂盛林荫中。

The genus contains seven species, distributed in subtropical and tropical Asia. Only one is recorded from China, confined to the South. Species of the genus are medium-sized dragonflies. Wings hyaline, no crossvein in triangle, hypertriangle and subtriangle, cubital space with one crossvein, anal loop 2-celled, anal triangle 4-celled. Male superior appendages with the tips curved downwards, the inferior appendage half length of the superiors with two branches strongly divergent.

Orientogomphus species inhabit montane streams but are rarely seen. Males exhibit territorial behavior by hovering close to water at twilight. They usually perch in shady forests during the daytime.

具突东方春蜓 雄
Orientogomphus armatus, male

东方春蜓属 头部正面观
Genus *Orientogomphus*, head in frontal view

具突东方春蜓
Orientogomphus armatus

东方春蜓属 雄性肛附器
Genus *Orientogomphus*, male appendages

具突东方春蜓 雄
Orientogomphus armatus, male

具突东方春蜓 *Orientogomphus armatus* Chao & Xu, 1987

具突东方春蜓 雄，海南
Orientogomphus armatus, male from Hainan

【形态特征】雄性上唇具1对黄斑，额横纹甚阔，中央间断；胸部黑色，背条纹与领条纹不相连，具甚小的肩前上点，合胸侧面第2条纹和第3条纹完整；腹部黑色，第1～8节具黄斑，上肛附器白色。雌性与雄性相似。【长度】体长52～58 mm，腹长 39～44 mm，后翅 30～38 mm。【栖息环境】海拔 1000 m以下的山区溪流。【分布】中国特有，分布于广东、海南、福建。【飞行期】5—9月。

[Identification] Male labrum with a pair of yellow spots, frons with a broad stripe interrupted medially. Thorax black, dorsal stripes and collar stripes not connecting, superior spots very small, second and third lateral stripes complete. Abdomen black, S1-S8 with yellow markings, superior appendages white. Female similar to male. **[Measurements]** Total length 52-58 mm, abdomen 39-44 mm, hind wing 30-38 mm. **[Habitat]** Montane streams below 1000 m elevation. **[Distribution]** Endemic to China, recorded from Guangdong, Hainan, Fujian. **[Flight Season]** May to September.

具突东方春蜓 雄，海南
Orientogomphus armatus, male from Hainan

副春蜓属 Genus *Paragomphus* Cowley, 1934

本属全球已知约50种，广布于亚洲、欧洲、非洲和大洋洲，在热带地区种类繁多。中国已知4种，分布于华南和西南地区。本属春蜓体中型；翅透明，三角室、上三角室和下三角室无横脉，基臀区具1条横脉，臀圈缺如，臀三角室4室；雄性腹部第8～9节侧下缘具片状突起，上肛附器较长，末端向下弯曲，下肛附器短，不及上肛附器长度的1/2。

本属蜻蜓栖息于较开阔具有浅滩的溪流，雄性经常停落在溪流边缘的沙滩上或者停落在水边的枝头上占据领地。

The genus contains about 50 species, widespread in Asia, Europe, Africa and Oceania, being most speciose in the tropics. Four are recorded from China, found in the South and Southwest. Species of the genus are medium-sized dragonflies. Wings hyaline, no crossvein in triangle, hypertriangle and subtriangle, cubital space with one crossvein, anal loop absent, anal triangle 4-celled. Male S8-S9 with leaf-like flaps at the lower lateral margin, superior appendages long, the tips curved downwards, the inferior appendage short and less than half length of the superiors.

Paragomphus species inhabit exposed streams with shallow beach. Territorial males usually perch on the beach or top of branches near water.

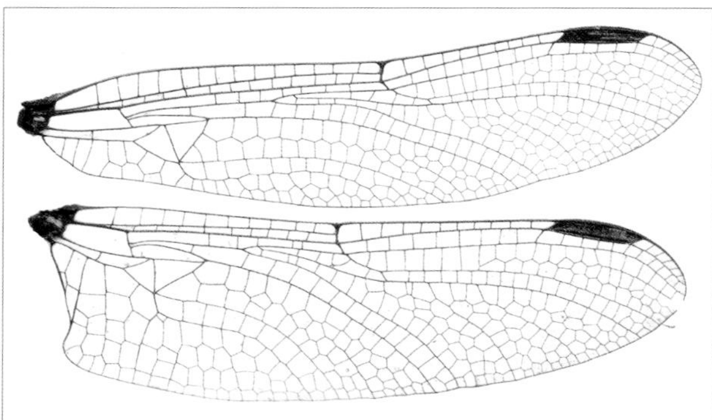

钩尾副春蜓，雄翅
Paragomphus capricornis, male wings

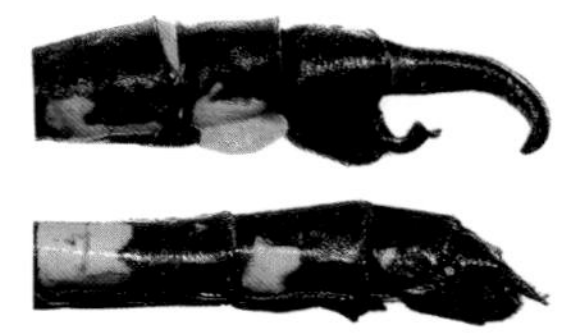

钩尾副春蜓
Paragomphus capricornis

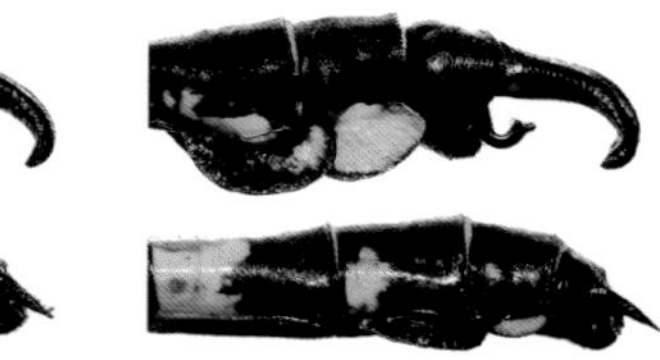

豹纹副春蜓
Paragomphus pardalinus

副春蜓属 腹部末端
Genus *Paragomphus*, distal abdominal segments

豹纹副春蜓 雄
Paragomphus pardalinus, male

钩尾副春蜓 *Paragomphus capricornis* (Förster, 1914)

钩尾副春蜓 雌，广东｜宋黎明 摄
Paragomphus capricorni, female from Guangdong | Photo by Liming Song

钩尾副春蜓 雄，云南（西双版纳）
Paragomphus capricorni, male from Yunnan (Xishuangbanna)

钩尾副春蜓 雄，广西
Paragomphus capricorni, male from Guangxi

【形态特征】雄性上唇具1对黄斑，额横纹甚阔，中央间断，后头黄色；胸部黑色，背条纹与领条纹不相连，具甚小的肩前上点，合胸侧面第2条纹和第3条纹合并；腹部黑色，第1～7节具黄斑，第8节侧下缘的片状突起较小，第9节的突起较大。雌性与雄性相似，腹部第8～9节无片状突起。【长度】体长 45～49 mm，腹长 34～37 mm，后翅 25～28 mm。【栖息环境】海拔 1000 m以下的溪流和河流。【分布】云南、广西、广东、香港、福建；缅甸、泰国、老挝、越南、新加坡、马来半岛。【飞行期】3—11月。

[Identification] Male labrum with a pair of yellow spots, frons with a broad stripe interrupted medially. Thorax black, dorsal stripes and collar stripes not connecting, superior spots very small, second and third lateral stripes combined. Abdomen black, S1-S7 with yellow markings, the leaf-like flap at the lower margin of S8 small, S9 large. Female similar to male, S8-S9 without leaf-like flap. [Measurements] Total length 45-49 mm, abdomen 34-37 mm, hind wing 25-28 mm. [Habitat] Streams and rivers below 1000 m elevation. [Distribution] Yunnan, Guangxi, Guangdong, Hong Kong, Fujian; Myanmar, Thailand, Laos, Vietnam, Singapore, Peninsular Malaysia. [Flight Season] March to November.

豹纹副春蜓 *Paragomphus pardalinus* Needham, 1942

【形态特征】雄性面部黑色具黄斑，上唇大面积黄白色，额横纹甚阔；胸部黑色，背条纹与领条纹不相连，具甚小的肩前上点，合胸侧面第2条纹和第3条纹合并；腹部黑色，第1~7节具黄斑，第8~9节侧缘的片状突起较大。雌性与雄性相似，第8~9节侧缘片状突起较小。【长度】体长 53~54 mm，腹长 39~40 mm，后翅 30~32 mm。【栖息环境】海拔 1000 m以下的溪流和河流。【分布】中国海南特有。【飞行期】3—11月。

[Identification] Male face black with yellow markings, labrum largely yellowish white, frons with a broad stripe. Thorax black, dorsal stripes and collar stripes not connecting, superior spots very small, second and third lateral stripes

豹纹副春蜓 雄，海南
Paragomphus pardalinus, male from Hainan

豹纹副春蜓 雌，海南
Paragomphus pardalinus, female from Hainan

豹纹副春蜓 交尾，海南 |莫善濂 摄
Paragomphus pardalinus, mating pair from Hainan | Photo by Shanlian Mo

combined. Abdomen black, S1-S7 with yellow markings, S8-S9 with large leaf-like flaps at the lower lateral margin. Female similar to male, S8-S9 with small leaf-like flaps. **[Measurements]** Total length 53-54 mm, abdomen 39-40 mm, hind wing 30-32 mm. **[Habitat]** Streams and rivers below 1000 m elevation. **[Distribution]** Endemic to Hainan of China. **[Flight Season]** March to November.

副春蜓属待定种 *Paragomphus* sp.

【形态特征】本种与钩尾副春蜓相似，但体型稍大，黄斑更发达，雄性后钩片的构造不同。【长度】雄性体长 51～55 mm，腹长 38～42 mm，后翅 28～30 mm。【栖息环境】海拔 500 m以下的溪流。【分布】云南（临沧）。【飞行期】6月。

[Identification] Similar to *P. capricornis* but body size larger, yellow markings more developed, shape of posterior hamulus different. **[Measurements]** Male total length 51-55 mm, abdomen 38-42 mm, hind wing 28-30 mm. **[Habitat]** Streams below 500 m elevation. **[Distribution]** Yunnan (Lincang). **[Flight Season]** June.

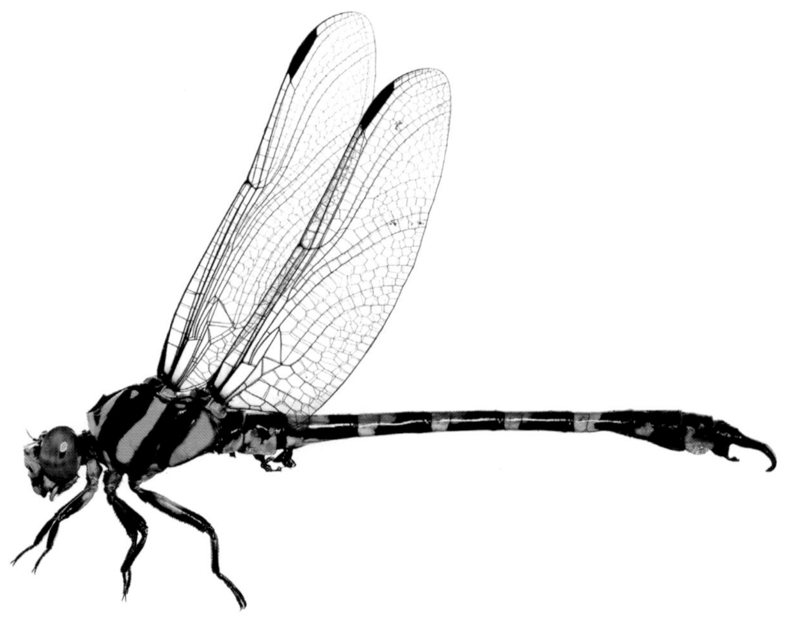

副春蜓属待定种 雄，云南（临沧）
Paragomphus sp., male from Yunnan (Lincang)

奇春蜓属 Genus *Perissogomphus* Laidlaw, 1922

本属全球已知仅有2种，分布于喜马拉雅山脉。中国已知2种，分布于云南和西藏。本属春蜓体中型，体色绿色，与日春蜓属的种类较相似，但本属雄性的肛附器与日春蜓属种类明显不同。

The genus contains only two species known from the Himalayas. Both are recorded from China, found in Yunnan and Tibet. Species of the genus are medium-sized, body green, the general habitus similar to species of genus *Nihonogomphus*, but the shape of male anal appendages different.

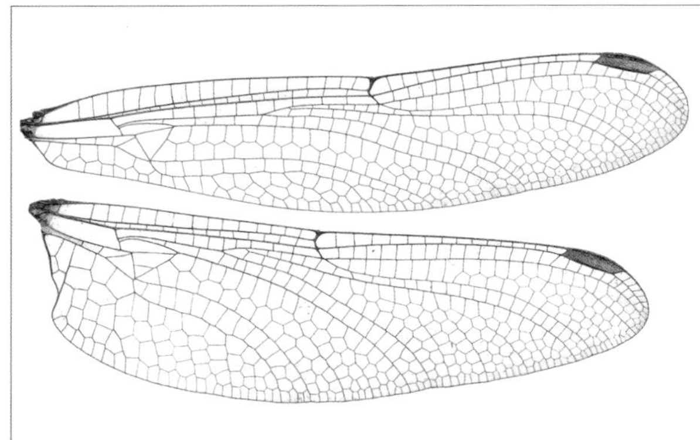

史蒂奇春蜓 雄翅
Perissogomphus stevensi, male wings

史蒂奇春蜓
Perissogomphus stevensi

奇春蜓属 雄性肛附器
Genus *Perissogomphus*, male anal appendages

史蒂奇春蜓 *Perissogomphus stevensi* Laidlaw, 1922

【形态特征】雄性复眼深绿色，面部大面积黑色，额横纹甚阔；胸部黑色，背条纹与领条纹相连，合胸侧面大面积绿色；腹部黑色具黄绿色条纹，肛附器黄色。【长度】雄性体长 55 mm，腹长 42 mm，后翅 37 mm。【栖息环境】山区溪流。【分布】西藏；印度、尼泊尔。【飞行期】7—9月。

[Identification] Male eyes dark green, face largely black, frons with a broad stripe. Thorax black, dorsal stripes and collar stripes connecting, sides largely green. Abdomen black with yellowish green markings, anal appendages yellow. **[Measurements]** Male total length 55 mm, abdomen 42 mm, hind wing 37 mm. **[Habitat]** Montane streams. **[Distribution]** Tibet; India, Nepal. **[Flight Season]** July to September.

史蒂奇春蜓 雄，西藏 | 吴超 摄
Perissogomphus stevensi, male from Tibet | Photo by Chao Wu

显春蜓属 Genus *Phaenandrogomphus* Lieftinck, 1964

本属全球已知7种，分布于亚洲的亚热带和热带区域。中国已知4种，分布于华南和西南地区。本属春蜓体中型；翅透明，三角室、上三角室和下三角室无横脉，基臀区具1条横脉，臀圈1室，臀三角室4室，臀三角室的基边甚斜；雄性肛附器发达，黄色，上肛附器甚长，末端尖锐并向下弯曲，下肛附器短于上肛附器，末端向上弯曲，侧面观末端几乎呈水平方向横截。阳茎具1对鞭。

本属蜻蜓栖息于山区溪流和河流。雄性经常停落在溪流边缘的枝头上或水面的大岩石上占据领地，并长时间定点低空悬停等待雌性。

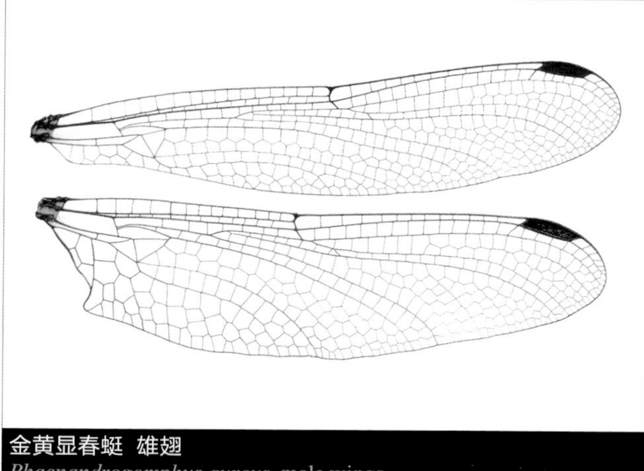

金黄显春蜓 雄翅
Phaenandrogomphus aureus, male wings

The genus contains seven species, distributed in subtropical and tropical Asia. Four species are recorded from China, found in the South and Southwest. Species of the genus are medium-sized; Wings hyaline, no crossvein in triangle, hypertriangle and subtriangle, cubital space with one crossvein, anal loop 1-celled, anal triangle 4-celled with the basal side oblique. Male anal appendages developed, color yellow, superior appendages long, the tips pointed and curved downwards, the inferior appendage shorter, the tips curved upwards. Penis with a pair of flagella.

Phaenandrogomphus species inhabit montane streams and rivers. Territorial males usually perch on top of branches or big rocks and often hover above water for long time.

金黄显春蜓 雄
Phaenandrogomphus aureus, male

金黄显春蜓
Phaenandrogomphus aureus

赵氏显春蜓
Phaenandrogomphus chaoi

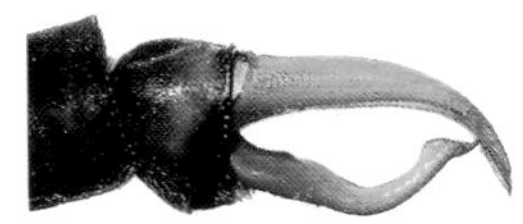

细尾显春蜓
Phaenandrogomphus tonkinicus

显春蜓属 雄性肛附器
Genus *Phaenandrogomphus*, male anal appendages

金黄显春蜓
Phaenandrogomphus aureus

赵氏显春蜓
Phaenandrogomphus chaoi

细尾显春蜓
Phaenandrogomphus tonkinicus

显春蜓属 雌性下生殖板
Genus *Phaenandrogomphus*, female vulvar lamina

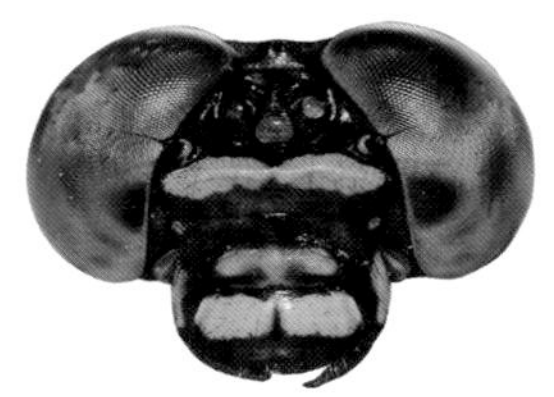

金黄显春蜓 雄
Phaenandrogomphus aureus, male

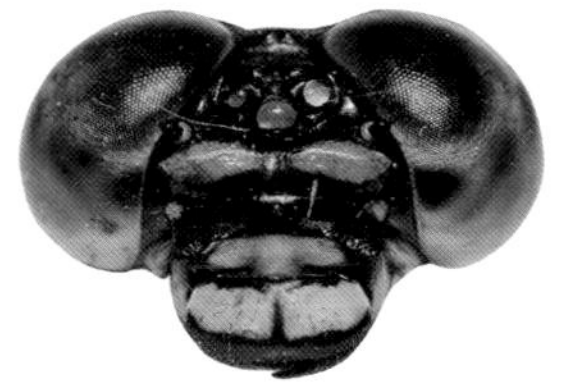

金黄显春蜓 雌
Phaenandrogomphus aureus, female

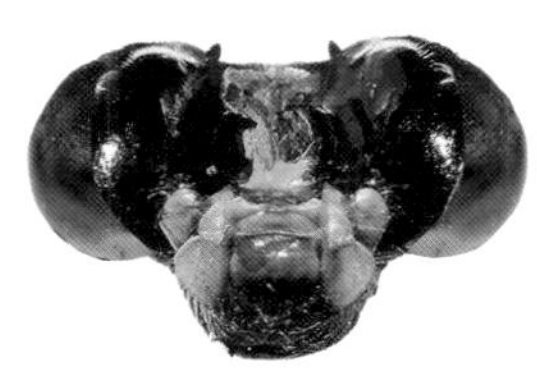

金黄显春蜓 雌
Phaenandrogomphus aureus, female

细尾显春蜓 雌
Phaenandrogomphus tonkinicus, female

显春蜓属 头部正面观
Genus *Phaenandrogomphus*, head in frontal view

显春蜓属 头部后面观
Genus *Phaenandrogomphus*, head in posterior view

金黄显春蜓 *Phaenandrogomphus aureus* (Laidlaw, 1922)

【形态特征】雄性上唇黄白色，额横纹甚阔；胸部黑色，背条纹与领条纹相连，肩前条纹甚长，合胸侧面第2条纹和第3条纹基本完整；腹部黑色，第1～8节具黄斑。雌性与雄性相似，后头后缘两侧具角状突起。【长度】体长49～52 mm，腹长 37～39 mm，后翅 29～34 mm。【栖息环境】海拔 1000 m以下的溪流和河流。【分布】云南（西双版纳、普洱）；印度、缅甸。【飞行期】4—10月。

[Identification] Male labrum yellowish white, frons with a broad stripe. Thorax black, dorsal stripes and collar stripes connecting, antehumeral stripes long, second and third lateral stripes complete. Abdomen black, S1-S8 with yellow markings. Female similar to male, hind margin of occiput with a pair of horns laterally. **[Measurements]** Total length 49-52 mm, abdomen 37-39 mm, hind wing 29-34 mm. **[Habitat]** Streams and rivers below 1000 m elevation. **[Distribution]** Yunnan (Xishuangbanna, Pu'er); India, Myanmar. **[Flight Season]** April to October.

金黄显春蜓 雄，云南（普洱）
Phaenandrogomphus aureus, male from Yunnan (Pu'er)

金黄显春蜓 雌，云南（普洱）
Phaenandrogomphus aureus, female from Yunnan (Pu'er)

金黄显春蜓 雄，云南（普洱）
Phaenandrogomphus aureus, male from Yunnan (Pu'er)

赵氏显春蜓 *Phaenandrogomphus chaoi* Zhu & Liang, 1994

【形态特征】雄性上唇黄白色，额横纹甚阔；胸部黑色，背条纹与领条纹相连，肩前上点甚小，合胸侧面第2条纹中央间断，第3条纹完整；腹部黑色，第1~8节具黄斑。雌性与雄性相似，后头后缘两侧具1对角状突起。本种已被作为细尾显春蜓的异名，但根据雌性下生殖板的构造，此处恢复其有效种的身份。【长度】体长 50~52 mm，腹长 38~40 mm，后翅 29~31 mm。【栖息环境】海拔 1000 m以下的山区溪流。【分布】中国特有，分布于广西、广东、福建。【飞行期】4—8月。

[Identification] Male labrum yellowish white, frons with a broad stripe. Thorax black, dorsal stripes and collar stripes connecting, superior spots very small, second lateral stripes largely interrupted medially, third lateral stripe complete. Abdomen black, S1-S8 with yellow markings. Female similar to male, hind margin of occiput with a pair of horns laterally. The species has been treated as synonym of *P. tonkinicus*, based on the female vulvar lamina structure it is considered as a valid species here. **[Measurements]** Total length 50-52 mm, abdomen 38-40 mm, hind wing 29-31 mm. **[Habitat]** Moutane streams below 1000 m elevation. **[Distribution]** Endemic to China, recorded from Guangxi, Guangdong, Fujian. **[Flight Season]** April to August.

赵氏显春蜓 雌，广东 | 宋黎明 摄
Phaenandrogomphus chaoi, female from Guangdong | Photo by Liming Song

赵氏显春蜓 雄，广西
Phaenandrogomphus chaoi, male from Guangxi

赵氏显春蜓 雄，广西
Phaenandrogomphus chaoi, male from Guangxi

细尾显春蜓 *Phaenandrogomphus tonkinicus* (Fraser, 1926)

细尾显春蜓 雄，云南（红河）
Phaenandrogomphus tonkinicus, male from Yunnan (Honghe)

细尾显春蜓 雌，云南（红河）| 莫善濂 摄
Phaenandrogomphus tonkinicus, female from Yunnan (Honghe) | Photo by Shanlian Mo

细尾显春蜓 雄，海南
Phaenandrogomphus tonkinicus, male from Hainan

【形态特征】雄性上唇黄白色，额横纹甚阔；胸部黑色，背条纹与领条纹相连，肩前条纹细而短或完全缺失，合胸侧面第2条纹和第3条纹基本完整；腹部黑色，第1～8节具黄斑。雌性与雄性相似，后头后缘两侧具较小的瘤状突起，海南个体胸部肩前条纹缺失，与云南东部个体不同。【长度】体长50～53 mm，腹长38～40 mm，后翅32～34 mm。【栖息环境】海拔 1000 m以下的溪流和河流。【分布】云南（红河）、海南；老挝、泰国、越南。【飞行期】4—10月。

[Identification] Male labrum largely yellowish white, frons with a broad stripe. Thorax black, dorsal stripes and collar stripes connecting, antehumeral stripes short and narrow or absent, second and third lateral stripes complete. Abdomen black, S1-S8 with yellow markings. Female similar to male, occipital margin with small tuberculiform prominences, individuals from Hainan without antehumeral stripes thus differing from individuals from the east of Yunnan.**[Measurements]** Total length 50-53 mm, abdomen 38-40 mm, hind wing 32-34 mm. **[Habitat]** Moutane streams and rivers below 1000 m elevation. **[Distribution]** Yunnan (Honghe), Hainan; Laos, Thailand, Vietnam. **[Flight Season]** April to October.

刀春蜓属 Genus *Scalmogomphus* Chao, 1990

本属全球已知5种，分布于喜马拉雅山脉和中国西南地区。中国已知5种。本属春蜓体中型；翅透明，三角室、上三角室和下三角室无横脉，基臀区具1条横脉，臀圈1室，臀三角室4室；雄性腹部第8～9节侧缘膨大，并具有红褐色或者浅黄色斑，肛附器黄褐色，上肛附器长，末端向内侧方弯曲，下肛附器稍短，具2个明显的齿状突起。

本属蜻蜓栖息于较开阔的溪流，一些种类可以在高山的寒冷环境生存。雄性经常停落在水面的大岩石或者水边的枝头上。

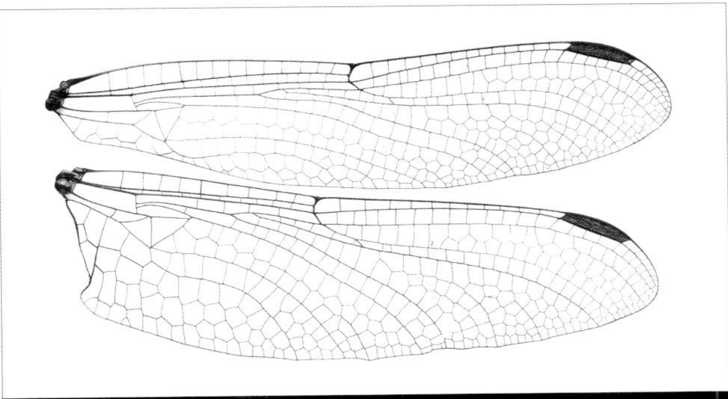

黄条刀春蜓 雄翅
Scalmogomphus bistrigatus, male wings

The genus contains five species known from the Himalayas and Southwest China. All of them are recorded from China. Species of the genus are medium-sized. Wings hyaline, no crossvein in triangle, hypertriangle and subtriangle, cubital space with one crossvein, anal loop 1-celled, anal triangle 4-celled. Male abdomen expanded at S8-S9, and tinted with reddish brown or pale yellow markings, anal appendages yellowish brown, superiors long with tips curved downwards, inferior shorter, with two teeth dorsally.

Scalmogomphus species inhabit exposed streams, some species live in high mountains. Territorial males usually perch on big rocks above water or the top of branches.

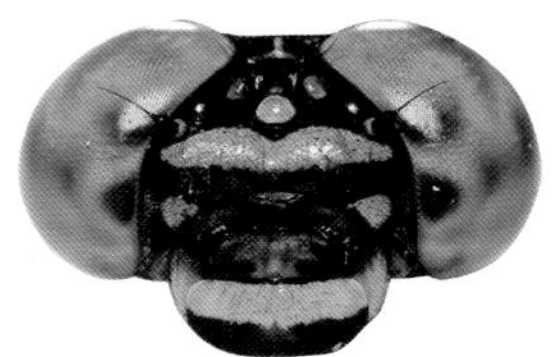

黄条刀春蜓 雄
Scalmogomphus bistrigatus, male

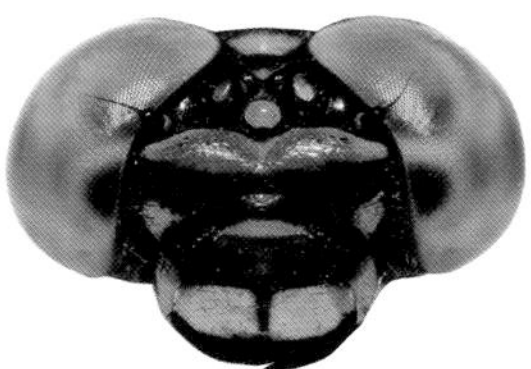

丁格刀春蜓 雄
Scalmogomphus dingavani, male

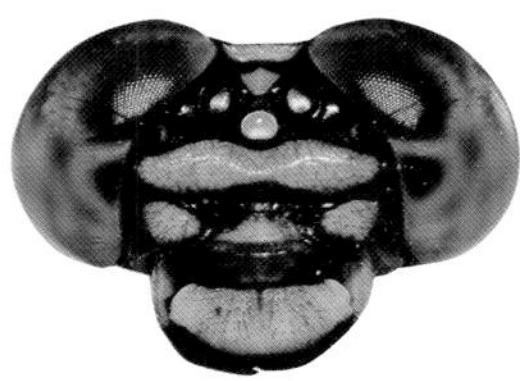

丁格刀春蜓 雌
Scalmogomphus dingavani, female

文山刀春蜓 雄
Scalmogomphus wenshanensis, male

刀春蜓属 头部正面观
Genus *Scalmogomphus*, head in frontal view

黄条刀春蜓 雄
Scalmogomphus bistrigatus, male

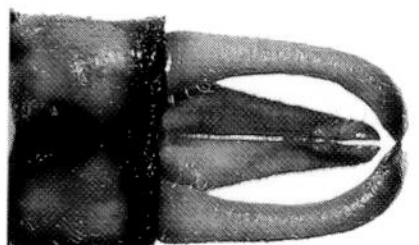

黄条刀春蜓
Scalmogomphus bistrigatus

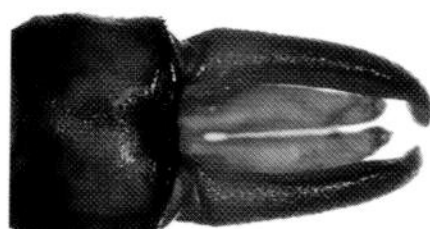

丁格刀春蜓
Scalmogomphus dingavani

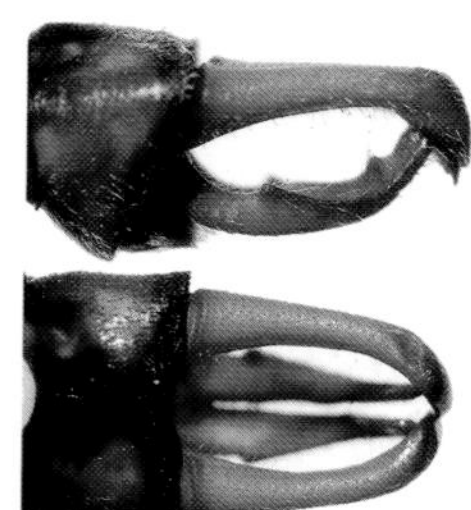

文山刀春蜓
Scalmogomphus wenshanensis

刀春蜓属 雄性肛附器
Genus *Scalmogomphus*, male anal appendages

黄条刀春蜓
Scalmogomphus bistrigatus

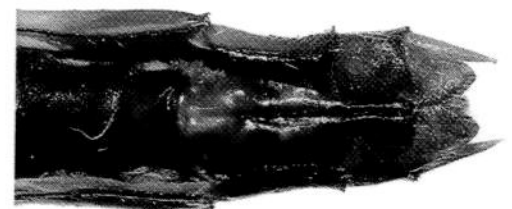

丁格刀春蜓
Scalmogomphus dingavani

刀春蜓属待定种1
Scalmogomphus sp. 1

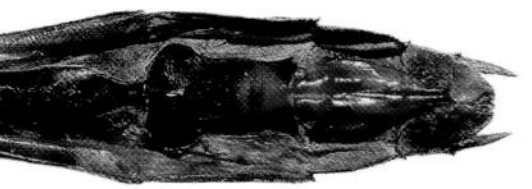

刀春蜓属待定种2
Scalmogomphus sp. 2

刀春蜓属 雌性下生殖板
Genus *Scalmogomphus*, female vulvar lamina

黄条刀春蜓 *Scalmogomphus bistrigatus* (Hagen, 1854)

【形态特征】雄性面部黑色具黄斑，上唇黄白色，额横纹甚阔；胸部黑色，背条纹与领条纹相连，肩前条纹细长，合胸侧面第2条纹和第3条纹完整；腹部黑色具黄斑。雌性与雄性相似，腹部末端未膨大。【长度】体长 51~53 mm，腹长 39~40 mm，后翅 32~34 mm。【栖息环境】海拔 1500~2500 m的山区开阔溪流。【分布】云南（大理、保山）；不丹、印度、尼泊尔。【飞行期】6—10月。

[Identification] Male face black with yellow markings, labrum yellowish white, frons with a broad stripe. Thorax black, dorsal stripes and collar stripes connecting, antehumeral stripes long and narrow, second and third

黄条刀春蜓 雄，云南（大理）
Scalmogomphus bistrigatus, male from Yunnan (Dali)

黄条刀春蜓 雌，云南（大理）
Scalmogomphus bistrigatus, female from Yunnan (Dali)

lateral stripes complete. Abdomen black with yellow markings. Female similar to male, distal abdomen not expanded. **[Measurements]** Total length 51-53 mm, abdomen 39-40 mm, hind wing 32-34 mm. **[Habitat]** Exposed montane streams at 1500-2500 m elevation. **[Distribution]** Yunnan (Dali, Baoshan); Bhutan, India, Nepal. **[Flight Season]** June to October.

黄条刀春蜓 雄，云南（大理）
Scalmogomphus bistrigatus, male from Yunnan (Dali)

丁格刀春蜓 *Scalmogomphus dingavani* (Fraser, 1924)

【形态特征】雄性面部黑色具黄斑，上唇黄白色，额横纹甚阔，后头黄色；胸部黑色，背条纹与领条纹相连，肩前条纹细长，合胸侧面第2条纹大面积缺失，第3条纹完整；腹部黑色具黄色和褐色斑。雌性与雄性相似，腹部具甚大的黄斑。【长度】体长 43～47 mm，腹长 32～35 mm，后翅 28～34 mm。【栖息环境】海拔 1000 m以下的山区溪流。【分布】云南（西双版纳、红河）；印度、缅甸、泰国、越南。【飞行期】9—11月。

丁格刀春蜓 雄，云南（红河）
Scalmogomphus dingavani, male from Yunnan (Honghe)

[Identification] Male face black with yellow markings, labrum largely yellowish white, frons with a broad stripe, occiput yellow. Thorax black, dorsal stripes and collar stripes connecting, antehumeral stripes long and narrow, second lateral stripe mostly absent, third lateral stripe complete. Abdomen black with yellow and brown markings. Female similar to male, abdomen with large yellow markings. **[Measurements]** Total length 43-47 mm, abdomen

32-35 mm, hind wing 28-34 mm. **[Habitat]** Montane streams below 1000 m elevation. **[Distribution]** Yunnan (Xishuangbanna, Honghe); India, Myanmar, Thailand, Vietnam. **[Flight Season]** September to November.

丁格刀春蜓 雄，云南（红河）
Scalmogomphus dingavani, male from Yunnan (Honghe)

丁格刀春蜓 雌，云南（红河）
Scalmogomphus dingavani, female from Yunnan (Honghe)

文山刀春蜓 *Scalmogomphus wenshanensis* Zhou, Zhou & Lu, 2005

文山刀春蜓 雄，云南（文山）
Scalmogomphus wenshanensis, male from Yunnan (Wenshan)

【形态特征】雄性面部黑色具黄斑，上唇黄白色，额横纹甚阔；胸部黑色，背条纹与领条纹相连，肩前条纹细长，合胸侧面第2条纹和第3条纹完整；腹部黑色具黄斑。本种与黄条刀春蜓相似，但身体的黄色斑纹不如后者发达，两者的肛附器稍有差异。【长度】雄性体长54～55 mm，腹长41～42 mm，后翅33～34 mm。【栖息环境】海拔1500～2500 m的山区开阔溪流。【分布】中国云南（红河、文山）特有。【飞行期】6—10月。

[Identification] Male face black with yellow markings, labrum yellowish white, frons with a broad stripe. Thorax black, dorsal stripes and collar stripes connecting, antehumeral stripes long and narrow, second and third lateral stripes complete. Abdomen black with yellow markings. Similar to *S. bistrigatus* but yellow markings less developed, male anal appendages a little different. **[Measurements]** Male total length 54-55 mm, abdomen 41-42 mm, hind wing 33-34 mm. **[Habitat]** Exposed montane streams at 1500-2500 m elevation. **[Distribution]** Endemic to Yunnan (Honghe, Wenshan) of China. **[Flight Season]** June to October.

刀春蜓属待定种1 *Scalmogomphus* sp. 1

刀春蜓属待定种1 雄，云南（西双版纳）
Scalmogomphus sp. 1, male from Yunnan (Xishuangbanna)

刀春蜓属待定种1 雌，云南（西双版纳）
Scalmogomphus sp. 1, female from Yunnan (Xishuangbanna)

【形态特征】雄性上唇具1对黄色斑，额横纹较细，中央间断；胸部黑色，背条纹与领条纹相连，肩前条纹细长，有时缺如，合胸侧面第2条纹中央间断，第3条纹完整；腹部黑色具黄色和褐色斑点。雌性与雄性相似，腹部黄斑更发达。【长度】体长 50～53 mm，腹长 37～41 mm，后翅 31～33 mm。【栖息环境】海拔 500～1000 m森林中的溪流。【分布】云南（西双版纳）。【飞行期】6—11月。

[Identification] Male labrum with a pair of yellow spots, frons with a narrow stripe and interrupted medially. Thorax black, dorsal stripes and collar stripes connecting, antehumeral stripes long and narrow, sometimes absent, second lateral stripe interrupted medially, third lateral stripes complete. Abdomen black with yellow and brown markings. Female similar to male, abdomen with larger yellow markings. **[Measurements]** Total length 50-53 mm, abdomen 37-41 mm, hind wing 31-33 mm. **[Habitat]** Streams in forest at 500-1000 m elevation. **[Distribution]** Yunnan (Xishuangbanna). **[Flight Season]** June to November.

刀春蜓属待定种2 *Scalmogomphus* sp. 2

【形态特征】雌性面部黑色具黄斑，上唇黄白色，额横纹甚阔；胸部黑色，背条纹与领条纹相连，肩前条纹细长，合胸侧面第2条纹和第3条纹完整；腹部黑色具黄色和褐色斑点，腹部第7～9节显著膨大。【长度】雌性体长50～51 mm，腹长 38～39 mm，后翅 32～33 mm。【栖息环境】海拔 500 m处森林中的开阔溪流。【分布】云南（临沧）。【飞行期】9月。

[Identification] Female face black with yellow markings, labrum yellowish white, frons with a broad stripe. Thorax black, dorsal stripes and collar stripes connecting, antehumeral stripes long and narrow, second and third lateral stripes complete. Abdomen black with yellow and brown markings, S7-S9 expanded laterally. **[Measurements]** Female total length 50-51 mm, abdomen 38-39 mm, hind wing 32-33 mm. **[Habitat]** Exposed streams in forest at 500 m elevation. **[Distribution]** Yunnan (Lincang). **[Flight Season]** September.

刀春蜓属待定种2 雌，云南（临沧）
Scalmogomphus sp. 2, female from Yunnan (Lincang)

邵春蜓属 Genus *Shaogomphus* Chao, 1984

本属全球已知3种，分布于亚洲的温带和亚热带区域。中国已知3种，主要分布于东北地区。本属春蜓体中型；翅透明，三角室、上三角室和下三角室无横脉，基臀区具1条横脉，臀圈缺如，臀三角室3～5室；雄性腹部第8～9节膨大，肛附器较短，上肛附器的两枝稍微分歧，向下弯曲，下肛附器约与上肛附器等长，两枝分歧。

本属蜻蜓栖息于较宽阔的溪流和河流。雄性经常停落在溪流边缘的树丛或水面的岩石上。雌性产卵前先是停在水边排出卵块，然后以快速的飞行点水方式将卵投入水中。

The genus contains three species known from temperate and subtropical Asia. Three species are recorded from China, mainly found in the Northeast. Species of the genus are medium-sized. Wings hyaline, no crossvein in triangle, hypertriangle and subtriangle, cubital space with one crossvein, anal loop absent, anal triangle 3- to 5-celled. Male S8-S9 expanded, anal appendages short, superiors slightly divergent and curved downwards, the inferior of the same length with the two branches divergent.

Shaogomphus species inhabit fairly wide streams and rivers. Males usually perch on tree limbs near shore or rocks above water. Females exude an egg mass beforehand and deposit it into water flying rapidly and dipping the abdomen tip.

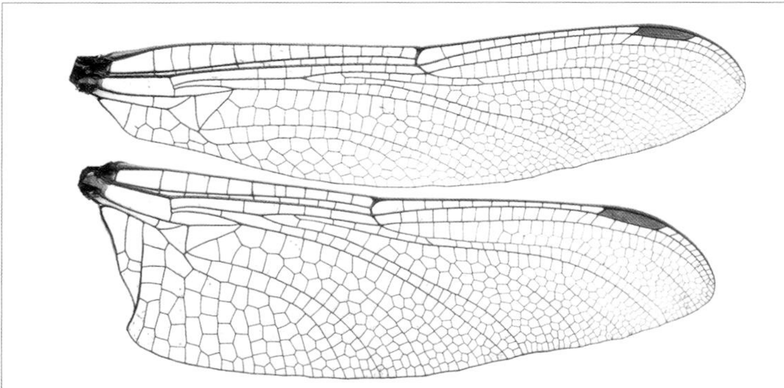

寒冷邵春蜓 雄翅
Shaogomphus postocularis, male wings

寒冷邵春蜓
Shaogomphus postocularis

邵春蜓属 雄性肛附器
Genus *Shaogomphus*, male anal appendages

寒冷邵春蜓 雄 | 莫善濂 摄
Shaogomphus postocularis, male | Photo by Shanlian Mo

寒冷邵春蜓 *Shaogomphus postocularis epophthalmus* (Selys, 1872)

寒冷邵春蜓 雄，黑龙江 | 莫善濂 摄
Shaogomphus postocularis, male from Heilongjiang | Photo by Shanlian Mo

【形态特征】雄性面部大面积黑色，额横纹甚阔，后头黄色；胸部黑色具灰白色条纹，背条纹与领条纹相连，肩前条纹细长，合胸侧面第2条纹上方间断，第3条纹完整；腹部黑色，第1～9节具灰白色斑。雌性黑色具黄色斑纹。【长度】体长 52～55 mm，腹长 36～38 mm，后翅 32～35 mm。【栖息环境】海拔 500 m以下的宽阔河流。【分布】黑龙江、吉林；朝鲜半岛、俄罗斯远东。【飞行期】5—7月。

[Identification] Male face largely black, frons with a broad stripe, occiput yellow. Thorax black with grayish white markings, dorsal stripes and collar stripes connecting, antehumeral stripes long and narrow, upper half of second lateral stripe absent, third lateral stripe complete. Abdomen black, S1-S9 with grayish white markings. Female black with yellow markings. **[Measurements]** Total length 52-55 mm, abdomen 36-38 mm, hind wing 32-35 mm. **[Habitat]** Large rivers below 500 m elevation. **[Distribution]** Heilongjiang, Jilin; Korean peninsular, Russian Far East. **[Flight Season]** May to July.

寒冷邵春蜓 雌，黑龙江 | 莫善濂 摄
Shaogomphus postocularis, female from Heilongjiang | Photo by Shanlian Mo

施氏邵春蜓 *Shaogomphus schmidti* (Asahina, 1956)

【形态特征】雄性面部大面积黑色，额横纹甚阔，后头黄色；胸部黑色具黄色条纹，背条纹与领条纹相连，肩前条纹细长，合胸侧面第2条纹上方间断，第3条纹完整；腹部黑色具黄色斑纹。雌性与雄性相似。【长度】腹长36~37 mm，后翅 30~32 mm。【栖息环境】海拔 500 m以下的宽阔河流。【分布】黑龙江；俄罗斯远东。【飞行期】5—7月。

[Identification] Male face largely black, frons with a broad stripe, occiput yellow. Thorax black with yellow markings, dorsal stripes and collar stripes connecting, antehumeral stripes long and narrow, second lateral stripe interrupted above, third lateral stripe complete. Abdomen black with abundant yellow markings. Female similar to male. **[Measurements]** Abdomen 36-37 mm, hind wing 30-32 mm. **[Habitat]** Large rivers below 500 m elevation. **[Distribution]** Heilongjiang; Russian Far East. **[Flight Season]** May to July.

施氏邵春蜓 雌，俄罗斯 | Oleg E. Kosterin 摄
Shaogomphus schmidti, female from Russia | Photo by Oleg E. Kosterin

施氏邵春蜓 雄，俄罗斯 | Oleg E. Kosterin 摄
Shaogomphus schmidti, male from Russia | Photo by Oleg E. Kosterin

施春蜓属 Genus *Sieboldius* Selys, 1854

本属全球已知8种，亚洲广布。中国已知6种，除西北地区全国广布。本属是体型巨大的春蜓，胸部巨大，头甚小；翅透明，三角室内具横脉，基臀区具2条横脉，臀圈4室，臀三角室3～5室；雄性肛附器不发达，上肛附器具齿状突起。

本属蜻蜓栖息于山区溪流。雄性经常停落在水面的大岩石上占据领地。雌性以定点悬停点水的方式产卵。

The genus contains eight species widespread in Asia. Six species are recorded from all of China, except from the Northwest. Species of the genus are largest among its family. The thorax is very robust but head small. Wings hyaline, triangle with crossveins, cubital space with two crossveins, anal loop 4-celled, anal triangle 3- to 5-celled. Male anal appendages less developed, superior appendages with teeth.

Sieboldius species inhabit montane streams. Territorial males usually perch on big rocks above water. Females lay eggs by hovering, and dipping the abdomen tip into water.

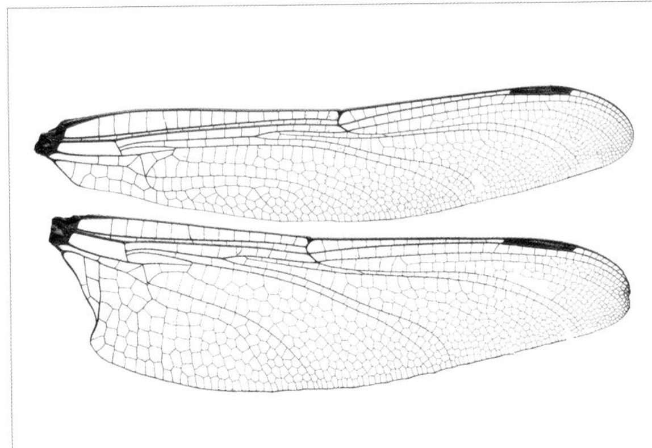

艾氏施春蜓 雄翅
Sieboldius albardae, male wings

亚力施春蜓 雄
Sieboldius alexanderi, male

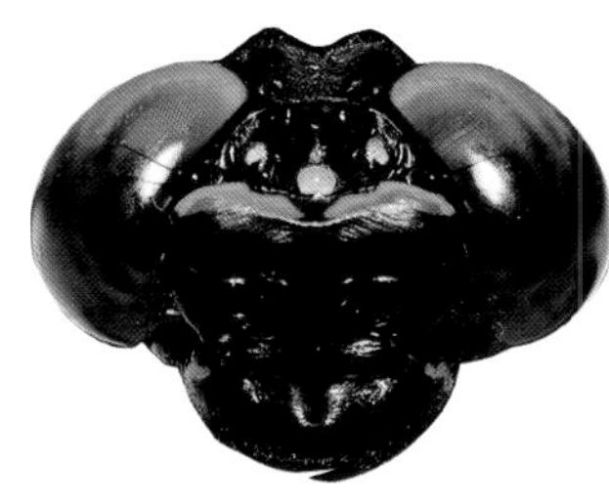

黑纹施春蜓 雄
Sieboldius nigricolor, male

施春蜓属 头部正面观
Genus *Sieboldius*, head in frontal view

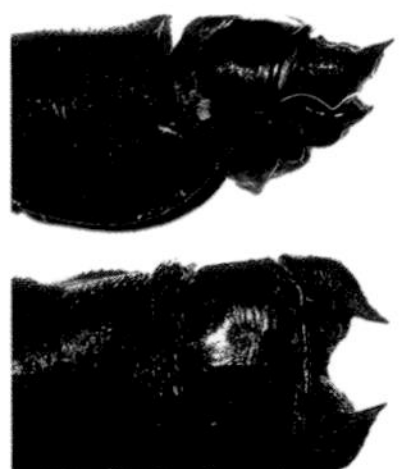

艾氏施春蜓
Sieboldius albardae

亚力施春蜓
Sieboldius alexanderi

折尾施春蜓
Sieboldius deflexus

黑纹施春蜓
Sieboldius nigricolor

施春蜓属 雄性肛附器
Genus *Sieboldius*, male anal appendages

黑纹施春蜓 雄
Sieboldius nigricolor, male

艾氏施春蜓 *Sieboldius albardae* Selys, 1886

【形态特征】雄性面部主要黑色，额横纹甚阔，侧单眼后方具1对矮突起，后头缘呈驼峰状隆起；胸部黑色具灰白色条纹，背条纹与领条纹相连，合胸侧面第2条纹和第3条纹完整并在下方合并；腹部黑色，第1～8节具灰白色斑。雌性黑色具黄斑，腹部较短。【长度】体长 78～81 mm，腹长 57～60 mm，后翅 46～49 mm。【栖息环境】海拔1000 m以下岩石丰富的开阔溪流。【分布】黑龙江、吉林、辽宁、北京、河北、山东；俄罗斯远东、朝鲜半岛、日本。【飞行期】6—9月。

[Identification] Male face largely black, frons with a broad stripe, vertex with a pair of short prominences, occipital margin bimodally raised. Thorax black with grayish white markings, dorsal stripes and collar stripes connecting, second and third lateral stripe complete and connecting ventrally. Abdomen black, S1-S8 with grayish white markings. Female black with yellow markings, abdomen shorter. **[Measurements]** Total length 78-81 mm, abdomen 57-60 mm, hind wing 46-49 mm. **[Habitat]** Exposed and rocky streams below 1000 m elevation. **[Distribution]** Heilongjiang, Jilin, Liaoning, Beijing, Hebei, Shandong; Russian Far East, Korean peninsula, Japan. **[Flight Season]** June to September.

艾氏施春蜓 雄，北京
Sieboldius albardae, male from Beijing

艾氏施春蜓 雌，山东
Sieboldius albardae, female from Shandong

艾氏施春蜓 雄，北京 | 吴超 摄
Sieboldius albardae, male from Beijing | Photo by Chao Wu

亚力施春蜓 *Sieboldius alexanderi* (Chao, 1955)

【形态特征】雄性面部主要黑色，额横纹甚阔，侧单眼后方具1对矮突起；胸部黑色具灰白色条纹，背条纹与领条纹不相连，合胸侧面第2条纹和第3条纹完整并在下方合并；腹部黑色，第1~9节具灰白色斑，上肛附器较长，端部具1个齿状突起，下肛附器甚短。雌性黑色具黄斑，后头缘稍微隆起。【长度】体长 86~91 mm，腹长 66~68 mm，后翅 56~60 mm。【栖息环境】海拔 1000 m以下的山区溪流。【分布】中国特有，分布于湖北、浙江、福建、江西、广西、广东、海南、香港。【飞行期】4—10月。

[Identification] Male face largely black, frons with a broad stripe, vertex with a pair of short prominences. Thorax black with grayish white markings, dorsal stripes and collar stripes not connecting, second and third lateral stripes complete and connecting ventrally. Abdomen black, S1-S9 with grayish white markings, superior appendages long with an apical tooth, inferior appendage short. Female black with yellow markings, occipital margin slightly raised. **[Measurements]** Total length 86-91 mm, abdomen 66-68 mm, hind wing 56-60 mm. **[Habitat]** Montane streams below 1000 m elevation. **[Distribution]** Endemic to China, recorded from Hubei, Zhejiang, Fujian, Jiangxi, Guangxi, Guangdong, Hainan, Hong Kong. **[Flight Season]** April to October.

亚力施春蜓 雌，广东 | 宋睿斌 摄
Sieboldius alexanderi, female from Guangdong | Photo by Ruibin Song

亚力施春蜓 雄，浙江
Sieboldius alexanderi, male from Zhejiang

亚力施春蜓 雄，浙江
Sieboldius alexanderi, male from Zhejiang

折尾施春蜓 *Sieboldius deflexus* (Chao, 1955)

【形态特征】雄性面部主要黑色，额横纹甚阔，侧单眼后方具1对锥形突起；胸部黑色具灰白色条纹，背条纹与领条纹不相连，合胸侧面第2条纹和第3条纹合并；腹部黑色，第1～8节具灰白色斑，第7节侧下缘中央具1个较短的瘤状突起，肛附器较短。雌性黑色具黄斑。【长度】体长 88～92 mm，腹长 65～67 mm，后翅 58～59 mm。【栖息环境】海拔 1000 m以下的开阔溪流。【分布】中国特有，分布于安徽、湖北、贵州、浙江、福建、广西、广东、台湾。【飞行期】5—10月。

[Identification] Male face mainly black, frons with a broad stripe, vertex with a pair of short prominences. Thorax black with grayish white markings, dorsal stripes and collar stripes not connecting, second and third lateral stripe combined. Abdomen black, S1-S8 with grayish white markings, lower margin of S7 with a short tuberculiform prominence medially, anal appendages short. Female black with yellow markings. **[Measurements]** Total length 88-92 mm, abdomen 65-67 mm, hind wing 58-59 mm. **[Habitat]** Exposed streams below 1000 m elevation. **[Distribution]** Endemic to China, recorded from Anhui, Hubei, Guizhou, Zhejiang, Fujian, Guangxi, Guangdong, Taiwan. **[Flight Season]** May to October.

折尾施春蜓 雄，湖北
Sieboldius deflexus, male from Hubei

折尾施春蜓 雌，广东
Sieboldius deflexus, female from Guangdong

黑纹施春蜓 *Sieboldius nigricolor* (Fraser, 1924)

黑纹施春蜓 雌，云南（西双版纳）
Sieboldius nigricolor, female from Yunnan (Xishuangbanna)

黑纹施春蜓 雄，云南（普洱）
Sieboldius nigricolor, male from Yunnan (Pu'er)

【形态特征】雄性面部主要黑色，额横纹甚阔，中央间断，头顶具1对锥形突起，后头缘隆起；胸部黑色，背条纹与领条纹不相连，合胸侧面第2条纹和第3条纹合并；腹部黑色，第1～8节具黄斑，上肛附器侧缘具1个齿状突起。雌性与雄性相似。【长度】体长 79～83 mm，腹长 59～62 mm，后翅 49～55 mm。【栖息环境】海拔 500～1500 m的开阔溪流。【分布】云南（西双版纳、普洱）；缅甸、泰国、老挝、越南。【飞行期】4—11月。

[Identification] Male face largely black, frons with a broad stripe interrupted medially, vertex with a pair of short prominences, occipital margin raised. Thorax black, dorsal stripes and collar stripes not connecting, second and third lateral stripes combined. Abdomen black, S1-S8 with yellow markings, superiors with a lateral spine. Female similar to male. **[Measurements]** Total length 79-83 mm, abdomen 59-62 mm, hind wing 49-55 mm. **[Habitat]** Exposed streams at 500-1500 m elevation. **[Distribution]** Yunnan (Xishuangbanna, Pu'er); Myanmar, Thailand, Laos, Vietnam. **[Flight Season]** April to November

新叶春蜓属 Genus *Sinictinogomphus* Fraser, 1939

本属全球已知仅1种，分布于亚洲，在中国广布。本属蜻蜓是体型较大的春蜓；翅透明，翅痣甚长，具支持脉，三角室3室，上三角室2～3室，前翅下三角室2室，后翅1～2室，臀圈3室，基臀区具2条横脉，后翅基缘呈钝角；腹部第8节背板侧缘极度扩大，雄性下肛附器较短。

本属蜻蜓栖息于各类静水水域和流速缓慢的溪流。雄性经常停在枝头上占据领地，并时而飞行巡逻。雄性护卫产卵。

The genus contains a single Asian species that is widespread in China. Species of the genus is rather large-sized. Wings hyaline, pterostigma long and braced, triangle 3 celled, hypertriangle 2- to 3-celled, subtriangle in fore wings 2-celled and 1- to 2-celled in hind wings, anal loop 3-celled, cubital space with two crossveins, tornus obtuse angled. Sides of S8 with leaf-like projection. Male inferior appendage short.

Sinictinogomphus species inhabit both standing water and slow flowing streams. Territorial males usually perch on top of branches, and sometimes patrol. Females lay eggs with male guarding around.

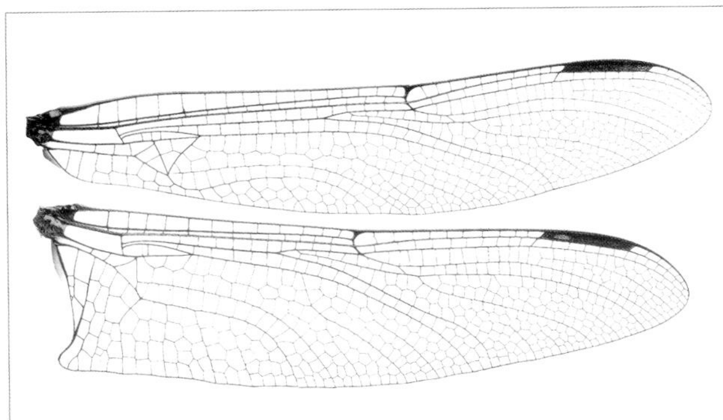

大团扇春蜓 雄翅
Sinictinogomphus clavatus, male wings

大团扇春蜓 雄
Sinictinogomphus clavatus, male

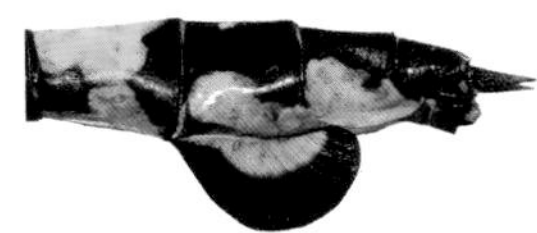

大团扇春蜓 雌
Sinictinogomphus clavatus, female

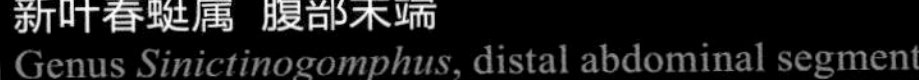

新叶春蜓属 腹部末端
Genus *Sinictinogomphus*, distal abdominal segments

大团扇春蜓 雄
Sinictinogomphus clavatus, male

大团扇春蜓 *Sinictinogomphus clavatus* (Fabricius, 1775)

大团扇春蜓 雌，广东 | 莫善濂 摄
Sinictinogomphus clavatus, female from Guangdong | Photo by Shanlian Mo

大团扇春蜓 雌，湖北
Sinictinogomphus clavatus, female from Hubei

大团扇春蜓 雄，海南
Sinictinogomphus clavatus, male from Hainan

【形态特征】雄性面部主要黄色，侧单眼后方具1对锥形突起；胸部黑色，背条纹与领条纹不相连，肩前条纹较长，合胸侧面第2条纹和第3条纹完整；腹部黑色具黄斑。雌性与雄性相似。【长度】体长 69～71 mm，腹长 51～55 mm，后翅 41～47 mm。【栖息环境】海拔 1500 m以下的池塘、水库和流速缓慢的溪流。【分布】除西北地区外全国广布；俄罗斯远东、朝鲜半岛、日本、尼泊尔、缅甸、泰国、柬埔寨、老挝、越南。【飞行期】3—11月。

[Identification] Male face mainly yellow, vertex with a pair of pyramidal prominences behind lateral ocelli. Thorax black, dorsal stripes and collar stripes not connecting, antehumeral stripes long, second and third lateral stripes complete. Abdomen black with yellow markings. Female similar to male. **[Measurements]** Total length 69-71 mm, abdomen 51-55 mm, hind wing 41-47 mm. **[Habitat]** Ponds, reservoirs and slow flowing streams below 1500 m elevation. **[Distribution]** Widespread in China except the Northwest; Russian Far East, Korean peninsula, Japan, Nepal, Myanmar, Thailand, Cambodia, Laos, Vietnam. **[Flight Season]** March to November.

华春蜓属 Genus *Sinogomphus* May, 1935

本属全球已知10余种，分布于中国、日本和越南。中国已知9种，分布于华中、华南和西南地区。本属蜻蜓是体中型的春蜓；翅透明，三角室、上三角室和下三角室无横脉，基臀区具1条横脉，臀圈缺如，臀三角室3室；雄性上肛附器的两枝在基方相距甚远，乳白色，手指状，基部腹方具齿状突起，下肛附器仅为上肛附器长度的1/2。阳茎具1对短鞭。

本属蜻蜓栖息于山区溪流。雄性经常停落在溪流边缘的树丛或水面的大岩石上。

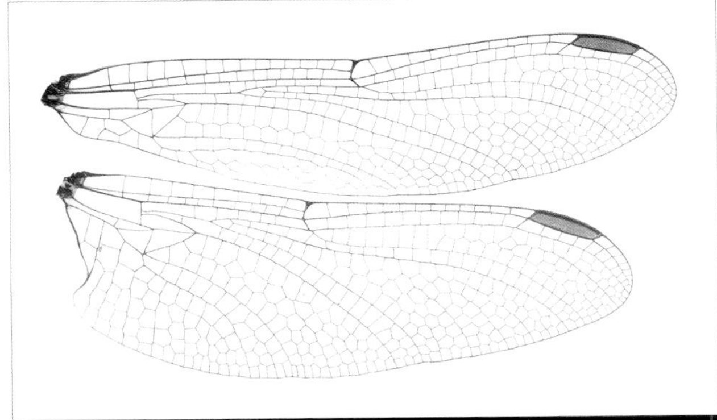

长角华春蜓，雄翅
Sinogomphus scissus, male wings

The genus contains over ten species known from China, Japan and Vietnam. Nine species are recorded from China, found in the Central, South and Southwest regions. Species of the genus are medium-sized. Wings hyaline, no crossvein in triangle, hypertriangle and subtriangle, cubital space with one crossvein, anal triangle 3-celled. Male superior appendages strongly separated at base, color white, finger-shaped, with ventral teeth present basally, the inferior appendage about half length of the superiors. Penis with a pair of short flagella.

Sinogomphus species inhabit montane streams. Males usually perch on trees near water or big rocks above water.

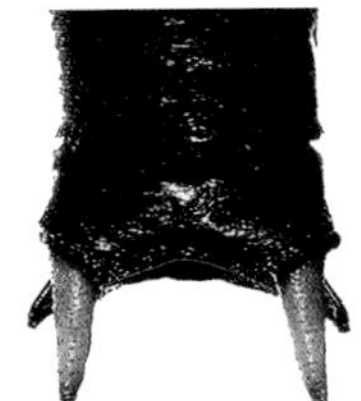

台湾华春蜓
Sinogomphus formosanus

长角华春蜓
Sinogomphus scissus

修氏华春蜓
Sinogomphus suensoni

华春蜓属 雄性肛附器
Genus *Sinogomphus*, male anal appendages

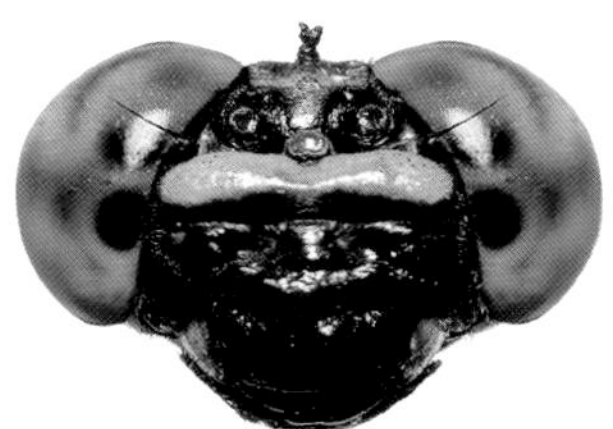

长角华春蜓 雌
Sinogomphus scissus, female

华春蜓属 头部正面观
Genus *Sinogomphus*, head in frontal view

长角华春蜓 雄 | 莫善濂 摄
Sinogomphus scissus, male | Photo by Shanlian Mo

台湾华春蜓 *Sinogomphus formosanus* Asahina, 1951

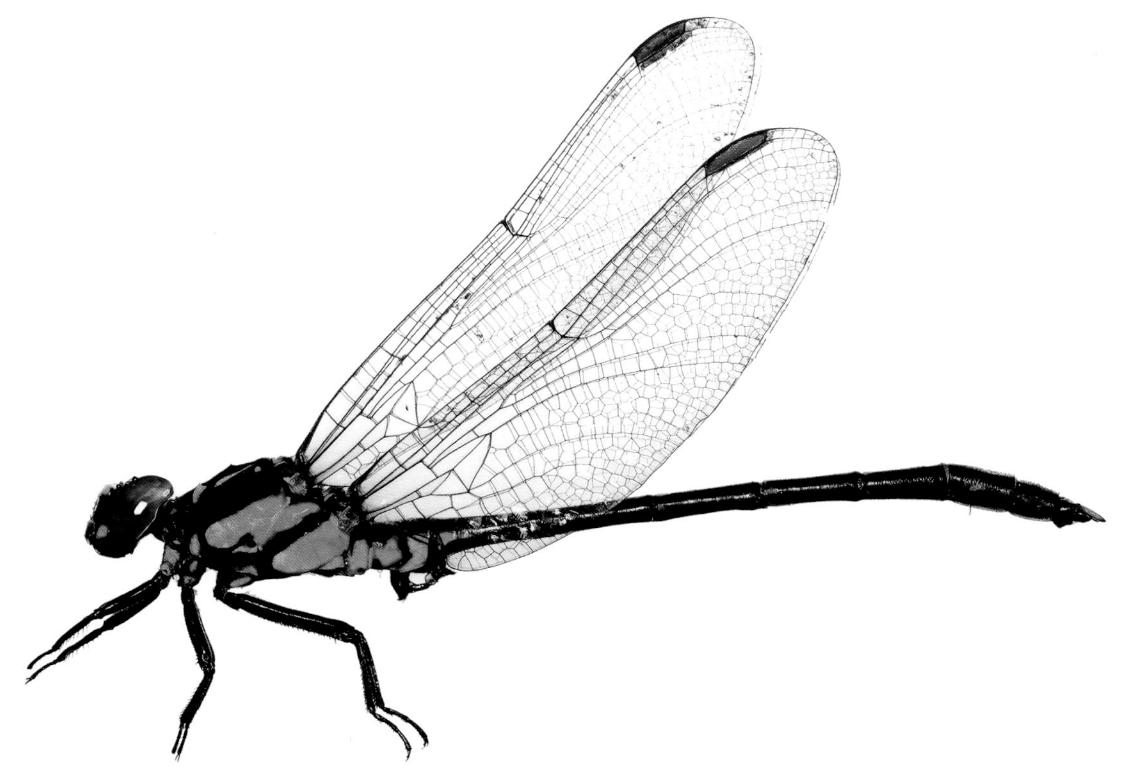

台湾华春蜓 雄，台湾
Sinogomphus formosanus, male from Taiwan

【形态特征】雄性面部主要黑色，额横纹甚阔；胸部黑色，背条纹与领条纹不相连，具甚小的肩前上点，合胸侧面第2条纹中央间断，第3条纹完整；腹部黑色，第1~8节具黄斑。雌性与雄性相似，头顶具1对角状突起。【长度】体长 44~47 mm，腹长 29~34 mm，后翅 26~28 mm。【栖息环境】海拔 2000 m以下森林中的小溪。【分布】中国台湾特有。【飞行期】3—8月。

[Identification] Male face mainly black, frons with a broad stripe. Thorax black, dorsal stripes and collar stripes not connecting, superior spots small, second lateral stripe interrupted medially, third lateral stripe complete. Abdomen black, S1-S8 with yellow markings. Female similar to male, vertex with a pair of horns. **[Measurements]** Total length 44-47 mm, abdomen 29-34 mm, hind wing 26-28 mm. **[Habitat]** Streams in forest below 2000 m elevation. **[Distribution]** Endemic to Taiwan of China. **[Flight Season]** March to August.

长角华春蜓 *Sinogomphus scissus* (McLachlan, 1896)

【形态特征】雄性面部主要黑色，额横纹甚阔；胸部黑色，背条纹与领条纹不相连，具甚小的肩前上点，合胸侧面第2条纹上方间断，第3条纹完整；腹部黑色，第1~7节具黄斑。雌性与雄性相似，后头中央具1个长角。【长度】体长 49~52 mm，腹长 37~40 mm，后翅 30~35 mm。【栖息环境】海拔 500~1500 m森林中的小溪。【分布】中国特有，分布于陕西、湖北、四川、贵州。【飞行期】6—9月。

[Identification] Male face mainly black, frons with a broad stripe. Thorax black, dorsal stripes and collar stripes not connecting, superior spots small, second lateral stripe interrupted above, third lateral stripe complete. Abdomen black, S1-S7 with yellow markings. Female similar to male, occipital margin with a long median horn. **[Measurements]** Total length 49-52 mm, abdomen 37-40 mm, hind wing 30-35 mm. **[Habitat]** Streams in forest at 500-1500 m elevation. **[Distribution]** Endemic to China, recorded from Shaanxi, Hubei, Sichuan, Guizhou. **[Flight Season]** June to September.

长角华春蜓 雄，湖北 | 莫善濂 摄
Sinogomphus scissus, male from Hubei | Photo by Shanlian Mo

修氏华春蜓 *Sinogomphus suensoni* (Lieftinck, 1939)

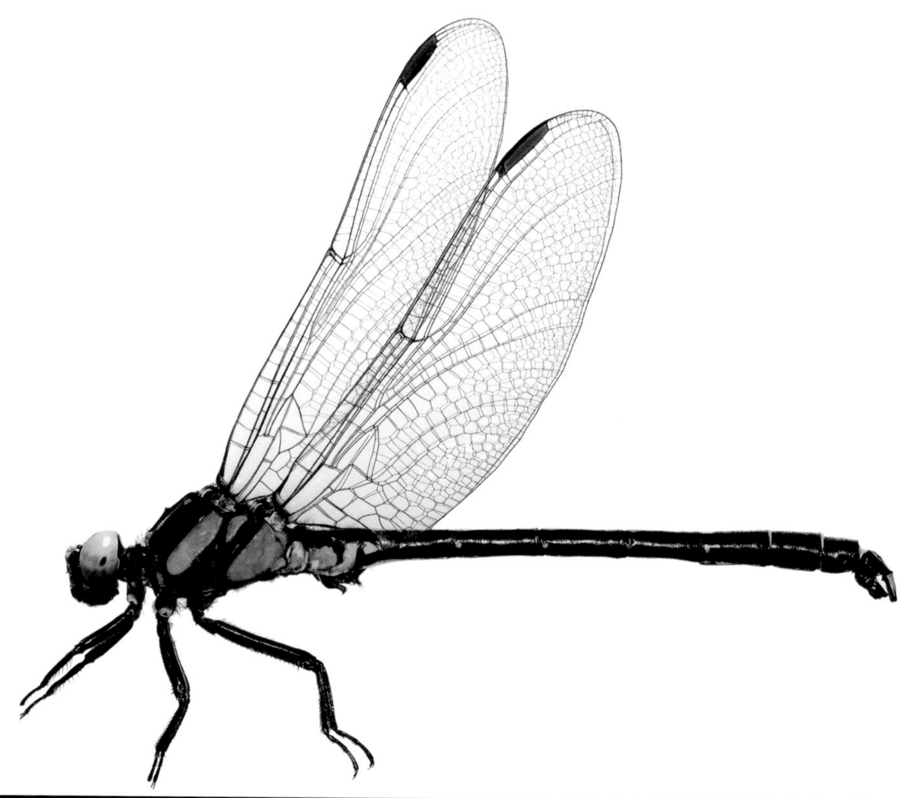

修氏华春蜓 雄，湖北
Sinogomphus suensoni, male from Hubei

【形态特征】雄性面部主要黑色，额横纹甚阔；胸部黑色，背条纹与领条纹不相连，具甚小的肩前上点，合胸侧面第2条纹和第3条纹完整并在下方合并；腹部黑色，第1~9节具黄斑。【长度】体长 48~51 mm，腹长 35~38 mm，后翅 30~34 mm。【栖息环境】海拔 1000~2000 m森林中的小溪。【分布】中国特有，分布于陕西、山西、湖北。【飞行期】6—8月。

[Identification] Male face mainly black, frons with a broad stripe. Thorax black, dorsal stripes and collar stripes not connecting, superior spots small, second and third lateral stripes complete and connecting below. Abdomen black, S1-S9 with yellow markings. **[Measurements]** Total length 48-51 mm, abdomen 35-38 mm, hind wing 30-34 mm. **[Habitat]** Streams in forest at 1000-2000 m elevation. **[Distribution]** Endemic to China, recorded from Shaanxi, Shanxi, Hubei. **[Flight Season]** June to August.

华春蜓属待定种 *Sinogomphus* sp.

【形态特征】雄性面部主要黑色，额横纹甚阔；胸部黑色，背条纹与领条纹不相连，具甚小的肩前上点，合胸侧面第2条纹和第3条纹完整；腹部黑色，第1~7节具黄斑。雌性与雄性相似，后头缘中央具1个长角。【长度】体长 47~48 mm，腹长35~37 mm，后翅30~33 mm。【栖息环境】海拔 2000~2500 m的山区小溪。【分布】云南（大理）。【飞行期】6—8月。

[Identification] Male face mainly black, frons with a broad stripe. Thorax black, dorsal stripes and collar stripes not connecting, superior spots small, second and third lateral stripes complete. Abdomen black, S1-S7 with yellow markings. Female similar to male, occipital margin with a long median horn. **[Measurements]** Total length 47-48 mm, abdomen 35-37 mm, hind wing 30-33 mm. **[Habitat]** Montane streams at 2000-2500 m elevation. **[Distribution]** Yunnan (Dali). **[Flight Season]** June to August.

华春蜓属待定种 雄，云南（大理）
Sinogomphus sp., male from Yunnan (Dali)

华春蜓属待定种 雌，云南（大理）
Sinogomphus sp., female from Yunnan (Dali)

华春蜓属待定种 雄，云南（大理）
Sinogomphus sp., male from Yunnan (Dali)

尖尾春蜓属 Genus *Stylogomphus* Fraser, 1922

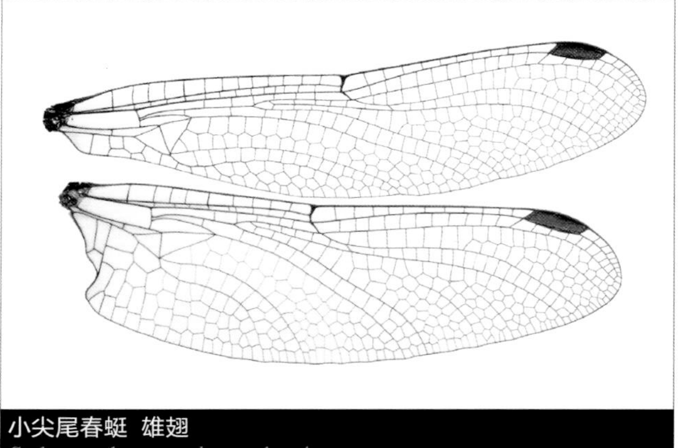

小尖尾春蜓 雄翅
Stylogomphus tantulus, male wings

本属全球已知10余种，广布于东亚、东南亚和北美洲东北部。中国已知8种，主要分布于华南和西南地区。本属蜻蜓是一类体小型的春蜓；翅透明，基方稍染琥珀色，三角室、上三角室和下三角室无横脉，基臀区具1条横脉，臀三角室3室；雄性上肛附器基方粗大，末端尖锐且显著弯曲，下肛附器黑色，中央凹陷；阳茎无鞭。

本属蜻蜓栖息于山区溪流和沟渠。雄性经常停落在水面的大岩石上占据领地，喜欢暴晒于太阳下。

The genus contains over ten species from eastern and southeastern Asia and northeastern part of North America. Eight species are recorded from China, mainly found in the South and Southwest. Species of the genus are small-sized. Wings hyaline, bases slightly tinted with amber brown, no crossvein in triangle, hypertriangle and subtriangle, cubital space with one crossvein, anal triangle 3-celled. Male superior appendages broad at base, tapering to the end, the tips pointed and curved, inferior appendage black with a median hollow. Penis without flagella.

Stylogomphus species inhabit montane streams and ditches. Males usually perch on exposed big rocks above water.

越中尖尾春蜓 雄
Stylogomphus annamensis, male

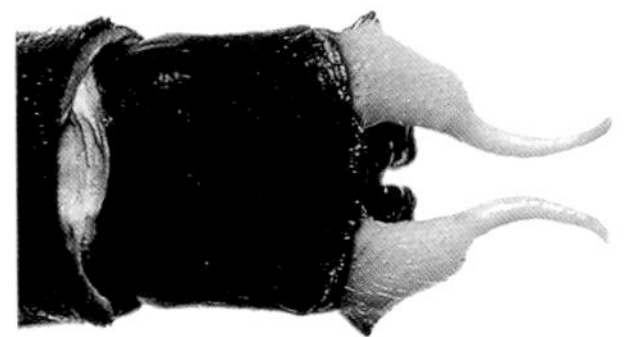

越中尖尾春蜓
Stylogomphus annamensis

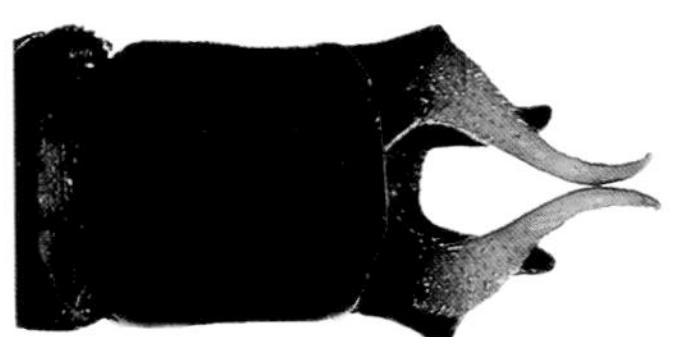

张氏尖尾春蜓
Stylogomphus changi

纯鎏尖尾春蜓
Stylogomphus chunliuae

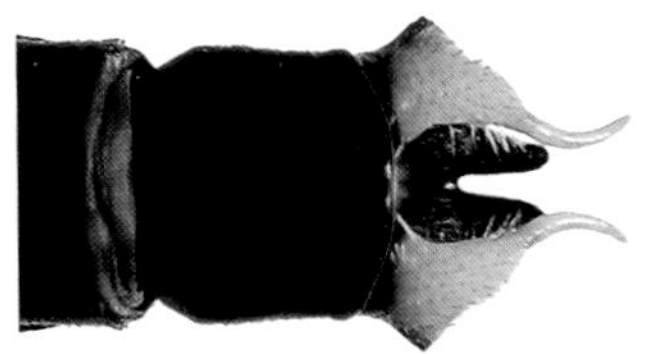

英格尖尾春蜓
Stylogomphus inglisi

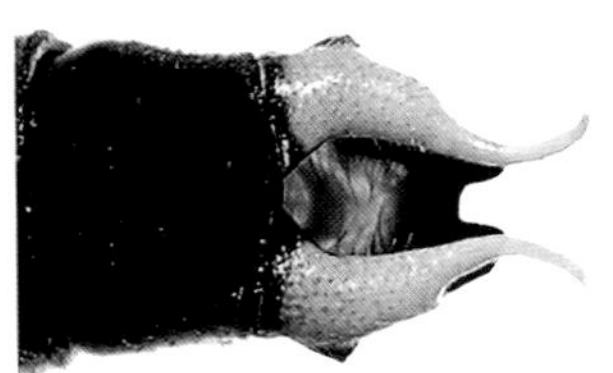

劳伦斯尖尾春蜓
Stylogomphus lawrenceae

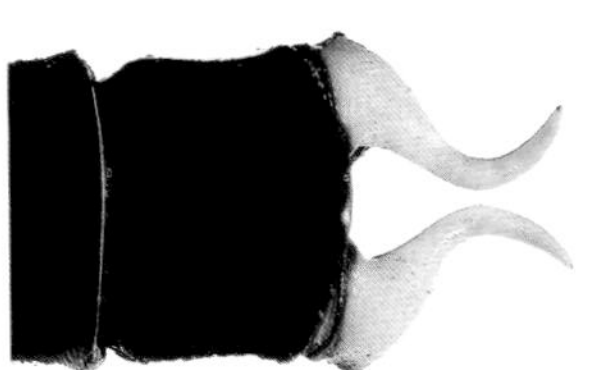

台湾尖尾春蜓
Stylogomphus shirozui

小尖尾春蜓
Stylogomphus tantulus

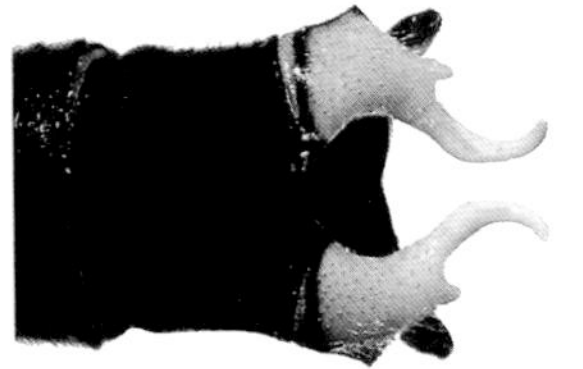

尖尾春蜓属待定种1
Stylogomphus sp. 1

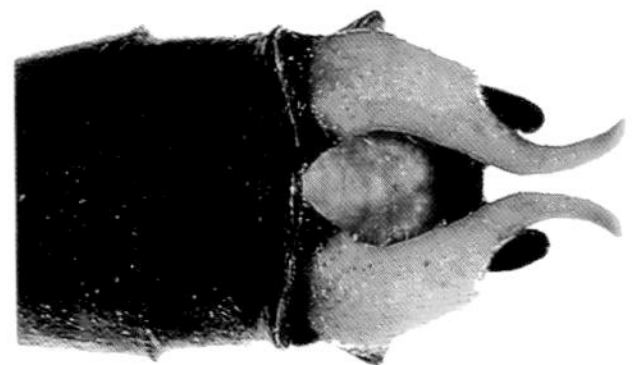

尖尾春蜓属待定种2
Stylogomphus sp. 2

尖尾春蜓属 雄性肛附器
Genus *Stylogomphus*, male anal appendages

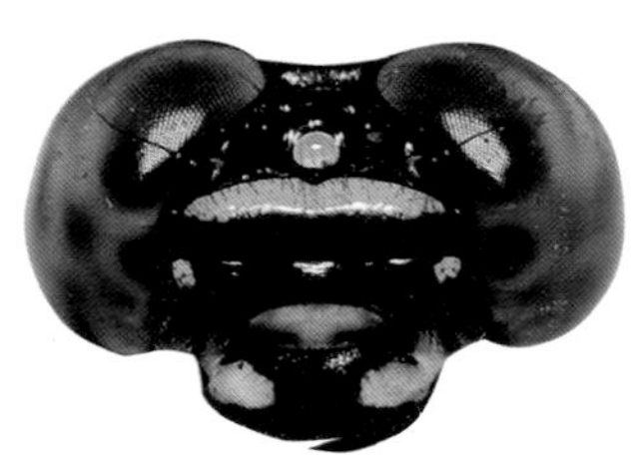

劳伦斯尖尾春蜓 雄
Stylogomphus lawrenceae, male

劳伦斯尖尾春蜓 雌
Stylogomphus lawrenceae, female

越中尖尾春蜓 雌
Stylogomphus annamensis, female

尖尾春蜓属 头部正面观
Genus *Stylogomphus*, head in frontal view

越中尖尾春蜓 *Stylogomphus annamensis* Kompier, 2017

【形态特征】雄性上唇具1对黄斑，额横纹甚阔；胸部黑色，背条纹与领条纹不相连，合胸侧面第2条纹和第3条纹完整并在中央合并；腹部黑色，第1~7节具黄斑。雌性与雄性相似，后头缘中央具1个“Y”形突起。【长度】体长39~43 mm，腹长 30~33 mm，后翅 24~25 mm。【栖息环境】海拔 1000 m以下的山区溪流和沟渠。【分布】云南（红河）、广西；越南。【飞行期】5—7月。

[Identification] Male labrum with a pair of yellow spots, frons with a broad stripe. Thorax black, dorsal stripes and collar stripes not connecting, second and third lateral stripes complete and conjoined medially. Abdomen black, S1-S7

越中尖尾春蜓 雌，广西
Stylogomphus annamensis, female from Guangxi

越中尖尾春蜓 雄，广西
Stylogomphus annamensis, male from Guangxi

越中尖尾春蜓 雄，云南（红河）
Stylogomphus annamensis, male from Yunnan (Honghe)

with yellow markings. Female similar to male, occiput margin with a Y-shaped prominence centrally. [Measurements] Male total length 39-43 mm, abdomen 30-33 mm, hind wing 24-25 mm. [Habitat] Montane streams and ditches below 1000 m elevation. [Distribution] Yunnan (Honghe), Guangxi; Vietnam. [Flight Season] May to July.

张氏尖尾春蜓 *Stylogomphus changi* Asahina, 1968

【形态特征】雄性上唇具1对黄斑，额横纹甚阔；胸部黑色，背条纹与领条纹不相连，合胸侧面第2条纹和第3条纹完整；腹部黑色，第1~2节具黄斑。【长度】体长 44~48 mm，腹长 33~38 mm，后翅 27~29 mm。【栖息环境】海拔 1500 m以下森林中的溪流。【分布】中国台湾特有。【飞行期】4—9月。

[Identification] Male labrum with a pair of yellow spots, frons with a broad stripe. Thorax black, dorsal stripes and collar stripes not connecting, second and third lateral stripes complete. Abdomen black, S1-S2 with yellow markings. [Measurements] Total length 44-48 mm, abdomen 33-38 mm, hind wing 27-29 mm. [Habitat] Streams in forest below 1500 m elevation. [Distribution] Endemic to Taiwan of China. [Flight Season] April to September.

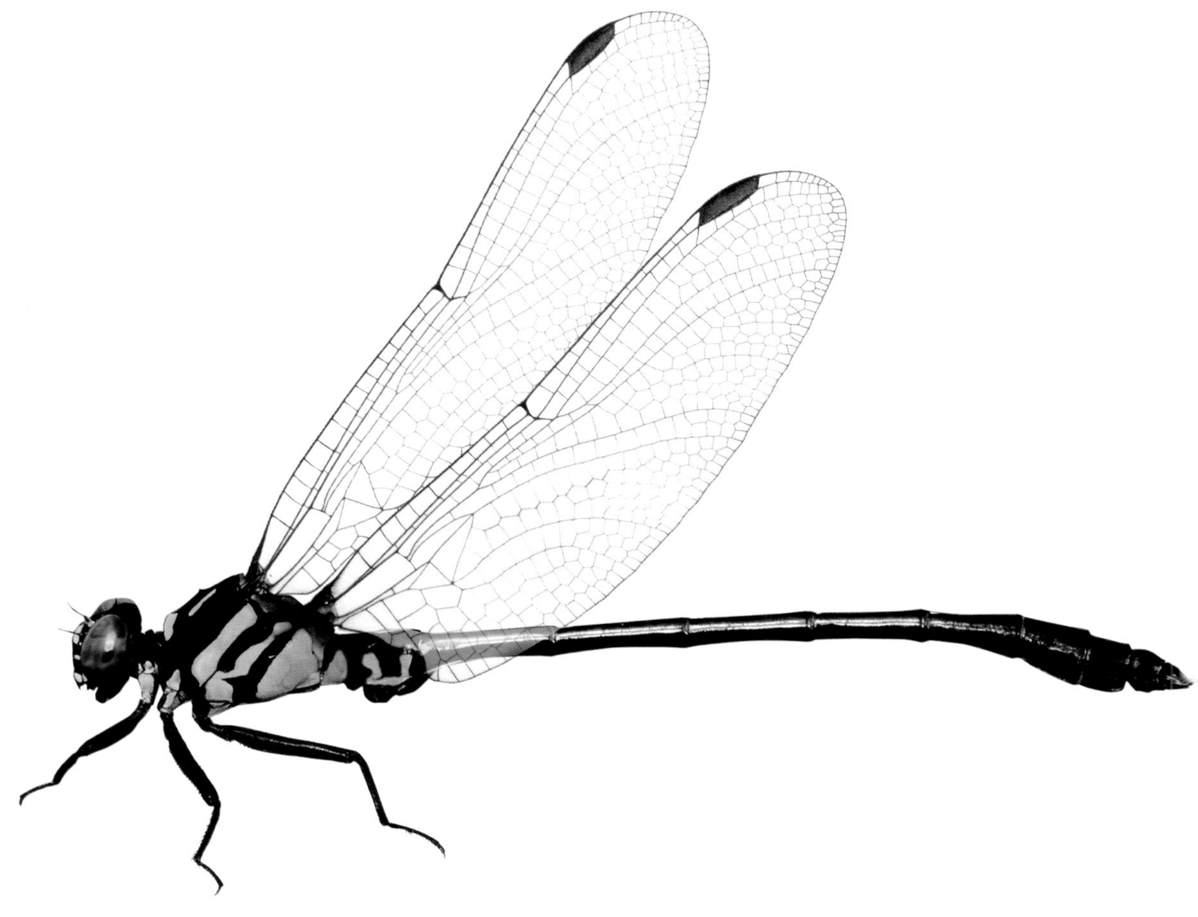

张氏尖尾春蜓 雄，台湾
Stylogomphus changi, male from Taiwan

纯鎏尖尾春蜓 *Stylogomphus chunliuae* Chao, 1954

【形态特征】雄性上唇具1对黄斑，额横纹甚阔；胸部黑色，背条纹与领条纹不相连，合胸侧面第2条纹和第3条纹完整，在中央处合并；腹部黑色，第1～3节具明显的黄斑。【长度】腹长 31～32 mm，后翅 25～28 mm。【栖息环境】海拔 1000 m以下森林中的溪流。【分布】福建、广东、海南、香港；越南的分布记录待确定。【飞行期】4—8月。

纯鎏尖尾春蜓 雄，香港 | 祁麟峰 摄
Stylogomphus chunliuae, male from Hong Kong | Photo by Mahler Ka

[Identification] Male labrum with a pair of yellow spots, frons with a broad stripe. Thorax black, dorsal stripes and collar stripes not connecting, second and third lateral stripes complete and conjoined medially. Abdomen black, S1-S3 with clear yellow spots. **[Measurements]** Abdomen 31-32 mm, hind wing 25-28 mm. **[Habitat]** Streams in forest below 1000 m elevation. **[Distribution]** Fujian, Guangdong, Hainan, Hong Kong; Records from Vietnam need confirmation. **[Flight Season]** April to August.

英格尖尾春蜓 *Stylogomphus inglisi* Fraser, 1922

【形态特征】雄性上唇具1对黄斑，额横纹甚阔；胸部黑色，背条纹与领条纹不相连，合胸侧面第2条纹和第3条纹完整；腹部黑色，第1～7节具黄斑。【长度】雄性体长 34～35 mm，腹长 26～27 mm，后翅 19～21 mm。【栖息环境】海拔 1000 m以下森林中的狭窄小溪。【分布】云南（德宏）；孟加拉国、印度、尼泊尔、缅甸。【飞行期】4—7月。

[Identification] Male labrum with a pair of yellow spots, frons with a broad stripe. Thorax black, dorsal stripes and collar stripes not connecting, second and third lateral stripes complete. Abdomen black, S1-S7 with yellow markings. **[Measurements]** Male total length 34-35 mm, abdomen 26-27 mm, hind wing 19-21 mm. **[Habitat]** Narrow streams in forest below 1000 m elevation. **[Distribution]** Yunnan (Dehong); Bangladesh, India, Nepal, Myanmar. **[Flight Season]** April to July.

英格尖尾春蜓 雄，云南（德宏）
Stylogomphus inglisi, male from Yunnan (Dehong)

劳伦斯尖尾春蜓 *Stylogomphus lawrenceae* Yang & Davies, 1996

劳伦斯尖尾春蜓 雄，云南（西双版纳）
Stylogomphus lawrenceae, male from Yunnan (Xishuangbanna)

劳伦斯尖尾春蜓 雌，云南（西双版纳）
Stylogomphus lawrenceae, female from Yunnan (Xishuangbanna)

劳伦斯尖尾春蜓 雄，云南（西双版纳）
Stylogomphus lawrenceae, male from Yunnan (Xishuangbanna)

【形态特征】雄性上唇具1对黄斑，额横纹甚阔；胸部黑色，背条纹与领条纹不相连，合胸侧面第2条纹和第3条纹完整；腹部黑色，第1~7节具黄斑。雌性黑色具黄斑，后头缘中央具1个鲸尾状突起。【长度】体长 33~37 mm，腹长 24~27 mm，后翅 20~22 mm。【栖息环境】海拔 1000 m以下森林中的狭窄小溪。【分布】云南（西双版纳、红河）；老挝。【飞行期】4—7月。

[Identification] Male labrum with a pair of yellow spots, frons with a broad stripe. Thorax black, dorsal stripes and collar stripes not connecting, second and third lateral stripes complete. Abdomen black, S1-S7 with yellow markings. Female black with yellow markings, occipital margin with a whale tail like prominence centrally. **[Measurements]** Total length 33-37 mm, abdomen 24-27 mm, hind wing 20-22 mm. **[Habitat]** Narrow streams in forest below 1000 m elevation. **[Distribution]** Yunnan (Xishuangbanna, Honghe); Laos. **[Flight Season]** April to July.

台湾尖尾春蜓 *Stylogomphus shirozui* Asahina, 1966

【形态特征】雄性上唇具1对黄斑，额横纹甚阔；胸部黑色，背条纹与领条纹不相连，合胸侧面第2条纹上方缺失，第3条纹“Y”形；腹部黑色，第1~7节具黄斑。【长度】体长 45~50 mm，腹长 32~37 mm，后翅 23~27 mm。【栖息环境】海拔 1000 m以下森林中的小溪。【分布】中国台湾；日本。【飞行期】3—9月。

[Identification] Male labrum with a pair of yellow spots, frons with a broad stripe. Thorax black, dorsal stripes and collar stripes not connecting, second lateral stripe absent above, third lateral stripe Y-shaped. Abdomen black, S1-S7 with yellow markings. [Measurements] Total length 45-50 mm, abdomen 32-37 mm, hind wing 23-27 mm. [Habitat] Streams in forest below 1000 m elevation. [Distribution] Taiwan of China; Japan. [Flight Season] March to September.

台湾尖尾春蜓 雄，台湾 | 嘎嘎 摄
Stylogomphus shirozui, male from Taiwan | Photo by Gaga

小尖尾春蜓 *Stylogomphus tantulus* Chao, 1954

【形态特征】雄性上唇大面积黄色，额横纹甚阔；胸部黑色，背条纹与领条纹不相连，合胸侧面第2条纹和第3条纹完整；腹部黑色，第1~7节具黄斑。雌性与雄性相似，但腹部较短。【长度】体长 42~43 mm，腹长 32~33 mm，后翅 23~27 mm。【栖息环境】海拔 1500 m以下森林中的开阔溪流。【分布】中国特有，分布于河南、贵州、浙江、江西、福建、广东。【飞行期】5—8月。

小尖尾春蜓 雌，贵州
Stylogomphus tantulus, female from Guizhou

[Identification] Male labrum largely yellow, frons with a broad stripe. Thorax black, dorsal stripes and collar stripes not connecting, second and third lateral stripes complete. Abdomen black, S1-S7 with yellow markings. Female similar to male, abdomen shorter. **[Measurements]** Total length 42-43 mm, abdomen 32-33 mm, hind wing 23-27 mm. **[Habitat]** Exposed streams in forest below 1500 m elevation. **[Distribution]** Endemic to China, recorded from Henan, Guizhou, Zhejiang, Jiangxi, Fujian, Guangdong. **[Flight Season]** May to August.

小尖尾春蜓 雄，贵州
Stylogomphus tantulus, male from Guizhou

尖尾春蜓属待定种1 *Stylogomphus* sp. 1

尖尾春蜓属待定种1 雄，云南（普洱）
Stylogomphus sp. 1, male from Yunnan (Pu'er)

【形态特征】雄性上唇具1对黄斑，额横纹甚阔；胸部黑色，背条纹与领条纹不相连，合胸侧面第2条纹和第3条纹完整；腹部黑色，第1~7节具黄斑。【长度】雄性体长 42 mm，腹长 32 mm，后翅 25 mm。【栖息环境】海拔1000~1500 m森林中的开阔溪流。【分布】云南（普洱）。【飞行期】5—7月。

[Identification] Male labrum with a pair of yellow spots, frons with a broad stripe. Thorax black, dorsal stripes and collar stripes not connecting, second and third lateral stripes complete. Abdomen black, S1-S7 with yellow markings. **[Measurements]** Male total length 42 mm, abdomen 32 mm, hind wing 25 mm. **[Habitat]** Exposed streams in forest at 1000-1500 m elevation. **[Distribution]** Yunnan (Pu'er). **[Flight Season]** May to July.

尖尾春蜓属待定种1 雄，云南（普洱）
Stylogomphus sp. 1, male from Yunnan (Pu'er)

尖尾春蜓属待定种2 *Stylogomphus* sp. 2

【形态特征】雄性上唇具1对黄斑，额横纹甚阔；胸部黑色，背条纹与领条纹不相连，合胸侧面第2条纹和第3条纹完整；腹部黑色，第1~7节具黄斑。雌性与雄性相似，后头缘具1对角状突起。【长度】体长 39~42 mm，腹长 30~32 mm，后翅 23~25 mm。【栖息环境】海拔 1000 m以下森林中的溪流。【分布】广东。【飞行期】3—7月。

[Identification] Male labrum with a pair of yellow spots, frons with a broad stripe. Thorax black, dorsal stripes and collar stripes not connecting, second and third lateral stripes complete. Abdomen black, S1-S7 with yellow markings.

尖尾春蜓属待定种2 雄，广东 | 宋睿斌 摄
Stylogomphus sp. 2, male from Guangdong | Photo by Ruibin Song

尖尾春蜓属待定种2 雌，广东 | 宋睿斌 摄
Stylogomphus sp. 2, female from Guangdong | Photo by Ruibin Song

尖尾春蜓属待定种2 雄，广东 | 宋睿斌 摄
Stylogomphus sp. 2, male from Guangdong | Photo by Ruibin Song

Female similar to male, occipital margin with a pair of horns. **[Measurements]** Total length 39-42 mm, abdomen 30-32 mm, hind wing 23-25 mm. **[Habitat]** Streams in forest below 1000 m elevation. **[Distribution]** Guangdong. **[Flight Season]** March to July.

扩腹春蜓属 Genus *Stylurus* Needham, 1897

本属全球已知约30种，分布于北美洲和欧亚大陆。中国已知10余种，全国广布。本属蜻蜓是体中型的春蜓；翅透明，三角室、上三角室和下三角室无横脉，基臀区具1条横脉，臀圈开放，臀三角室3室；雄性腹部第7~9节显著膨大，肛附器构造简单，下肛附器与上肛附器的长度和分歧角度都几乎相同；阳茎具1对短鞭。

本属蜻蜓比较偏爱大型河流和溪流。雄性经常停落在河岸带茂盛的植被丛中，时而飞到河面低空定点悬停。

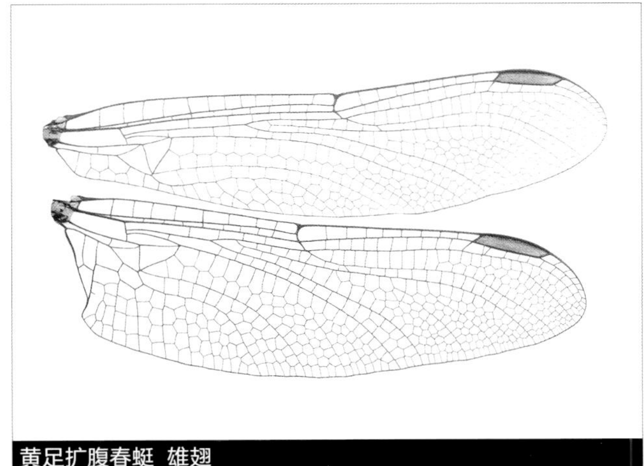

黄足扩腹春蜓 雄翅
Stylurus flavipes, male wings

The genus contains about 30 species distributed in North America and Eurasia. Over ten species are recorded from China, widespread throughout the country. Species of the genus are medium-sized. Wings hyaline, no crossvein in triangle, hypertriangle and subtriangle, cubital space with one crossvein, anal loop open, anal triangle 3-celled. Male S7-S9 expanded laterally, anal appendages simple, the superiors and inferior appendage of the same length and divergent in same angle. Penis with a pair of flagella.

Stylurus species inhabit large rivers and streams. Males usually seen in the marginal vegetation and sometimes hover above water.

长节扩腹春蜓
Stylurus amicus

黄角扩腹春蜓
Stylurus flavicornis

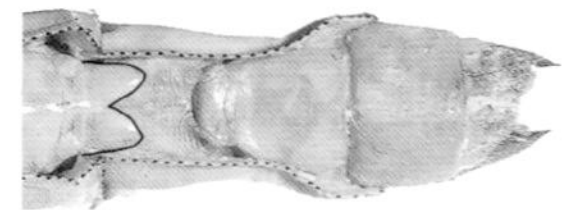

黄足扩腹春蜓
Stylurus flavipes

扩腹春蜓属 雌性下生殖板
Genus *Stylurus*, female vulvar lamina

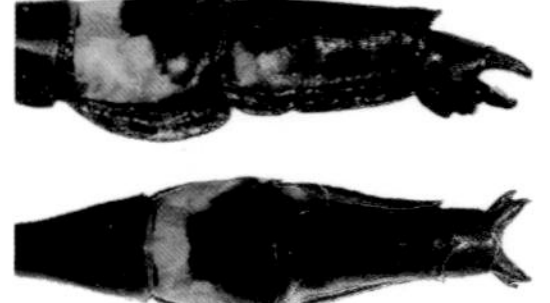

长节扩腹春蜓
Stylurus amicus

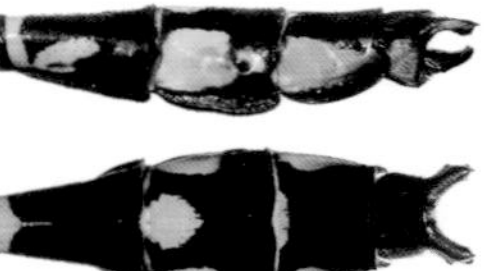

黄角扩腹春蜓
Stylurus flavicornis

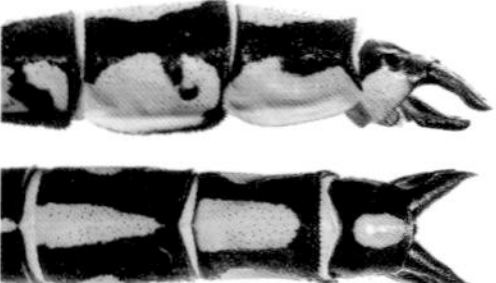

黄足扩腹春蜓
Stylurus flavipes

南宁扩腹春蜓
Stylurus nanningensis

扩腹春蜓属 雄性肛附器
Genus *Stylurus*, male anal appendages

长节扩腹春蜓 雄
Stylurus amicus, male

长节扩腹春蜓 *Stylurus amicus* (Needham, 1930)

【形态特征】雄性上唇具1条黄纹，额横纹甚阔；胸部黑色，背条纹与领条纹不相连，肩前条纹较长，合胸侧面第2条纹和第3条纹完整；腹部黑色具黄斑，第7～9节背板侧缘扩大，第9节较长。雌性与雄性相似。【长度】体长64～70 mm，腹长 48～55 mm，后翅 38～43 mm。【栖息环境】海拔 1000 m以下的开阔溪流和宽阔河流。【分布】四川、贵州、福建、广西、广东、海南；越南。【飞行期】3—7月。

[Identification] Male labrum with a yellow stripe, frons with a broad stripe. Thorax black, dorsal stripes and collar stripes not connecting, antehumeral stripes long, second and third lateral stripes complete. Abdomen black with yellow markings, S7-S9 expanded laterally, S9 long. Female similar to male. **[Measurements]** Total length 64-70 mm, abdomen 48-55 mm, hind wing 38-43 mm. **[Habitat]** Exposed streams and large rivers below 1000 m elevation. **[Distribution]** Sichuan, Guizhou, Fujian, Guangxi, Guangdong, Hainan; Vietnam. **[Flight Season]** March to July.

长节扩腹春蜓 雄，海南
Stylurus amicus, male from Hainan

长节扩腹春蜓 雌，广东 | 宋睿斌 摄
Stylurus amicus, female from Guangdong | Photo by Ruibin Song

长节扩腹春蜓 雄，海南
Stylurus amicus, male from Hainan

黄角扩腹春蜓 *Stylurus flavicornis* (Needham, 1931)

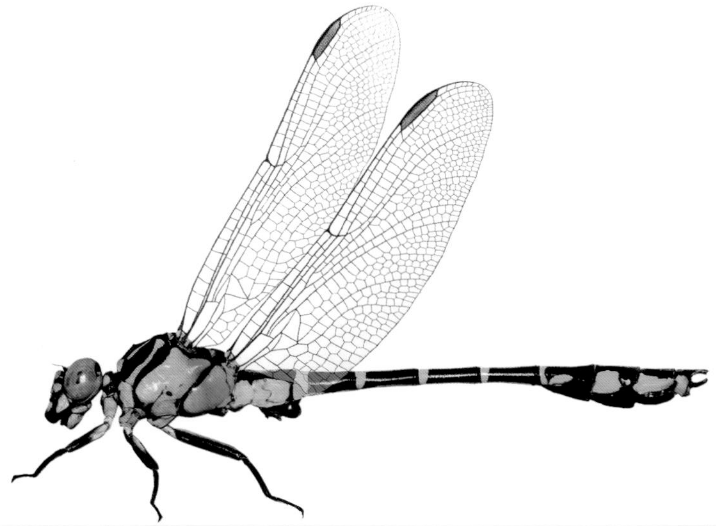

黄角扩腹春蜓 雄，黑龙江
Stylurus flavicornis, male from Heilongjiang

黄角扩腹春蜓 雌，黑龙江
Stylurus flavicornis, female from Heilongjiang

【形态特征】雄性面部黑色具黄斑，上唇和前唇基黄色，额横纹甚阔；胸部黑色，背条纹与领条纹不相连，肩前条纹较长，合胸侧面第2条纹缺失，第3条纹完整；腹部黑色具黄斑，第7～9节背板侧缘扩大。雌性与雄性相似，腹部末端未膨大。【长度】体长 55～64 mm，腹长 40～45 mm，后翅 32～37 mm。【栖息环境】海拔 500 m以下的宽阔河流。【分布】中国特有，分布于黑龙江、福建。【飞行期】6—8月。

[Identification] Male face black with yellow markings, labrum and anteclypeus yellow, frons with a broad stripe. Thorax black, dorsal stripes and collar stripes not connecting, antehumeral stripes long, second lateral stripe absent, third lateral stripe complete. Abdomen black with yellow markings, S7-S9 expanded laterally. Female similar to male, abdominal tip not expanded. [Measurements] Total length 55-64 mm, abdomen 40-45 mm, hind wing 32-37 mm. [Habitat] Large rivers below 500 m elevation. [Distribution] Endemic to China, recorded from Heilongjiang, Fujian. [Flight Season] June to August.

黄足扩腹春蜓 *Stylurus flavipes* (Charpentier, 1825)

【形态特征】雄性面部主要黄色，头顶黑色；胸部背条纹与领条纹不相连，肩前条纹较长，合胸侧面第2条纹上方缺失，第3条纹完整；腹部黑色具丰富的黄斑。雌性与雄性相似。【长度】体长 46~51 mm，腹长 33~37 mm，后翅 30~32 mm。【栖息环境】海拔 500 m以下的宽阔河流。【分布】黑龙江、吉林、陕西；欧洲至中国东北地区广布。【飞行期】6—8月。

[Identification] Male face mainly yellow, vertex black. Thorax with dorsal stripes and collar stripes not connecting, antehumeral stripes long, second lateral stripe absent above, third lateral stripe complete. Abdomen black with abundant yellow markings. Female similar to male. [Measurements] Total length 46-51 mm, abdomen 33-37 mm, hind wing 30-32 mm. [Habitat] Large rivers below 500 m elevation. [Distribution] Heilongjiang, Jilin, Shaanxi; Widespread from Europe to Northeast China. [Flight Season] June to August.

黄足扩腹春蜓 雄，吉林 | 金洪光 摄
Stylurus flavipes, male from Jilin | Photo by Hongguang Jin

黄足扩腹春蜓 雌，黑龙江 | 莫善濂 摄
Stylurus flavipes, female from Heilongjiang | Photo by Shanlian Mo

克雷扩腹春蜓 *Stylurus kreyenbergi* (Ris, 1928)

【形态特征】雄性面部主要黑色，上唇和后头黄色；胸部黑色，背条纹与领条纹不相连，肩前条纹较长，合胸侧面第2条纹和第3条纹完整，有时第2条纹上方缺失；腹部黑色具黄斑，第7～9节背板侧缘扩大。【长度】雄性腹长 40 mm，后翅 30 mm。【栖息环境】海拔 1000 m以下的开阔溪流和宽阔河流。【分布】山东、四川、浙江、江西、广东、香港；越南。【飞行期】4—8月。

[Identification] Male face mainly black, labrum and occiput yellow. Thorax black, dorsal stripes and collar stripes not connecting, antehumeral stripes long, second and third lateral stripes complete, second stripes sometimes absent above. Abdomen black with yellow markings, S7-S9 expanded laterally. **[Measurements]** Male abdomen 40 mm, hind wing 30 mm. **[Habitat]** Exposed streams and large rivers below 1000 m elevation. **[Distribution]** Shandong, Sichuan, Zhejiang, Jiangxi, Guangdong, Hong Kong; Vietnam. **[Flight Season]** April to August.

克雷扩腹春蜓 雄，广东 | 吴宏道 摄
Stylurus kreyenbergi, male from Guangdong | Photo by Hongdao Wu

南宁扩腹春蜓 *Stylurus nanningensis* Liu, 1985

【形态特征】雄性上唇具1对黄斑，额横纹甚阔；胸部黑色，背条纹与领条纹不相连，肩前条纹较长，合胸侧面第2条纹和第3条纹完整；腹部黑色具黄斑，第7～9节背板侧缘扩大。雌性与雄性相似。【长度】体长 55～60 mm，腹长 42～45 mm，后翅 36～38 mm。【栖息环境】海拔 500 m以下的开阔溪流和宽阔河流。【分布】中国特有，分布于广西、广东。【飞行期】5—8月。

[Identification] Male labrum with a pair of yellow spots, frons with a broad stripe. Thorax black, dorsal stripes and collar stripes not connecting, antehumeral stripes long, second and third lateral stripes complete. Abdomen black with yellow markings, S7-S9 expanded laterally. Female similar to male. **[Measurements]** Total length 55-60 mm, abdomen 42-45 mm, hind wing 36-38 mm. **[Habitat]** Exposed streams and large rivers below 500 m elevation. **[Distribution]** Endemic to China, recorded from Guangxi, Guangdong. **[Flight Season]** May to August.

南宁扩腹春蜓 雄，广东 | 宋睿斌 摄
Stylurus nanningensis, male from Guangdong | Photo by Ruibin Song

南宁扩腹春蜓 雌，广东 | 宋睿斌 摄
Stylurus nanningensis, female from Guangdong | Photo by Ruibin Song

棘尾春蜓属 Genus *Trigomphus* Bartenev, 1911

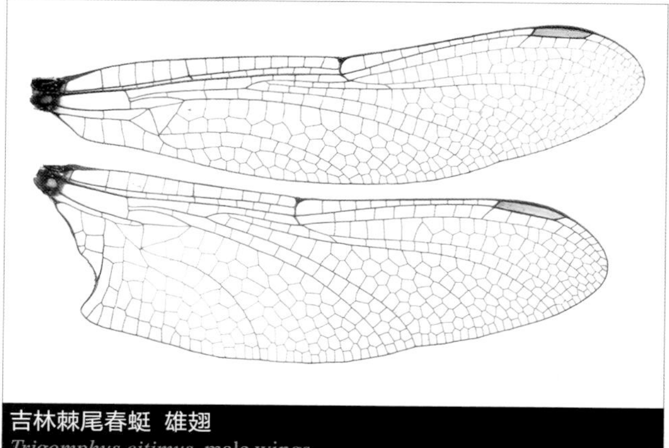

吉林棘尾春蜓 雄翅
Trigomphus citimus, male wings

本属全球已知约15种，主要分布于亚洲的温带和亚热带区域。中国已知10种，除西北地区全国广布。本属蜻蜓是体中型的春蜓；翅透明，三角室、上三角室和下三角室无横脉，基臀区具1~2条横脉，臀圈开放；本属雄性上肛附器大面积白色或灰白色，内缘和外缘具齿状突起，下肛附器稍短于上肛附器，两枝分歧显著。

本属蜻蜓多栖息于水草茂盛的湿地。雄性经常停落在叶片或者地面上占据领地。

The genus contains about 15 species distributed in temperate and subtropical Asia. Ten species are recorded from China, widespread throughout the country except from the Northwest. Species of the genus are medium-sized. Wings hyaline, no crossvein in triangle, hypertriangle and subtriangle, cubital space with 1-2 crossveins, anal loop open. Male superior appendages largely white or grayish white, the inner and outer margin with teeth, inferior appendage slightly shorter, two branches divergent.

Trigomphus species usually inhabit well vegetated wetlands. Territorial males usually perch on the leaves or ground.

野居棘尾春蜓
Trigomphus agricola

黄唇棘尾春蜓
Trigomphus beatus

棘尾春蜓属 雄性肛附器
Genus *Trigomphus*, male anal appendages

野居棘尾春蜓 交尾
Trigomphus agricola, mating pair

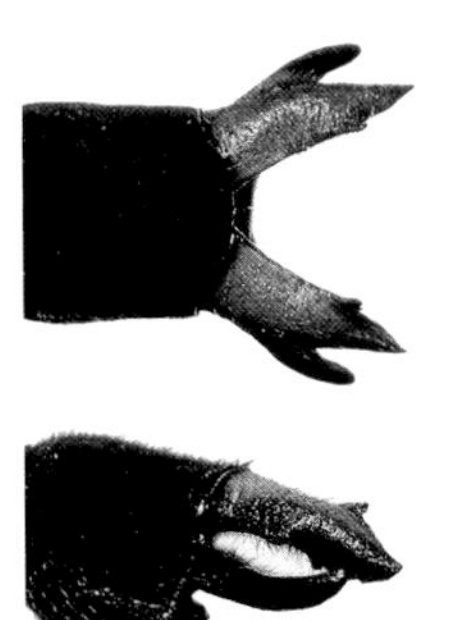

吉林棘尾春蜓
Trigomphus citimus

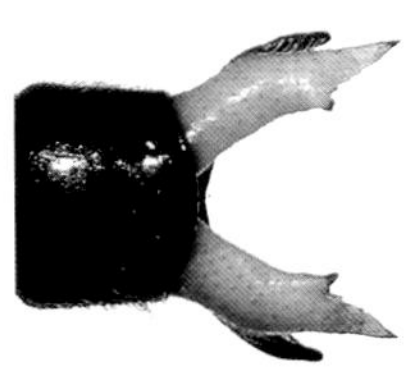

海南棘尾春蜓
Trigomphus hainanensis

净棘尾春蜓
Trigomphus lautus

棘尾春蜓属 雄性肛附器
Genus *Trigomphus*, male anal appendages

野居棘尾春蜓 *Trigomphus agricola* (Ris, 1916)

【形态特征】雄性复眼蓝色，面部大面积黄色，头顶黑色；胸部黑色，背条纹与领条纹相连，肩前条纹细长，有时间断，合胸侧面第2条纹大面积缺失，第3条纹完整；腹部黑色具黄白色斑，上肛附器上面白色，下面黑色，下肛附器黑色。雌性黑色具黄斑。【长度】体长 42~45 mm，腹长 31~33 mm，后翅 24~26 mm。【栖息环境】海拔 500 m 以下的池塘和流速缓慢的溪流。【分布】中国特有，湖北、安徽、江苏、浙江、福建。【飞行期】3—5月。

野居棘尾春蜓 雄，湖北
Trigomphus agricola, male from Hubei

[Identification] Male eyes blue, face largely yellow, vertex black. Thorax black, dorsal stripes and collar stripes connecting, antehumeral stripes long and sometimes interrupted, second lateral stripe largely absent, third lateral stripe complete. Abdomen black with yellowish white markings, superior appendages white above and black below, inferior appendage black. Female black with yellow markings. **[Measurements]** Total length 42-45 mm, abdomen 31-33 mm, hind wing 24-26 mm. **[Habitat]** Ponds and slow flowing streams below 500 m elevation. **[Distribution]** Endemic to China, recorded from Hubei, Anhui, Jiangsu, Zhejiang, Fujian. **[Flight Season]** March to May.

野居棘尾春蜓 雌，湖北
Trigomphus agricola, female from Hubei

黄唇棘尾春蜓 *Trigomphus beatus* Chao, 1954

黄唇棘尾春蜓 雄，湖北
Trigomphus beatus, male from Hubei

【形态特征】雄性复眼蓝色，上唇白色，额横纹甚阔；胸部黑色，背条纹与领条纹相连，具较小的肩前上点，合胸侧面第2条纹上方缺失，下方与第3条纹合并，第3条纹完整；腹部黑色具黄斑，上肛附器基方1/2白色，下肛附器黑色。雌性与雄性相似。【长度】体长 41～43 mm，腹长 29～32 mm，后翅 23～24 mm。【栖息环境】海拔 500 m以下的池塘和水库。【分布】中国特有，湖北、湖南、福建、广西。【飞行期】3—4月。

[Identification] Male eyes blue, labrum white, frons with a broad stripe. Thorax black, dorsal stripes and collar stripes connecting, superior spots small, second lateral stripe absent above but connecting with complete third lateral stripe below. Abdomen black with yellow markings, superior appendages with complete basal half white, inferior appendage black. Female similar to male. **[Measurements]** Total length 41-43 mm, abdomen 29-32 mm, hind wing 23-24 mm. **[Habitat]** Ponds and reservoirs below 500 m elevation. **[Distribution]** Endemic to China, recorded from Hubei, Hunan, Fujian, Guangxi. **[Flight Season]** March to April.

黄唇棘尾春蜓 雌，湖北
Trigomphus beatus, female from Hubei

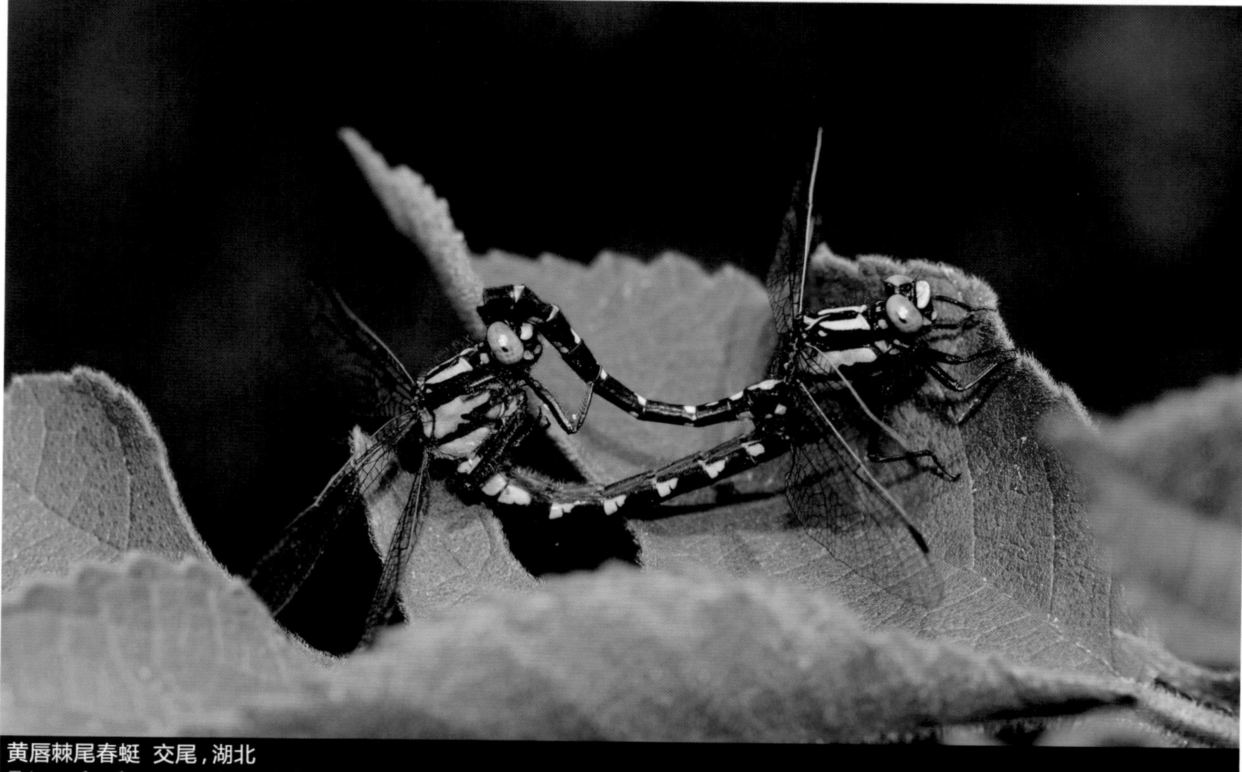

黄唇棘尾春蜓 交尾，湖北
Trigomphus beatus, mating pair from Hubei

吉林棘尾春蜓 *Trigomphus citimus* (Needham, 1931)

【形态特征】雄性复眼绿色，上唇黄色，额横纹甚阔；胸部黑色，背条纹与领条纹相连，具肩前上点和甚细的肩前下条纹，合胸侧面第2条纹大面积缺失，第3条纹完整；腹部黑色，第1～8节具灰白色斑，上肛附器上面白色，下面黑色，下肛附器黑色。雌性黑色具黄斑。【长度】体长 44～48 mm，腹长 32～35 mm，后翅 26～29 mm。【栖息环境】海拔 500 m以下的池塘和流速缓慢的溪流。【分布】黑龙江、吉林；俄罗斯远东、朝鲜半岛、日本。【飞行期】5—7月。

[Identification] Male eyes green, labrum yellow, frons with a broad stripe. Thorax black, dorsal stripes and collar stripes connecting, superior spots and narrow antehumeral stripes present, second lateral stripe largely absent, third lateral stripe complete. Abdomen black, S1-S8 with grayish white markings, superior appendages white above and black below, inferior appendage black. Female black with yellow markings. **[Measurements]** Total length 44-48 mm, abdomen 32-35 mm, hind wing 26-29 mm. **[Habitat]** Ponds and slow flowing streams below 500 m elevation. **[Distribution]** Heilongjiang, Jilin; Russian Far East, Korean peninsula, Japan. **[Flight Season]** May to July.

吉林棘尾春蜓 雄，黑龙江 | 莫善濂 摄
Trigomphus citimus, male from Heilongjiang | Photo by Shanlian Mo

吉林棘尾春蜓 雌，黑龙江 | 莫善濂 摄
Trigomphus citimus, female from Heilongjiang | Photo by Shanlian Mo

吉林棘尾春蜓 交尾，黑龙江 | 莫善濂 摄
Trigomphus citimus, mating pair from Heilongjiang | Photo by Shanlian Mo

海南棘尾春蜓 *Trigomphus hainanensis* Zhang & Tong, 2009

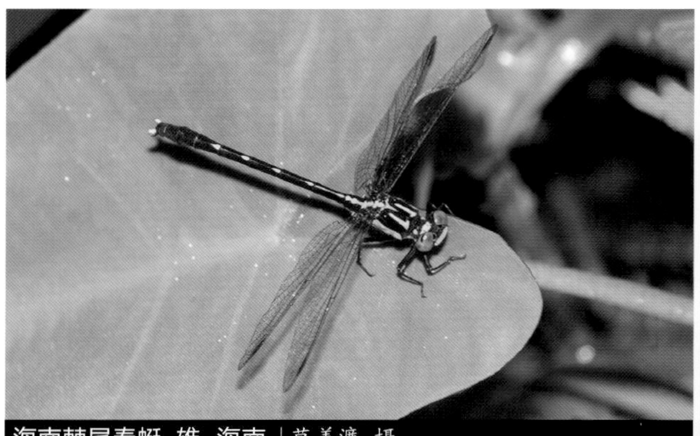

海南棘尾春蜓 雄，海南 | 莫善濂 摄
Trigomphus hainanensis, male from Hainan | Photo by Shanlian Mo

海南棘尾春蜓 雌，海南 | 莫善濂 摄
Trigomphus hainanensis, female from Hainan | Photo by Shanlian Mo

海南棘尾春蜓 雄，海南 | 莫善濂 摄
Trigomphus hainanensis, male from Hainan | Photo by Shanlian Mo

【形态特征】雄性复眼绿色，上唇黄色，中央具1个黑斑，额横纹甚阔；胸部黑色，背条纹与领条纹相连，具甚小的肩前上点，合胸侧面第2条纹上方缺失，第3条纹完整；腹部黑色，第1~7节具黄斑，上肛附器白色，下肛附器黑色。雌性与雄性相似，上唇黑色具1对小黄斑，后头缘具1个长角。【长度】体长 39~43mm，腹长 29~33 mm，后翅 24~26 mm。【栖息环境】海拔 1000 m以下森林中的小型池塘。【分布】中国海南特有。【飞行期】3—4月。

[Identification] Male eyes green, labrum yellow with a median black spot, frons with a broad stripe. Thorax black, dorsal stripes and collar stripes connecting, superior spots small, second lateral stripe absent above, third lateral stripe complete. Abdomen black, S1-S7 with yellow markings, superior appendages white, inferior appendage black. Female similar to male, labrum with a pair of small yellow spots, occipital margin with a long horn. **[Measurements]** Total length 39-43 mm, abdomen 29-33 mm, hind wing 24-26 mm. **[Habitat]** Small ponds in forest below 1000 m elevation. **[Distribution]** Endemic to Hainan of China. **[Flight Season]** March to April.

净棘尾春蜓 *Trigomphus lautus* (Needham, 1931)

【形态特征】雄性复眼绿色，上唇具1对黄斑，额横纹甚阔；胸部黑色，背条纹与领条纹相连，具甚小的肩前上点，合胸侧面第2条纹和第3条纹缺失；腹部黑色，第1～7节具黄斑，上肛附器白色，下肛附器黑色。雌性与雄性相似。【长度】体长 44～48 mm，腹长 33～37 mm，后翅 24～27 mm。【栖息环境】海拔 500 m以下的沼泽地、池塘和水库。【分布】中国特有，分布于福建、广东。【飞行期】3—5月。

[Identification] Male eyes green, labrum with a pair of yellow spots, frons with a broad stripe. Thorax black, dorsal stripes and collar stripes connecting, superior spots small, second and third lateral stripes absent. Abdomen black, S1-S7 with yellow markings, superior appendages white, inferior appendage black. Female similar to male. **[Measurements]** Total length 44-48 mm, abdomen 33-37 mm, hind wing 24-27 mm. **[Habitat]** Marshes, ponds and reservoirs below 500 m elevation. **[Distribution]** Endemic to China, recorded from Fujian, Guangdong. **[Flight Season]** March to May.

净棘尾春蜓 雄，广东 | 宋黎明 摄
Trigomphus lautus, male from Guangdong | Photo by Liming Song

净棘尾春蜓 雌，广东 | 吴宏道 摄
Trigomphus lautus, female from Guangdong | Photo by Hongdao Wu

黑足棘尾春蜓 *Trigomphus nigripes* (Selys, 1887)

【形态特征】雄性复眼绿色，面部大面积黑色，上唇黄色，额横纹甚阔；胸部黑色，背条纹与领条纹、具肩前上点相连，合胸侧面第2条纹大面积缺失，第3条纹完整；腹部黑色具灰白色斑。雌性黑色具黄斑。【长度】不详。【栖息环境】海拔 500 m以下的湿地。【分布】吉林、辽宁；俄罗斯远东、朝鲜半岛。【飞行期】5—7月。

[Identification] Male eyes green, face largely black, labrum yellow, frons with a broad stripe. Thorax black, dorsal stripes connecting with collar stripes and superior spots, second lateral stripe largely absent, third lateral stripe complete. Abdomen black with grayish white markings. Female black with yellow markings. **[Measurements]** Unknown. **[Habitat]** Wetlands below 500 m elevation. **[Distribution]** Jilin, Liaoning; Russian Far East, Korean peninsula. **[Flight Season]** May to July.

黑足棘尾春蜓 雄，俄罗斯 | Oleg E. Kosterin 摄
Trigomphus nigripes, male from Russia | Photo by Oleg E. Kosterin

黑足棘尾春蜓 雌，俄罗斯 | Oleg E. Kosterin 摄
Trigomphus nigripes, female from Russia | Photo by Oleg E. Kosterin

斯氏棘尾春蜓 *Trigomphus svenhedini* Sjöstedt, 1932

【形态特征】雄性复眼绿色，上唇具1对黄斑，额横纹甚阔；胸部黑色，背条纹与领条纹相连，具甚小的肩前上点，合胸侧面第2条纹和第3条纹缺失；腹部黑色具黄斑。雌性与雄性相似。【长度】腹长33～37 mm，后翅29～31 mm。【栖息环境】沼泽地、池塘和水库。【分布】中国四川特有。【飞行期】4—5月。

[Identification] Male eyes green, labrum with a pair of yellow spots, frons with a broad stripe. Thorax black, dorsal stripes and collar stripes connecting, superior spots small, second and third lateral stripes absent. Abdomen black with yellow markings. Female similar to male. **[Measurements]** Abdomen 33-37 mm, hind wing 29-31 mm. **[Habitat]** Marshes, ponds and reservoirs. **[Distribution]** Endemic to Sichuan of China. **[Flight Season]** April to May.

斯氏棘尾春蜓 雄，四川 | 陈尽 摄
Trigomphus svenhedini, male from Sichuan | Photo by Jin Chen

3 裂唇蜓科 Family Chlorogomphidae

裂唇蜓被认为是一类较古老的蜻蜓类群，仅在亚洲分布，东洋界为分布中心。目前全世界已知3属50余种，但仍将有大量的新种被发现。本科中国已知3属30余种，在中国南方分布广泛。本科体大型，身体黑褐色具黄色条纹；头部正面观呈椭圆形，两复眼在头顶稍微分离，额高度隆起；翅宽阔，许多种类雌性的翅具有2种色型，一种是大面积透明，另一种则是染有大面积的黄色、橙色、黑色和白色斑纹；腹部细长。

本科栖息于植被茂盛的深山因此不容易遇见。其中最显著的蝴蝶裂唇蜓，也是全世界体型最大的蜻蜓之一，身体色彩十分艳丽。蝴蝶裂唇蜓的发现使得整个裂唇蜓家族在全世界备受关注。

裂唇蜓十分依赖茂盛的森林，栖息于具有一定海拔高度的清澈溪流。海拔500~2500 m的清澈山区溪流是它们比较偏爱的环境。雄性的裂唇蜓具有显著的领域行为，主要有2种方式，一种是以低空慢速飞行在短距离内巡逻，有时它们会非常接近水面，有时则距离水面有一定高度；另一种是长距离巡逻，例如蝴蝶裂唇蜓可以沿着几千米的溪流巡飞来寻找雌性。裂唇蜓的稚虫期较长，通常需要2年或者更长的时间才能发育成熟。未熟的成虫经常可见在峡谷和溪流上空翱翔。

Chlorogomphidae is regarded as one of the ancient dragonfly groups, the family is only distributed in Asia, being most speciose in the Oriental region. Over 50 species within three genera are known and no doubt more remain to be discovered. Over 30 species among three genera have been recorded from the southern part of China. They are usually large, black with yellow markings; head in frontal view oblong, eyes slightly separated above, the frons strongly developed; wings broad. Females of many species exist in two color morphs, one with largely hyaline wings, the other patterned with yellow, orange, black and white; the abdomen is narrow and long.

Species of the family occur in very dense forests and are not so easy to see. The largest and most famous species, *Chlorogomphus papilio* Ris, 1927, one of the largest anisopterans known, is a very colorful dragonfly. This magnificent species has attracted intense attention among current researchers specializing in this family.

All members of the family depend on forests. They prefer streams with a moderate gradient ranging from 500-2500 m elevation. Males display two types of mate locating behavior. In the first case they patrol slowly along a well defined territory, flying just above the water or somewhat higher depending on the species. In the second case, found in *C. papilio*, males search far and wide, travelling long distances to locate females. Larval growth takes at least two years. Juveniles are often seen flying very high in forest gaps and along margins.

斑翅裂唇蜓 雌

Chlorogomphus (*Orogomphus*) *usudai*, female

戴维裂唇蜓 雌

Chlorogomphus (*Neorogomphus*) *daviesi*, female

雄性的蝴蝶裂唇蜓沿着宽阔的溪流巡飞寻找雌性 | 宋睿斌 摄
A male *Chlorogomphus papilio* travelling along a wide stream in order to find a female | Photo by Ruibin Song

雌性的蝴蝶裂唇蜓在溪流边缘浅水潭插秧式产卵 | 宋睿斌 摄
A female *Chlorogomphus papilio* oviposting in a shallow pool of the stream with the body erected, | Photo by Ruibin Song

雄性的长鼻裂唇蜓沿着阴暗的小溪飞行巡视领地
A male *Chlorogomphus nasutus* establishing its territory by patrolling along a shady stream

雌性的长鼻裂唇蜓先在水面飞行排出卵块，然后插秧式产卵于溪流的浅滩
A female *Chlorogomphus nasutus* flying around above water and exuding egg mass, then depositing it into the shallow portion of the stream with the body erected

裂唇蜓属 Genus *Chlorogomphus* Selys, 1854

本属全球已知40余种。中国已知20余种。本属的主要特点是胸部侧面的黄色条纹出现在后胸前侧板，有时后胸后侧板后缘具黄色斑纹。本属被分成很多亚属，其中至少有5个亚属已经在中国发现。

The genus contains over 40 species and more than 20 are recorded from China. The main character for the genus is the presence of thoracic yellow stripes on metepisternum, in some species the ventral part of metepimeron have yellow markings. The genus is divided into many subgenera, and at least five of them have been found in China.

蝴蝶裂唇蜓 交尾 | 宋睿斌 摄
Chlorogomphus (*Aurorachlorus*) *papilio*, mating pair | Photo by Ruibin Song

蝶裂唇蜓亚属 Subgenus *Aurorachlorus* Carle, 1995

本亚属仅包含1种形态特殊且著名的蜻蜓，蝴蝶裂唇蜓。本种巨型蜻蜓，翅展巨大，两性的翅基方都染有黄褐相间的大块色斑，翅端褐色，臀圈十分发达。雄性的肛附器很短，容易与其他亚属的种类区分。

蝴蝶裂唇蜓栖息于森林中的宽阔溪流。它们的飞行能力很强，可以在不扇动翅的情况下滑翔很长一段距离，雄性可以沿着溪流巡飞长达数千米的距离寻找配偶。

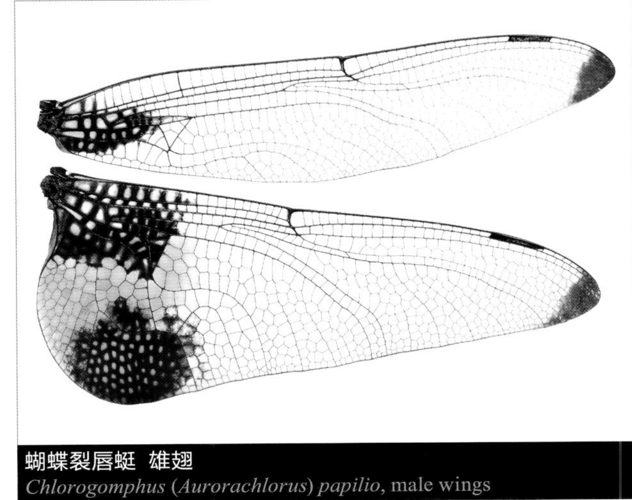

蝴蝶裂唇蜓 雄翅
Chlorogomphus (*Aurorachlorus*) *papilio*, male wings

The subgenus contains a single, remarkable and famous species, *Chlorogomphus* (*Aurorachlorus*) *papilio* Ris, 1927. It is a huge species with very large wingspan, wings in both sexes are tinted with yellow and brown bands basally, wing tips brown, anal loop well developed. The male anal appendages are short and differ from other congeners in the genus.

C. papilio inhabits wide streams in forest. Their flight ability is excellent and they can fly across a long distances without flapping their wings, males usually travel miles for searching females.

蝴蝶裂唇蜓 雄
Chlorogomphus (*Aurorachlorus*) *papilio*, male

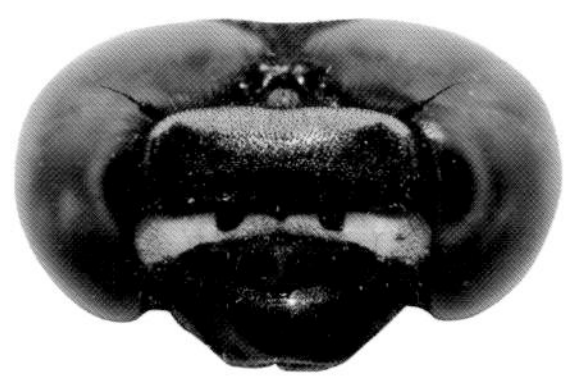

蝴蝶裂唇蜓 雄
Chlorogomphus (*Aurorachlorus*) *papilio*, male

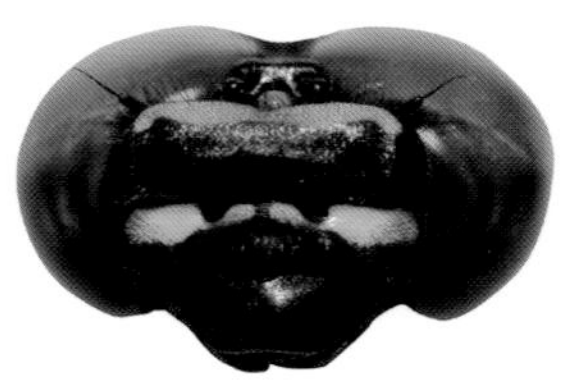

蝴蝶裂唇蜓 雌
Chlorogomphus (*Aurorachlorus*) *papilio*, female

蝶裂唇蜓亚属 头部正面观
Subgenus *Aurorachlorus*, head in frontal view

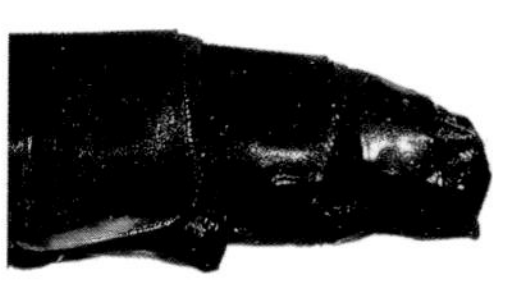

蝴蝶裂唇蜓 雌
Chlorogomphus (*Aurorachlorus*) *papilio*, female

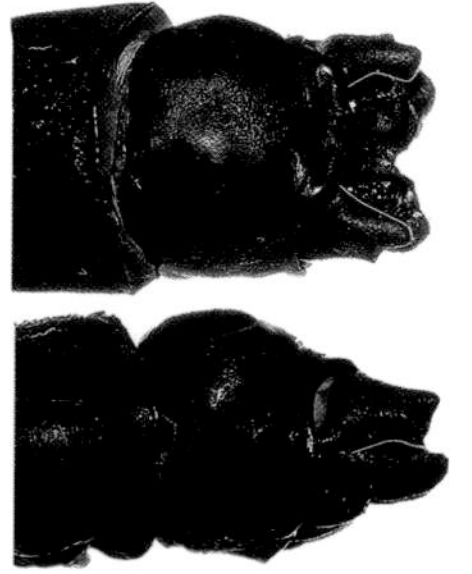

蝴蝶裂唇蜓 雄
Chlorogomphus (*Aurorachlorus*) *papilio*, male

蝶裂唇蜓亚属 雄性肛附器和雌性下生殖板
Subgenus *Aurorachlorus*, male anal appendages and female vulvar lamina

蝴蝶裂唇蜓 *Chlorogomphus* (*Aurorachlorus*) *papilio* Ris, 1927

【形态特征】雄性复眼墨绿色；胸部具肩前条纹，合胸侧面具1条甚阔的黄色条纹，翅染有淡褐色，翅端具深褐色斑，后翅基部具黄褐色相间的色斑；腹部黑色，第2～7节具黄斑。雌性翅的色斑更发达，黑斑伸达翅结处；腹部第2～6节具黄斑。【长度】体长 81～88 mm，腹长 58～63 mm，后翅 63～73 mm。【栖息环境】海拔 1500 m以下森林中的宽阔溪流。【分布】华中、华南、西南广布；越南。【飞行期】4—9月。

[Identification] Male eyes dark green. Thorax with antehumeral stripes and a broad yellow stripe laterally, wings tinted with pale brown, tips with dark brown spots, hind wing bases with yellow and brown markings. Abdomen black,

蝴蝶裂唇蜓 雄，贵州
Chlorogomphus (*Aurorachlorus*) *papilio*, male from Guizhou

S2-S7 with yellow spots. Female wings with more extensive markings with black reaching the nodus. S2-S6 with yellow spot. **[Measurements]** Total length 81-88 mm, abdomen 58-63 mm, hind wing 63-73 mm. **[Habitat]** Wide streams in forest below 1500 m elevation. **[Distribution]** Widespread in the Central, South and Southwest regions; Vietnam. **[Flight Season]** April to September.

蝴蝶裂唇蜓 雌，贵州
Chlorogomphus (*Aurorachlorus*) *papilio*, female from Guizhou

蝴蝶裂唇蜓 雄，贵州
Chlorogomphus (*Aurorachlorus*) *papilio*, male from Guizhou

蝴蝶裂唇蜓 雌，贵州
Chlorogomphus (*Aurorachlorus*) *papilio*, female from Guizhou

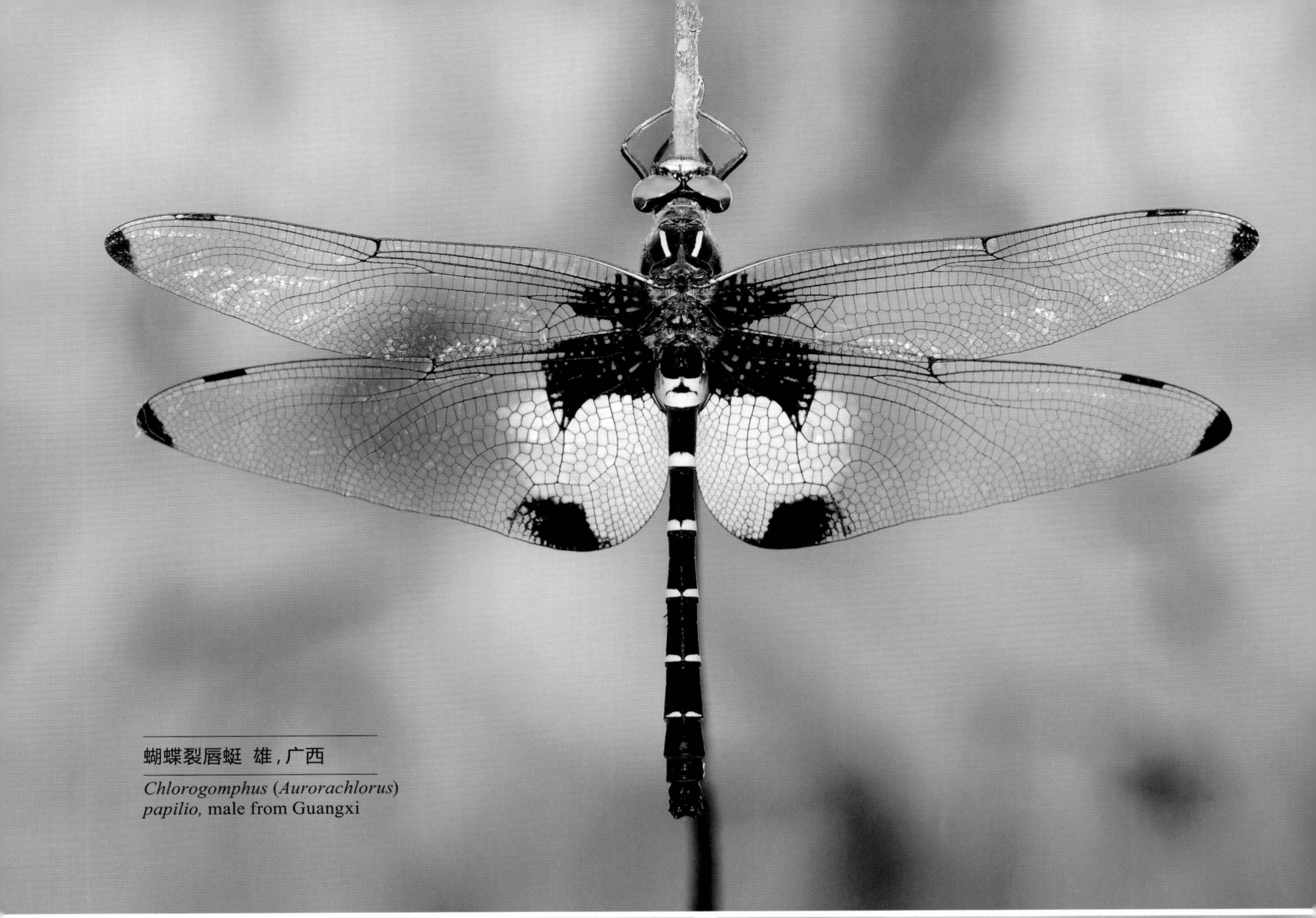

蝴蝶裂唇蜓 雄，广西

Chlorogomphus (*Aurorachlorus*) *papilio,* male from Guangxi

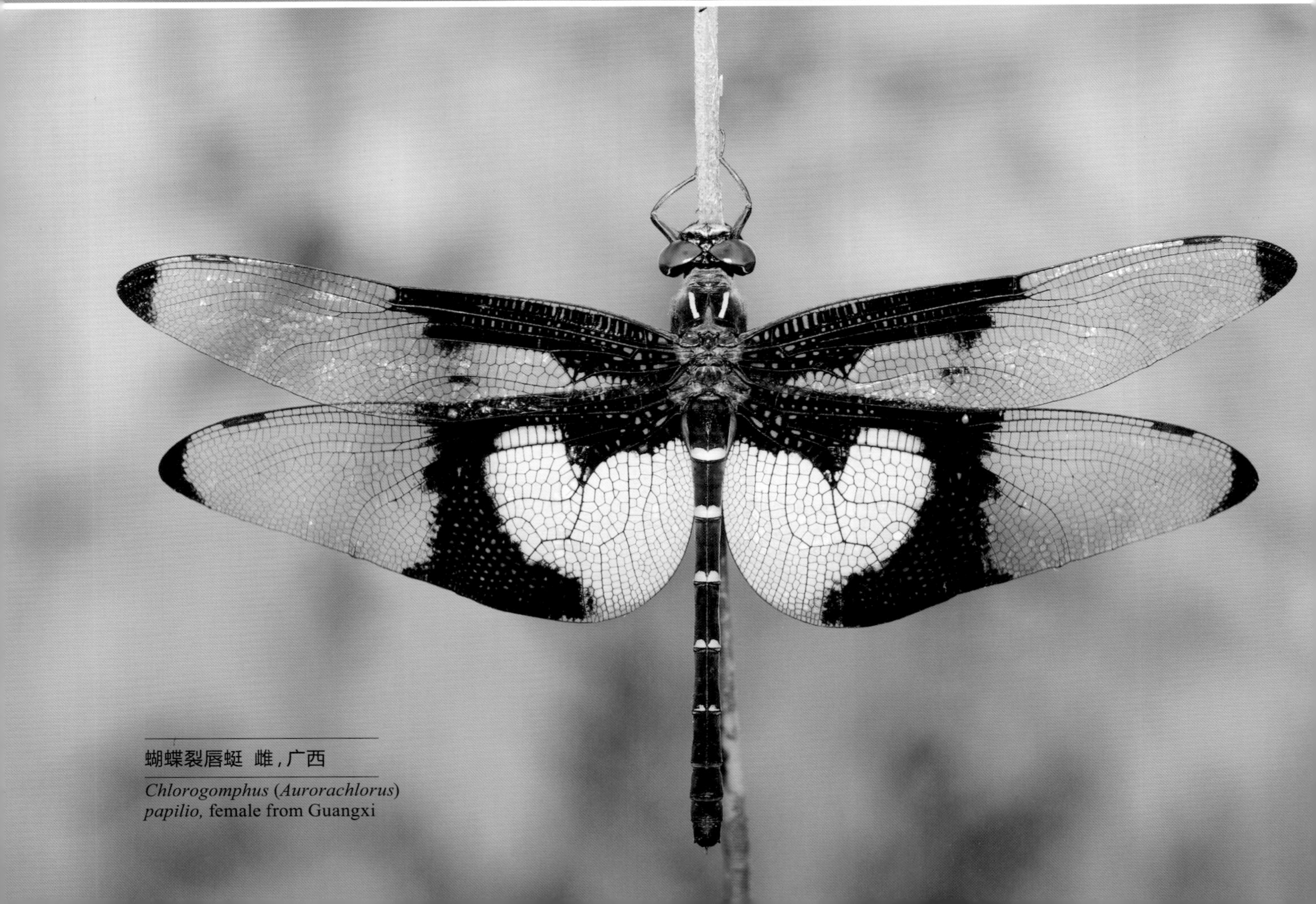

蝴蝶裂唇蜓 雌，广西

Chlorogomphus (*Aurorachlorus*) *papilio,* female from Guangxi

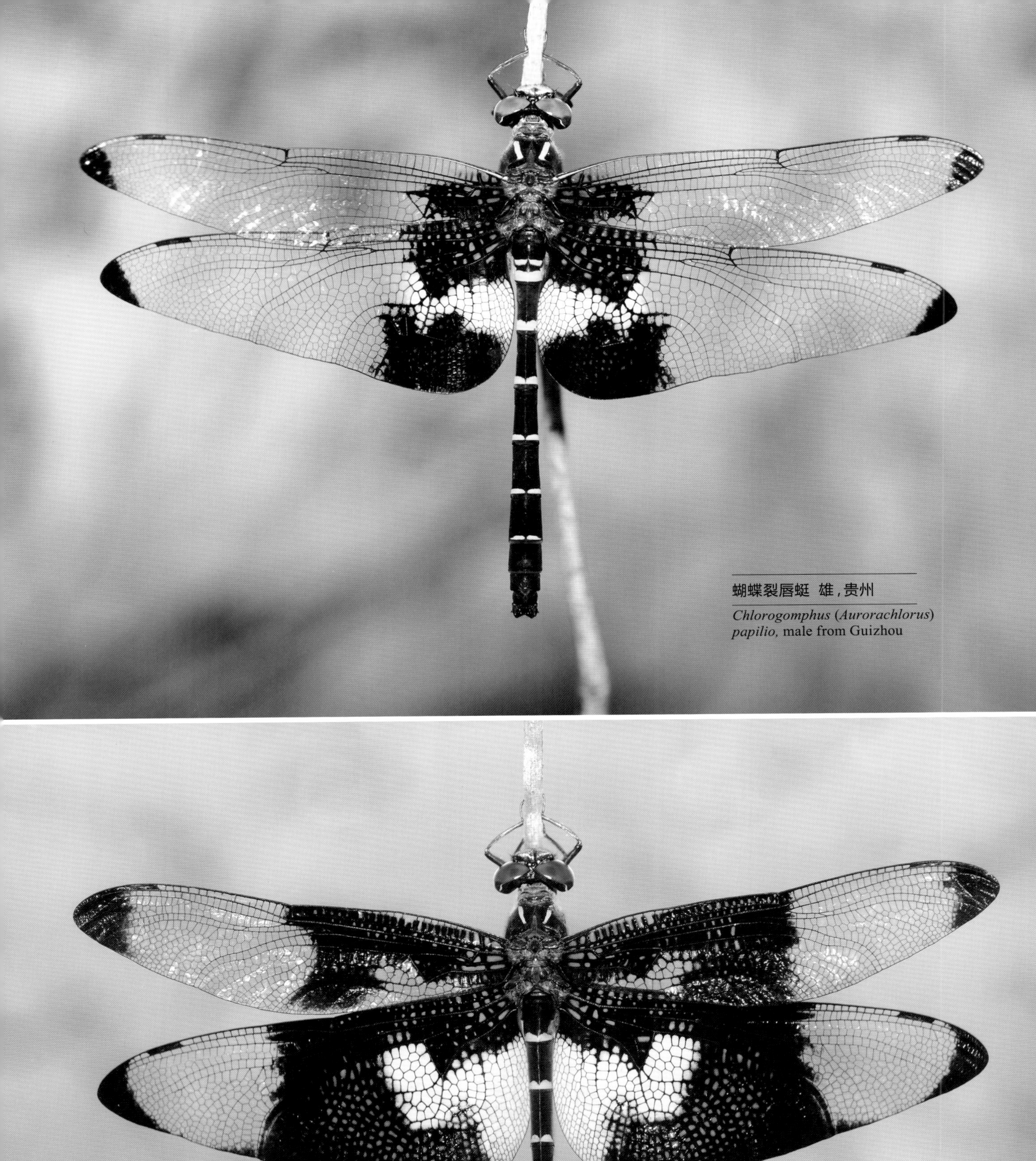

蝴蝶裂唇蜓 雄，贵州

Chlorogomphus (*Aurorachlorus*) *papilio,* male from Guizhou

蝴蝶裂唇蜓 雌，贵州

Chlorogomphus (*Aurorachlorus*) *papilio,* female from Guizhou

金翅裂唇蜓亚属 Subgenus *Neorogomphus* Carle, 1995

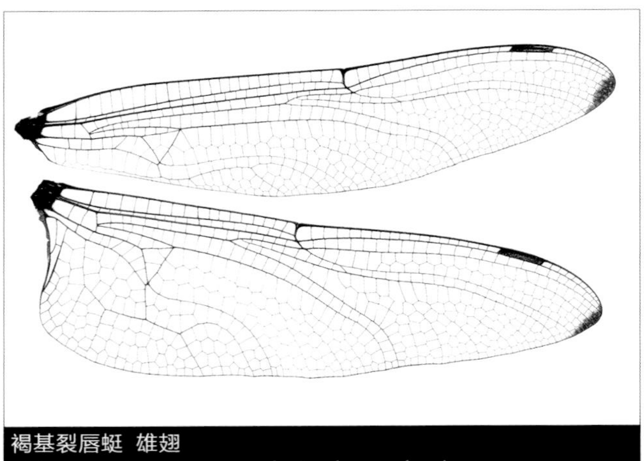

褐基裂唇蜓 雄翅
Chlorogomphus (*Neorogomphus*) *yokoii*, male wings

本亚属全球已知6种，仅分布于中国、越南、老挝和泰国地区。中国已知4种，分布于云南和广西。雄性翅透明，端部具小褐斑，腹部第6～9节向体侧方显著膨大，第10节末端呈角状突起并向下弯曲。雌性多型，透翅型翅大面积透明，仅基方具色斑，金翅型翅大面积金褐色。本亚属可以通过雄性肛附器的构造和雌性翅的色彩区分。

本亚属栖息于茂盛森林中的狭窄溪流。它们喜欢在森林的空旷地和峡谷中飞行。雄性经常在狭窄的溪流上方来回飞行，或者在森林中长距离穿梭寻找配偶。它们会飞到树枝附近寻找吊挂在森林中休息的雌性，这与某些溪栖的蜓科种类相似。

The subgenus contains six species, confined to China, Vietnam, Laos and Thailand. Four species are recorded from China, distributed in Yunnan and Guangxi. Males have hyaline wings with small patches of brown on the wing tips, S6-S9 expanded laterally, the end of S10 protruded and curved downwards. Females are polymorphic, hyaline winged morph with wings largely hyaline and dark markings at bases, golden winged morph with wings largely golden brown.

黄翅裂唇蜓 雄
Chlorogomphus (*Neorogomphus*) *auratus*, male

Species of the subgenus can be distinguished by the male anal appendages and female wing color pattern.

Species of this subgenus inhabit narrow streams in dense forest. They are usually seen fly high above the valley. Males usually patrol above narrow streams, or travel across the forest searching for females. Sometimes the male will check the trees looking for vertically perched female, a similar behavior shared by some species of Aeshnidae.

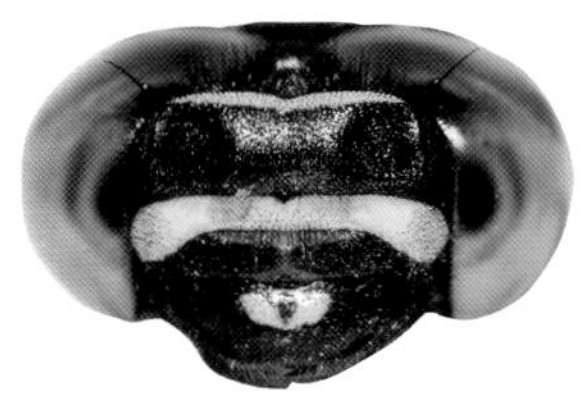

戴维裂唇蜓 雄
Chlorogomphus (*Neorogomphus*) *daviesi*, male

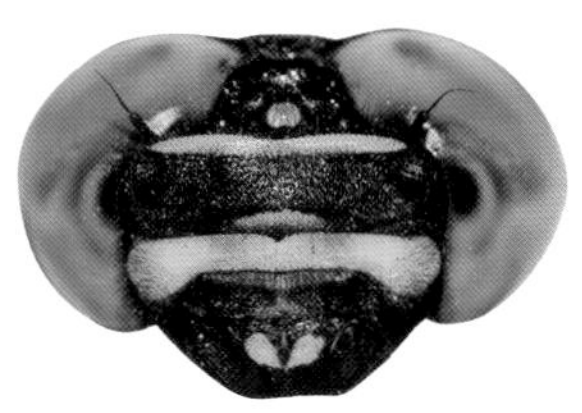

戴维裂唇蜓 雌
Chlorogomphus (*Neorogomphus*) *daviesi*, female

金翅裂唇蜓亚属 头部正面观
Subgenus *Neorogomphus*, head in frontal view

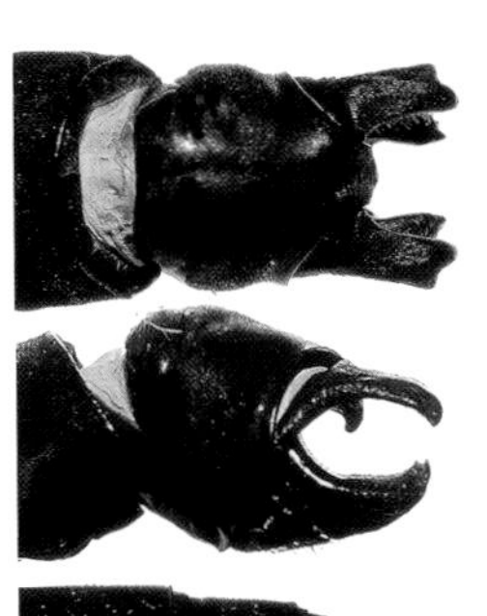

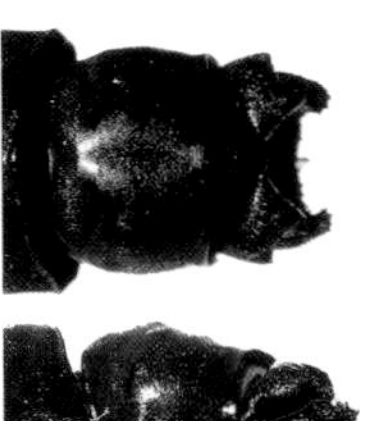

金翅裂唇蜓
Chlorogomphus (*Neorogomphus*) *canhvang*

戴维裂唇蜓
Chlorogomphus (*Neorogomphus*) *daviesi*

褐基裂唇蜓
Chlorogomphus (*Neorogomphus*) *yokoii*

金翅裂唇蜓亚属待定种
Chlorogomphus (*Neorogomphus*) sp.

金翅裂唇蜓亚属 雄性肛附器和雌性下生殖板
Subgenus *Neorogomphus*, male anal appendages and female vulvar lamina

金翅裂唇蜓 *Chlorogomphus* (*Neorogomphus*) *canhvang* Kompier & Karube, 2018

【形态特征】雄性面部黑色具黄色条纹，上唇中央具1个U形黄斑；胸部肩前条纹较细，肩条纹稍阔，合胸侧面具1条甚阔的黄色条纹；腹部黑色，第1~7节具黄斑。雌性多型，金翅型翅金色，端部具较大褐斑；褐翅型翅基方1/2褐色，端方1/2透明。【长度】体长 70~77 mm，腹长 52~57 mm，后翅 46~56 mm。【栖息环境】1500 m以下森林中的渗流地、狭窄溪流和小型瀑布。【分布】云南（红河）；越南。【飞行期】4—7月。

[Identification] Male face black with yellow markings, labrum with a U-shaped yellow spot centrally. Thorax with narrow antehumeral stripes and boarder humeral stripes, laterally with a broad yellow stripe. Abdomen black, S1-

S7 with yellow spots. Female polymorphic, golden winged morph with wings golden and extensive brown tips. Brown winged morph with wings basal half brown and apical half hyaline. **[Measurements]** Total length 70-77 mm, abdomen 52-57 mm, hind wing 46-56 mm. **[Habitat]** Seepages, narrow streams and small waterfalls below 1500 m elevation. **[Distribution]** Yunnan (Honghe); Vietnam. **[Flight Season]** April to July.

金翅裂唇蜓 雌，褐翅型，云南（红河）
Chlorogomphus (*Neorogomphus*) *canhvang*, female, brown winged from Yunnan (Honghe)

金翅裂唇蜓 雄，云南（红河）
Chlorogomphus (*Neorogomphus*) *canhvang*, male from Yunnan (Honghe)

金翅裂唇蜓 雄，云南（红河）| 莫善濂 摄
Chlorogomphus (*Neorogomphus*) *canhvang*, male from Yunnan (Honghe) | Photo by Shanlian Mo

金翅裂唇蜓 雌，褐翅型，云南（红河）
Chlorogomphus (*Neorogomphus*) *canhvang*, female, brown winged from Yunnan (Honghe)

金翅裂唇蜓 雌，金翅型，云南（红河） | 莫善濂 摄
Chlorogomphus (*Neorogomphus*) *canhvang*, female, golden winged from Yunnan (Honghe) | Photo by Shanlian Mo

黄翅裂唇蜓 *Chlorogomphus* (*Neorogomphus*) *auratus* Martin, 1910

【形态特征】本种与金翅裂唇蜓十分相似，但雄性的下肛附器构造不同。本种下肛附器的两枝末端向内侧显著钩曲，中央形成1个圆形空腔，而金翅裂唇蜓雄性下肛附器的两枝几乎平行。两者的雌性则较难区分。【长度】体长72～80 mm，腹长 52～61 mm，后翅 50～58 mm。【栖息环境】1000 m以下森林中的溪流和小型瀑布。【分布】广西；越南。【飞行期】5—7月。

[Identification] Similar to *C. canhvang* but male anal appendages different. The inferior appendage of this species with tips of two branches curved inwards, forming a median circular cavity, but males of *C. canhvang* with the two branches of inferior appendage almost parallel. Females are difficult to separate. **[Measurements]** Total length 72-80 mm, abdomen 52-61 mm, hind wing 50-58 mm. **[Habitat]** Streams and small waterfalls in forest below 1000 m elevation. **[Distribution]** Guangxi; Vietnam. **[Flight Season]** May to July.

黄翅裂唇蜓 雄，广西
Chlorogomphus (*Neorogomphus*) *auratus*, male from Guangxi

黄翅裂唇蜓 雌，广西
Chlorogomphus (*Neorogomphus*) *auratus*, female from Guangxi

黄翅裂唇蜓 雄，广西
Chlorogomphus (*Neorogomphus*) *auratus*, male from Guangxi

黄翅裂唇蜓 雌，广西
Chlorogomphus (*Neorogomphus*) *auratus*, female from Guangxi

戴维裂唇蜓 *Chlorogomphus* (*Neorogomphus*) *daviesi* Karube, 2001

戴维裂唇蜓 雄，云南（临沧）
Chlorogomphus (*Neorogomphus*) *daviesi*, male from Yunnan (Lincang)

戴维裂唇蜓 雌，金翅型，云南（临沧）
Chlorogomphus (*Neorogomphus*) *daviesi*, female, golden winged from Yunnan (Lincang)

【形态特征】雄性面部黑色具黄色条纹，上唇中央具1个U形黄斑；胸部肩前条纹较细，肩条纹稍阔，合胸侧面具1条甚阔的黄色条纹；腹部黑色，第1~8节具黄斑。雌性多型，金翅型翅金色，端部具小褐斑；透翅型翅大面积透明，基方具橙褐斑。【长度】体长 68~71 mm，腹长 51~53 mm，后翅 45~52 mm。【栖息环境】海拔 500 m左右森林中的狭窄溪流。【分布】中国云南（临沧）特有。【飞行期】4—6月。

[Identification] Male face black with yellow markings, labrum with a U-shaped yellow spot centrally. Thorax with narrow antehumeral stripes and broader humeral stripes, laterally with a broad yellow stripe. Abdomen black, S1-

S8 with yellow spots. Female polymorphic, golden winged morph has golden wings largely golden with small brown tips; hyaline winged morph has largely hyaline wings with basal orange brown spots. **[Measurements]** Total length 68-71 mm, abdomen 51-53 mm, hind wing 45-52 mm. **[Habitat]** Narrow streams in forest at about 500 m elevation. **[Distribution]** Endemic to Yunnan (Lincang) of China. **[Flight Season]** April to June.

戴维裂唇蜓 雌，金翅型，云南（临沧）
Chlorogomphus (*Neorogomphus*) *daviesi*, female, golden winged from Yunnan (Lincang)

戴维裂唇蜓 雌，透翅型，云南（临沧）
Chlorogomphus (*Neorogomphus*) *daviesi*, female, hyaline winged from Yunnan (Lincang)

戴维裂唇蜓 雄，云南（临沧）
Chlorogomphus (Neorogomphus) daviesi, male from Yunnan (Lincang)

褐基裂唇蜓 *Chlorogomphus* (*Neorogomphus*) *yokoii* Karube, 1995

【形态特征】雄性面部黑色具黄纹，上唇中央具1个U形黄斑；胸部具较细的肩条纹和肩前条纹，合胸侧面具1条甚阔的黄色条纹；腹部黑色，第1~7节基方具黄斑。雌性多型，金翅型翅金褐色具黑褐色斑；透翅型翅大面积透明，基方具黑褐色斑。【长度】体长 73~79 mm，腹长 55~60 mm，后翅 47~55 mm。【栖息环境】海拔 1500 m以下森林中的狭窄溪流和渗流地。【分布】云南（西双版纳、红河、普洱）；泰国、老挝。【飞行期】4—9月。

[Identification] Male face black with yellow markings, labrum with a U-shaped yellow spot centrally. Thorax with narrow antehumeral and humeral stripes, laterally with a broad yellow stripe. Abdomen black, S1-S7 with yellow spots. Female polymorphic, golden winged morph has golden brown wings brown with blackish brown spots; hyaline winged morph has largely hyaline wings with basal blackish brown spots. **[Measurements]** Total length 73-79 mm, abdomen 55-60 mm, hind wing 47-55 mm. **[Habitat]** Narrow streams and seepages in forest below 1500 m elevation. **[Distribution]** Yunnan (Xishuangbanna, Honghe, Pu'er); Thailand, Laos. **[Flight Season]** April to September.

褐基裂唇蜓 雄，云南（普洱）
Chlorogomphus (*Neorogomphus*) *yokoii*, male from Yunnan (Pu'er)

褐基裂唇蜓 雌，透翅型，云南（普洱）
Chlorogomphus (*Neorogomphus*) *yokoii*, female, hyaline winged from Yunnan (Pu'er)

褐基裂唇蜓 雌，金翅型，云南（红河）
Chlorogomphus (*Neorogomphus*) *yokoii*, female, golden winged from Yunnan (Honghe)

褐基裂唇蜓 雄，云南（普洱）
Chlorogomphus (*Neorogomphus*) *yokoii*, male from Yunnan (Pu'er)

褐基裂唇蜓 雌，透翅型，云南（普洱）
Chlorogomphus (*Neorogomphus*) *yokoii*, female, hyaline winged from Yunnan (Pu'er)

褐基裂唇蜓 雌，金翅型，云南（红河）
Chlorogomphus (*Neorogomphus*) *yokoii*, female, golden winged from Yunnan (Honghe)

金翅裂唇蜓亚属待定种 *Chlorogomphus* (*Neorogomphus*) sp.

【形态特征】雄性面部黑色具黄斑，上唇中央具1个U形黄斑；胸部具黄色的肩条纹和肩前条纹，合胸侧面具1条甚阔的黄色条纹；腹部黑色，第1~7节基方具黄斑。雌性多型，金翅型翅金褐色具黑褐色斑；透翅型翅大面积透明，基方具黑褐色斑。【长度】体长 72~77 mm，腹长 55~58 mm，后翅 50~57 mm。【栖息环境】海拔 500~1500 m 森林中的狭窄溪流。【分布】云南（普洱）。【飞行期】4—6月。

[Identification] Male face black with yellow markings, labrum with a U-shaped yellow spot centrally. Thorax with yellow antehumeral and humeral stripes, laterally with a broad yellow stripe. Abdomen black, S1-S7 with yellow spots. Female polymorphic, golden winged morph has golden brown wings with blackish brown spots. Hyaline winged morph has largely hyaline wings with basal blackish brown spots. **[Measurements]** Total length 72-77 mm, abdomen 55-58 mm, hind wing 50-57 mm. **[Habitat]** Narrow streams in forest at 500-1500 m elevation. **[Distribution]** Yunnan (Pu'er). **[Flight Season]** April to June.

金翅裂唇蜓亚属待定种 雄，云南（普洱）
Chlorogomphus (*Neorogomphus*) sp., male from Yunnan (Pu'er)

金翅裂唇蜓亚属待定种 雄，云南（普洱）
Chlorogomphus (*Neorogomphus*) sp., male from Yunnan (Pu'er)

金翅裂唇蜓亚属待定种 雌，金翅型，云南（普洱）
Chlorogomphus (*Neorogomphus*) sp., female, golden winged from Yunnan (Pu'er)

金翅裂唇蜓亚属待定种 雌，透翅型，云南（普洱）
Chlorogomphus (*Neorogomphus*) sp., female, hyaline winged from Yunnan (Pu'er)

金翅裂唇蜓亚属待定种 雌，金翅型，云南（普洱）
Chlorogomphus (*Neorogomphus*) sp., female, golden winged from Yunnan (Pu'er)

金翅裂唇蜓亚属待定种 雌，透翅型，云南（普洱）
Chlorogomphus (*Neorogomphus*) sp., female, hyaline winged from Yunnan (Pu'er)

山裂唇蜓亚属 Subgenus *Orogomphus* Selys, 1878

本亚属中国已知5种，主要分布在华南地区。雄性的上肛附器平直，侧缘腹面常具1个刺状突起，下肛附器分成两枝，腹面观呈三角形。雄性的翅透明，端部常具小褐斑，雌性多型。中国已知种可以通过雄性肛附器的构造和雌性翅的色彩区分。

本亚属栖息于茂盛森林中的狭窄溪流，喜欢在森林的空旷地和峡谷的高空飞行，与金翅裂唇蜓亚属的种类有很多相似的行为。雄性通常在狭窄的溪流上方巡飞。

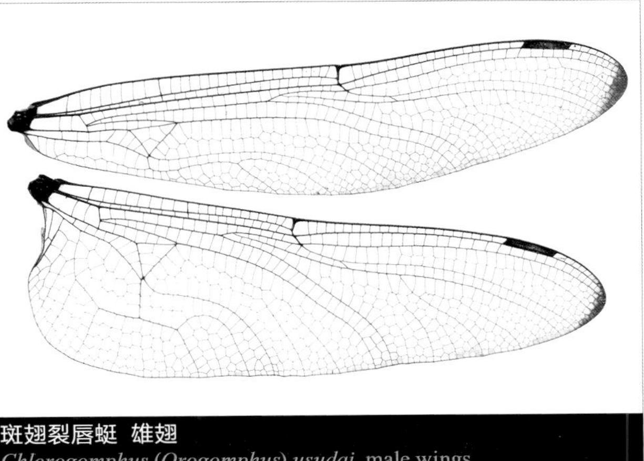

斑翅裂唇蜓 雄翅
Chlorogomphus (*Orogomphus*) *usudai*, male wings

Five species of the subgenus have been recorded from China, mainly distributed in the South. Males have straight superior appendages with a ventro-lateral spine, the inferior appendage has a deep median hollow thus forming two branches that is triangular in lateral view. Male wings hyaline with small brown tips, female polymorphic. The Chinese species can be distinguished by the male anal appendages and female wing color pattern.

斑翅裂唇蜓 雄
Chlorogomphus (*Orogomphus*) *usudai*, male

Species of this subgenus inhabit narrow streams in dense forest where they are usually seen flying high above valleies sharing similar behavior patterns with species of subgenus *Neorogomphus*. Males usually patrol above narrow streams.

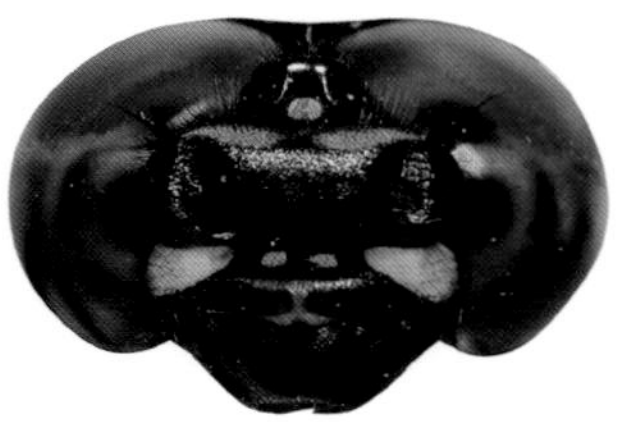

斑翅裂唇蜓 雄
Chlorogomphus (*Orogomphus*) *usudai*, male

斑翅裂唇蜓 雌
Chlorogomphus (*Orogomphus*) *usudai*, female

山裂唇蜓亚属 头部正面观
Subgenus *Orogomphus*, head in frontal view

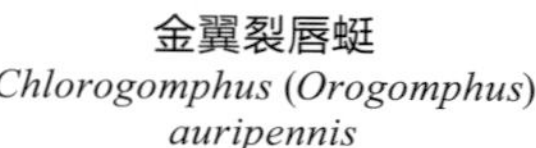

金翼裂唇蜓
Chlorogomphus (*Orogomphus*) *auripennis*

短痣裂唇蜓
Chlorogomphus (*Orogomphus*) *brevistigma*

李氏裂唇蜓
Chlorogomphus (*Orogomphus*) *risi*

斑翅裂唇蜓
Chlorogomphus (*Orogomphus*) *usudai*

山裂唇蜓亚属 雄性肛附器
Subgenus *Orogomphus*, male anal appendages

金翼裂唇蜓 *Chlorogomphus* (*Orogomphus*) *auripennis* Zhang & Cai, 2014

【形态特征】雄性面部黑色，后唇基和上额黄色；胸部肩前条纹甚细，肩条纹稍阔，合胸侧面具1条黄色条纹；腹部黑色，第1~6节具黄斑。雌性多型，金翅型翅基方1/2大面积金褐色；透翅型翅稍染浅褐色，翅结处具黄褐色带。【长度】体长 74~81 mm，腹长 56~61 mm，后翅 53~58 mm。【栖息环境】海拔 2000 m以下森林中的渗流地、沟渠和狭窄溪流。【分布】中国特有，分布于福建、广东。【飞行期】3—7月。

[Identification] Male face black, postclypeus and top of frons yellow. Thorax with narrow antehumeral stripes and boarder humeral stripes, laterally with a broad yellow stripe. Abdomen black, S1-S6 with yellow spots. Female polymorphic, the golden winged morph with basal half of wings largely golden brown; hyaline winged morph has wings slightly tinted with light brown with yellowish brown bands under nodus. **[Measurements]** Total length 74-81 mm, abdomen 56-61 mm, hind wing 53-58 mm. **[Habitat]** Seepages, ditches and narrow shady streams in forest below 2000 m elevation. **[Distribution]** Endemic to China, recorded from Fujian, Guangdong. **[Flight Season]** March to July.

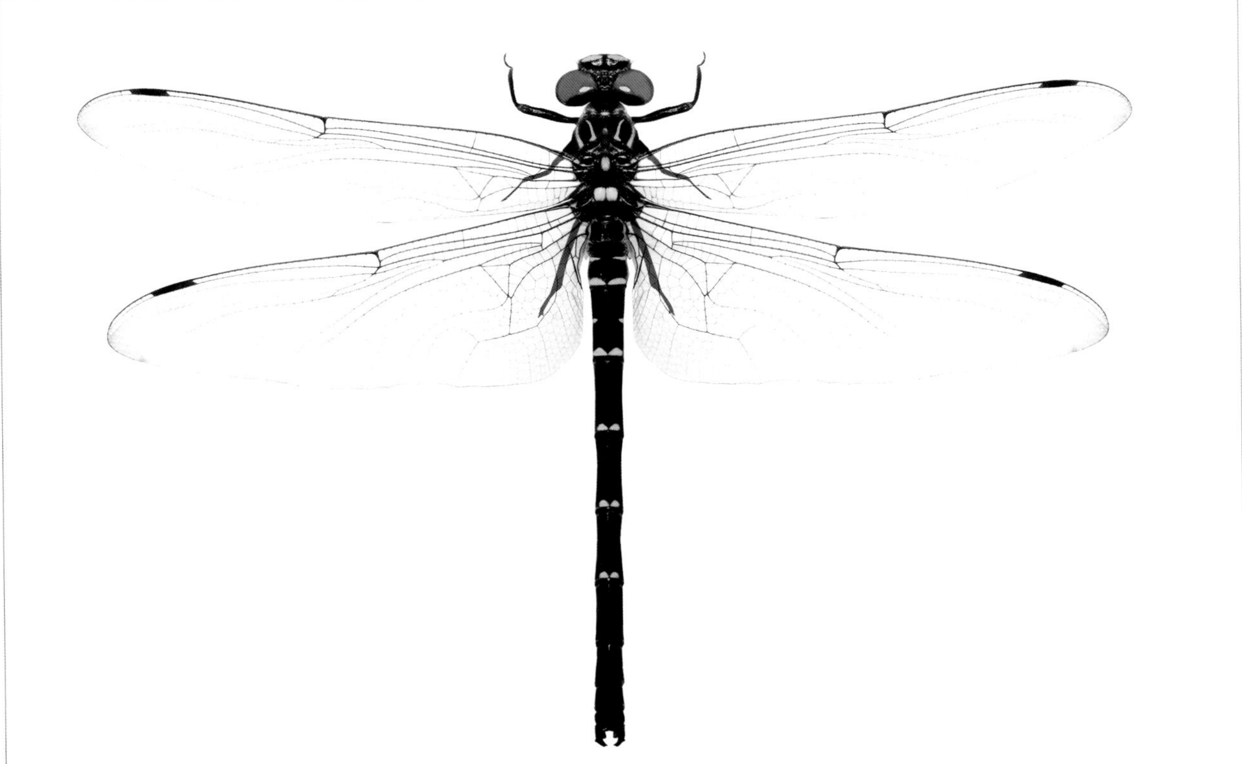

金翼裂唇蜓 雄，广东
Chlorogomphus (*Orogomphus*) *auripennis*, male from Guangdong

金翼裂唇蜓 雌，透翅型，广东
Chlorogomphus (*Orogomphus*) *auripennis*, female, hyaline winged from Guangdong

金翼裂唇蜓 雌，金翅型，广东
Chlorogomphus (*Orogomphus*) *auripennis*, female, golden winged from Guangdong

短痣裂唇蜓 *Chlorogomphus* (*Orogomphus*) *brevistigma* Oguma, 1926

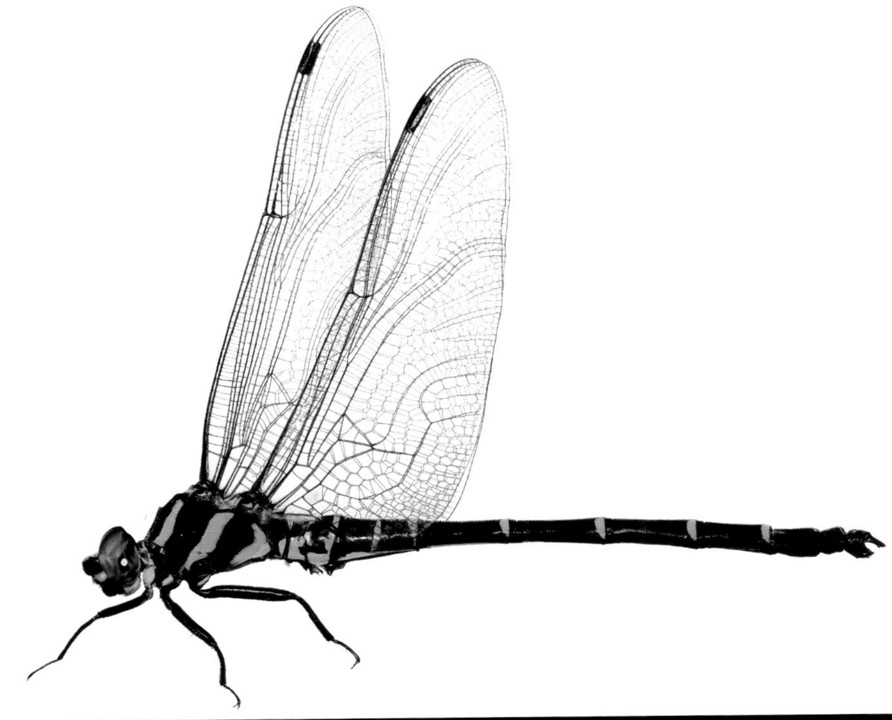

短痣裂唇蜓 雄，台湾
Chlorogomphus (*Orogomphus*) *brevistigma*, male from Taiwan

【形态特征】雄性面部黑色，后唇基和上额黄色；胸部肩前条纹甚细，肩条纹宽阔，合胸侧面具2条黄色条纹，翅端无褐斑；腹部黑色，第1~7节具黄斑。【长度】雄性体长 80~83 mm，腹长 61~64 mm，后翅 50~53 mm。【栖息环境】海拔 2000 m以下森林中的小溪。【分布】中国台湾特有。【飞行期】5—9月。

[Identification] Male face black, postclypeus and top of frons yellow. Thorax with narrow antehumeral stripes and broad humeral stripes, laterally with two broad yellow stripes, wings lacking apical brown spots. Abdomen black, S1-S7 with yellow spots. **[Measurements]** Male total length 80-83 mm, abdomen 61-64 mm, hind wing 50-53 mm. **[Habitat]** Streams in forest below 2000 m elevation. **[Distribution]** Endemic to Taiwan of China. **[Flight Season]** May to September.

李氏裂唇蜓 *Chlorogomphus* (*Orogomphus*) *risi* Chen, 1950

【形态特征】雄性面部黑色，后唇基和上额黄色；胸部具黄色的肩条纹和肩前条纹，合胸侧面具2条黄色条纹；腹部黑色，第1~7节具黄斑，第7节后方具1对甚大的黄斑。雌性多型，透翅型翅大面积透明仅端部具褐斑；斑翅型翅基方具甚大的黑褐色斑。【长度】体长 77~80 mm，腹长 60~61 mm，后翅 49~55 mm。【栖息环境】海拔 1500 m以下森林中的小溪。【分布】中国台湾特有。【飞行期】4—9月。

[Identification] Male face black, postclypeus and top of frons yellow. Thorax with yellow antehumeral and humeral stripes, laterally with two yellow stripes. Abdomen black, S1-S7 with yellow spots, S7 with a pair of large yellow spots apically. Female polymorphic, hyaline winged morph has largely hyaline wings with apical brown spots;

spotted winged morph has wings with large basal blackish brown markings. **[Measurements]** Total length 77-80 mm, abdomen 60-61 mm, hind wing 49-55 mm. **[Habitat]** Streams in forest below 1500 m elevation. **[Distribution]** Endemic to Taiwan of China. **[Flight Season]** April to September.

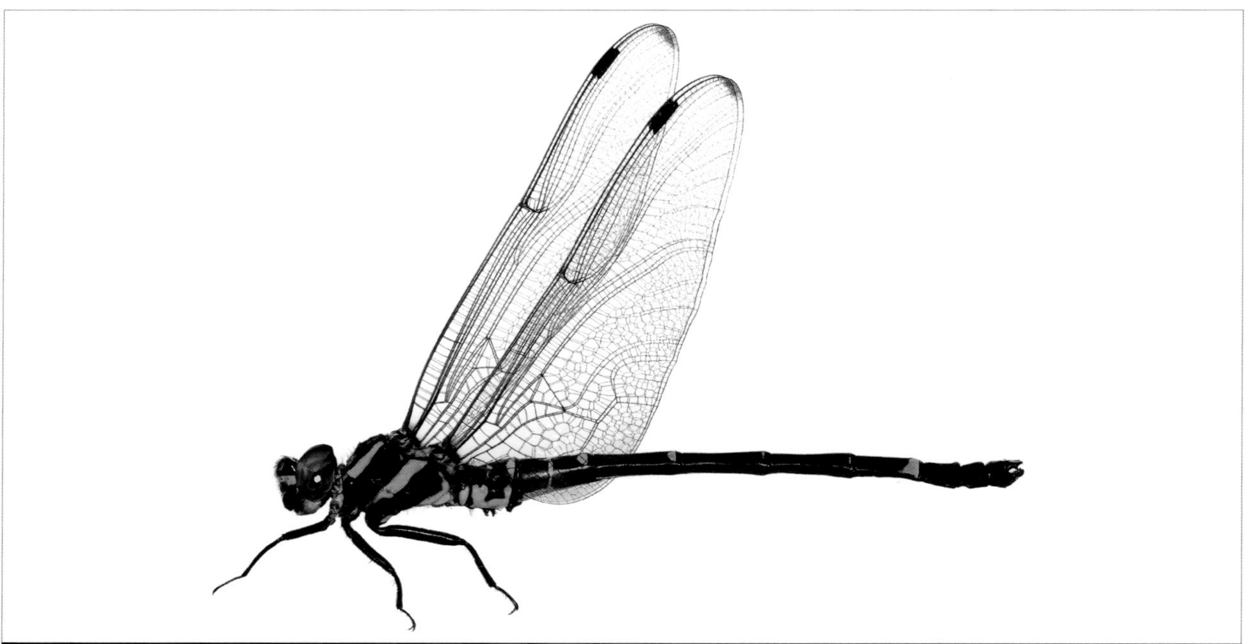

李氏裂唇蜓 雄，台湾
Chlorogomphus (*Orogomphus*) *risi*, male from Taiwan

李氏裂唇蜓 雌，台湾
Chlorogomphus (*Orogomphus*) *risi*, female from Taiwan

斑翅裂唇蜓 *Chlorogomphus* (*Orogomphus*) *usudai* Ishida, 1996

【形态特征】雄性面部黑色，后唇基具2对黄色斑点，上额黄色；胸部具黄色的肩条纹和肩前条纹，合胸侧面具2条黄色条纹；腹部黑色，第1～6节具黄斑。雌性多型，斑翅型翅基方和端方具发达的黑褐色斑；透翅型仅在翅端部具较大的黑褐色斑。【长度】体长 67～70 mm，腹长 52～55 mm，后翅 46～52 mm。【栖息环境】海拔 500～1500 m森林中的狭窄溪流、沟渠和小型瀑布。【分布】中国海南特有。【飞行期】4—7月。

[Identification] Male face black, postclypeus with two pairs of yellow spots, top of frons yellow. Thorax with yellow antehumeral and humeral stripes, laterally with two yellow stripes. Abdomen black, S1-S6 with yellow spots.

斑翅裂唇蜓 雄，海南
Chlorogomphus (*Orogomphus*) *usudai*, male from Hainan

Female polymorphic, spotted winged morph has wings with basal half and wing tips blackish brown; hyaline winged morph has wings with blackish brown spots apically. **[Measurements]** Total length 67-70 mm, abdomen 52-55 mm, hind wing 46-52 mm. **[Habitat]** Narrow streams, ditches and small waterfalls in forest at 500-1500 m elevation. **[Distribution]** Endemic to Hainan of China. **[Flight Season]** April to July.

斑翅裂唇蜓 雌，斑翅型，海南
Chlorogomphus (*Orogomphus*) *usudai*, female, spotted winged from Hainan

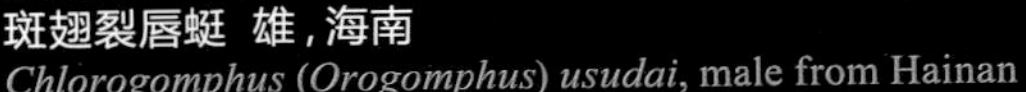

斑翅裂唇蜓 雄，海南
Chlorogomphus (*Orogomphus*) *usudai*, male from Hainan

斑翅裂唇蜓 雌，透翅型，海南
Chlorogomphus (*Orogomphus*) *usudai*, female, hyaline winged from Hainan

斑翅裂唇蜓 雌，透翅型，海南
Chlorogomphus (*Orogomphus*) *usudai*, female, hyaline winged from Hainan

斑翅裂唇蜓 雌，斑翅型，海南
Chlorogomphus (*Orogomphus*) *usudai*, female, spotted winged from Hainan

华裂唇蜓亚属 Subgenus *Sinorogomphus* Carle, 1995

本亚属中国已知10余种，在南方分布较广泛。本亚属的显著特征是腹部极为细长；雄性的上肛附器通常具1个刺状的侧突，下肛附器中央缺刻；绝大多数种类的翅透明，少数种类的雌性翅具色彩。

本亚属栖息于具有一定海拔高度的山区溪流。雄性会以慢速的低空飞行在溪流上方巡逻和占据领地。

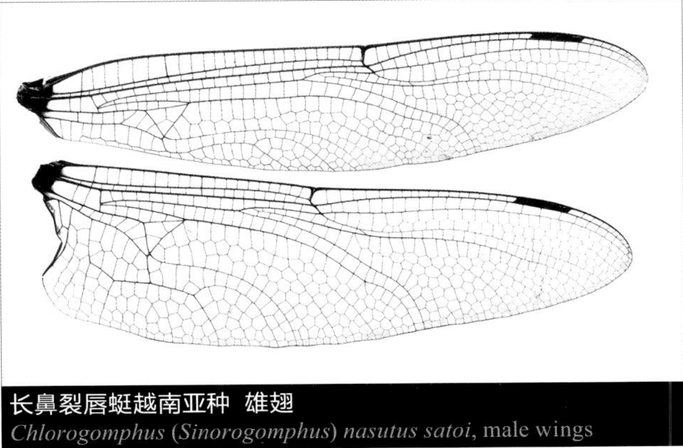

长鼻裂唇蜓越南亚种 雄翅
Chlorogomphus (*Sinorogomphus*) *nasutus satoi*, male wings

Over ten species of the subgenus have been found in China, widespread in the southern part. Species of the subgenus are characterized by the extremely long abdomen. Male superior appendages with a lateral spine, inferior appendage with a median hollow. Most species have hyaline wings but some females have colorful wings.

Species of this subgenus inhabit montane streams at moderate altitudes. Territorial males usually fly slowly above water.

铃木裂唇蜓 雌
Chlorogomphus (*Sinorogomphus*) *suzukii*, female

细腹裂唇蜓
Chlorogomphus (*Sinorogomphus*) *gracilis*

老挝裂唇蜓
Chlorogomphus (*Sinorogomphus*) *hiten*

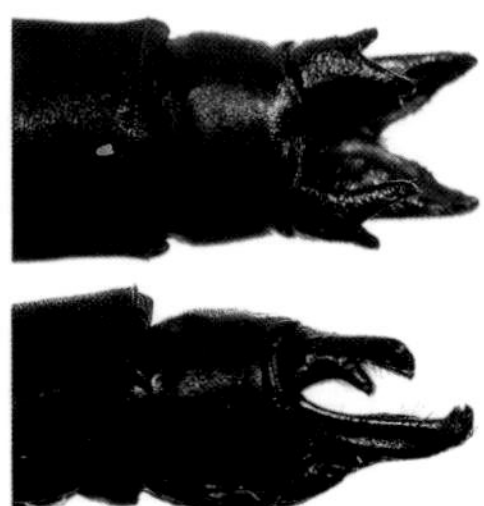

褐翅裂唇蜓
Chlorogomphus (*Sinorogomphus*) *infuscatus*

长腹裂唇蜓
Chlorogomphus (*Sinorogomphus*) *kitawakii*

中越裂唇蜓
Chlorogomphus (*Sinorogomphus*) *sachiyoae*

山裂唇蜓
Chlorogomphus (*Sinorogomphus*) *shanicus*

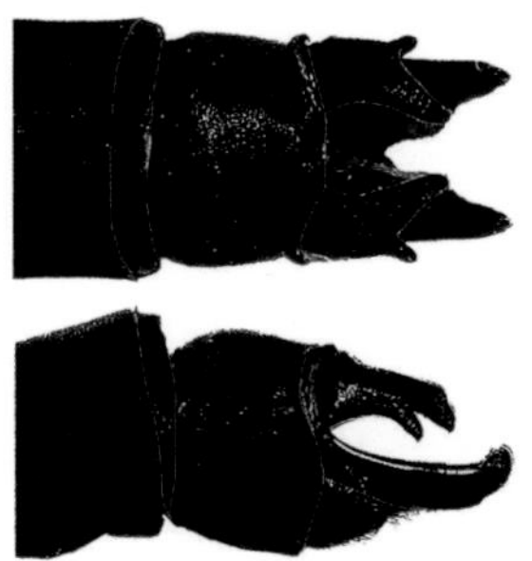

铃木裂唇蜓
Chlorogomphus (*Sinorogomphus*) *suzukii*

华裂唇蜓亚属待定种1
Chlorogomphus (*Sinorogomphus*) sp. 1

华裂唇蜓亚属待定种2
Chlorogomphus (*Sinorogomphus*) sp. 2

朴氏裂唇蜓
Chlorogomphus (*Sinorogomphus*) *piaoacensis*

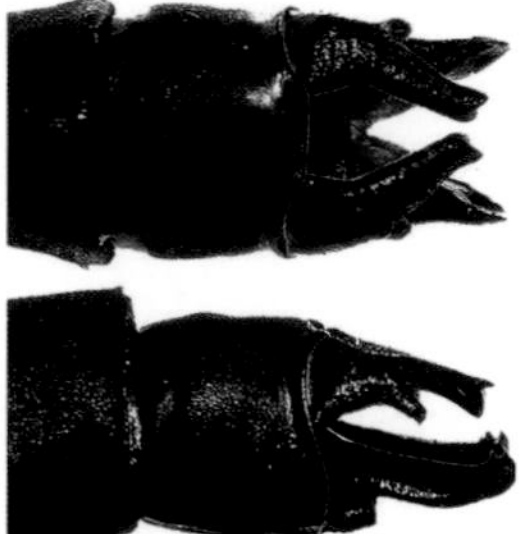

长鼻裂唇蜓指名亚种
Chlorogomphus (*Sinorogomphus*) *nasutus nasutus*

长鼻裂唇蜓越南亚种
Chlorogomphus (*Sinorogomphus*) *nasutus satoi*

华裂唇蜓亚属 雄性肛附器
Subgenus *Sinorogomphus*, male anal appendages

细腹裂唇蜓
Chlorogomphus (*Sinorogomphus*) *gracilis*

老挝裂唇蜓
Chlorogomphus (*Sinorogomphus*) *hiten*

褐翅裂唇蜓
Chlorogomphus (*Sinorogomphus*) *infuscatus*

长腹裂唇蜓
Chlorogomphus (*Sinorogomphus*) *kitawakii*

朴氏裂唇蜓
Chlorogomphus (*Sinorogomphus*) *piaoacensis*

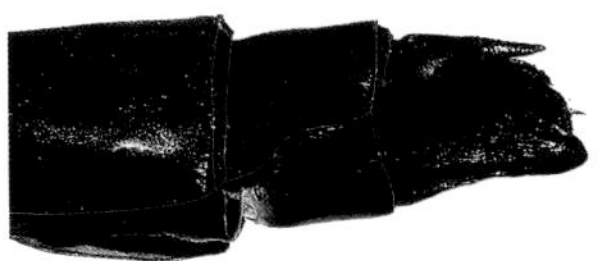

山裂唇蜓
Chlorogomphus (*Sinorogomphus*) *shanicus*

华裂唇蜓亚属 雌性下生殖板
Subgenus *Sinorogomphus*, female vulvar lamina

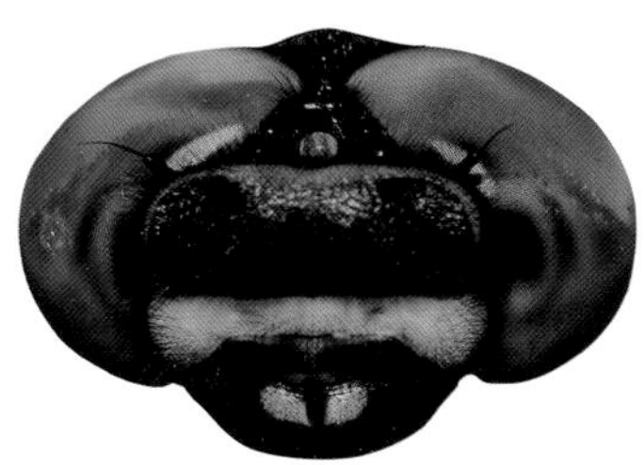

老挝裂唇蜓 雄
Chlorogomphus (*Sinorogomphus*) *hiten*, male

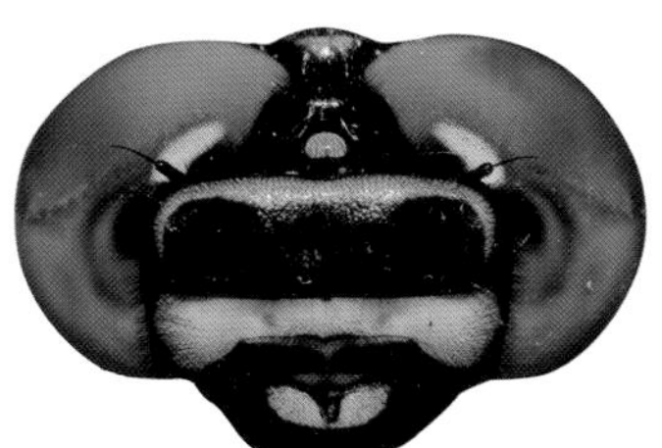

老挝裂唇蜓 雌
Chlorogomphus (*Sinorogomphus*) *hiten*, female

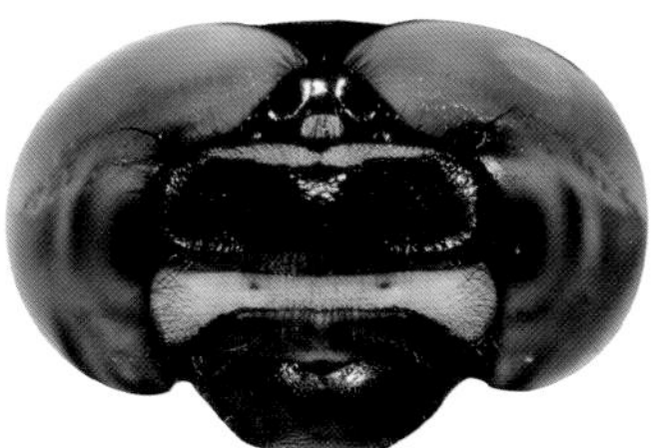

长腹裂唇蜓 雄
Chlorogomphus (*Sinorogomphus*) *kitawakii*, male

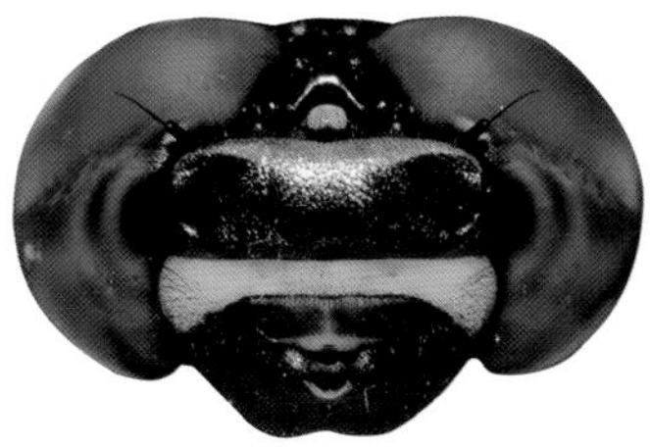

长腹裂唇蜓 雌
Chlorogomphus (*Sinorogomphus*) *kitawakii*, female

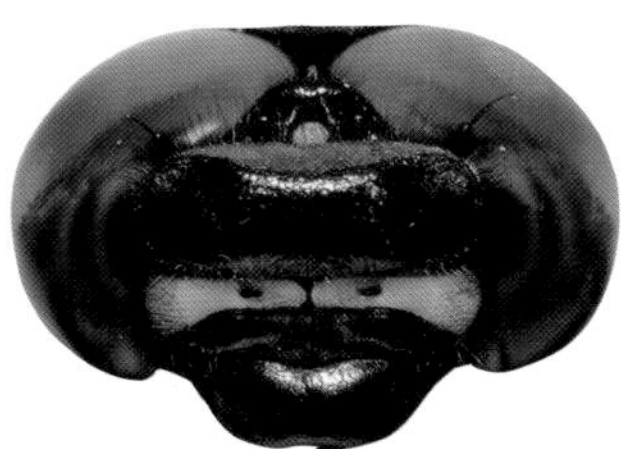

山裂唇蜓 雄
Chlorogomphus (*Sinorogomphus*) *shanicus*, male

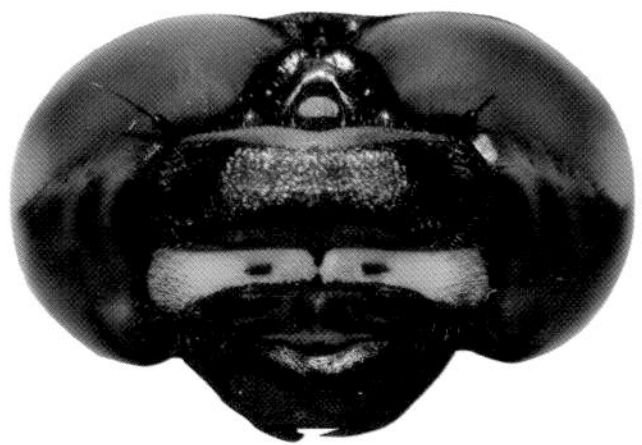

山裂唇蜓 雌
Chlorogomphus (*Sinorogomphus*) *shanicus*, female

华裂唇蜓亚属 头部正面观
Subgenus *Sinorogomphus*, head in frontal view

细腹裂唇蜓 *Chlorogomphus* (*Sinorogomphus*) *gracilis* Wilson & Reels, 2001

【形态特征】雄性面部黑色，后唇基和上额黄色；胸部肩前条纹和肩条纹较宽阔，合胸侧面具1条甚阔的黄色条纹；腹部黑色，第1~7节具黄斑。雌性与雄性相似但更粗壮。【长度】体长 91~93 mm，腹长 73~75 mm，后翅 46~51 mm。【栖息环境】海拔 500~1000 m森林中的小溪。【分布】中国海南特有。【飞行期】4—6月。

[Identification] Male face black, postclypeus and top of frons yellow. Thorax with broad antehumeral and humeral stripes, laterally with a broad yellow stripe. Abdomen black, S1-S7 with yellow spots. Female similar to male but stouter. **[Measurements]** Total length 91-93 mm, abdomen 73-75 mm, hind wing 46-51 mm. **[Habitat]** Streams in forest at 500-1000 m elevation. **[Distribution]** Endemic to Hainan of China. **[Flight Season]** April to June.

细腹裂唇蜓 雌，海南
Chlorogomphus (*Sinorogomphus*) *gracilis*, female from Hainan

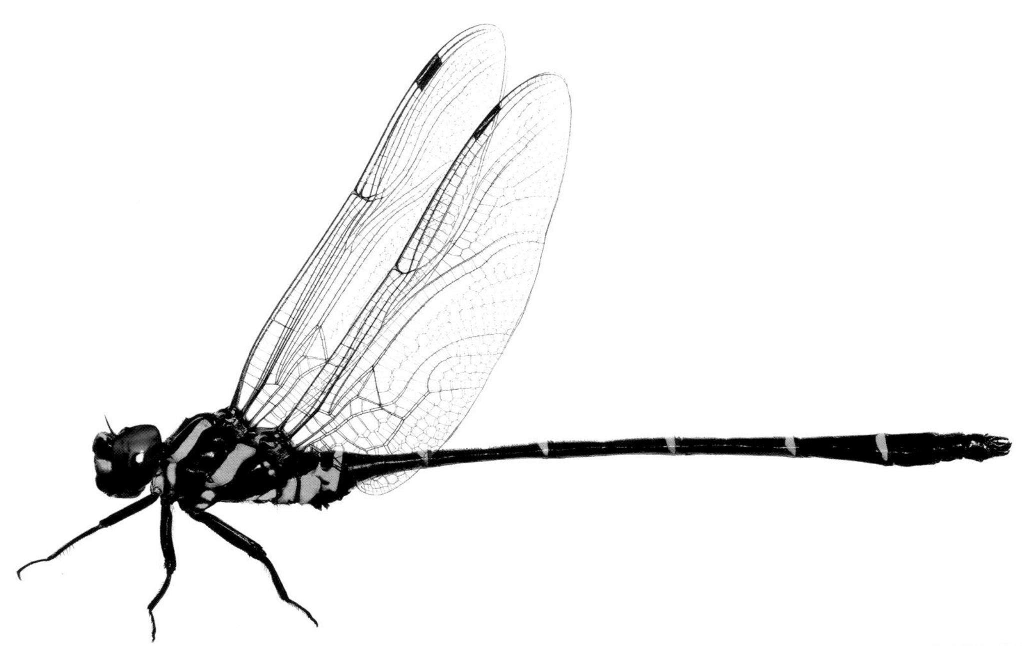

细腹裂唇蜓 雄，海南
Chlorogomphus (*Sinorogomphus*) *gracilis*, male from Hainan

老挝裂唇蜓 *Chlorogomphus* (*Sinorogomphus*) *hiten* (Sasamoto, Yokoi & Teramoto, 2011)

老挝裂唇蜓 雄，云南（西双版纳）
Chlorogomphus (*Sinorogomphus*) *hiten*, male from Yunnan (Xishuangbanna)

【形态特征】雄性面部黑色，上唇具1对黄斑，后唇基和上额黄色；胸部具黄色的肩前条纹和肩条纹，合胸侧面具1条甚阔的黄色条纹；腹部黑色，第1~7节具黄斑。雌性与雄性相似，有时胸部侧面具2~3条黄条纹。【长度】体长75~77 mm，腹长 59~60 mm，后翅 42~47 m 【栖息环境】海拔 500~1500 m森林中的小溪。【分布】云南（西双版纳、红河）；老挝、越南。【飞行期】4—7月。

老挝裂唇蜓 雌，云南（西双版纳）
Chlorogomphus (*Sinorogomphus*) *hiten*, female from Yunnan (Xishuangbanna)

[Identification] Male face black, labrum with a pair of yellow spots, postclypeus and top of frons yellow. Thorax with yellow antehumeral and humeral stripes, laterally with a broad yellow stripe. Abdomen black, S1-S7 with yellow spots. Female similar to male, laterally of thorax with 2-3 yellow stripes occasionally. **[Measurements]** Total length 75-77 mm, abdomen 59-60 mm, hind wing 42-47 mm. **[Habitat]** Streams in forest at 500-1500 m elevation. **[Distribution]** Yunnan (Xishuangbanna, Honghe); Laos, Vietnam. **[Flight Season]** April to July.

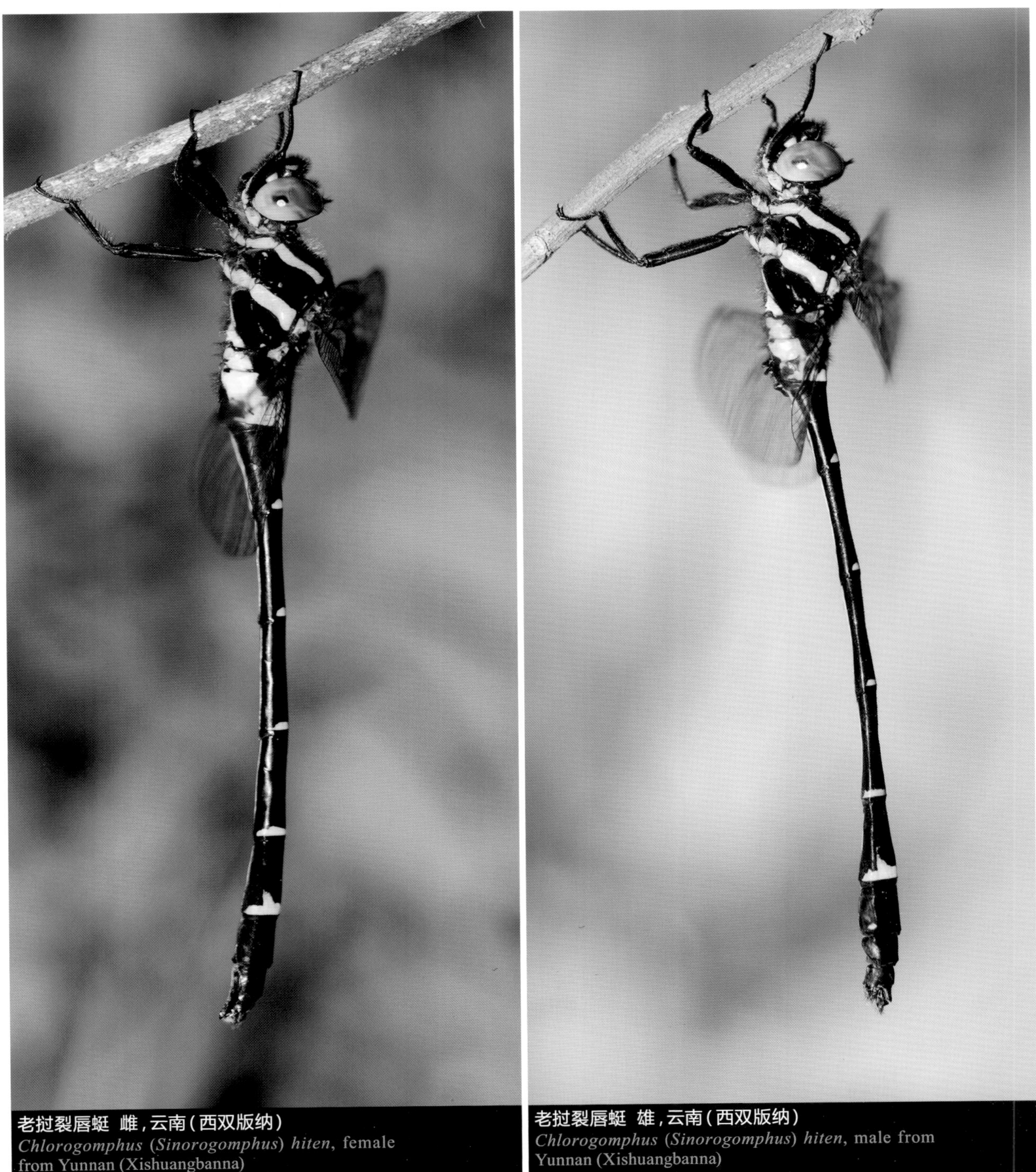

老挝裂唇蜓 雌，云南（西双版纳）
Chlorogomphus (*Sinorogomphus*) *hiten*, female from Yunnan (Xishuangbanna)

老挝裂唇蜓 雄，云南（西双版纳）
Chlorogomphus (*Sinorogomphus*) *hiten*, male from Yunnan (Xishuangbanna)

褐翅裂唇蜓 *Chlorogomphus* (*Sinorogomphus*) *infuscatus* Needham, 1930

【形态特征】雄性面部黑色，后唇基和上额黄色；胸部具黄色的肩前条纹和肩条纹，合胸侧面具1条甚阔的黄色条纹，翅基方具较小的黑褐色斑；腹部黑色，第1~7节具黄斑。雌性翅黑褐色，有时整个翅黑褐色，有时仅基方1/2黑褐色。【长度】体长 84~91 mm，腹长 65~71 mm，后翅 47~56 mm。【栖息环境】海拔 1000~1500 m森林中的溪流。【分布】中国特有，分布于四川、重庆、贵州、湖北。【飞行期】6—8月。

[Identification] Male face black, postclypeus and top of frons yellow. Thorax with yellow antehumeral and humeral stripes, laterally with a broad yellow stripe, wings hyaline with small blackish brown spots at wing bases. Abdomen black, S1-S7 with yellow spots. Female wings tinted with blackish brown, sometimes wings wholly blackish brown, sometimes only basal half of wings blackish brown. **[Measurements]** Total length 84-91 mm, abdomen 65-71 mm, hind wing 47-56 mm. **[Habitat]** Streams in forest at 1000-1500 m elevation. **[Distribution]** Endemic to China, recorded from Sichuan, Chongqing, Guizhou, Hubei. **[Flight Season]** June to August.

褐翅裂唇蜓 雄，重庆
Chlorogomphus (*Sinorogomphus*) *infuscatus*, male from Chongqing

褐翅裂唇蜓 雄，重庆
Chlorogomphus (*Sinorogomphus*) *infuscatus*,
male from Chongqing

褐翅裂唇蜓 雌，重庆
Chlorogomphus (*Sinorogomphus*) *infuscatus*,
female from Chongqing

褐翅裂唇蜓 雌，重庆
Chlorogomphus (*Sinorogomphus*) *infuscatus*, female from Chongqing

褐翅裂唇蜓 雌，贵州
Chlorogomphus (*Sinorogomphus*) *infuscatus*, female from Guizhou

长腹裂唇蜓 *Chlorogomphus* (*Sinorogomphus*) *kitawakii* Karube, 1995

【形态特征】雄性面部黑色，后唇基和上额黄色，额呈锥状向体前方突起；胸部肩前条纹和肩条纹较宽阔，合胸侧面具1条甚阔的黄色条纹；腹部黑色，第1~6节具黄斑，第6节黄斑较大。雌性与雄性相似但更粗壮。【长度】体长96~102 mm，腹长 77~81 mm，后翅 47~52 mm。【栖息环境】海拔 1000 m以下森林中的溪流。【分布】中国特有，分布于福建、广东、广西。【飞行期】4—8月。

长腹裂唇蜓 雄，广东 | 宋睿斌 摄
Chlorogomphus (*Sinorogomphus*) *kitawakii*, male from Guangdong | Photo by Ruibin Song

长腹裂唇蜓 雌，广东 | 宋睿斌 摄
Chlorogomphus (*Sinorogomphus*) *kitawakii*, female from Guangdong | Photo by Ruibin Song

[Identification] Male face black, postclypeus and top of frons yellow, triangular antefrons protruding anteriorly. Thorax with broad antehumeral and humeral stripes, laterally with a broad yellow stripe. Abdomen black, S1-S6 with yellow spots, S6 with large yellow spots. Female similar to male but stouter. **[Measurements]** Total length 96-102 mm, abdomen 77-81 mm, hind wing 47-52 mm. **[Habitat]** Streams in forest below 1000 m elevation. **[Distribution]** Endemic to China, recorded from Fujian, Guangdong, Guangxi. **[Flight Season]** April to August.

长腹裂唇蜓 雄，广东 | 宋睿斌 摄
Chlorogomphus (Sinorogomphus) kitawakii, male from Guangdong | Photo by Ruibin Song

长腹裂唇蜓 雌，广东 | 宋睿斌 摄
Chlorogomphus (Sinorogomphus) kitawakii, female from Guangdong | Photo by Ruibin Song

武夷裂唇蜓 *Chlorogomphus* (*Sinorogomphus*) *montanus* (Chao, 1999)

【形态特征】本种与长鼻裂唇蜓相似，但雌性后翅基方具褐斑。【长度】雌性体长 88~93 mm，腹长 67~75mm，后翅 60 mm。【栖息环境】1500 m以下森林中的溪流。【分布】中国特有，分布于湖北、福建、广东。【飞行期】5—8月。

武夷裂唇蜓　雌，湖北
Chlorogomphus (*Sinorogomphus*) *montanus*, female from Hubei

[Identification] Similar to *C. nasutus*, but female hind wings with the basal brown marking. **[Measurements]** Female total length 88-93 mm, abdomen 67-75 mm, hind wing 60 mm. **[Habitat]** Streams in forest below 1500 m elevation. **[Distribution]** Endemic to China, recorded from Hubei, Fujian, Guangdong. **[Flight Season]** May to August.

长鼻裂唇蜓指名亚种 *Chlorogomphus* (*Sinorogomphus*) *nasutus nasutus* Needham, 1930

【形态特征】雄性面部黑色，后唇基和上额黄色，额向体前方呈锥状突起；胸部肩前条纹和肩条纹较宽阔，合胸侧面具1条甚阔的黄色条纹；腹部黑色，第1~6节具黄斑，第6节黄斑较大。雌性与雄性相似，翅基方有时具褐斑。【长度】体长 88~93 mm，腹长 67~73 mm，后翅 52~58 mm。【栖息环境】海拔 2000 m以下森林中的狭窄溪流和渗流地。【分布】四川、贵州、湖北、湖南、浙江、福建、广东、广西；越南。【飞行期】3—9月。

[Identification] Male face black, postclypeus and top of frons yellow, triangular antefrons protruding anteriorly. Thorax with broad antehumeral and humeral stripes, laterally with a broad yellow stripe. Abdomen black, S1-S6 with yellow spots, S6 with large yellow spots. Female similar to male, sometimes wing bases with brown spots. **[Measurements]** Total length 88-93 mm, abdomen 67-73 mm, hind wing 52-58 mm. **[Habitat]** Narrow streams and seepages in forest below 2000 m elevation. **[Distribution]** Sichuan, Guizhou, Hubei, Hunan, Zhejiang, Fujian, Guangdong, Guangxi; Vietnam. **[Flight Season]** March to September.

长鼻裂唇蜓指名亚种　雄，广西
Chlorogomphus (*Sinorogomphus*) *nasutus nasutus*, male from Guangxi

长鼻裂唇蜓指名亚种　雌，广西
Chlorogomphus (*Sinorogomphus*) *nasutus nasutus*, female from Guangxi

长鼻裂唇蜓指名亚种　雌，广西
Chlorogomphus (*Sinorogomphus*) *nasutus nasutus*, female from Guangxi

长鼻裂唇蜓指名亚种 雄，广西
Chlorogomphus (Sinorogomphus) nasutus nasutus, male from Guangxi

长鼻裂唇蜓越南亚种 *Chlorogomphus (Sinorogomphus) nasutus satoi* Asahina, 1995

【形态特征】雄性面部黑色，后唇基和上额黄色，额向体前方呈锥状突起；胸部肩前条纹和肩条纹较宽阔，合胸侧面具1条甚阔的黄色条纹；腹部黑色，第1～6节具黄斑，第6节黄斑较大。雌性与雄性相似但更粗壮。【长度】体长79～88 mm，腹长 60～69 mm，后翅 48～51 mm。【栖息环境】海拔 500～1500 m森林中的狭窄溪流和渗流地。【分布】云南（红河）；越南。【飞行期】4—6月。

[Identification] Male face black, postclypeus and top of frons yellow, triangular antefrons protruding anteriorly. Thorax with broad antehumeral and humeral stripes, laterally with a broad yellow stripe. Abdomen black, S1-S6 with yellow spots, S6 with large yellow spots. Female similar to male but stouter. **[Measurements]** Total length 79-88 mm, abdomen 60-69 mm, hind wing 48-51 mm. **[Habitat]** Narrow streams and seepages in forest at 500-1500 m elevation. **[Distribution]** Yunnan (Honghe); Vietnam. **[Flight Season]** April to June.

长鼻裂唇蜓越南亚种 雄，云南（红河）
Chlorogomphus (*Sinorogomphus*) *nasutus satoi*, male from Yunnan (Honghe)

长鼻裂唇蜓越南亚种 雌，云南（红河）
Chlorogomphus (*Sinorogomphus*) *nasutus satoi*, female from Yunnan (Honghe)

长鼻裂唇蜓越南亚种 雄，云南（红河）
Chlorogomphus (*Sinorogomphus*) *nasutus satoi*, male from Yunnan (Honghe)

朴氏裂唇蜓 *Chlorogomphus* (*Sinorogomphus*) *piaoacensis* Karube, 2013

【形态特征】雄性面部黑色，后唇基和上额黄色；胸部具肩前条纹和肩条纹，合胸侧面具1条甚阔的黄色条纹；腹部黑色，第1~7节具黄斑。雌性与雄性相似，有时翅基方具黑褐色斑。【长度】体长 70~80 mm，腹长 58~63 mm，后翅 43~48 mm。【栖息环境】海拔 500~1000 m森林中的狭窄小溪。【分布】云南（红河）；越南。【飞行期】4—7月。

[Identification] Male face black, postclypeus and top of frons yellow. Thorax with yellow antehumeral and humeral stripes, laterally with a broad yellow stripe. Abdomen black, S1-S7 with yellow spots. Female similar to male, wing bases occasionally with dark brown spots. **[Measurements]** Total length 70-80 mm, abdomen 58-63 mm, hind wing 43-48 mm. **[Habitat]** Narrow streams in forest at 500-1000 m elevation. **[Distribution]** Yunnan (Honghe); Vietnam. **[Flight Season]** April to July.

朴氏裂唇蜓 雌，云南（红河）
Chlorogomphus (*Sinorogomphus*) *piaoacensis*, female from Yunnan (Honghe)

朴氏裂唇蜓 雄，云南（红河）
Chlorogomphus (*Sinorogomphus*) *piaoacensis*, male from Yunnan (Honghe)

朴氏裂唇蜓 雄，云南（红河）
Chlorogomphus (*Sinorogomphus*) *piaoacensis*, male from Yunnan (Honghe)

朴氏裂唇蜓 雌，云南（红河）
Chlorogomphus (*Sinorogomphus*) *piaoacensis*, female from Yunnan (Honghe)

中越裂唇蜓 *Chlorogomphus* (*Sinorogomphus*) *sachiyoae* Karube, 1995

中越裂唇蜓 雌，广西
Chlorogomphus (*Sinorogomphus*) *sachiyoae*, female from Guangxi

中越裂唇蜓 雄，广西
Chlorogomphus (*Sinorogomphus*) *sachiyoae*, male from Guangxi

【形态特征】雄性面部黑色，后唇基和上额黄色，额呈锥状向体前方突起；胸部肩前条纹和肩条纹较宽阔，合胸侧面具1条甚阔的黄色条纹；腹部黑色，第1～6节具黄斑，第6节黄斑较大。雌性与雄性相似但更粗壮。本种与长腹裂唇蜓相似，但腹部的比例明显不同，中越裂唇蜓的腹部明显短于后者。【长度】体长 86～92 mm，腹长 68～73 mm，后翅 46～50 mm。【栖息环境】海拔 1000 m以下森林中的溪流。【分布】云南（红河）、广西；越南。【飞行期】4—8月。

中越裂唇蜓 雌，广西
Chlorogomphus (Sinorogomphus) sachiyoae, female from Guangxi

中越裂唇蜓 雄，广西
Chlorogomphus (Sinorogomphus) sachiyoae, male from Guangxi

[Identification] Male face black, postclypeus and top of frons yellow, triangular antefrons protruding anteriorly. Thorax with broad antehumeral and humeral stripes, laterally with a broad yellow stripe. Abdomen black, S1-S6 with yellow spots, S6 with large yellow spots. Female similar to male but stouter. Similar to *C. kitawakii*, but abdominal proportion is clearly different, abdomen much shorter in *C. sachiyoae*. **[Measurements]** Total length 86-92 mm, abdomen 68-73 mm, hind wing 46-50 mm. **[Habitat]** Streams in forest below 1000 m elevation. **[Distribution]** Yunnan (Honghe), Guangxi; Vietnam. **[Flight Season]** April to August.

山裂唇蜓 *Chlorogomphus* (*Sinorogomphus*) *shanicus* Wilson, 2002

【形态特征】雄性面部黑色，后唇基和上额黄色；胸部具黄色的肩条纹和肩前条纹，合胸侧面具1条甚阔的黄色条纹；腹部黑色，第1～6节具小黄斑，第7节具较大的黄斑。雌性与雄性相似但更粗壮。【长度】体长 85～88 mm，腹长 66～69 mm，后翅 46～52 mm。【栖息环境】海拔 500～1000 m森林中的溪流。【分布】中国特有，分布于广东、湖南。【飞行期】4—8月。

[Identification] Male face black, postclypeus and top of frons yellow. Thorax with yellow antehumeral and humeral stripes, laterally with a broad yellow stripe. Abdomen black, S1-S6 with small yellow spots, S7 with large yellow spots. Female similar to male but stouter. **[Measurements]** Total length 85-88 mm, abdomen 66-69 mm, hind wing 46-52 mm. **[Habitat]** Streams in forest at 500-1000 m elevation. **[Distribution]** Endemic to China, recorded from Guangdong, Hunan. **[Flight Season]** April to August.

山裂唇蜓 雄，广东 | 宋睿斌 摄
Chlorogomphus (*Sinorogomphus*) *shanicus*, male from Guangdong | Photo by Ruibin Song

山裂唇蜓 雌，广东 | 宋睿斌 摄
Chlorogomphus (*Sinorogomphus*) *shanicus*, female from Guangdong | Photo by Ruibin Song

山裂唇蜓 雄，广东 | 宋睿斌 摄
Chlorogomphus (*Sinorogomphus*) *shanicus*, male from Guangdong | Photo by Ruibin Song

山裂唇蜓 雌，广东 | 宋睿斌 摄
Chlorogomphus (*Sinorogomphus*) *shanicus*, female from Guangdong | Photo by Ruibin Song

铃木裂唇蜓 *Chlorogomphus* (*Sinorogomphus*) *suzukii* (Oguma, 1926)

【形态特征】雄性面部黑色，后唇基和上额黄色；胸部具甚阔的肩条纹和肩前条纹，合胸侧面具1条甚阔的黄色条纹；腹部黑色，第1~7节具黄斑。雌性与雄性相似但更粗壮。此处将侗族裂唇蜓作为本种的异名。【长度】体长81~90 mm，腹长 63~70 mm，后翅 45~55 mm。【栖息环境】海拔 1500 m以下森林中的溪流。【分布】中国特有，分布于山东、河南、四川、贵州、湖北、浙江、福建、台湾。【飞行期】5—9月。

[Identification] Male face black, postclypeus and top of frons yellow. Thorax with broad antehumeral and humeral stripes, laterally with a broad yellow stripe. Abdomen black, S1-S7 with yellow spots. Female similar to male but stouter. *C. tunti* Needham, 1930 is regarded as a synonym of this species here. [Measurements] Total length 81-90 mm, abdomen 63-70 mm, hind wing 45-55 mm. [Habitat] Streams in forest below 1500 m elevation. [Distribution] Endemic to China, recorded from Shandong, Henan, Sichuan, Guizhou, Hubei, Zhejiang, Fujian, Taiwan. [Flight Season] May to September.

铃木裂唇蜓 雄，湖北
Chlorogomphus (*Sinorogomphus*) *suzukii*, male from Hubei

铃木裂唇蜓 雄，湖北
Chlorogomphus (*Sinorogomphus*) *suzukii*, male from Hubei

铃木裂唇蜓 雌，贵州
Chlorogomphus (*Sinorogomphus*) *suzukii*, female from Guizhou

铃木裂唇蜓 雌，贵州
Chlorogomphus (*Sinorogomphus*) *suzukii*, female from Guizhou

叶尾裂唇蜓 *Chlorogomphus* (*Sinorogomphus*) *urolobatus* Chen, 1950

叶尾裂唇蜓 雌，福建 | 安起迪 摄
Chlorogomphus (*Sinorogomphus*) *urolobatus*, female from Fujian | Photo by Qidi An

【形态特征】本种与长鼻裂唇蜓、武夷裂唇蜓的体型和身体条纹较近似。这3个种的雌性在翅的色彩上有较明显的差异。本种雌性与武夷裂唇蜓更相似，但前后翅均具褐色斑，武夷裂唇蜓仅后翅具明显的褐色斑，而长鼻裂唇蜓仅在翅基方具甚小的橙褐色斑或缺失。本种雄性的翅末端具褐斑，下肛附器中央具1个甚大的中央缺刻，可以与近似的长鼻裂唇蜓区分，武夷裂唇蜓的雄性未知。【长度】雌性体长 83~91 mm，腹长 65~75 mm，后翅 50~57 mm。【栖息环境】海拔1500 m以下森林中的溪流。【分布】中国特有，分布于浙江、福建、广东。【飞行期】5—8月。

[Identification] Similar to *C. nasutus* and *C. montanus* in size and body markings. Wing color is clearly different in females of the three species. This species is more similar to *C. montanus*, but both wing bases with brown spots, in female of *C. montanus* the brown spots only present on hind wings. Female of *C. nasutus* with or without very small orange brown on wing bases. Male of *C. urolobatus* can be distinguished from *C. nasutus* by the small brown tips of wings as well as the deep medial cavity of the inferior appendage. Male of *C. montanus* is still unknown. **[Measurements]** Female total length 83-91 mm, abdomen 65-75 mm, hind wing 50-57 mm. **[Habitat]** Streams in forest below 1500 m elevation. **[Distribution]** Endemic to China, recorded from Zhejiang, Fujian, Guangdong. **[Flight Season]** May to August.

华裂唇蜓亚属待定种1 *Chlorogomphus* (*Sinorogomphus*) sp. 1

【形态特征】雄性面部黑色，上唇具1对黄斑，后唇基和上额黄色；胸部具黄色的肩前条纹和肩条纹，合胸侧面具1条甚阔的黄色条纹；腹部黑色，第1~8节具黄斑。【长度】雄性体长 77 mm，腹长 61 mm，后翅 41 mm。【栖息环境】海拔 500 m处森林中的小溪。【分布】云南（普洱）。【飞行期】4—6月。

华裂唇蜓属待定种1 雄，云南（普洱）
Chlorogomphus (*Sinorogomphus*) sp. 1, male from Yunnan (Pu'er)

[Identification] Male face black, labrum with a pair of yellow spots, postclypeus and top of frons yellow. Thorax with yellow antehumeral and humeral stripes, laterally with a broad yellow stripe. Abdomen black, S1-S8 with yellow spots. **[Measurements]** Male total length 77 mm, abdomen 61 mm, hind wing 41 mm. **[Habitat]** Streams in forest at 500 m elevation. **[Distribution]** Yunnan (Pu'er). **[Flight Season]** April to June.

华裂唇蜓亚属待定种2 *Chlorogomphus* (*Sinorogomphus*) sp. 2

【形态特征】雄性面部黑色，上唇具1对黄斑，后唇基和上额黄色；胸部具黄色的肩前条纹和肩条纹，合胸侧面具1条甚阔的黄色条纹和2条较细的黄色条纹；腹部黑色，第1～8节具黄斑。雌性与雄性相似。【长度】体长 71～80 mm，腹长 56～64 mm，后翅 39～46 mm。【栖息环境】海拔 500～1500 m森林中的小溪。【分布】云南（普洱、临沧）。【飞行期】4—6月。

华裂唇蜓亚属待定种2 雄，云南（普洱）
Chlorogomphus (*Sinorogomphus*) sp. 2, male from Yunnan (Pu'er)

华裂唇蜓亚属待定种2 雌，云南（普洱）
Chlorogomphus (*Sinorogomphus*) sp. 2, female from Yunnan (Pu'er)

华裂唇蜓亚属待定种2 雄，云南（普洱）
Chlorogomphus (*Sinorogomphus*) sp. 2, male from Yunnan (Pu'er)

华裂唇蜓亚属待定种2 雌，云南（普洱）
Chlorogomphus (*Sinorogomphus*) sp. 2, female from Yunnan (Pu'er)

[Identification] Male face black, labrum with a pair of yellow spots, postclypeus and top of frons yellow. Thorax with yellow antehumeral and humeral stripes, laterally with a broad yellow stripe and two narrow yellow stripes. Abdomen black, S1-S8 with yellow spots. Female similar to male. **[Measurements]** Total length 71-80 mm, abdomen 56-64 mm, hind wing 39-46 mm. **[Habitat]** Streams in forest at 500-1500 m elevation. **[Distribution]** Yunnan (Pu'er, Lincang). **[Flight Season]** April to June.

华裂唇蜓亚属待定种3 *Chlorogomphus* (*Sinorogomphus*) sp. 3

【形态特征】本种与铃木裂唇蜓相似，但体型稍小，翅具褐斑，腹部的黄斑较小。【长度】体长 77～81 mm，腹长 57～62 mm，后翅 48～53 mm。【栖息环境】海拔 1000 m以下森林中的溪流。【分布】浙江。【飞行期】5—7月。

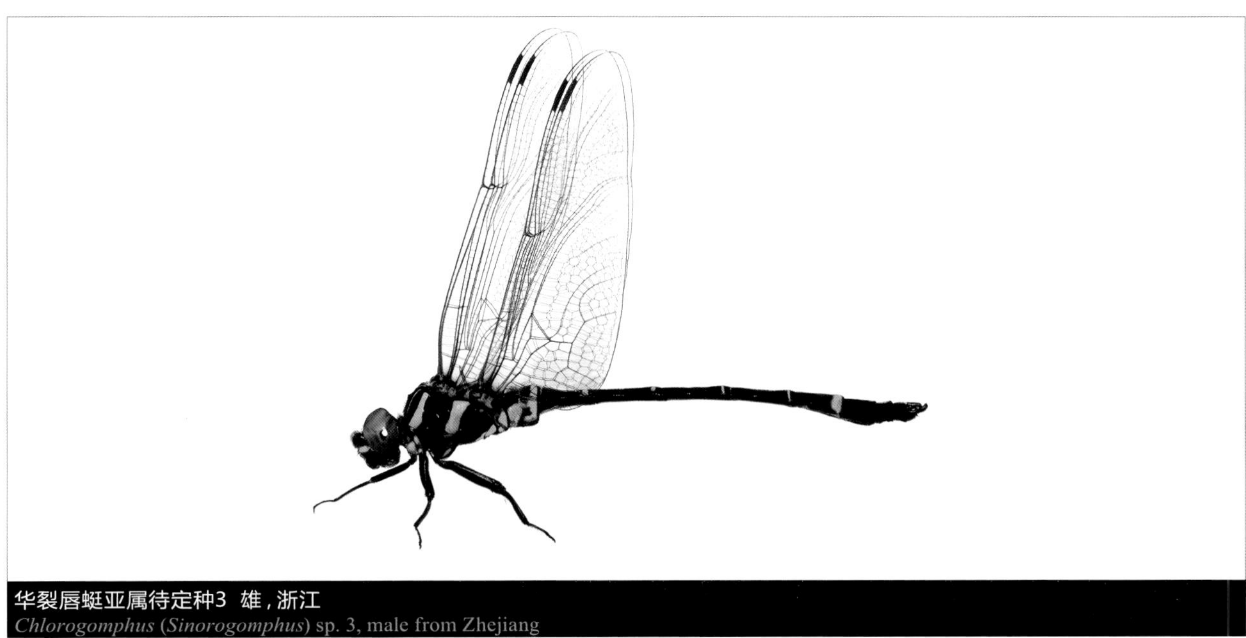

华裂唇蜓亚属待定种3 雄，浙江
Chlorogomphus (*Sinorogomphus*) sp. 3, male from Zhejiang

[Identification] Similar to *C. suzukii* but smaller, wings with brown markings, abdomen with smaller yellow stripes. [Measurements] Total length 77-81 mm, abdomen 57-62 mm, hind wing 48-53 mm. [Habitat] Streams in forest below 1000 m elevation. [Distribution] Zhejiang. [Flight Season] May to July.

华裂唇蜓亚属待定种3 雌，浙江
Chlorogomphus (*Sinorogomphus*) sp. 3, female from Zhejiang

华裂唇蜓亚属待定种4 *Chlorogomphus* (*Sinorogomphus*) sp. 4

【形态特征】本种与铃木裂唇蜓相似，但身体比例不同，腹部较短，腹部第2节侧面以及第7节背面黄斑的形状不同。【长度】雄性体长 77~84 mm，腹长 58~63 mm，后翅 45~48 mm。【栖息环境】海拔 1000 m以下森林中的溪流。【分布】广西；越南。【飞行期】4—7月。

[Identification] Similar to *C. suzukii* but with different body proportions and shorter abdomen, yellow markings on side of S2 and dorsum of S7 of different shape. **[Measurements]** Male total length 77-84 mm, abdomen 58-63 mm, hind wing 45-48 mm. **[Habitat]** Streams in forest below 1000 m elevation. **[Distribution]** Guangxi; Vietnam. **[Flight Season]** April to July.

华裂唇蜓亚属待定种4 雄，广西
Chlorogomphus (*Sinorogomphus*) sp. 4, male from Guangxi

华裂唇蜓亚属待定种4 雌，广西
Chlorogomphus (*Sinorogomphus*) sp. 4, female from Guangxi

华裂唇蜓亚属待定种4 雄，广西
Chlorogomphus (*Sinorogomphus*) sp. 4, male from Guangxi

华裂唇蜓亚属待定种4 雌，广西
Chlorogomphus (*Sinorogomphus*) sp. 4, female from Guangxi

花裂唇蜓亚属 Subgenus *Petaliorogomphus* Karube, 2013

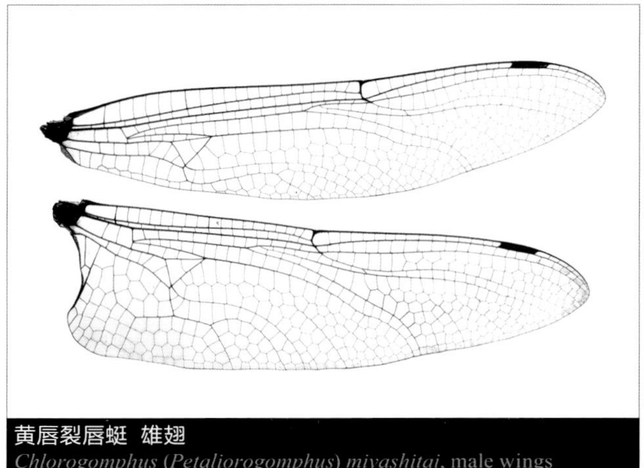

黄唇裂唇蜓 雄翅
Chlorogomphus (*Petaliorogomphus*) *miyashitai*, male wings

本亚属已知仅2种，分布于中国西南、老挝和越南。中国已知1种，仅分布于云南。本亚属翅透明；雄性腹部第7~9节向侧面膨大，上肛附器平直，腹方具齿，下肛附器侧面观在基方1/2处膨大，背面观时其端部的两个分枝为三角形。

本亚属栖息于山区溪流。在云南它们通常生活在海拔1000 m左右的茂盛森林。雄性会以慢速的低空飞行沿溪流巡飞，与华裂唇蜓亚属的行为相似。

The subgenus contains only two species distributed in Southwest China, Laos and Vietnam. Only one described species is recorded from Yunnan, China. Species of the subgenus have hyaline wings. Male S7-S9 expanded laterally, superior appendages straight with ventral teeth, inferior appendage expanded at basal half in lateral view, the two distal branches triangular in dorsal view.

Species of this subgenus inhabit montane streams where they occur along mountain streams in Yunnan at about 1000 m elevation. Males patrol low above streams for territory, similar in behavior to subgenus *Sinorogomphus*.

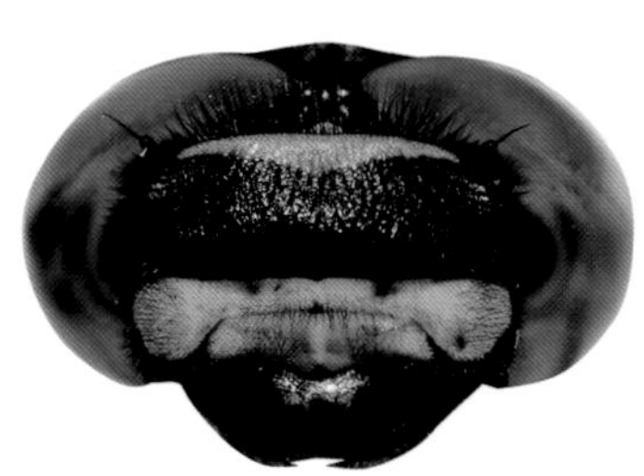

黄唇裂唇蜓 雄
Chlorogomphus (*Petaliorogomphus*) *miyashitai*, male

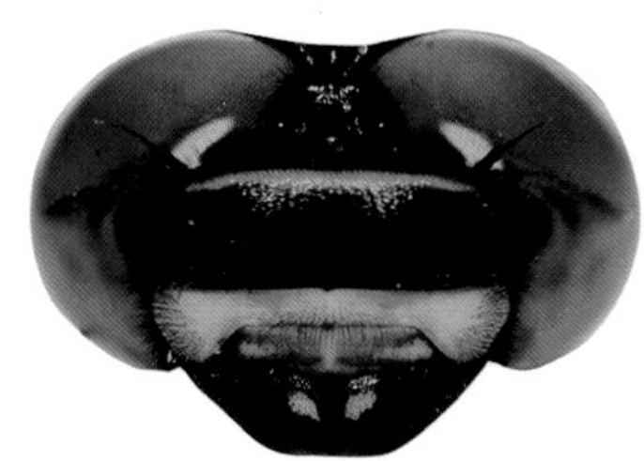

花裂唇蜓亚属待定种 雌
Chlorogomphus (*Petaliorogomphus*) sp., female

花裂唇蜓亚属 头部正面观
Subgenus *Petaliorogomphus*, head in frontal view

黄唇裂唇蜓 雄
Chlorogomphus (*Petaliorogomphus*) *miyashitai*, male

黄唇裂唇蜓
Chlorogomphus (*Petaliorogomphus*) *miyashitai*

花裂唇蜓亚属 雄性肛附器
Subgenus *Petaliorogomphus*, male anal appendages

黄唇裂唇蜓
Chlorogomphus (*Petaliorogomphus*) *miyashitai*

花裂唇蜓亚属 雌性下生殖板
Subgenus *Petaliorogomphus*, female vulvar lamina

黄唇裂唇蜓 *Chlorogomphus* (*Petaliorogomphus*) *miyashitai* Karube, 1995

【形态特征】雄性面部黑色，上唇具黄斑，唇基和上额黄色；胸部具黄色的肩前条纹和肩条纹，合胸侧面具1条甚阔的黄条纹和2条较细的黄条纹；腹部黑色，第1~7节具黄斑。雌性与雄性相似，翅基方染有琥珀色斑。【长度】体长 72~81 mm，腹长 55~63 mm，后翅 41~50 mm。【栖息环境】海拔 800~1500 m森林中的溪流。【分布】云南（普洱）；老挝。【飞行期】4—6月。

[Identification] Male face black, labrum with yellow spots, clypeus and top of frons yellow. Thorax with yellow antehumeral and humeral stripes, laterally with a broad yellow stripe and two narrow yellow stripes. Abdomen black,

黄唇裂唇蜓 雄，云南（普洱）
Chlorogomphus (*Petaliorogomphus*) *miyashitai*, male from Yunnan (Pu'er)

S1-S7 with yellow spots. Female similar to male, wing bases with amber spots. **[Measurements]** Total length 72-81 mm, abdomen 55-63 mm, hind wing 41-50 mm. **[Habitat]** Streams in forest at 800-1500 m elevation. **[Distribution]** Yunnan (Pu'er); Laos. **[Flight Season]** April to June.

黄唇裂唇蜓 雄，云南（普洱）
Chlorogomphus (Petaliorogomphus) miyashitai, male from Yunnan (Pu'er)

黄唇裂唇蜓 雌，云南（普洱）
Chlorogomphus (*Petaliorogomphus*) *miyashitai*, female from Yunnan (Pu'er)

花裂唇蜓亚属待定种 *Chlorogomphus* (*Petaliorogomphus*) sp.

花裂唇蜓亚属待定种 雌，云南（德宏）
Chlorogomphus (*Petaliorogomphus*) sp., female from Yunnan (Dehong)

【形态特征】雌性面部黑色，上唇具黄斑，唇基和上额黄色；胸部具黄色的肩前条纹和肩条纹，合胸侧面具1条甚阔的黄条纹和2条较细的黄条纹，翅基方具琥珀色斑；腹部黑色，第1~7节具黄斑。【长度】雌性体长 76 mm，腹长 60 mm，后翅 48 mm。【栖息环境】海拔 800~1200 m森林中的溪流。【分布】云南（德宏）。【飞行期】3—6月。

[Identification] Female face black, labrum with yellow spots, clypeus and top of frons yellow. Thorax with yellow antehumeral and humeral stripes, laterally with a broad yellow stripe and two narrow yellow stripes, wing bases with amber spots. Abdomen black, S1-S7 with yellow spots. **[Measurements]** Female total length 76 mm, abdomen 60 mm, hind wing 48 mm. **[Habitat]** Streams in forest at 800-1200 m elevation. **[Distribution]** Yunnan (Dehong). **[Flight Season]** March to June.

花裂唇蜓亚属待定种 雌，云南（德宏）
Chlorogomphus (*Petaliorogomphus*) sp., female from Yunnan (Dehong)

凹尾裂唇蜓属 Genus *Chloropetalia* Carle, 1995

赛丽裂唇蜓
Chloropetalia selysi

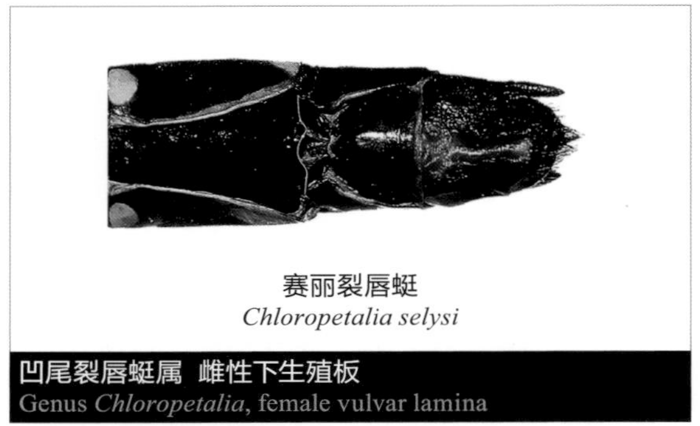

凹尾裂唇蜓属 雌性下生殖板
Genus *Chloropetalia*, female vulvar lamina

本属全球已知4种，记录于印度、越南、中国和印度尼西亚的高海拔山区。中国仅知1种，分布于云南西部的高山环境。本属体型稍小于本科其他属种类，面部甚阔而隆起，身体黑色具黄色条纹，翅透明。胸部侧面具有2条宽阔的黄色条纹，分别位于中胸后侧板和后胸后侧板，这与裂唇蜓属种类不同。

The genus contains four species confined to mountainous regions of India, Vietnam, China and Indonesia. Only one species is recorded from China, found in the high mountains from the west of Yunnan. Species of the genus is smaller than other genera of chlorogomphids, face very broad and protruding, body black with yellow markings, wings hyaline. The side of thorax has two broad yellow stripes on mesepimeron and metepimeron respectively, which in this respect differs from species of genus *Chlorogomphus*.

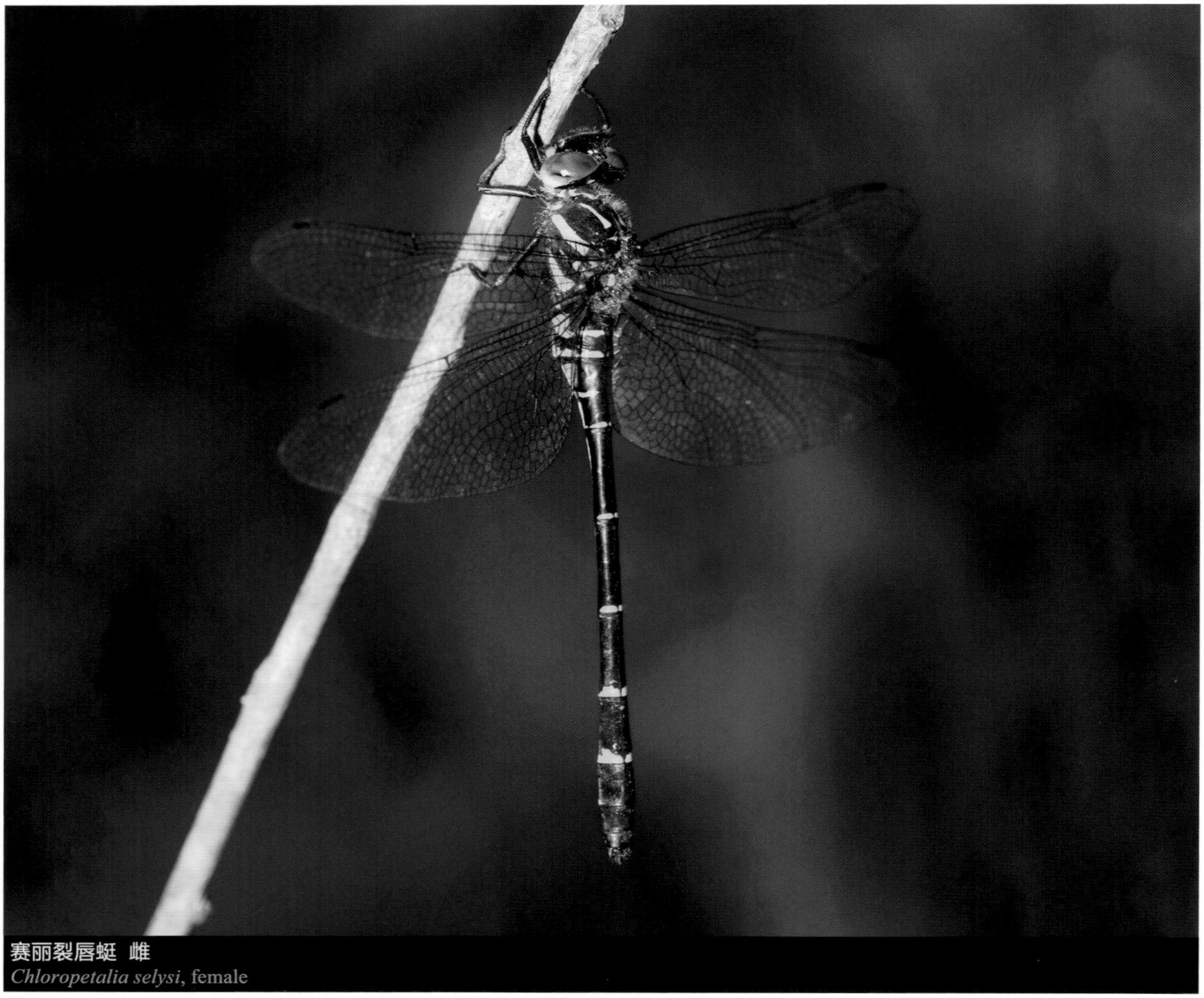

赛丽裂唇蜓 雌
Chloropetalia selysi, female

赛丽裂唇蜓 *Chloropetalia selysi* (Fraser, 1929)

赛丽裂唇蜓 雌，云南（大理）
Chloropetalia selysi, female from Yunnan (Dali)

【形态特征】雌性面部黑色，上唇具黄斑，后唇基和上额黄色；胸部仅有肩前条纹，合胸侧面具2条甚阔的黄色条纹，后胸前侧板具2~3个小黄斑；腹部黑色，第1~8节具黄斑。【长度】雌性体长 71 mm，腹长 56 mm，后翅 42 mm。【栖息环境】海拔 2000~2500 m森林中的溪流。【分布】云南（大理）；印度、尼泊尔。【飞行期】5—7月。

[Identification] Female face black, labrum with yellow spots, postclypeus and top of frons yellow. Thorax with only antehumeral stripes, laterally with two broad yellow stripes, metepisternum with 2-3 small yellow spots. Abdomen black, S1-S8 with yellow spots. **[Measurements]** Female total length 71 mm, abdomen 56 mm, hind wing 42 mm. **[Habitat]** Stream in forest at 2000-2500 m elevation. **[Distribution]** Yunnan (Dali); India, Nepal. **[Flight Season]** May to July.

楔尾裂唇蜓属 Genus *Watanabeopetalia* Karube, 2002

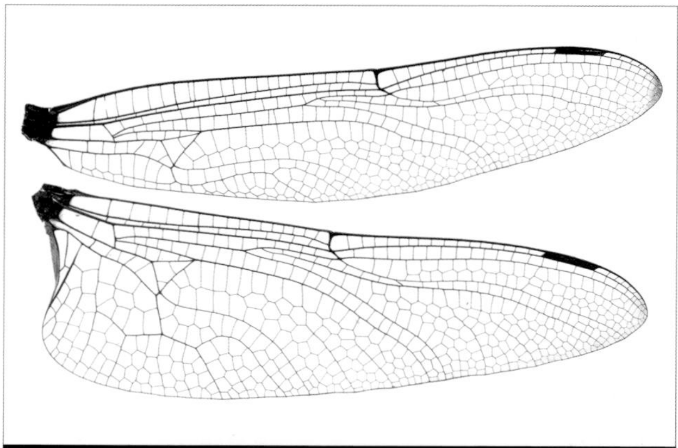

U纹裂唇蜓 雄翅
Watanabeopetalia (*Matsumotopetalia*) *usignata*, male wings

本属全球已知5种，分布于印度、尼泊尔、中国西南至越南北部地区。此处将高翔裂唇蜓移入该属，中国已知2种，都属于马特楔尾裂唇蜓亚属。本属与凹尾裂唇蜓属种类相似，但雄性肛附器、阳茎和雌性下生殖板的构造均不同。凹尾裂唇蜓属和楔尾裂唇蜓属翅的基室仅具1条横脉，而在裂唇蜓属则具2条或者更多条横脉。

本属栖息于茂盛森林中的溪流，通常生活在海拔较高的山区。雄性具有领域行为，会沿着溪流短距离巡飞。

The genus contains five species from India, Nepal, southwestern China and northern Vietnam. *Chloropetalia soarer* Wilson, 2002 is transferred here, so two described species are now recorded from China both belonging to the newly erected subgenus, *Matsumotopetalia* Karube, 2013. Species of the genus are similar to *Chloropetalia* species but differ by the structure of male anal appendages, penis and female vulvar lamina. Species of *Chloropetalia* and

U纹裂唇蜓 雄
Watanabeopetalia (*Matsumotopetalia*) *usignata*, male

Watanabeopetalia possess only one crossvein in median space instead of two or more in *Chlorogomphus*.

Watanabeopetalia species inhabit moderately high montane streams. Territorial males exhibit patrol above streams for a short distance.

高翔裂唇蜓
Watanabeopetalia (*Matsumotopetalia*) *soarer*

楔尾裂唇蜓属 雌性下生殖板
Genus *Watanabeopetalia*, female vulvar lamina

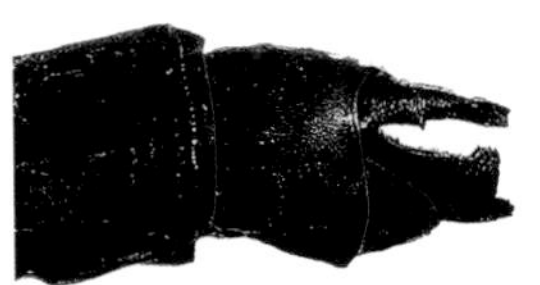

U纹裂唇蜓
Watanabeopetalia (*Matsumotopetalia*) *usignata*

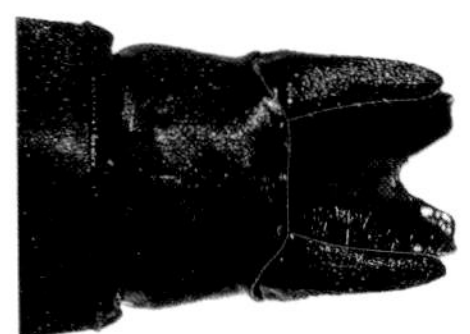

楔尾裂唇蜓属待定种
Watanabeopetalia (*Matsumotopetalia*) sp.

楔尾裂唇蜓属 雄性肛附器
Genus *Watanabeopetalia*, male anal appendages

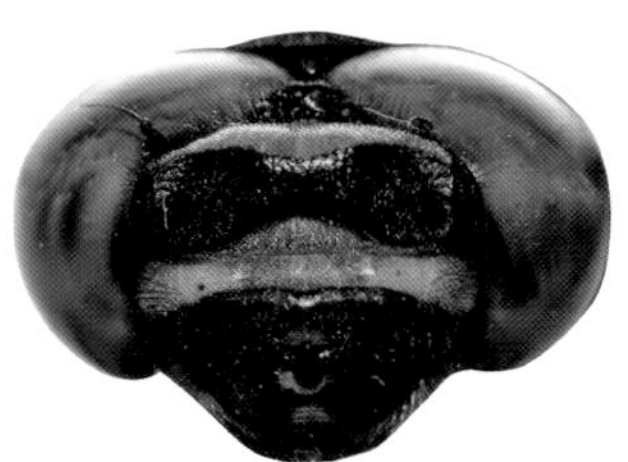

U纹裂唇蜓 雄
Watanabeopetalia (Matsumotopetalia) usignata, male

U纹裂唇蜓 雌
Watanabeopetalia (Matsumotopetalia) usignata, female

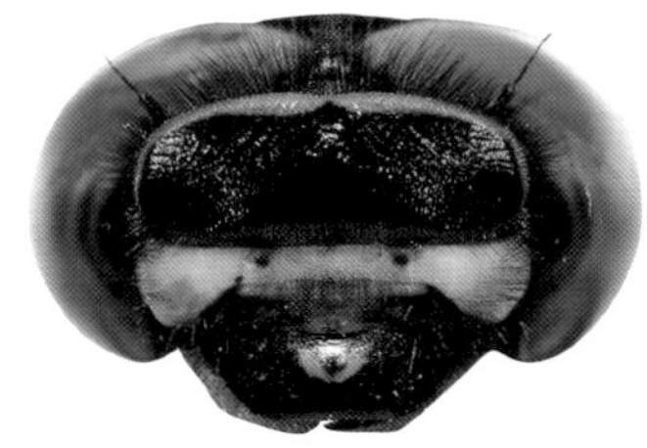

楔尾裂唇蜓属待定种 雄
Watanabeopetalia (*Matsumotopetalia*) sp., male

楔尾裂唇蜓属 头部正面观
Genus *Watanabeopetalia*, head in frontal view

高翔裂唇蜓 *Watanabeopetalia (Matsumotopetalia) soarer* (Wilson, 2002)

高翔裂唇蜓 雌，广东
Watanabeopetalia (Matsumotopetalia) soarer, female from Guangdong

【形态特征】雌性面部黑色，后唇基和上额黄色；胸部仅有肩前条纹，合胸侧面具2条甚阔的黄色条纹，翅透明，前翅染有浅褐色；腹部黑色，第1~7节具黄斑。【长度】雌性体长 78 mm，腹长 61 mm，后翅 53 mm。【栖息环境】海拔 500~2000 m森林中的溪流。【分布】中国广东特有。【飞行期】5—7月。

[Identification] Female face black, postclypeus and top of frons yellow. Thorax only with antehumeral stripes, laterally with two broad yellow stripes, wings hyaline, fore wings tinted with light brown. Abdomen black, S1-S7 with yellow spots. **[Measurements]** Female total length 78 mm, abdomen 61 mm, hind wing 53 mm. **[Habitat]** Streams in forest at 500-2000 m elevation. **[Distribution]** Endemic to Guangdong of China. **[Flight Season]** May to July.

U纹裂唇蜓 *Watanabeopetalia (Matsumotopetalia) usignata* (Chao, 1999)

【形态特征】雄性面部黑色，上唇有时具1个“U”形黄斑，后唇基和上额黄色；胸部仅有肩前条纹，合胸侧面具2条甚阔的黄色条纹，翅透明，翅端具甚小的褐斑；腹部黑色，第1~7节具黄斑。雌性与雄性相似，翅染浅褐色。【长度】体长 78~80 mm，腹长 60~62 mm，后翅 47~50 mm。【栖息环境】海拔 500~2000 m森林中的溪流。【分布】陕西、湖北、四川、贵州；越南。【飞行期】5—9月。

[Identification] Male face black, labrum sometimes with a U-shaped yellow spot, postclypeus and top of frons yellow. Thorax with only antehumeral stripes, sides with two broad yellow stripes, wings hyaline with small brown spots apically. Abdomen black, S1-S7 with yellow spots. Female similar to male, wings tinted with light brown.

[Measurements] Total length 78-80 mm, abdomen 60-62 mm, hind wing 47-50 mm. **[Habitat]** Streams in forest at 500-2000 m elevation. **[Distribution]** Shaanxi, Hubei, Sichuan, Guizhou; Vietnam. **[Flight Season]** May to September.

U纹裂唇蜓 雄，贵州
Watanabeopetalia (Matsumotopetalia) usignata, male from Guizhou

U纹裂唇蜓 雌，贵州
Watanabeopetalia (Matsumotopetalia) usignata, female from Guizhou

U纹裂唇蜓 雄，贵州
Watanabeopetalia (*Matsumotopetalia*) *usignata*, male from Guizhou

楔尾裂唇蜓属待定种 *Watanabeopetalia* (*Matsumotopetalia*) sp.

楔尾裂唇蜓属待定种 雄，云南（普洱）
Watanabeopetalia (*Matsumotopetalia*) sp., male from Yunnan (Pu'er)

楔尾裂唇蜓属待定种 雄，云南（普洱）
Watanabeopetalia (*Matsumotopetalia*) sp., male from Yunnan (Pu'er)

【形态特征】雄性面部黑色，上唇具1个“U”形黄斑，后唇基和上额黄色，额向体前方呈锥状突起；胸部仅有肩前条纹，合胸侧面具2条甚阔的黄色条纹，翅透明；腹部黑色，第1~8节具黄斑。本种与越南分布的长者裂唇蜓相似，但肛附器构造不同。【长度】雄性体长 79 mm，腹长 61 mm，后翅 44 mm。【栖息环境】海拔 1000 m处森林中的溪流。【分布】云南（普洱）。【飞行期】5—6月。

[Identification] Male face black, labrum with a U-shaped yellow spot, postclypeus and top of frons yellow, triangular antefrons protruding anteriorly. Thorax only with antehumeral stripes, laterally with two broad yellow stripes, wings hyaline. Abdomen black, S1-S8 with yellow spots. Similar to *Watanabeopetalia* (*Matsumotopetalia*) *ojisan* Karube, 2013 from Vietnam but male anal appendages different. **[Measurements]** Male total length 79 mm, abdomen 61 mm, hind wing 44 mm. **[Habitat]** Streams in forest at 1000 m elevation. **[Distribution]** Yunnan (Pu'er). **[Flight Season]** May to June.

4 大蜓科 Family Cordulegastridae

本科全球已知3属50余种，广泛分布于全北界（另有2种分布于新热带界北部）和东洋界。本科是体大型至巨型的昆虫，有些种类的雌性体长超过10 cm。复眼在头顶仅稍微分离，额隆起较高，上颚发达；身体以黑色为主，具鲜艳的黄色条纹、环纹和斑点；翅狭长而透明；雌性产卵管突出并超出腹部末端。

本科蜻蜓栖息于茂盛森林中的溪流和沟渠，偏爱狭窄而浅的泥沙溪流。成虫的飞行能力很强，是具有游荡行为的蜻蜓，经常远离溪流出没在空旷地和山脉顶峰。雄性的大蜓会沿着小溪慢速低空飞行以寻找雌性。本属雌性具特殊的产卵行为——插秧式产卵。它们身体直立，重复插秧动作将卵埋藏于水底的泥土中。一次产卵可以插秧多达500余次，持续时间10 min以上。

This family contains over 50 species in three genera, widespread within the Holarctic (two species are recorded from the northern Neotropical region) and Oriental regions. Species of the family are large or giant insects with the female of some species measuring more than 10 cm in total length. Their eyes are slightly separated above, frons strongly developed, the mouth with large mandibles. The ground color is usually black with yellow markings in the form of stripes, rings or spots throughout. Wings long and narrow. Female vulvar scale projects beyond the end of abdomen.

Species of the family inhabit streams and ditches in dense forests, they prefer narrow and shallow montane streams with muddy substrates. Adults are strong flyers and may be seen away from streams where they fly along pastures or on mountain tops. Males slowly course up down streams along a definite beat searching for females. The method of oviposition is highly characteristic, the tip of the abdomen being inserted into the substrate while the abdomen is held vertically as the insect hovers. During oviposition females repeatedly insert eggs into the muddy substrate in a series of up and down movements. Females were observed ovipositing over 500 times at a site for more than ten minutes.

金斑圆臀大蜓 雌

Anotogaster klossi, female

圆臀大蜓属 Genus *Anotogaster* Selys, 1854

本属全球已知约15种，以东洋界为分布中心，古北界已知种较少。中国已知10余种。本属的多数种类形态上十分相似，是分类学上最难的类群，需要借助分子手段才能完全确认。本属已经发现了大量未被描述的新物种。

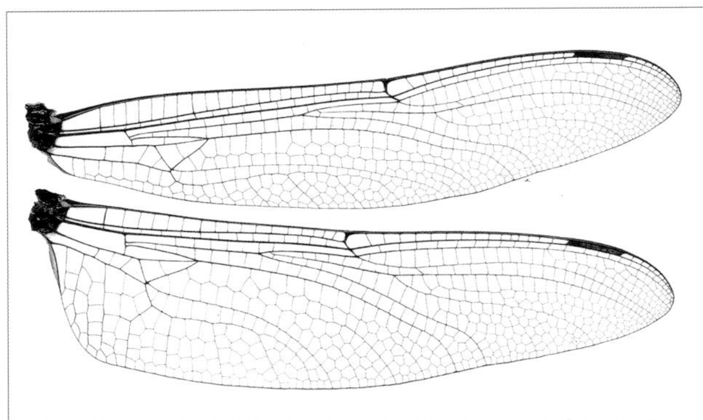
双斑圆臀大蜓　雄翅
Anotogaster kuchenbeiseri, male wings

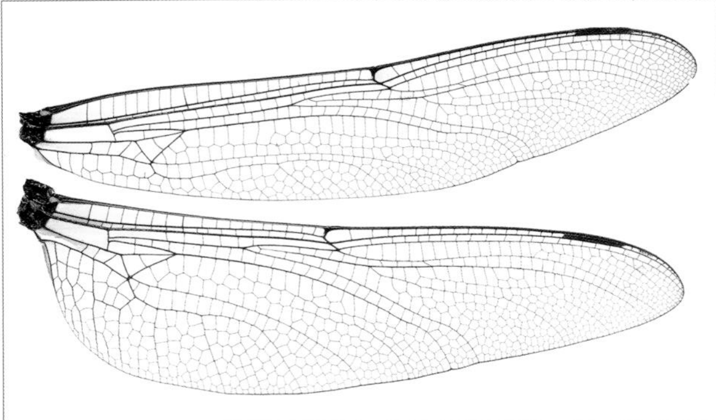
双斑圆臀大蜓　雌翅
Anotogaster kuchenbeiseri, female wings

本属多数种类体大型至巨型，身体黑色具黄色斑纹，胸部肩前条纹宽阔，合胸侧面具2条宽阔的黄色条纹；翅透明，部分种类雌性的翅染有琥珀色或金色。雄性肛附器是较重要的分类特征，此外，面部是本属重要的识别特征，包括额横纹的有无和宽度、上颚外方色彩、上唇的黄斑大小等方面。圆臀大蜓属的雄性腹部第2节没有耳状突，可以与本科的大蜓属和角臀大蜓属相区分。

双斑圆臀大蜓　雌，产卵
Anotogaster kuchenbeiseri, female, laying eggs

本属蜻蜓栖息于狭窄而浅的山区溪流和小型沟渠。雄性的圆臀大蜓经常在山路和溪流间来回飞行寻找配偶，有时它们会沿着小溪以慢速低空飞行巡视，但通常不会在一个地点停留很久。在阳光充足炎热的午后，雄性经常停落在小溪边缘的树枝上，暴露在太阳下。雌性则是以本科特有的插秧式产卵，将卵埋藏于水底的泥土中。由于体型巨大，圆臀大蜓的捕食对象很多，包括同类，一些种类具有捕食蜜蜂的喜好，它们经常游窜于蜂箱附近。

The genus contains about 15 species mainly distributed in Oriental region, a few are recorded from the Palaearctic region. Over ten species are recorded from China. The general habitus of most species is very similar, making them a difficult group for taxonomic studies. Molecular studies must be performed in order to correctly identification. Many undescribed species have await description.

赵氏圆臀大蜓 雄
Anotogaster chaoi, male

赵氏圆臀大蜓 雌
Anotogaster chaoi, female

黑额圆臀大蜓 雄
Anotogaster gigantica, male

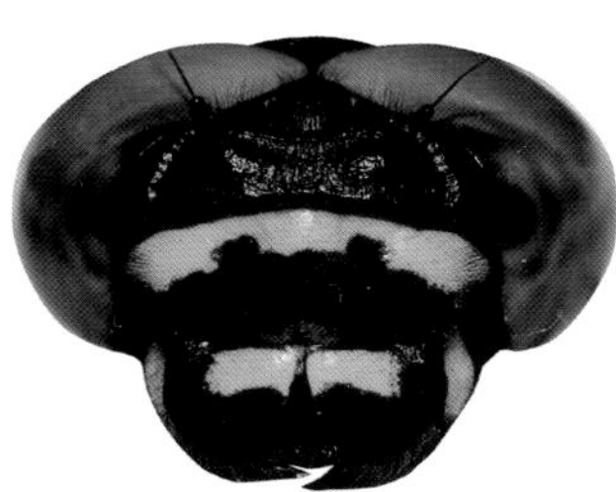

黑额圆臀大蜓 雌
Anotogaster gigantica, female

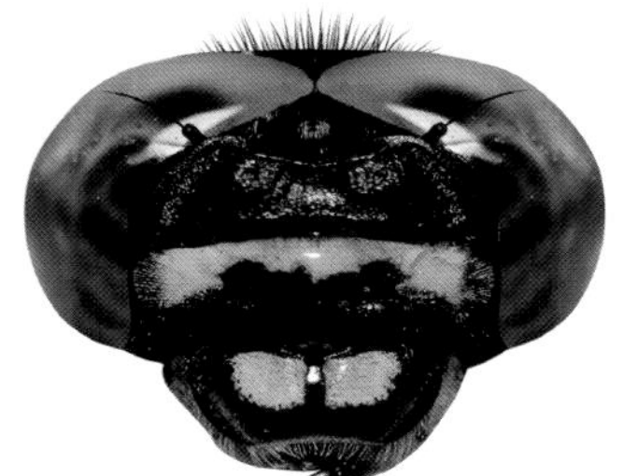

格氏圆臀大蜓 雄
Anotogaster gregoryi, male

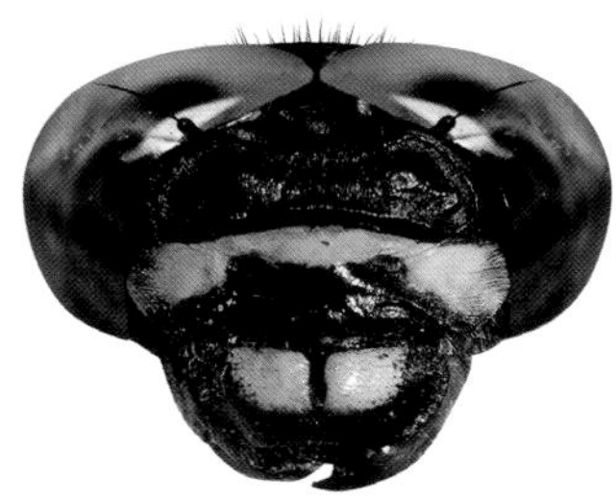

格氏圆臀大蜓 雌
Anotogaster gregoryi, female

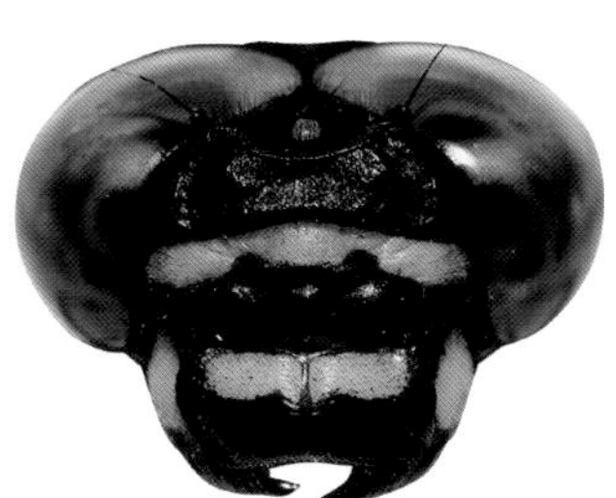

金斑圆臀大蜓 雄
Anotogaster klossi, male

金斑圆臀大蜓 雌
Anotogaster klossi, female

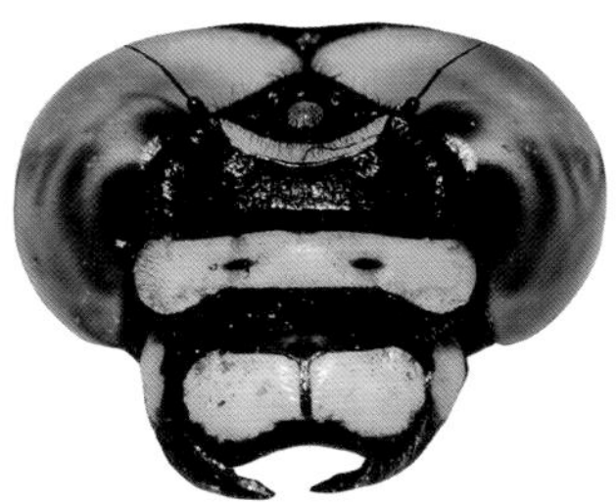

双斑圆臀大蜓 雄
Anotogaster kuchenbeiseri, male

圆臀大蜓属 头部正面观
Genus *Anotogaster*, head in frontal view

Most species of the genus are large to giant insects, body usually black with yellow markings, thorax with broad antehumeral stripes, laterally with two broad yellow stripes. Wings hyaline, in some species female wing bases tinted with amber or golden. Male anal appendages is a key for identification as well as detailed characters of face, including the presence and width of the yellow stripes on top of frons, color at the base of mandibles and the shape of spots on labrum. The auricle is absent in male *Anotogaster,* but is present in *Cordulegaster* and *Neallogaster.*

Anotogaster species prefer narrow and shallow montane streams and ditches. Males usually patrol among the paths and streams, sometimes they approach water for finding female by flying back and forth along the streams, but will not remain in an area for any great length of time. Males often perch pendently on exposed portions during periods of sunshine. Female repeatedly deposit the eggs into the substrate. Since they are very large, they prey on many insects, including their own kind. Some species prefer bees for food, and they usually stay around beehive.

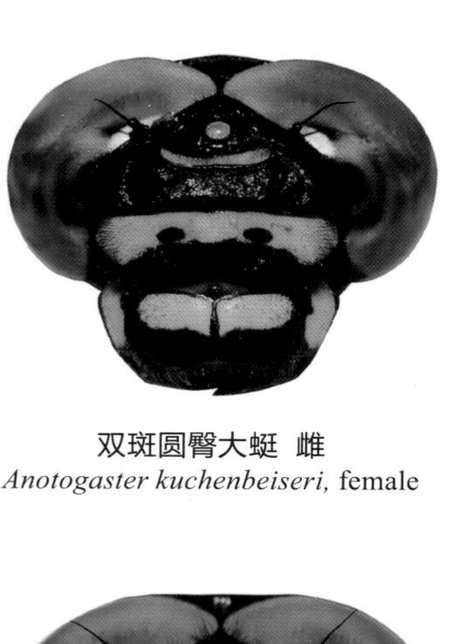

双斑圆臀大蜓 雌
Anotogaster kuchenbeiseri, female

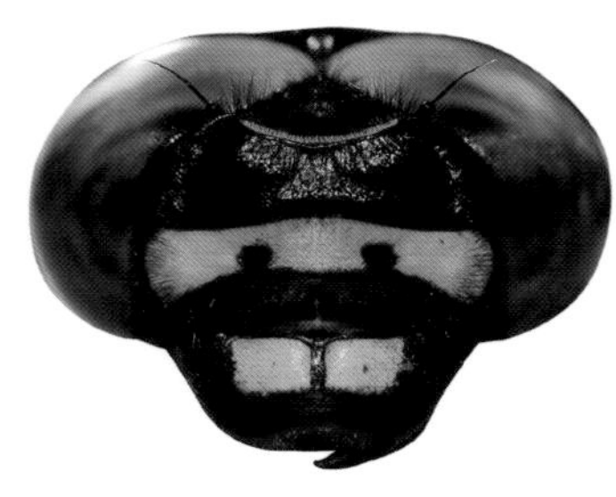

细纹圆臀大蜓 雄
Anotogaster myosa, male

细纹圆臀大蜓 雌
Anotogaster myosa, female

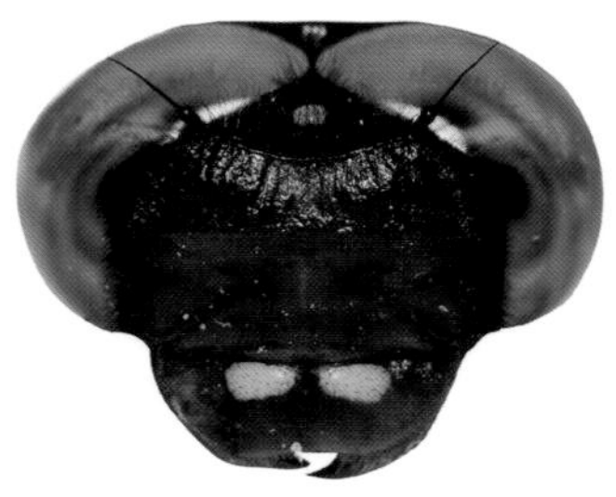

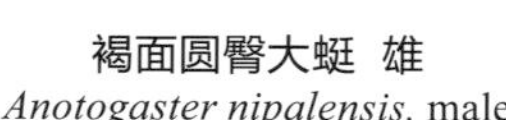

褐面圆臀大蜓 雄
Anotogaster nipalensis, male

褐面圆臀大蜓 雌
Anotogaster nipalensis, female

清六圆臀大蜓 雄
Anotogaster sakaii, male

清六圆臀大蜓 雌
Anotogaster sakaii, female

萨帕圆臀大蜓 雄
Anotogaster sapaensis, male

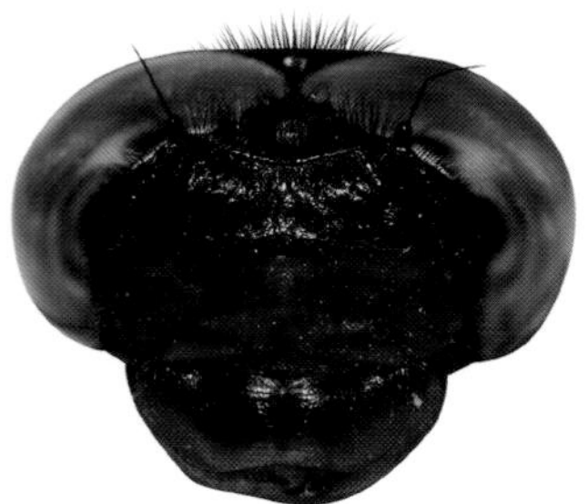

萨帕圆臀大蜓 雌
Anotogaster sapaensis, female

圆臀大蜓属 头部正面观
Genus *Anotogaster*, head in frontal view

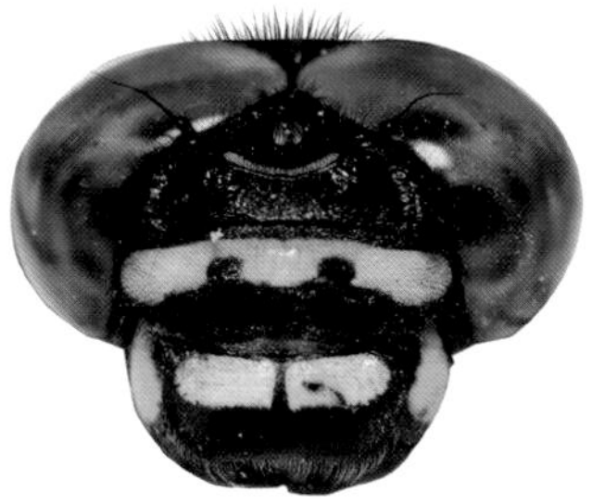

巨圆臀大蜓 雄
Anotogaster sieboldii, male

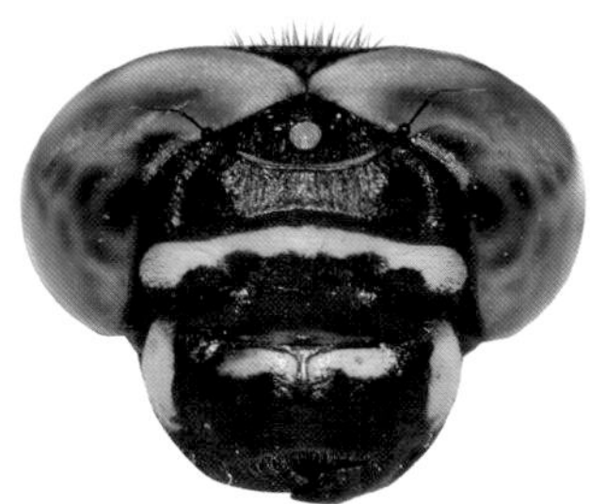

巨圆臀大蜓 雌
Anotogaster sieboldii, female

圆臀大蜓属待定种1 雄
Anotogaster sp. 1, male

圆臀大蜓属待定种1 雌
Anotogaster sp. 1, female

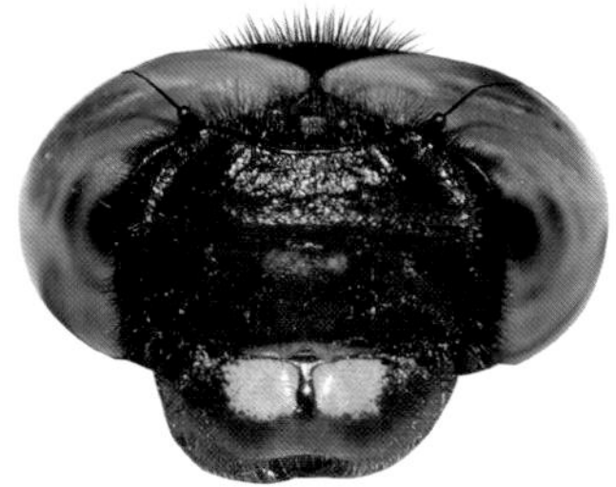

圆臀大蜓属待定种2 雄
Anotogaster sp. 2, male

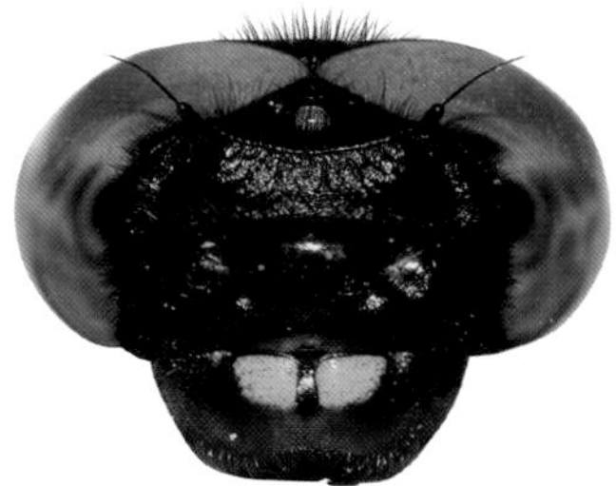

圆臀大蜓属待定种2 雌
Anotogaster sp. 2, female

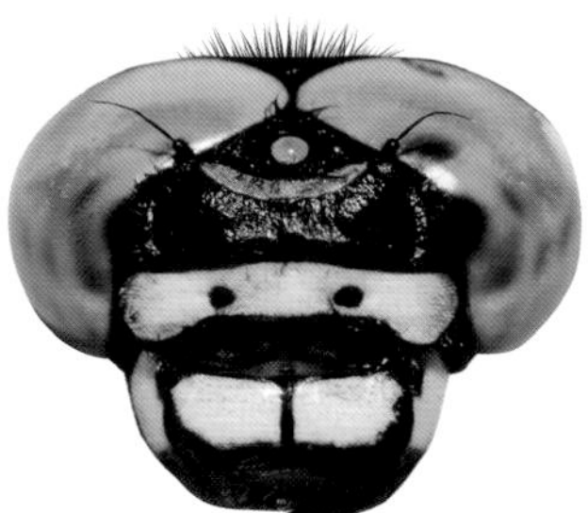

圆臀大蜓属待定种3 雄
Anotogaster sp. 3, male

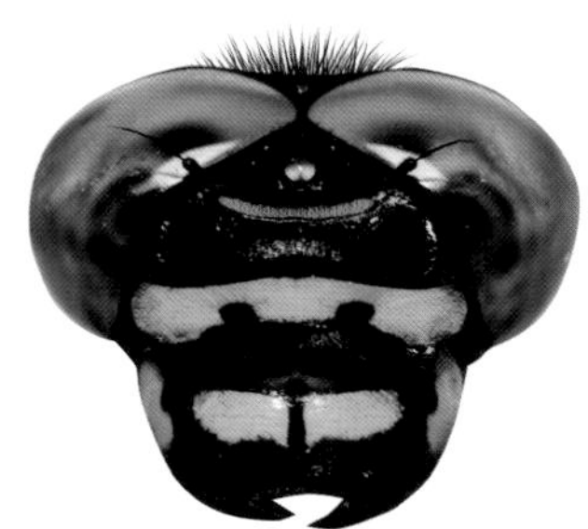

圆臀大蜓属待定种3 雌
Anotogaster sp. 3, female

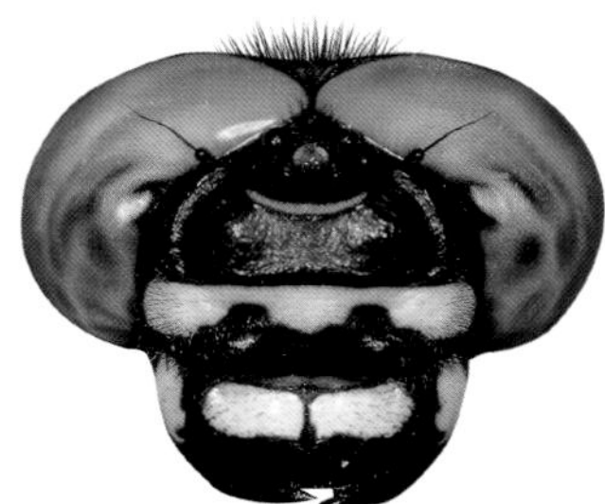

圆臀大蜓属待定种6 雄
Anotogaster sp. 6, male

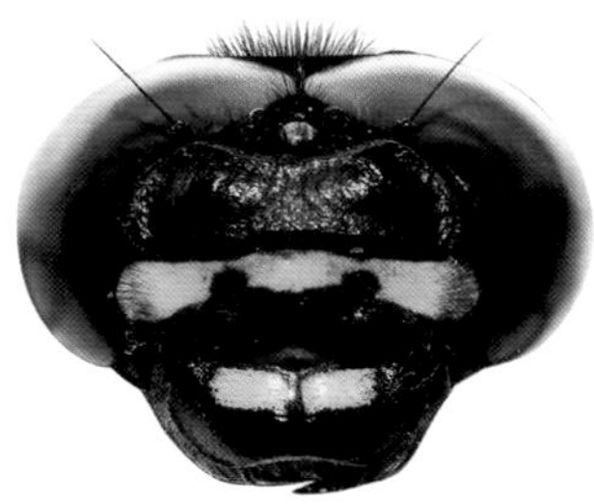

圆臀大蜓属待定种7 雄
Anotogaster sp. 7, male

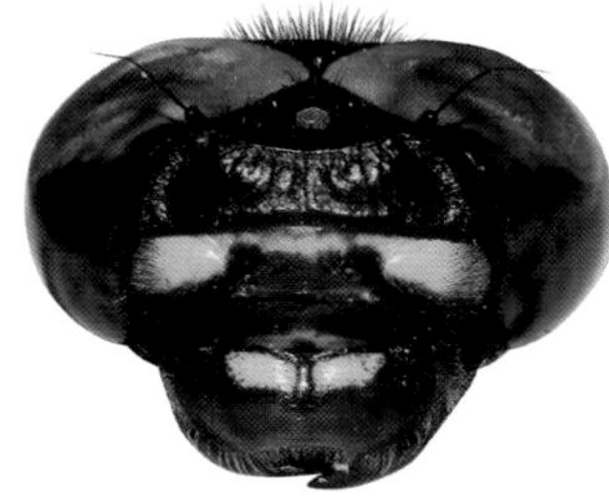

圆臀大蜓属待定种7 雌
Anotogaster sp. 7, female

圆臀大蜓属待定种8 雄
Anotogaster sp. 8, male

圆臀大蜓属 头部正面观
Genus *Anotogaster*, head in frontal view

赵氏圆臀大蜓
Anotogaster chaoi

黑额圆臀大蜓
Anotogaster gigantica

格氏圆臀大蜓
Anotogaster gregoryi

金斑圆臀大蜓
Anotogaster klossi

双斑圆臀大蜓
Anotogaster kuchenbeiseri

细纹圆臀大蜓
Anotogaster myosa

褐面圆臀大蜓
Anotogaster nipalensis

清六圆臀大蜓
Anotogaster sakaii

萨帕圆臀大蜓
Anotogaster sapaensis

巨圆臀大蜓（中国）
Anotogaster sieboldii (China)

巨圆臀大蜓（日本）
Anotogaster sieboldii (Japan)

圆臀大蜓属待定种1
Anotogaster sp. 1

圆臀大蜓属待定种2
Anotogaster sp. 2

圆臀大蜓属待定种3
Anotogaster sp. 3

圆臀大蜓属待定种4
Anotogaster sp. 4

圆臀大蜓属待定种5
Anotogaster sp. 5

圆臀大蜓属待定种6
Anotogaster sp. 6

圆臀大蜓属待定种7
Anotogaster sp. 7

圆臀大蜓属　雄性肛附器
Genus *Anotogaster*, male appendages

圆臀大蜓属待定种8
Anotogaster sp. 8

赵氏圆臀大蜓
Anotogaster chaoi

圆臀大蜓属 雄性肛附器
Genus *Anotogaster*, male appendages

圆臀大蜓属 雌性产卵管
Genus *Anotogaster*, female ovipositor

格氏圆臀大蜓
Anotogaster gregoryi

金斑圆臀大蜓
Anotogaster klossi

双斑圆臀大蜓
Anotogaster kuchenbeiseri

细纹圆臀大蜓
Anotogaster myosa

清六圆臀大蜓
Anotogaster sakaii

萨帕圆臀大蜓
Anotogaster sapaensis

巨圆臀大蜓（中国）
Anotogaster sieboldii (China)

圆臀大蜓属待定种1
Anotogaster sp. 1

圆臀大蜓属待定种2
Anotogaster sp. 2

圆臀大蜓属待定种3
Anotogaster sp. 3

圆臀大蜓属待定种5
Anotogaster sp. 5

圆臀大蜓属待定种6
Anotogaster sp. 6

圆臀大蜓属待定种7
Anotogaster sp. 7

圆臀大蜓属待定种8
Anotogaster sp. 8

黑额圆臀大蜓
Anotogaster gigantica

圆臀大蜓属 雌性产卵管
Genus *Anotogaster*, female ovipositor

赵氏圆臀大蜓 *Anotogaster chaoi* Zhou, 1998

【形态特征】雄性复眼翠绿色，上唇具1对甚大黄斑，上颚外方褐色，后唇基黄色，额具1条黄色横纹甚阔；胸部肩前条纹较长；腹部黑色，第1～9节具宽阔的黄斑。雌性体型更大，翅脉金色，腹部具更多的黄色斑点。【长度】体长 80～88 mm，腹长 60～67 mm，后翅 47～51 mm。【栖息环境】海拔 2000～3000 m森林中的渗流地、狭窄小溪和沟渠。【分布】云南（昆明、红河、大理、保山、德宏）；本种在越南的分布记录待确定。【飞行期】5—10月。

[Identification] Male eyes emerald, labrum with a pair of large yellow spots, base of mandible brown, postclypeus yellow, top of frons with a broad yellow stripe. Thorax with long antehumeral stripes. Abdomen black, S1-S9 with broad yellow stripes. Female larger, wing venation golden, abdomen with more extensive yellow spots. [Measurements] Total length 80-88 mm, abdomen 60-67 mm, hind wing 47-51 mm. [Habitat] Seepages, narrow streams and ditches in forest at 2000-3000 m elevation. [Distribution] Yunnan (Kunming, Honghe, Dali, Baoshan, Dehong); Records from Vietnam need confirmation. [Flight Season] May to October.

赵氏圆臀大蜓 雄，云南（大理）
Anotogaster chaoi, male from Yunnan (Dali)

赵氏圆臀大蜓 雌，云南（大理）
Anotogaster chaoi, female from Yunnan (Dali)

黑额圆臀大蜓 *Anotogaster gigantica* Fraser, 1924

【形态特征】雄性复眼蓝绿色，上唇具1对甚大黄斑，上颚外方黄色，后唇基黄色，额全黑色；胸部肩前条纹短；腹部黑色，第2～9节具较细的黄斑。雌性体型更大，上唇的黄斑稍小，腹部黄斑更丰富。【长度】体长 87～102 mm，腹长 66～79 mm，后翅 51～58 mm。【栖息环境】海拔 1000～2000 m森林中的渗流地、狭窄小溪和沟渠。【分布】云南（西双版纳、红河、普洱）；缅甸、越南。【飞行期】6—12月。

[Identification] Male eyes bluish green, labrum with a pair of large yellow spots, base of mandible yellow, postclypeus yellow, frons entirely black. Thorax with short antehumeral stripes. Abdomen black, S2-S9 with narrow yellow spots. Female larger, labrum with smaller yellow spots, abdomen with more extensive yellow spots. **[Measurements]** Total length 87-102 mm, abdomen 66-79 mm, hind wing 51-58 mm. **[Habitat]** Seepages, narrow streams and ditches in forest at 1000-2000 m elevation. **[Distribution]** Yunnan (Xishuangbanna, Honghe, Pu'er); Myanmar, Vietnam. **[Flight Season]** June to December.

黑额圆臀大蜓 雄，云南（红河）
Anotogaster gigantica, male from Yunnan (Honghe)

黑额圆臀大蜓 雌，云南（普洱）
Anotogaster gigantica, female from Yunnan (Pu'er)

黑额圆臀大蜓 雄，云南（红河）
Anotogaster gigantica, male from Yunnan (Honghe)

黑额圆臀大蜓 雌，云南（普洱）
Anotogaster gigantica, female from Yunnan (Pu'er)

格氏圆臀大蜓 *Anotogaster gregoryi* Fraser, 1924

【形态特征】雄性复眼翠绿色，上唇具1对较大黄斑，上颚外方褐色，后唇基黄色，额全黑色；胸部肩前条纹短；腹部黑色，第2～9节具黄斑。雌性体型更大，翅基方具琥珀色斑，腹部的黄色条纹更丰富。【长度】体长 82～100 mm，腹长 62～73 mm，后翅 44～57 mm。【栖息环境】海拔 1000～2500 m森林中的渗流地、狭窄小溪和沟渠。【分布】云南（大理、临沧、普洱、西双版纳、红河）、贵州；印度、尼泊尔、泰国、老挝、越南。【飞行期】5—8月。

[Identification] Male eyes emerald, labrum with a pair of fairly large yellow spots, base of mandible brown, postclypeus yellow, frons entirely black. Thorax with short antehumeral stripes. Abdomen black, S2-S9 with yellow stripes. Female larger, wing bases with amber markings, abdomen with broader yellow stripes. **[Measurements]** Total length 82-100 mm, abdomen 62-73 mm, hind wing 44-57 mm. **[Habitat]** Seepages, narrow streams and ditches in forest at 1000-2500 m elevation. **[Distribution]** Yunnan (Dali, Lincang, Pu'er, Xishuangbanna, Honghe), Guizhou; India, Nepal, Thailand, Laos, Vietnam. **[Flight Season]** May to August.

格氏圆臀大蜓 雄，云南（普洱）
Anotogaster gregoryi, male from Yunnan (Pu'er)

格氏圆臀大蜓 雌，云南（红河）
Anotogaster gregoryi, female from Yunnan (Honghe)

格氏圆臀大蜓 雄，云南（普洱）
Anotogaster gregoryi, male from Yunnan (Pu'er)

金斑圆臀大蜓 *Anotogaster klossi* Fraser, 1919

金斑圆臀大蜓 雄，广西
Anotogaster klossi, male from Guangxi

金斑圆臀大蜓 雌，广西
Anotogaster klossi, female from Guangxi

【形态特征】雄性复眼翠绿色或蓝绿色，上唇具1对黄斑，上颚外方黄色，后唇基大面积黄色，额黑色；胸部肩前条纹逗号形状；腹部黑色，第2～8节具黄斑，第2～3节有时大面积黄色。雌性多型，橙色型腹部第2～8节大面积金黄色；黑色型第2～8节黑色具黄斑。本种模式产地为越南南部，雌性模式标本腹部大面积橙色。本种在中国台湾和日本的记录很可能是误定。【长度】体长 93～116 mm，腹长 73～89 mm，后翅 54～67 mm。【栖息环境】海拔2000 m以下森林中的狭窄小溪、渗流地和沟渠。【分布】云南（红河）、广西；越南。【飞行期】5—8月。

[Identification] Male eyes emerald or bluish green, labrum with a pair of yellow spots, base of mandible yellow, postclypeus largely yellow, frons black. Thorax with comma-shaped antehumeral stripes. Abdomen black, S2-S8 with

金斑圆臀大蜓 雌，广西
Anotogaster klossi, female from Guangxi

金斑圆臀大蜓 雄，云南（红河）
Anotogaster klossi, male from Yunnan (Honghe)

yellow markings, sometimes S2-S3 largely yellow. Female polymorphic, the orange morph S2-S8 largely orange yellow; the black morph S2-S8 with yellow markings. Type locality is in southern Vietnam, female type specimens possess orange abdomen. Records from Taiwan of China and Japan may be incorrect. **[Measurements]** Male total length 93-116 mm, abdomen 73-89 mm, hind wing 54-67 mm. **[Habitat]** Narrow streams, seepages and ditches in forest below 2000 m elevation. **[Distribution]** Yunnan (Honghe), Guangxi; Vietnam. **[Flight Season]** May to August.

双斑圆臀大蜓 *Anotogaster kuchenbeiseri* (Förster, 1899)

【形态特征】雄性复眼翠绿色，上唇具1对甚大黄斑，上颚外方黄色，后唇基黄色，额横纹甚阔；胸部肩前条纹甚阔；腹部黑色，第2～9节具宽阔的黄色条纹。雌性体型更大，翅基方具琥珀色斑。【长度】体长 80～95 mm，腹长 60～73 mm，后翅 46～50 mm。【栖息环境】海拔 1500 m以下森林中的狭窄小溪和沟渠。【分布】中国特有，分布于北京、山西、陕西、河南、湖北、四川。【飞行期】6—9月。

[Identification] Male eyes emerald, labrum with a pair of large yellow spots, base of mandible yellow, postclypeus yellow, top of frons with a broad yellow stripe. Thorax with broad antehumeral stripes. Abdomen black, S2-S9 with broad yellow stripes. Female larger, wing bases with amber markings. [Measurements] Total length 80-95 mm, abdomen 60-73 mm, hind wing 46-50 mm. [Habitat] Narrow streams and ditches in forest below 1500 m elevation. [Distribution] Endemic to China, recorded from Beijing, Shanxi, Shaanxi, Henan, Hubei, Sichuan. [Flight Season] June to September.

双斑圆臀大蜓 雄，北京 | 安起迪 摄
Anotogaster kuchenbeiseri, male from Beijing | Photo by Qidi An

双斑圆臀大蜓 雌，北京
Anotogaster kuchenbeiseri, female from Beijing

双斑圆臀大蜓 雄，北京
Anotogaster kuchenbeiseri, male from Beijing

细纹圆臀大蜓 *Anotogaster myosa* Needham, 1930

【形态特征】本种与格氏圆臀大蜓非常近似，但有时具甚细的额横纹，腹部条纹比后者窄许多。【长度】体长84~91 mm，腹长 63~70 mm，后翅 48~56 mm。【栖息环境】海拔 500~2000 m森林中的渗流地、狭窄小溪和沟渠。【分布】中国特有，分布于湖北、四川。【飞行期】6—9月。

[Identification] The species is similar to *A. gregoryi*, but the top of frons sometimes with narrower yellow stripe, abdomen with narrower stripes. [Measurements] Total length 84-91 mm, abdomen 63-70 mm, hind wing 48-56 mm. [Habitat] Seepages, narrow streams and ditches in forest at 500-2000 m elevation. [Distribution] Endemic to China, recorded from Hubei, Sichuan. [Flight Season] June to September.

细纹圆臀大蜓 雄，湖北
Anotogaster myosa, male from Hubei

细纹圆臀大蜓 雄，湖北
Anotogaster myosa, male from Hubei

细纹圆臀大蜓 雌，湖北
Anotogaster myosa, female from Hubei

细纹圆臀大蜓 雌，湖北
Anotogaster myosa, female from Hubei

褐面圆臀大蜓 *Anotogaster nipalensis* (Selys, 1854)

褐面圆臀大蜓 雄，云南（德宏）
Anotogaster nipalensis, male from Yunnan (Dehong)

褐面圆臀大蜓 雌，云南（德宏）
Anotogaster nipalensis, female from Yunnan (Dehong)

【形态特征】雄性复眼翠绿色，面部深褐色，上唇具1对较小的黄斑；胸部肩前条纹甚短；腹部黑色，第2～8节具黄色条纹。雌性体型更大，上唇的黄斑更小，翅基方具琥珀色斑。【长度】体长 83～85 mm，腹长 64～65 mm，后翅 48～50 mm。【栖息环境】海拔 1000～2000 m森林中的渗流地、狭窄小溪和沟渠。【分布】云南（德宏）、西藏；不丹、印度、尼泊尔、缅甸。【飞行期】10—12月。

[Identification] Male eyes emerald, face dark brown, labrum with a pair of small yellow spots. Thorax with short antehumeral stripes. Abdomen black, S2-S8 with yellow stripes. Female larger, labrum with smaller yellow spots, wing bases with amber spots. **[Measurements]** Total length 83-85 mm, abdomen 64-65 mm, hind wing 48-50 mm. **[Habitat]** Seepages, narrow streams and ditches in forest at 1000-2000 m elevation. **[Distribution]** Yunnan (Dehong), Tibet; Bhutan, India, Nepal, Myanmar. **[Flight Season]** October to December.

褐面圆臀大蜓 雄，云南（德宏）
Anotogaster nipalensis, male from Yunnan (Dehong)

褐面圆臀大蜓 雌，云南（德宏）
Anotogaster nipalensis, female from Yunnan (Dehong)

清六圆臀大蜓 *Anotogaster sakaii* Zhou, 1988

【形态特征】本种与格氏圆臀大蜓相似，但体型较大，胸部和腹部的黄色条纹更细，而且随着年纪增长条纹变成灰白色，但格氏圆臀大蜓的黄色条纹不会随年纪而变化。【长度】体长 93～108 mm，腹长 71～83 mm，后翅 53～63 mm。【栖息环境】海拔 1000～2000 m森林中的渗流地、狭窄小溪和沟渠。【分布】中国特有，分布于贵州、浙江、福建、湖北、湖南、广东；本种在越南的分布记录为误定。【飞行期】6—8月。

[Identification] This spcies is similar to *A. gregoryi*, but size larger, yellow stripes on thorax and abdomen narrower, yellow stripes turn to grayish white when aged. In *A. gregoryi* color of these yellow stripes remain yellow despite with age. [Measurements] Total length 93-108 mm, abdomen 71-83 mm, hind wing 53-63 mm. [Habitat] Seepages, narrow streams and ditches in forest at 1000-2000 m elevation. [Distribution] Endemic to China, recorded from Guizhou, Zhejiang, Fujian, Hubei, Hunan, Guangdong; Records from Vietnam are based on incorrect identification. [Flight Season] June to August.

清六圆臀大蜓 雄，贵州
Anotogaster sakaii, male from Guizhou

清六圆臀大蜓 雌，贵州
Anotogaster sakaii, female from Guizhou

清六圆臀大蜓 雄，贵州
Anotogaster sakaii, male from Guizhou

萨帕圆臀大蜓 *Anotogaster sapaensis* Karube, 2012

【形态特征】本种与褐面圆臀大蜓相似，但体型更大，上唇无黄斑。【长度】体长 89~102 mm，腹长 68~80 mm，后翅 51~58 mm。【栖息环境】海拔 1000~2000 m森林中的渗流地、狭窄小溪和沟渠。【分布】云南（红河）；越南。【飞行期】6—8月。

萨帕圆臀大蜓 雄，云南（红河）
Anotogaster sapaensis, male from Yunnan (Honghe)

[Identification] The species is similar to *A. nipalensis*, but size larger, labrum without yellow spots. **[Measurements]** Total length 89-102 mm, abdomen 68-80 mm, hind wing 51-58 mm. **[Habitat]** Seepages, narrow streams and ditches in forest at 1000-2000 m elevation. **[Distribution]** Yunnan (Honghe); Vietnam. **[Flight Season]** June to August.

萨帕圆臀大蜓 雄，云南（红河）
Anotogaster sapaensis, male from Yunnan (Honghe)

萨帕圆臀大蜓 雌，云南（红河）
Anotogaster sapaensis, female from Yunnan (Honghe)

萨帕圆臀大蜓 雌，云南（红河）
Anotogaster sapaensis, female from Yunnan (Honghe)

巨圆臀大蜓 *Anotogaster sieboldii* (Selys, 1854)

巨圆臀大蜓 雌，广东 | 宋睿斌 摄
Anotogaster sieboldii, female from Guangdong | Photo by Ruibin Song

巨圆臀大蜓 雄，日本 | 梁嘉景 摄
Anotogaster sieboldii, male from Japan | Photo by Kenneth Leung

【形态特征】雄性复眼翠绿色，上唇具1对较大的黄斑，上颚外方黄色，后唇基黄色，额横纹较窄；胸部肩前条纹较长；腹部黑色，第2~8节具黄色条纹。雌性体型更大，翅基方具琥珀色斑。中国个体与日本个体在雄性肛附器、雌性产卵管长度等方面都有差异。【长度】体长 87~107 mm，腹长 67~82 mm，后翅 53~65 mm。【栖息环境】海拔 1500 m以下森林中的渗流地、狭窄小溪和沟渠。【分布】重庆、贵州、湖北、江西、广东；日本、朝鲜半岛、俄罗斯远东。【飞行期】4—11月。

[Identification] Male eyes emerald, labrum with a pair of fairly large yellow spots, base of mandible yellow, postclypeus yellow, top of frons with a narrow yellow stripe. Thorax with fairly long antehumeral stripes. Abdomen black, S2-S8 with yellow stripes. Female larger, wing bases with amber spots. Populations from China differs from those

巨圆臀大蜓 雄，贵州
Anotogaster sieboldii, male from Guizhou

巨圆臀大蜓 雌，湖北
Anotogaster sieboldii, female from Hubei

巨圆臀大蜓 雄，湖北
Anotogaster sieboldii, male from Hubei

in Japan in some characters including male anal appendages and length of female ovipositor. **[Measurements]** Total length 87-107 mm, abdomen 67-82 mm, hind wing 53-65 mm. **[Habitat]** Seepages, narrow streams and ditches in forest below 1500 m elevation. **[Distribution]** Chongqing, Guizhou, Hubei, Jiangxi, Guangdong; Japan, Korean peninsula, Russian Far East. **[Flight Season]** April to November.

圆臀大蜓属待定种1 *Anotogaster* sp. 1

【形态特征】本种与金斑圆臀大蜓相似，但体型较小，复眼蓝色，身体黄色条纹较窄；雄性肛附器和雌性产卵管长度也不同。【长度】体长 85~98 mm，腹长 64~75 mm，后翅 52~61 mm。【栖息环境】海拔 1000 m以下森林中的渗流地、狭窄小溪和沟渠。【分布】云南（西双版纳）。【飞行期】6—10月。

[Identification] The species is similar to *A. klossi*, but size smaller, eyes blue, body with yellow stripes narrower. Male anal appendages and length of female ovipositor also different. **[Measurements]** Total length 85-98 mm, abdomen 64-75 mm, hind wing 52-61 mm. **[Habitat]** Seepages, narrow streams and ditches in forest below 1000 m elevation. **[Distribution]** Yunnan (Xishuangbanna). **[Flight Season]** June to October.

圆臀大蜓属待定种1 雄，云南（西双版纳）
Anotogaster sp. 1, male from Yunnan (Xishuangbanna)

圆臀大蜓属待定种1 雌，云南（西双版纳）
Anotogaster sp. 1, female from Yunnan (Xishuangbanna)

圆臀大蜓属待定种1 雄，云南（西双版纳）
Anotogaster sp. 1, male from Yunnan (Xishuangbanna)

圆臀大蜓属待定种1 雌，云南（西双版纳）
Anotogaster sp. 1, female from Yunnan (Xishuangbanna)

圆臀大蜓属待定种2 *Anotogaster* sp. 2

【形态特征】本种与褐面圆臀大蜓相似，但体型更小，面部黑色。【长度】体长 73~80 mm，腹长 56~62 mm，后翅 42~47 mm。【栖息环境】海拔 1500 m以下森林中的渗流地、狭窄小溪和沟渠。【分布】云南（德宏）。【飞行期】9—12月。

圆臀大蜓属待定种2 雄，云南（德宏）
Anotogaster sp. 2, male from Yunnan (Dehong)

圆臀大蜓属待定种2 雌，云南（德宏）
Anotogaster sp. 2, female from Yunnan (Dehong)

圆臀大蜓属待定种2 雄，云南（德宏）
Anotogaster sp. 2, male from Yunnan (Dehong)

[Identification] The species is similar to *A. nipalensis*, but size smaller, face black. **[Measurements]** Total length 73-80 mm, abdomen 56-62 mm, hind wing 42-47 mm. **[Habitat]** Seepages, narrow streams and ditches in forest below 1500 m elevation. **[Distribution]** Yunnan (Dehong). **[Flight Season]** September to December.

圆臀大蜓属待定种2 雌，云南（德宏）
Anotogaster sp. 2, female from Yunnan (Dehong)

圆臀大蜓属待定种3 *Anotogaster* sp. 3

圆臀大蜓属待定种3 雄，湖北
Anotogaster sp. 3, male from Hubei

圆臀大蜓属待定种3 雄，山东
Anotogaster sp. 3, male from Shandong

圆臀大蜓属待定种3 雌，山东
Anotogaster sp. 3, female from Shandong

【形态特征】本种与双斑圆臀大蜓相似，但体型更大，雄性肛附器和雌性产卵管长度不同。【长度】体长86~97 mm，腹长 65~73 mm，后翅 52~59 mm。【栖息环境】海拔 1000 m以下森林中的溪流和沟渠。【分布】山东、江苏、湖北、安徽、浙江。【飞行期】6—9月。

[Identification] The species is similar to *A. kuchenbeiseri*, but size larger, male anal appendages and length of female ovipositor different. **[Measurements]** Total length 86-97 mm, abdomen 65-73 mm, hind wing 52-59 mm. **[Habitat]** Streams and ditches in forest below 1000 m elevation. **[Distribution]** Shandong, Jiangsu, Hubei, Anhui, Zhejiang. **[Flight Season]** June to September.

圆臀大蜓属待定种4 *Anotogaster* sp. 4

【形态特征】本种与金斑圆臀大蜓相似，但身体黄色条纹较窄。【长度】雄性体长 97 mm，腹长 76 mm，后翅 57 mm。【栖息环境】海拔 1000 m左右森林中的狭窄小溪。【分布】海南。【飞行期】7—9月。

[Identification] The species is similar to *A. klossi*, but body with yellow stripes narrower. **[Measurements]** Male total length 97 mm, abdomen 76 mm, hind wing 57 mm. **[Habitat]** Narrow streams at about 1000 m elevation. **[Distribution]** Hainan. **[Flight Season]** July to September.

圆臀大蜓属待定种4 雄，海南
Anotogaster sp. 4, male from Hainan

圆臀大蜓属待定种5 *Anotogaster* sp. 5

圆臀大蜓属待定种5 雄，云南（红河）
Anotogaster sp. 5, male from Yunnan (Honghe)

圆臀大蜓属待定种5 雄，云南（红河）
Anotogaster sp. 5, male from Yunnan (Honghe)

圆臀大蜓属待定种5 雌，云南（红河）
Anotogaster sp. 5, female from Yunnan (Honghe)

【形态特征】本种与赵氏圆臀大蜓近似，但额横纹和腹部条纹更窄，雌性产卵管更长，两者的分布区域重叠。【长度】体长 81~90 mm，腹长 62~68 mm，后翅 46~53 mm。【栖息环境】海拔 1500~2500 m森林中的狭窄溪流。【分布】云南（红河）；越南。【飞行期】2—7月。

[Identification] The species is similar to *A. chaoi*, but the top of frons and abdomen with narrower stripes, female ovipositor longer. This species and *A.chaoi* are partially sympatric. **[Measurements]** Total length 81-90 mm, abdomen 62-68 mm, hind wing 46-53 mm. **[Habitat]** Narrow streams in forest at 1500-2500 m elevation. **[Distribution]** Yunnan (Honghe); Vietnam. **[Flight Season]** February to July.

圆臀大蜓属待定种5 雌，云南（红河）
Anotogaster sp. 5, female from Yunnan (Honghe)

圆臀大蜓属待定种6 *Anotogaster* sp. 6

【形态特征】本种与巨圆臀大蜓相似，但雄性上肛附器在背面观时不见齿状突起，而巨圆臀大蜓在亚基方可见齿状突起；本种雌性的产卵管明显短于巨圆臀大蜓。【长度】体长 90～115 mm，腹长 69～90 mm，后翅 53～64 mm。【栖息环境】海拔 2000 m以下森林中的狭窄溪流和沟渠。【分布】中国台湾。【飞行期】4—11月。

[Identification] The species is similar to *A. sieboldii*, the sub-basal teeth of male superior appendages is not visible in dorsal view but can be seen in male *A. sieboldii*; female of this species has much shorter ovipositor. **[Measurements]** Total length 90-115 mm, abdomen 69-90 mm, hind wing 53-64 mm. **[Habitat]** Narrow streams and ditches in forest below 2000 m elevation. **[Distribution]** Taiwan of China. **[Flight Season]** April to November.

圆臀大蜓属待定种6 雄，台湾 | 嘎嘎 摄
Anotogaster sp. 6, male from Taiwan | Photo by Gaga

圆臀大蜓属待定种6 雌，台湾 | 嘎嘎 摄
Anotogaster sp. 6, female from Taiwan | Photo by Gaga

圆臀大蜓属待定种7 *Anotogaster* sp. 7

【形态特征】本种与清六圆臀大蜓相似，但本种体型更粗壮，复眼蓝色，雄性肛附器构造、雌性身体黄条纹的大小和形状完全不同。【长度】体长 93～105 mm，腹长 70～82 mm，后翅 54～62 mm。【栖息环境】海拔 2000 m 以下森林中的狭窄溪流和沟渠。【分布】云南（红河）；越南。【飞行期】4—7月。

[Identification] The species is similar to *A. sakaii*, but body stouter, eyes blue, male anal appendages, size and shape of yellow stripes in females different. [Measurements] Total length 93-105 mm, abdomen 70-82 mm, hind wing 54-62 mm. [Habitat] Narrow streams and ditches in forest below 2000 m elevation. [Distribution] Yunnan (Honghe); Vietnam. [Flight Season] April to July.

圆臀大蜓属待定种7 雄，云南（红河）
Anotogaster sp. 7, male from Yunnan (Honghe)

圆臀大蜓属待定种7 雌，云南（红河）
Anotogaster sp. 7, female from Yunnan (Honghe)

圆臀大蜓属待定种7 雄，云南（红河）
Anotogaster sp. 7, male from Yunnan (Honghe)

圆臀大蜓属待定种8 *Anotogaster* sp. 8

圆臀大蜓属待定种8 雄，广东
Anotogaster sp. 8, male from Guangdong

圆臀大蜓属待定种8 雌，广东
Anotogaster sp. 8, female from Guangdong

【形态特征】本种与金斑圆臀大蜓相似，但身体黄色条纹更窄，雌性差异更显著；雄性肛附器构造也不相同。【长度】体长 96~115 mm，腹长 74~85 mm，后翅 57~67 mm。【栖息环境】海拔 2000 m以下森林中的渗流地、狭窄小溪和沟渠。【分布】浙江、福建、湖南、广东。【飞行期】4—10月。

[Identification] The species is similar to *A. klossi*, but the body with narrower yellow stripes, more visible in female. Male anal appendages also different. **[Measurements]** Total length 96-115 mm, abdomen 74-85 mm, hind wing 57-67 mm. **[Habitat]** Seepages, narrow streams and ditches in forest below 2000 m elevation. **[Distribution]** Zhejiang, Fujian, Hunan, Guangdong. **[Flight Season]** April to October.

圆臀大蜓属待定种8 雄，广东
Anotogaster sp. 8, male from Guangdong

圆臀大蜓属待定种8 雌，广东
Anotogaster sp. 8, female from Guangdong

角臀大蜓属 Genus *Neallogaster* Cowley, 1934

本属和圆臀大蜓属相似，但本属目前的分类学研究不够完善。现归入大蜓属的部分种类可能会被移入该属，此处将中国所有大蜓属物种都归入本属中，中国大约有8种，但很多种类的身份需要更深入的研究才可以确定，包括云南西部记录的种类，仍需要和印度、尼泊尔地区的标本进一步比对。本属蜻蜓体型较大，身体黑色或褐色具显著的黄色条纹；额比圆臀大蜓属隆起更为显著；翅透明。和圆臀大蜓属相比，本属雄性后翅基方略呈角状，具明显的臀三角室，腹部具耳状突起。

本属蜻蜓喜欢较寒冷的气候，有些种类可以生活在海拔4000 m以上的高山环境。高海拔角臀大蜓的稚虫可能要经历多年才能发育成熟。雄性常沿溪流巡飞，有时在水面上方慢速低空飞行。雌性在溪流的浅滩或渗流地插秧式产卵。

As with *Anotogaster*, the taxonomy within this genus is problematic. Some species now assigned to genus *Cordulegaster* may be transferred to this genus. Here all the Chinese *Cordulegaster* species are regarded as *Neallogaster*, about eight species are recorded from China, but many of them need further study, including those known from the west of Yunnan. A comparison of our species with specimens from India and Nepal could reveal their true status. Species of the genus are fairly large-sized dragonflies, generally black or brown with yellow markings; frons more protruding than in *Anotogaster*; wings hyaline. Compared with *Anotogaster*, the hind wing tornus slightly angled, hind wings possess a distinct anal triangle, auricles present.

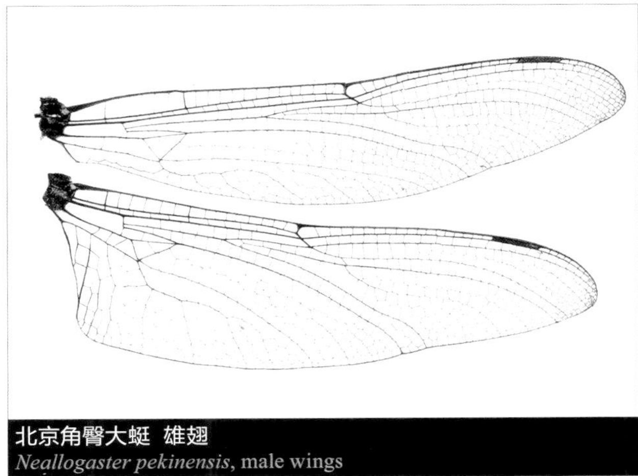

北京角臀大蜓 雄翅
Neallogaster pekinensis, male wings

Neallogaster species prefer colder climates, some species occur in high mountains above 4000 m elevation. The larval growth in these high altitude habitats may last for several years. Males usually patrol along streams, sometimes they fly low and slow above water. Female oviposit at the shallow beach of streams or seepages.

北京角臀大蜓 雌
Neallogaster pekinensis, female

云南角臀大蜓
Neallogaster annandalei

北京角臀大蜓
Neallogaster pekinensis

角臀大蜓属 雌性产卵管
Genus *Neallogaster*, female ovipositor

云南角臀大蜓
Neallogaster annandalei

浅色角臀大蜓
Neallogaster hermionae

褐面角臀大蜓
Neallogaster latifrons

北京角臀大蜓
Neallogaster pekinensis

角臀大蜓属 雄性肛附器
Genus *Neallogaster*, male appendages

云南角臀大蜓 雄
Neallogaster annandalei, male

云南角臀大蜓 雌
Neallogaster annandalei, female

浅色角臀大蜓 雄
Neallogaster hermionae, male

褐面角臀大蜓 雄
Neallogaster latifrons, male

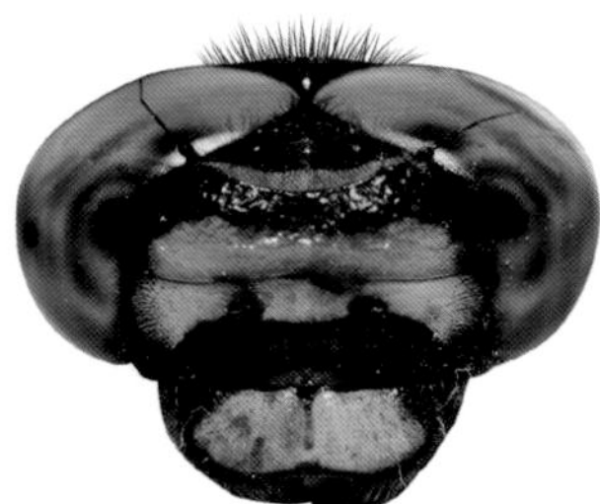

北京角臀大蜓 雄
Neallogaster pekinensis, male

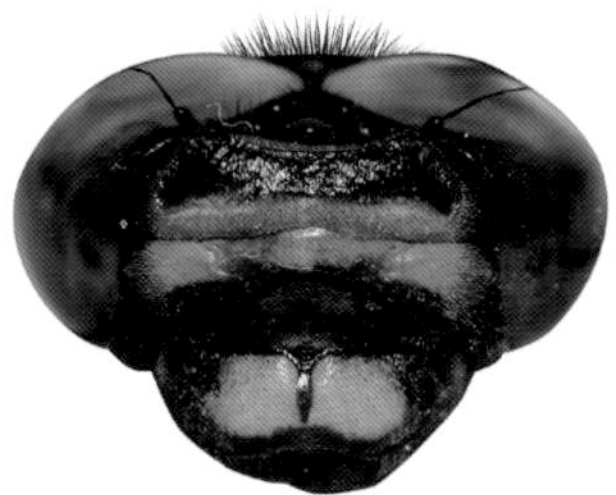

北京角臀大蜓 雌
Neallogaster pekinensis, female

角臀大蜓属 头部正面观
Genus *Neallogaster*, head in frontal view

云南角臀大蜓 *Neallogaster annandalei* (Fraser, 1924)

云南角臀大蜓 雄，云南（大理）
Neallogaster annandalei, male from Yunnan (Dali)

【形态特征】雄性上唇中央具1个甚大的黄斑，前唇基黑色，后唇基褐色两侧具黄斑，额褐色具1条黄色额横纹；胸部黑色，肩前条纹甚阔，合胸侧面具2条甚阔的黄色条纹；腹部黑色具黄斑。雌性体型更大且粗壮，翅淡琥珀色。【长度】体长 70~76 mm，腹长 53~58 mm，后翅 40~48 mm。【栖息环境】海拔 2500~3000 m森林中的溪流。【分布】中国云南（大理、迪庆）特有。【飞行期】5—7月。

[Identification] Male labrum with a large yellow spot centrally, anteclypeus black, postclypeus brown with lateral yellow spots, frons brown with a yellow stripe. Thorax black with broad antehumeral stripes, laterally with two broad yellow stripes. Abdomen black with yellow spots. Female larger and stouter, wings tinted with light amber. **[Measurements]** Total length 70-76 mm, abdomen 53-58 mm, hind wing 40-48 mm. **[Habitat]** Streams in forest at 2500-3000 m elevation. **[Distribution]** Endemic to Yunnan (Dali, Diqing) of China. **[Flight Season]** May to July.

云南角臀大蜓 雌，云南（大理）
Neallogaster annandalei, female from Yunnan (Dali)

浅色角臀大蜓 *Neallogaster hermionae* (Fraser, 1927)

浅色角臀大蜓 雄，云南（大理）
Neallogaster hermionae, male from Yunnan (Dali)

【形态特征】雄性面部主要淡黄色，额浅褐色；胸部褐色，肩前条纹甚阔，合胸侧面具2条甚阔的黄色条纹；腹部褐色，第2～9节具黄斑，耳状突黄色。【长度】雄性体长 70～74 mm，腹长 53～55 mm，后翅 40～41 mm。【栖息环境】海拔 2000～3000 m森林中的溪流。【分布】云南（大理）；印度、尼泊尔。【飞行期】4—6月。

[Identification] Male face largely pale yellow, frons pale brown. Thorax brown with broad antehumeral stripes, laterally with two broad yellow stripes. Abdomen brown, S2-S9 with yellow spots, auricle yellow. **[Measurements]** Male total length 70-74 mm, abdomen 53-55 mm, hind wing 40-41 mm. **[Habitat]** Streams in forest at 2000-3000 m elevation. **[Distribution]** Yunnan (Dali); India, Nepal. **[Flight Season]** April to June.

浅色角臀大蜓 雄，云南（大理）
Neallogaster hermionae, male from Yunnan (Dali)

褐面角臀大蜓 *Neallogaster latifrons* (Selys, 1878)

【形态特征】雄性面部大面积褐色，上唇中央具1对黄斑；胸部黑色，肩前条纹甚阔，合胸侧面具2条甚阔的黄色条纹，后胸前侧板具小黄斑；腹部黑色，第2~10节具黄斑。【长度】雄性体长 75 mm，腹长 57 mm，后翅 42 mm。【栖息环境】海拔 2500~3000 m森林中的溪流。【分布】云南（大理）；印度、尼泊尔。【飞行期】6—12月。

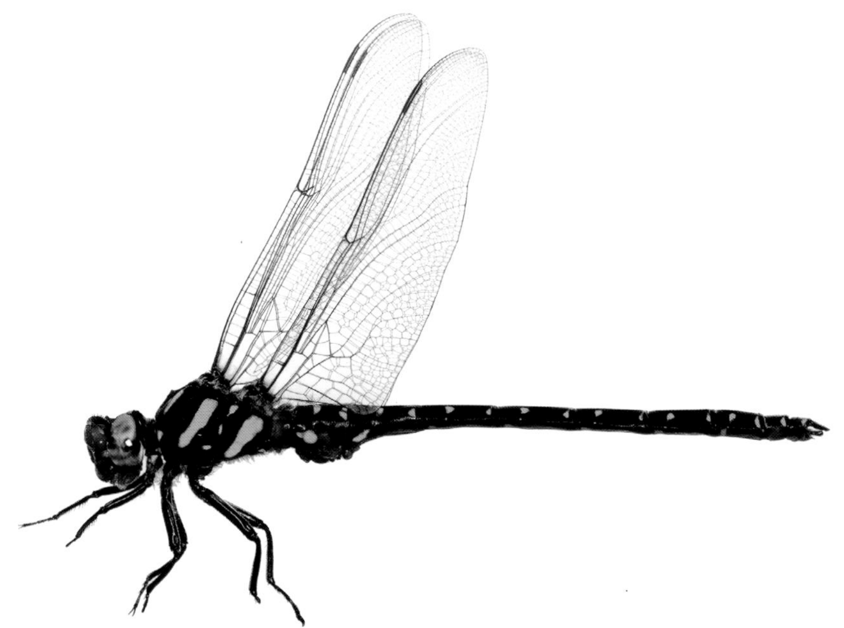

褐面角臀大蜓 雄，云南（大理）
Neallogaster latifrons, male from Yunnan (Dali)

[Identification] Male face largely brown, labrum with a pair of yellow spots centrally. Thorax black with broad antehumeral stripes, laterally with two broad yellow stripes, metepisternum with small yellow spots. Abdomen black, S2-S10 with yellow spots. **[Measurements]** Male total length 75 mm, abdomen 57 mm, hind wing 42 mm. **[Habitat]** Streams in forest at 2500-3000 m elevation. **[Distribution]** Yunnan (Dali); India, Nepal. **[Flight Season]** June to December.

北京角臀大蜓 *Neallogaster pekinensis* (Selys, 1886)

【形态特征】雄性面部黑色具黄斑，上唇中央具1个甚大的黄斑，后唇基和前额的下半部黄色，额具1个“T”形黄纹；胸部肩前条纹甚阔，合胸侧面具2条甚阔的黄色条纹；腹部黑色具黄斑。雌性体型更大且粗壮。【长度】体长 71~80 mm，腹长 54~62 mm，后翅 44~50 mm。【栖息环境】海拔 500~1500 m森林中的狭窄小溪和沟渠。【分布】中国特有，分布于北京、四川。【飞行期】5—8月。

[Identification] Male face black with yellow markings, labrum with a large yellow spot centrally, postclypeus and lower half of antefrons yellow, top of frons with a yellow T-mark. Thorax with broad antehumeral stripes, laterally with two broad yellow stripes. Abdomen black with yellow spots. Female larger and stouter. **[Measurements]** Total length 71-80 mm, abdomen 54-62 mm, hind wing 44-50 mm. **[Habitat]** Narrow streams and ditches in forest at 500-1500 m elevation. **[Distribution]** Endemic to China, recorded from Beijing, Sichuan. **[Flight Season]** May to August.

北京角臀大蜓 雄，北京
Neallogaster pekinensis, male from Beijing

北京角臀大蜓 雌，北京
Neallogaster pekinensis, female from Beijing

北京角臀大蜓 雄，北京
Neallogaster pekinensis, male from Beijing

5 伪蜻科 Family Corduliidae

本科全球已知20属超过150种，世界性分布。中国已知5属10余种，全国广布。本科蜻蜓体中型，复眼亮绿色并在头顶相交，如同绿宝石，很多种类身体具金属光泽；翅大面积透明，基室无横脉，前翅的三角室2室，前翅的基臀区具1条横脉，后翅的基臀区具1~2条横脉，臀圈靴状。

本科多数种类栖息于池塘、湖泊和水潭等静水环境，少数种类生活在流速缓慢的溪流。本科在中国北方较常见，在南方则隐蔽于高海拔山区。雄性具有显著的领域行为，经常靠近水面来回飞行或者间歇性悬停。

This family contains 20 genera with over 150 species distributed all over the world. Over ten species in five genera are widely distributed in China. Species of the family are medium-sized, eyes brilliant green that meet above, like green jewels, many species have brilliant metallic bodies. Wings largely hyaline, median space without crossvein, in fore wings the triangle 2-celled and the cubital space has one crossvein, in hind wings the cubital space has one or two crossveins and the anal loop boot-shaped.

Most species of the family inhabit standing water such as ponds, lakes or pools, a few others are found along slow flowing streams. They are comnonly seen in the northern part of China with those occurring in the southern region confined to high mountain elevations. Males exhibit territorial behavior by flying low above water or hovering at various intervals.

高山半伪蜻 雄

Hemicordulia edai, male

缘斑毛伪蜻 雄

Epitheca marginata, male

伪蜻属 Genus *Cordulia* Leach, 1815

本属全球已知3种，1种分布于北美洲，2种分布于欧亚大陆的温带区域。中国已知1种，分布于东北地区。本属蜻蜓体中型，身体墨绿色具金属光泽；翅透明，后翅的基臀区具1条横脉，后翅基方略呈角状。

本属蜻蜓栖息于静水环境。雄性沿池塘边缘快速飞行并时而悬停。

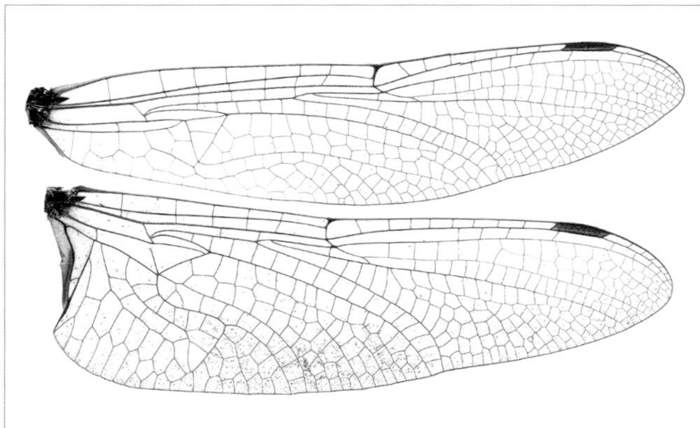

青铜伪蜻 雄翅
Cordulia amurensis, male wings

青铜伪蜻
Cordulia amurensis

伪蜻属 雄性肛附器
Genus *Cordulia*, male anal appenages

The genus contains three species, one from North America and the other two widespread in the temperate zone of Eurasia. Species of the genus are medium-sized, body metallic dark green. Wings hyaline, cubital space in hind wings with one crossvein, tornus slightly angled.

Cordulia species inhabit standing water. Male usually seen flying rapidly along the margin of ponds, interrupted by short intervals of hovering.

青铜伪蜻 交尾 | 金洪光 摄
Cordulia amurensis, mating pair | Photo by Hongguang Jin

青铜伪蜻 *Cordulia amurensis* Selys, 1887

青铜伪蜻 雌，黑龙江 | 莫善濂 摄
Cordulia amurensis, female from Heilongjiang | Photo by Shanlian Mo

青铜伪蜻 雄，黑龙江 | 莫善濂 摄
Cordulia amurensis, male from Heilongjiang | Photo by Shanlian Mo

青铜伪蜻 雄，黑龙江 | 莫善濂 摄
Cordulia amurensis, male from Heilongjiang | Photo by Shanlian Mo

【形态特征】雄性复眼绿色；面部和胸部墨绿色具金属光泽；腹部黑色，第2节耳状突下方具1个黄褐色斑，第3节侧面具1条白色条纹。雌性与雄性色彩相似，腹部第1～3节膨大。【长度】体长 47～55 mm，腹长 30～39 mm，后翅 29～35 mm。【栖息环境】海拔 500 m以下的湿地。【分布】黑龙江、吉林；西伯利亚东部及日本。【飞行期】5—7月。

[Identification] Male eyes green. Face and thorax metallic dark green. Abdomen black, S2 with a yellowish brown spot under the auricle, S3 with a lateral white stripe. Female similar to male, S1-S3 expanded. **[Measurements]** Total length 47-55 mm, abdomen 30-39 mm, hind wing 29-35 mm. **[Habitat]** Wetlands below 500 m elevation. **[Distribution]** Heilongjiang, Jilin; East Siberia and Japan. **[Flight Season]** May to July.

毛伪蜻属 Genus *Epitheca* Burmeister, 1839

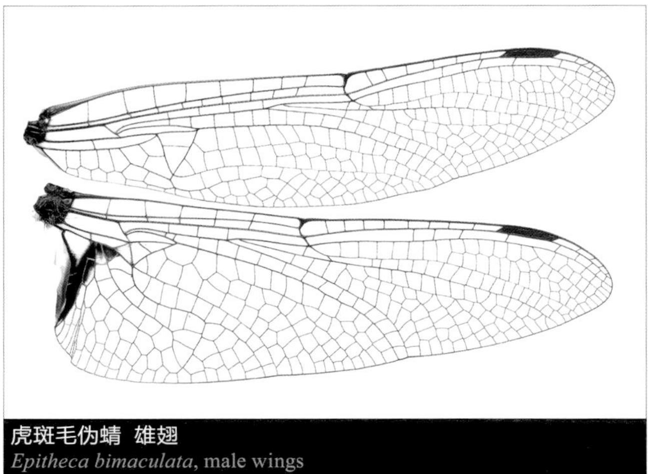
虎斑毛伪蜻 雄翅
Epitheca bimaculata, male wings

本属全球已知12种，多数种类在北美洲分布，欧亚大陆仅有2种，均分布于中国。本属蜻蜓是本科中体型稍大且粗壮的种类，身体缺乏金属光泽，体色较暗但具有显著的黄色条纹；翅透明，前翅的三角室2~3室，后翅三角室通常2室，后翅的基臀区具1~2条横脉，后翅基方略呈角状。中国分布的2种毛伪蜻可以通过身体色彩和肛附器形状区分。虎斑毛伪蜻的后翅基方具较大的黑褐色斑，而缘斑毛伪蜻没有此色彩。

本属蜻蜓栖息于静水环境。雄性通常在池塘中央悬停飞行却很少靠岸。雌性产卵时先停落在池塘边排出卵块，然后将腹部翘起并携带卵块飞行，最后将卵块投在水面漂浮的水草上。

The genus contains 12 species, most from North America with only two has been found in Eurasia both of which are recorded from China. They are moderate large-sized and robust species, but they lack the metallic colouration and are usually dark with yellow markings. Wings hyaline, in fore wings the triangle 2- to 3-celled, in hing wings usually 2-celled, the cubital space in hind wings with 1-2 crossveins, tornus slightly angled. The two species from China can be distinguished by the body color pattern and male anal appendages. Hind wing of *E. bimaculata* has a large blackish brown spot basally, which is not present in *E. marginata*.

缘斑毛伪蜻 雌
Epitheca marginata, female

Epitheca species frequent standing water. Males usually hover over the center of the water but seldom approach the margin. Females perch near ponds and exude an egg mass first, then carrying it with the abdominal tip curving before depositing it on floating plants.

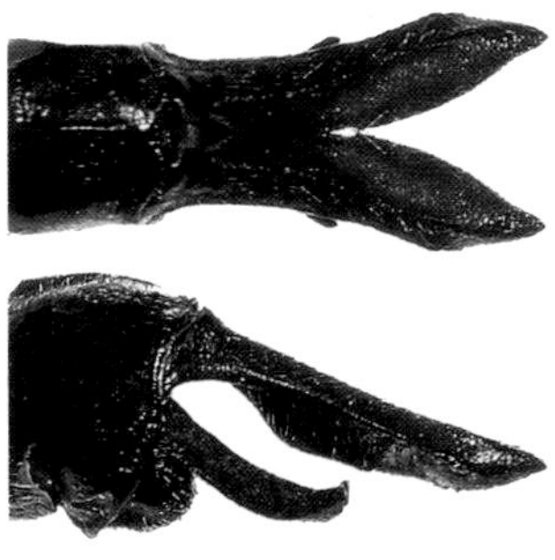

虎斑毛伪蜻
Epitheca bimaculata

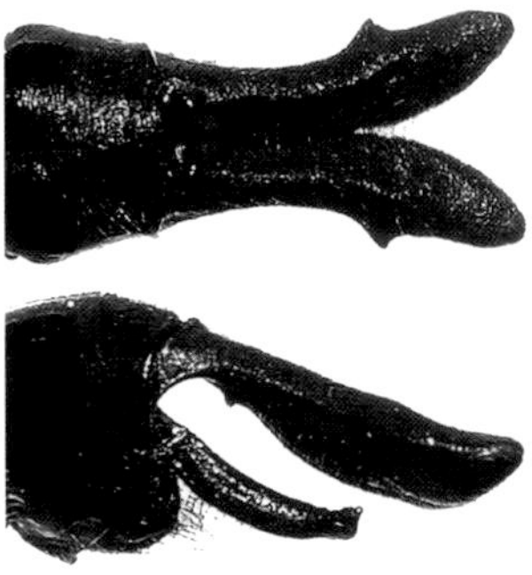

缘斑毛伪蜻
Epitheca marginata

毛伪蜻属 雄性肛附器
Genus *Epitheca*, male anal appenages

虎斑毛伪蜻 *Epitheca bimaculata* (Charpentier, 1825)

虎斑毛伪蜻 雄，吉林 | 金洪光 摄
Epitheca bimaculata, male from Jilin | Photo by Hongguang Jin

【形态特征】雄性复眼深绿色，面部黄褐色；胸部黑色具黄褐色条纹，后翅基方具1个黑褐色斑；腹部黑色，第1～8节侧缘具黄色条纹。雌性与雄性相似但腹部具更宽阔的黄色条纹。【长度】体长 55～65 mm，腹长 37～43 mm，后翅 36～44 mm。【栖息环境】海拔 500 m以下的池塘和水库。【分布】黑龙江、吉林、辽宁；从欧洲经西伯利亚至日本广布。【飞行期】5—8月。

[Identification] Male eyes dark green, face yellowish brown. Thorax black with yellowish brown stripes, hind wing base with a blackish brown spot. Abdomen black, S1-S8 with lateral yellow stripes. Female similar to male but abdomen with broader yellow markings. **[Measurements]** Total length 55-65 mm, abdomen 37-43 mm, hind wing 36-44 mm. **[Habitat]** Ponds and reservoirs below 500 m elevation. **[Distribution]** Heilongjiang, Jilin, Liaoning; Widespread from Europe throughout Siberia to Japan. **[Flight Season]** May to August.

虎斑毛伪蜻 雌，吉林 | 金洪光 摄
Epitheca bimaculata, female from Jilin | Photo by Hongguang Jin

虎斑毛伪蜻 雄，黑龙江 | 莫善濂 摄
Epitheca bimaculata, male from Heilongjiang | Photo by Shanlian Mo

缘斑毛伪蜻 *Epitheca marginata* (Selys, 1883)

缘斑毛伪蜻 雄，湖北
Epitheca marginata, male from Hubei

【形态特征】雄性复眼蓝绿色，面部黄褐色；胸部黑色具黄褐色条纹，翅稍染淡褐色；腹部黑色，第1～8节具黄斑。雌性多型，透翅型翅透明，斑翅型的翅前缘具1条黑带，从翅基方伸达翅端。【长度】体长 52～54 mm，腹长 36～38 mm，后翅 36～39 mm。【栖息环境】海拔 1500 m以下的池塘和水库。【分布】北京、河北、山东、山西、江苏、安徽、湖北、贵州；朝鲜半岛、日本。【飞行期】3—6月。

[Identification] Male eyes bluish green, face yellowish brown. Thorax black with yellowish brown stripes, wings slightly tinted with brown. Abdomen black, S1-S8 with yellow spots. Female polymorphic, the hyaline winged morph with wings hyaline, and the spotted winged morph with a black costal stripe, extending from the base to tip. **[Measurements]** Total length 52-54 mm, abdomen 36-38 mm, hind wing 36-39 mm. **[Habitat]** Ponds and reservoirs below 1500 m elevation. **[Distribution]** Beijing, Hebei, Shandong, Shanxi, Jiangsu, Anhui, Hubei, Guizhou; Korean peninsula, Japan. **[Flight Season]** March to June.

缘斑毛伪蜻 雌，透翅型，湖北
Epitheca marginata, female, hyaline winged morph from Hubei

缘斑毛伪蜻 雌，斑翅型，湖北
Epitheca marginata, female, spotted winged morph from Hubei

缘斑毛伪蜻 雄，湖北
Epitheca marginata, male from Hubei

半伪蜻属 Genus *Hemicordulia* Selys, 1870

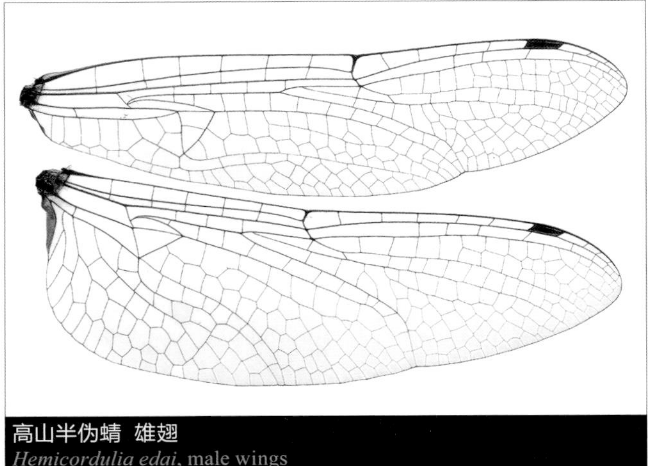

高山半伪蜻 雄翅
Hemicordulia edai, male wings

本属全球已知40余种，分布于亚洲、非洲和大洋洲。中国已知2种，分布于华南和西南地区。本属蜻蜓体中型，身体色彩较暗具金属光泽，有时具有显著的黄色条纹；翅透明，后翅三角室1室，后翅的基臀区具1条横脉，后翅基方略呈圆弧形。

本属蜻蜓栖息于水草茂盛的湿地和流速缓慢的溪流。雄性在池塘边缘和具有挺水植物或漂浮水草的区域飞行，通常飞行速度较快并经常短时定点悬停。之前认为本属种类仅在高山环境分布，但广东发现的1种半伪蜻可以在低海拔出没。

The genus contains over 40 species distributed in Asia, Africa and Oceania. Two species are recorded from China, found in the South and Southwest. Species of the genus are medium-sized, body fundamentally dark with metallic coloring, sometimes with yellow markings. Wings hyaline, triangle in hind wings 1-celled, cubital space with one crossvein, tornus slightly rounded.

高山半伪蜻 交尾
Hemicordulia edai, mating pair

Hemicordulia species inhabit well vegetated wetlands and slow flowing streams. Males fly along the margin where emergent plants or floating plants grow, they usually fly fast interrupted by short periods of hovering. The genus was previously believed to be confined to high mountain habitats but a species from Guangdong occurs in lowland.

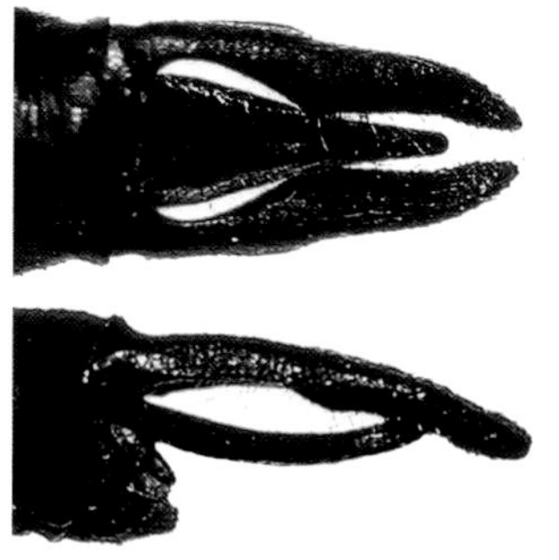

高山半伪蜻
Hemicordulia edai

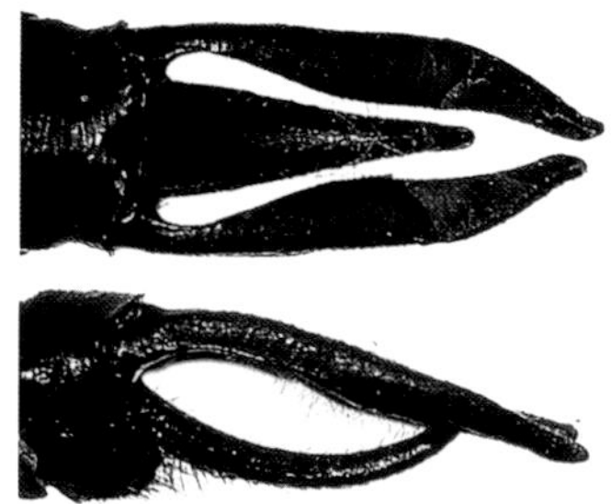

岷峨半伪蜻
Hemicordulia mindana nipponica

半伪蜻属待定种
Hemicordulia sp.

半伪蜻属 雄性肛附器
Genus *Hemicordulia*, male anal appenages

高山半伪蜻 *Hemicordulia edai* Karube & Katatani, 2012

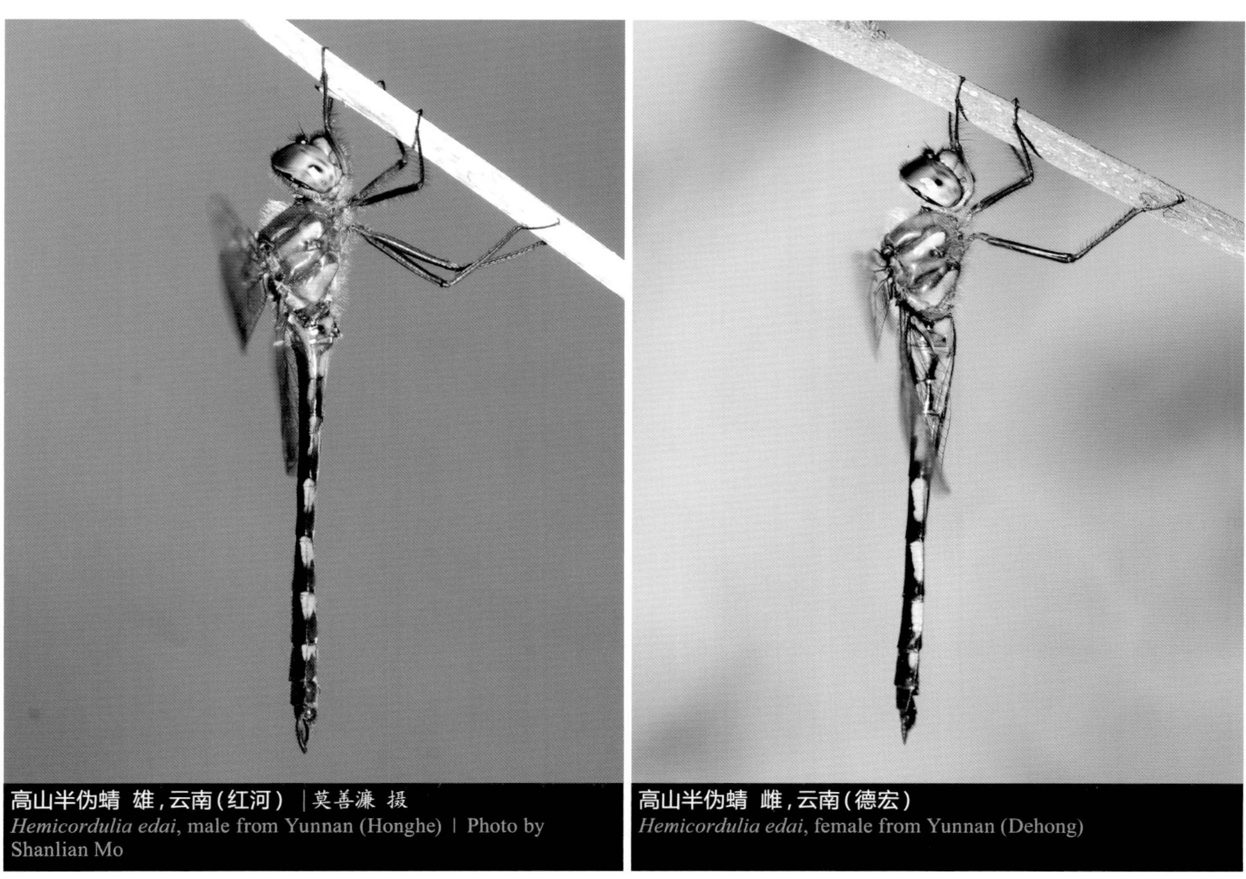

高山半伪蜻 雄，云南（红河）｜莫善濂 摄
Hemicordulia edai, male from Yunnan (Honghe) | Photo by Shanlian Mo

高山半伪蜻 雌，云南（德宏）
Hemicordulia edai, female from Yunnan (Dehong)

【形态特征】雄性复眼蓝绿色，面部黄褐色，上额黑色；胸部深绿色具金属光泽和黄褐色条纹；腹部黑色，第2~8节侧缘具黄斑。雌性与雄性相似。【长度】体长 48~50 mm，腹长 34~37 mm，后翅 32~34 mm。【栖息环境】海拔 1500~3000 m的湿地。【分布】云南（红河、大理、保山、德宏）、贵州；老挝、越南。【飞行期】4—8月。

[Identification] Male eyes bluish green, face yellowish brown, top of frons black. Thorax metallic dark green with yellowish brown stripes. Abdomen black, S2-S8 with lateral yellow spots. Female similar to male. **[Measurements]** Total length 48-50 mm, abdomen 34-37 mm, hind wing 32-34 mm. **[Habitat]** Wetlands at 1500-3000 m elevation. **[Distribution]** Yunnan (Honghe, Dali, Baoshan, Dehong), Guizhou; Laos, Vietnam. **[Flight Season]** April to August.

高山半伪蜻 雄，云南（保山）
Hemicordulia edai, male from Yunnan (Baoshan)

高山半伪蜻 雌，云南（德宏）
Hemicordulia edai, female from Yunnan (Dehong)

岷峨半伪蜻 *Hemicordulia mindana nipponica* Asahina, 1980

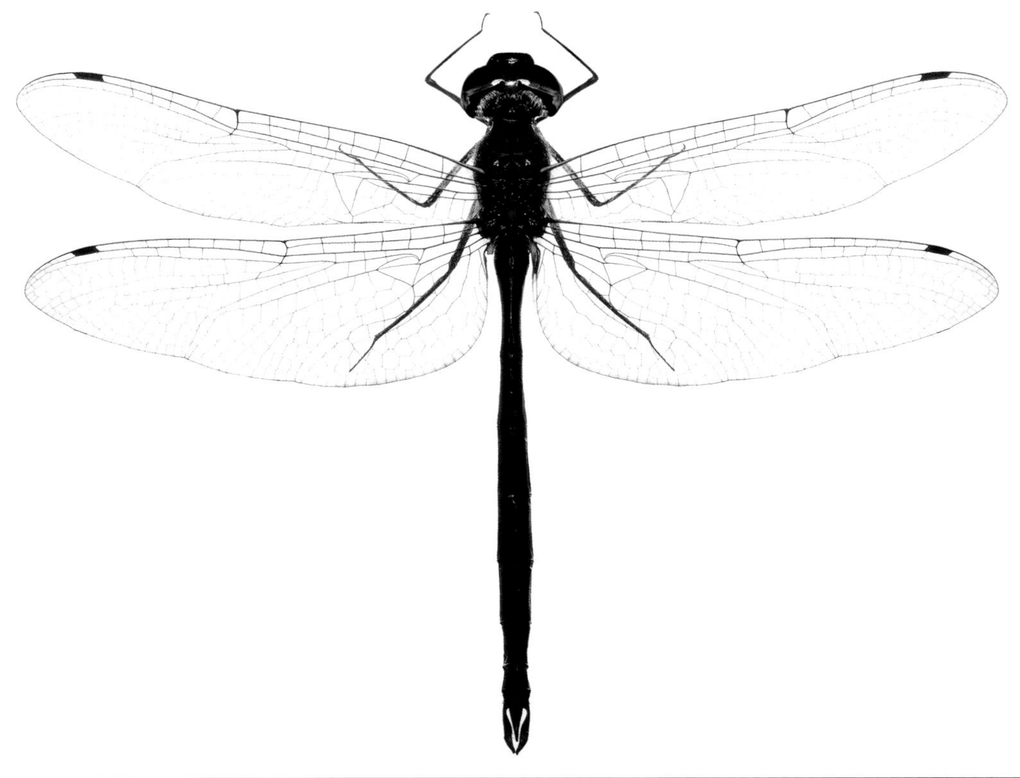

岷峨半伪蜻 雄，台湾
Hemicordulia mindana nipponica, male from Taiwan

【形态特征】雄性复眼深绿色，面部浅褐色，额黑绿色具金属光泽；胸部墨绿色具金属光泽和不清晰的深褐色条纹；腹部黑色。【长度】雄性体长 53 mm，腹长 39 mm，后翅 33 mm。【栖息环境】海拔 500 m以下的溪流、池塘和湖泊。【分布】中国台湾；日本。【飞行期】4—8月。

[Identification] Male eyes dark green, face light brown, frons metallic blackish green. Thorax metallic dark green with indistinct dark brown stripes. Abdomen black. **[Measurements]** Male total length 53 mm, abdomen 39 mm, hind wing 33 mm. **[Habitat]** Streams, ponds and lakes below 500 m elevation. **[Distribution]** Taiwan of China; Japan. **[Flight Season]** April to August.

半伪蜻属待定种 *Hemicordulia* sp.

【形态特征】雄性复眼深绿色，面部黄褐色，额黑色；胸部墨绿色具金属光泽和深褐色条纹；腹部有时全黑色，有时第2~7节侧缘具甚小的褐色斑。雌性与雄性相似。【长度】体长 48~50 mm，腹长 35~36 mm，后翅 32~33 mm。【栖息环境】海拔 1000 m以下的沟渠、流速缓慢的小溪和池塘。【分布】广西、广东。【飞行期】5—8月。

[Identification] Male eyes dark green, face yellowish brown, frons black. Thorax metallic dark green with dark brown stripes. Abdomen sometimes entirely black, sometimes with small brown spots on S2-S7. Female similar to male.

[Measurements] Total length 48-50 mm, abdomen 35-36 mm, hind wing 32-33 mm. **[Habitat]** Ditches, slowing flowing streams and ponds below 1000 m elevation. **[Distribution]** Guangxi, Guangdong. **[Flight Season]** May to August.

半伪蜻属待定种 雄，广东 | 莫善濂 摄
Hemicordulia sp., male from Guangdong | Photo by Shanlian Mo

半伪蜻属待定种 交尾，广东 | 莫善濂 摄
Hemicordulia sp., mating pair from Guangdong | Photo by Shanlian Mo

褐伪蜻属 Genus *Procordulia* Martin, 1907

本属全球已知18种，分布于亚洲和大洋洲。中国已知仅1种，分布于云南。本属与半伪蜻属相似，但雄性具有甚小的耳状突，而半伪蜻属雄性无此构造。

本属蜻蜓栖息于森林中的池塘。行为与半伪蜻属相似。

The genus contains 18 species distributed in Asia and Oceania. Only one species is recorded from China, found in Yunnan. Species of the genus are similar to those of *Hemicordulia*, but males have small auricles which are absent in *Hemicordulia*.

Procordulia species inhabit ponds in forest. The behavior is similar to species of *Hemicordulia*.

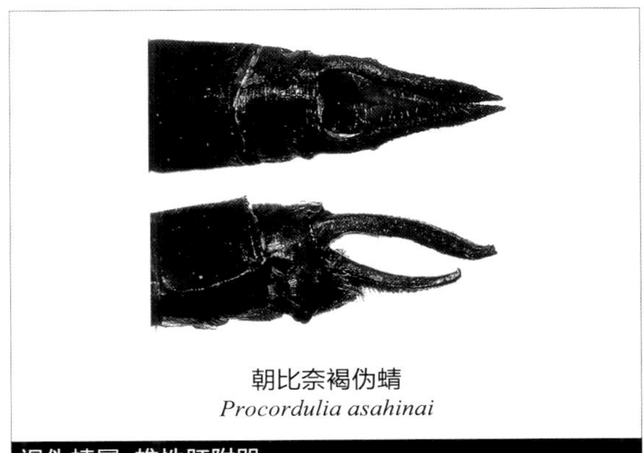

朝比奈褐伪蜻
Procordulia asahinai

褐伪蜻属 雄性肛附器
Genus *Procordulia*, male anal appenages

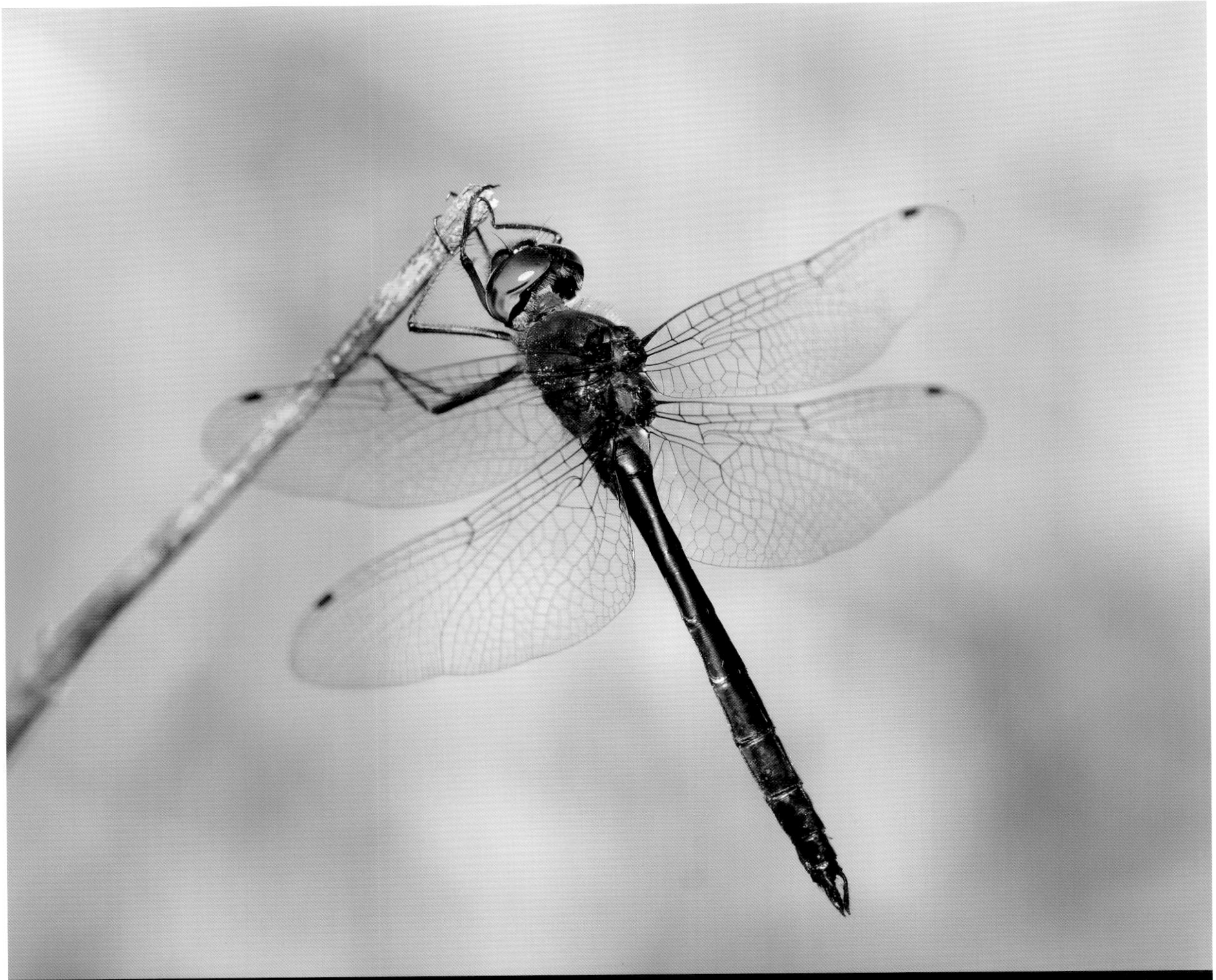

朝比奈褐伪蜻 雄
Procordulia asahinai, male

朝比奈褐伪蜻 *Procordulia asahinai* Karube, 2007

【形态特征】雄性复眼深绿色，面部深褐色，上额具1条黑色横纹；胸部墨绿色具金属光泽和深褐色条纹；腹部黑色，第1～9节侧缘具深褐色斑。雌性与雄性相似。【长度】体长 52～53 mm，腹长 38～39 mm，后翅 36～38 mm。【栖息环境】海拔 1500～2000 m 森林中的池塘。【分布】云南（普洱、红河）；越南。【飞行期】6—10月。

[Identification] Male eyes dark green, face dark brown, top of frons with a black stripe. Thorax metallic dark green with dark brown stripes. Abdomen black, S1-S9 with lateral dark brown spots. Female similar to male. **[Measurements]** Total length 52-53 mm, abdomen 38-39 mm, hind wing 36-38 mm. **[Habitat]** Ponds in forest at 1500-2000 m elevation. **[Distribution]** Yunnan (Pu'er, Honghe); Vietnam. **[Flight Season]** June to October.

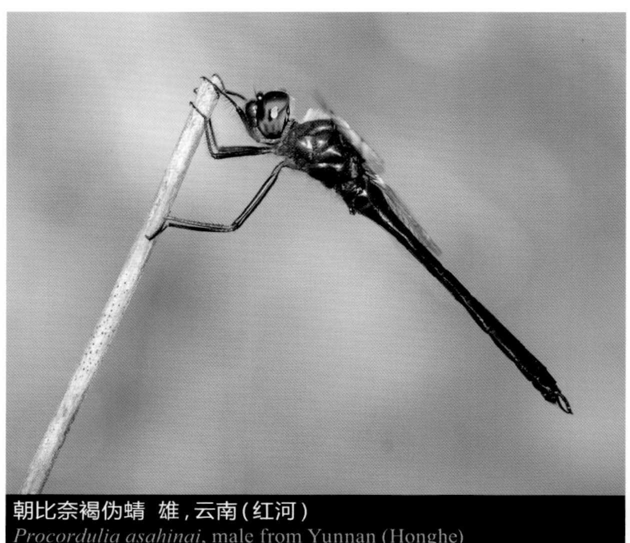

朝比奈褐伪蜻 雄，云南（红河）
Procordulia asahinai, male from Yunnan (Honghe)

朝比奈褐伪蜻 雌，云南（红河）
Procordulia asahinai, female from Yunnan (Honghe)

朝比奈褐伪蜻 雄，云南（红河）
Procordulia asahinai, male from Yunnan (Honghe)

朝比奈褐伪蜻 雌，云南（红河）
Procordulia asahinai, female from Yunnan (Honghe)

金光伪蜻属 Genus *Somatochlora* Selys, 1871

本属全球已知40余种，分布于北美洲和欧亚大陆的温带地区。中国已知10种，广布全国，在北方较常见，在南方仅栖息于高山环境。本属蜻蜓体中型，复眼亮绿色如同绿宝石，身体墨绿色具金属光泽；翅透明，后翅三角室通常1室，后翅的基臀区具2条横脉，后翅基方略呈角状。本属雄性可以通过肛附器的构造区分，雌性可以通过下生殖板的构造区分。

本属蜻蜓栖息于各类静水环境和流速缓慢的溪流。行为与半伪蜻属相似。

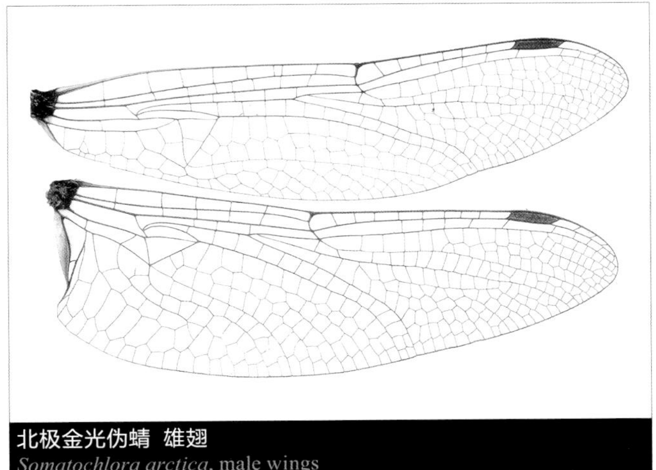

北极金光伪蜻 雄翅
Somatochlora arctica, male wings

The genus contains over 40 species distributed in North America and the temperate zone of Eurasia. Ten species are recorded from China, widespread throughout the country, commonly seen in the north but in the south they are confined to mountains. Species of the genus are medium-sized, eyes brilliant iridescent green, body metallic dark green. Wings hyaline, triangle in hind wings usually 1-celled, cubital space with two crossveins, tornus slightly angled. Males can be distinguished by the anal appendages and females by the vulvar lamina.

Somatochlora species inhabit several kinds of standing water habitats as well as slow flowing streams. Their behavior is similar to species of *Hemicordulia*.

日本金光伪蜻 雄 | 齐天博 摄
Somatochlora exuberata, male | Photo by Tianbo Qi

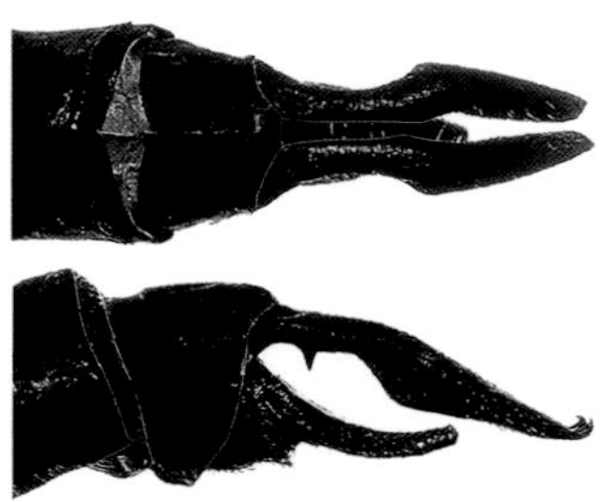

山西金光伪蜻
Somatochlora shanxiensis

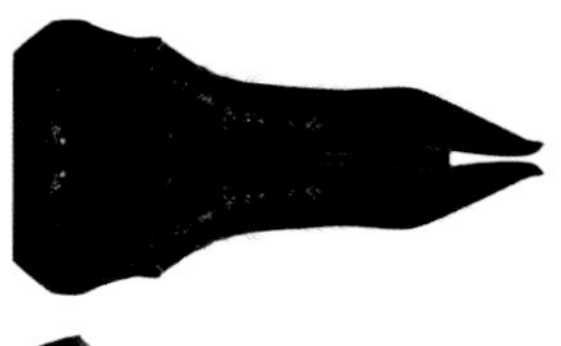

神农金光伪蜻
Somatochlora shennong

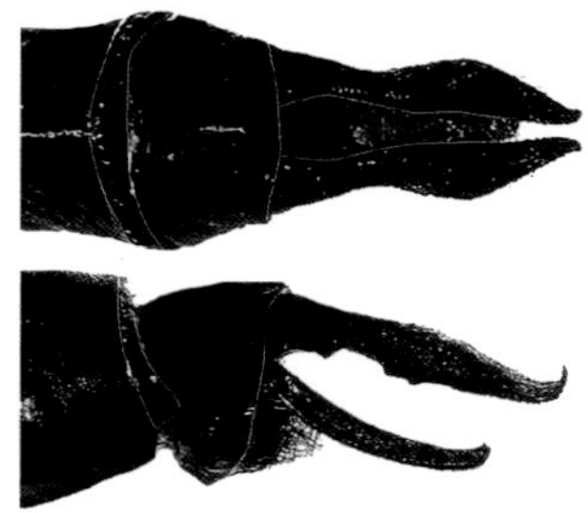

台湾金光伪蜻
Somatochlora taiwana

金光伪蜻属 雄性肛附器
Genus *Somatochlora*, male anal appenages

高地金光伪蜻
Somatochlora alpestris

北极金光伪蜻
Somatochlora arctica

绿金光伪蜻
Somatochlora dido

格氏金光伪蜻
Somatochlora graeseri

日本金光伪蜻
Somatochlora exuberata

金光伪蜻属 雄性肛附器
Genus *Somatochlora*, male anal appenages

神农金光伪蜻
Somatochlora shennong

台湾金光伪蜻
Somatochlora taiwana

北极金光伪蜻
Somatochlora arctica

日本金光伪蜻
Somatochlora exuberata

山西金光伪蜻
Somatochlora shanxiensis

金光伪蜻属 雌性下生殖板
Genus *Somatochlora*, female vulvar lamina

高地金光伪蜻 *Somatochlora alpestris* (Selys, 1840)

高地金光伪蜻 雄，芬兰 | Sami Karjalainen 摄
Somatochlora alpestris, male from Finland | Photo by Sami Karjalainen

高地金光伪蜻 雌，芬兰 | Sami Karjalainen 摄
Somatochlora alpestris, female from Finland | Photo by Sami Karjalainen

【形态特征】雄性面部主要黑色；胸部墨绿色具金属光泽；腹部黑色，第2节和第3节后缘具白色细条纹。雌性与雄性相似，但腹部更粗壮。【长度】体长 45~50 mm，腹长 30~36 mm，后翅 30~35 mm。【栖息环境】海拔500~1000 m的沼泽地和小型池塘。【分布】黑龙江；从欧洲经西伯利亚至日本广布。【飞行期】6—9月。

[Identification] Male face mainly black. Thorax metallic dark green. Abdomen black, S2-S3 with fine white stripes posteriorly. Female similar to male but the abdomen stouter. **[Measurements]** Total length 45-50 mm, abdomen 30-36 mm, hind wing 30-35 mm. **[Habitat]** Marshes and small ponds at 500-1000 m elevation. **[Distribution]** Heilongjiang; Widespread from Europe throughout Siberia to Japan. **[Flight Season]** June to September.

北极金光伪蜻 *Somatochlora arctica* (Zetterstedt, 1840)

【形态特征】雄性前唇基黄色，额黄色具1个甚大的黑色斑；胸部墨绿色具金属光泽；腹部黑色，第2节侧面具黄色斑纹。雌性腹部粗壮，第2~3节具黄斑。【长度】体长 45~52 mm，腹长 30~38 mm，后翅 27~35 mm。【栖息环境】海拔 1000 m以下的渗流地和沼泽地。【分布】黑龙江；从欧洲经西伯利亚至日本广布。【飞行期】6—9月。

北极金光伪蜻 雄，芬兰 | Sami Karjalainen 摄
Somatochlora arctica, male from Finland | Photo by Sami Karjalainen

北极金光伪蜻 交尾，芬兰 | Sami Karjalainen 摄
Somatochlora arctica, mating pair from Finland | Photo by Sami Karjalainen

[Identification] Male anteclypeus yellow, frons with a large black spot. Thorax metallic dark green. Abdomen black, S2 with lateral yellow stripes. Female abdomen stouter, S2-S3 with yellow markings. **[Measurements]** Total length 45-52 mm, abdomen 30-38 mm, hind wing 27-35 mm. **[Habitat]** Seepages and marshes below 1000 m elevation. **[Distribution]** Heilongjiang; Widespread from Europe throughout Siberia to Japan. **[Flight Season]** June to September.

绿金光伪蜻 *Somatochlora dido* Needham, 1930

【形态特征】雄性前唇基黄色，额黑绿色具金属光泽；胸部墨绿色具金属光泽，侧面具2个黄斑；腹部黑色，第2节和第3节具黄斑。【长度】雄性腹长 34 mm，后翅 32 mm。【栖息环境】海拔 1000~2500 m的湿地。【分布】中国四川特有。【飞行期】7—9月。

[Identification] Male anteclypeus yellow, frons metallic blackish green. Thorax metallic dark green, laterally with two yellow spots. Abdomen black, S2-S3 with yellow spots. **[Measurements]** Male abdomen 34 mm, hind wing 32 mm. **[Habitat]** Wetlands at 1000-2000 m elevation. **[Distribution]** Endemic to Sichuan of China. **[Flight Season]** July to September.

日本金光伪蜻 *Somatochlora exuberata* Bartenev, 1910

【形态特征】雄性前唇基黄色，额黄色具1个甚大的黑色斑；胸部墨绿色具金属光泽；腹部黑色，第1节侧面具1个黄斑，第2节侧面后缘具黄色细纹，第3节侧面基方具黄斑。雌性与雄性相似，下生殖板甚长，伸向体下方。【长度】体长 51~55 mm，腹长 37~41 mm，后翅 36~38 mm。【栖息环境】海拔 1000 m以下流速缓慢的小溪和池塘。【分布】黑龙江、吉林、辽宁、北京；朝鲜半岛、日本、西伯利亚。【飞行期】6—9月。

日本金光伪蜻 雄，黑龙江 | 莫善濂 摄
Somatochlora exuberata, male from Heilongjiang | Photo by Shanlian Mo

日本金光伪蜻 雄，黑龙江 | 莫善濂 摄
Somatochlora exuberata, male from Heilongjiang | Photo by Shanlian Mo

日本金光伪蜻 雌，黑龙江 | 莫善濂 摄
Somatochlora exuberata, female from Heilongjiang | Photo by Shanlian Mo

[Identification] Male anteclypeus yellow, frons yellow with large black spot. Thorax metallic dark green. Abdomen black, S1 with a yellow spot laterally, S2 with lateral yellow stripes posteriorly, S3 with basal yellow spots laterally. Female similar to male, the vulvar lamina long and projecting ventrally. **[Measurements]** Total length 51-55 mm, abdomen 37-41 mm, hind wing 36-38 mm. **[Habitat]** Slow flowing streams and ponds below 1000 m elevation. **[Distribution]** Heilongjiang, Jilin, Liaoning, Beijing; Korean peninsula, Japan, Siberia. **[Flight Season]** June to September.

格氏金光伪蜻 *Somatochlora graeseri* Selys, 1887

【形态特征】雄性前唇基黄色，额黄色具1个甚大的黑色斑；胸部墨绿色具金属光泽；腹部黑色，第1节侧面具1个黄斑，第2节侧面具黄色细纹，第3节侧面基方具1对甚大黄斑。雌性与雄性相似，但腹部较粗壮。【长度】体长 49～57 mm，腹长 35～41 mm，后翅 34～38 mm。【栖息环境】海拔 1000 m以下的池塘。【分布】黑龙江、吉林、辽宁、北京；朝鲜半岛、日本、俄罗斯远东。【飞行期】6—9月。

[Identification] Male anteclypeus yellow, frons yellow with a large black spot. Thorax metallic dark green. Abdomen black, S1 with a yellow spot laterally, S2 with lateral yellow stripes posteriorly, S3 with a pair of basal yellow spots laterally. Female similar to male but abdomen stouter. **[Measurements]** Total length 49-57 mm, abdomen 35-41 mm, hind wing 34-38 mm. **[Habitat]** Ponds below 1000 m elevation. **[Distribution]** Heilongjiang, Jilin, Liaoning, Beijing; Korean peninsula, Japan, Russsian Far East. **[Flight Season]** June to September.

格氏金光伪蜻 雄，北京 | 陈炜 摄
Somatochlora graeseri, male from Beijing | Photo by Wei Chen

格氏金光伪蜻 雌，吉林 | 金洪光 摄
Somatochlora graeseri, female from Jilin | Photo by Hongguang Jin

格氏金光伪蜻 雄，北京 | 陈炜 摄
Somatochlora graeseri, male from Beijing | Photo by Wei Chen

凝翠金光伪蜻 *Somatochlora metallica* (Vander Linden, 1825)

凝翠金光伪蜻 雄，芬兰 | Matti Hämäläinen 摄
Somatochlora metallica, male from Finland | Photo by Matti Hämäläinen

凝翠金光伪蜻 雌，芬兰 | Sami Karjalainen 摄
Somatochlora metallica, female from Finland | Photo by Sami Karjalainen

【形态特征】雄性前唇基黄色，额黄色具1个甚大的黑色斑；胸部墨绿色具金属光泽；腹部黑色，第1~3节具较小的黄斑。雌性与雄性相似，下生殖板极长，伸向体下方。【长度】体长 50~55 mm，腹长 37~44 mm，后翅 34~38 mm。【栖息环境】池塘、水库周边以及流速缓慢的小溪。【分布】黑龙江；从欧洲经西伯利亚至中国东北广布。【飞行期】6—9月。

[Identification] Male anteclypeus yellow, frons yellow with a large black spot. Thorax metallic dark green. Abdomen black, S1-S3 with very small yellow spots. Female similar to male, the vulvar lamina extremely long and projecting ventrally. **[Measurements]** Total length 50-55 mm, abdomen 37-44 mm, hind wing 34-38 mm. **[Habitat]** Ponds, reservoirs and slow flowing streams. **[Distribution]** Heilongjiang; Widespread from Europe throughout Siberia to Northeast China. **[Flight Season]** June to September.

山西金光伪蜻 *Somatochlora shanxiensis* Zhu & Zhang, 1999

【形态特征】雄性前唇基黄色，额黄色具1个甚大的黑色斑；胸部墨绿色具金属光泽；腹部黑色，第1节侧面具1个黄斑，第2节侧面后缘具黄色细纹，第3节侧面基方具1对甚大的黄斑。雌性与雄性相似，但腹部较粗壮。【长度】体长 52~55 mm，腹长 36~39 mm，后翅 38~40 mm。【栖息环境】海拔 1500~3000 m的湿地。【分布】中国特有，分布于山西、湖北。【飞行期】6—9月。

[Identification] Male anteclypeus yellow, frons yellow with a large black spot. Thorax metallic dark green. Abdomen black, S1 with a yellow spot laterally, S2 with lateral yellow stripes posteriorly, S3 with a pair of basal yellow spots laterally. Female similar to male but the abdomen stouter. **[Measurements]** Total length 52-55 mm, abdomen 36-39 mm, hind wing 38-40 mm. **[Habitat]** Wetlands at 1500-3000 m elevation. **[Distribution]** Endemic to China, recorded from Shanxi, Hubei. **[Flight Season]** June to September.

山西金光伪蜻 雄，湖北
Somatochlora shanxiensis, male from Hubei

山西金光伪蜻 雄，湖北 | 莫善濂 摄
Somatochlora shanxiensis, male from Hubei | Photo by Shanlian Mo

山西金光伪蜻 雌，湖北 | 莫善濂 摄
Somatochlora shanxiensis, female from Hubei | Photo by Shanlian Mo

神农金光伪蜻 *Somatochlora shennong* Zhang, Vogt & Cai, 2014

【形态特征】雄性前唇基黄色，额黄色具1个甚大的黑色斑；胸部墨绿色具金属光泽，侧面具2条黄色条纹；腹部黑色，第1节侧面具1个大黄斑，第2节侧面后缘具黄色细纹，第3节侧面基方具1对甚大的黄斑。雌性与雄性相似，但腹部较粗壮，下生殖板半圆形，稍微向下突出。【长度】体长 45~50 mm，腹长 32~37 mm，后翅 29~34 mm。【栖息环境】海拔 1500~2000 m的沼泽、沟渠和渗流地。【分布】中国特有，分布于湖北、广西。【飞行期】6—9月。

[Identification] Male anteclypeus yellow, frons yellow with a large black spot. Thorax metallic dark green, laterally with two yellow stripes. Abdomen black, S1 with a large yellow spot laterally, S2 with lateral yellow stripes posteriorly, S3 with a pair of basal yellow spots laterally. Female similar to male but the abdomen stouter, vulvar lamina semicircle-shaped, slightly projecting ventrally. **[Measurements]** Total length 45-50 mm, abdomen 32-37 mm, hind wing 29-34 mm. **[Habitat]** Marshes, ditches and seepages at 1500-2000 m elevation. **[Distribution]** Endemic to China, recorded from Hubei, Guangxi. **[Flight Season]** June to September.

神农金光伪蜻 交尾，湖北 | 莫善濂 摄
Somatochlora shennong, mating pair from Hubei | Photo by Shanlian Mo

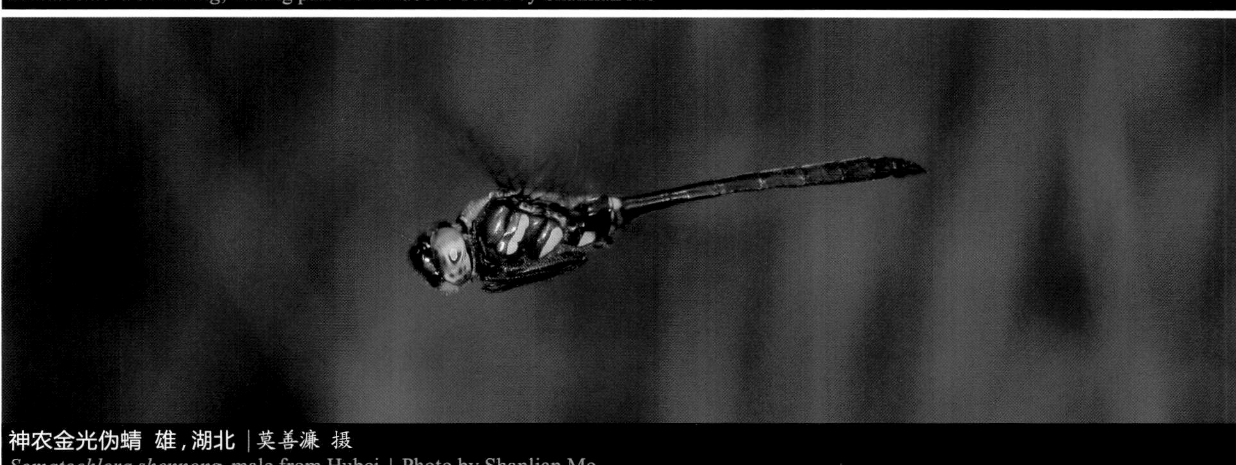

神农金光伪蜻 雄，湖北 | 莫善濂 摄
Somatochlora shennong, male from Hubei | Photo by Shanlian Mo

台湾金光伪蜻 *Somatochlora taiwana* Inoue & Yokota, 2001

【形态特征】雄性前唇基黄色，额黑绿色具金属光泽；胸部墨绿色具金属光泽，后胸后侧板具1个黄斑；腹部黑色，第1~3节具黄斑。雌性与雄性相似。【长度】体长 48~50 mm，腹长 34~37 mm，后翅 32~33 mm。【栖息环境】海拔 1500~2500 m的湿地。【分布】中国特有，分布于广东、台湾。【飞行期】7—9月。

[Identification] Male anteclypeus yellow, frons metallic blackish green. Thorax metallic dark green, metepimeron with a yellow spot. Abdomen black, S1-S3 with yellow spots. Female similar to male. **[Measurements]** Total length 48-50 mm, abdomen 34-37 mm, hind wing 32-33 mm. **[Habitat]** Wetlands at 1500-2500 m elevation. **[Distribution]** Endemic to China, recorded from Guangdong, Taiwan. **[Flight Season]** July to September.

台湾金光伪蜻 雌，广东
Somatochlora taiwana, female from Guangdong

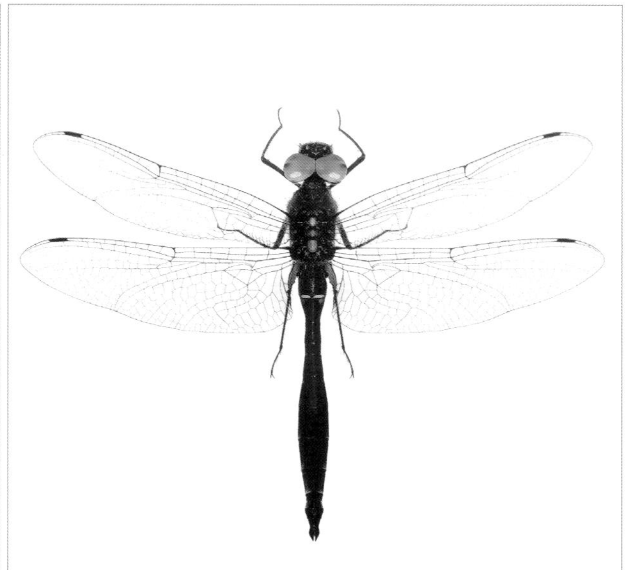

台湾金光伪蜻 雄，台湾
Somatochlora taiwana, male from Taiwan

台湾金光伪蜻 雌，广东
Somatochlora taiwana, female from Guangdong

6 大伪蜻科 Family Macromiidae

本科已知4属120余种，广布于欧亚大陆、澳新界、北美洲和非洲。中国已知2属20余种，全国广布。本科是一类中至大型的蜻蜓；复眼较大，具有如同宝石般的蓝色和绿色光泽，在头顶有很长的一段交汇；身体黑色或墨绿色具黄色条纹，许多种类具金属光泽，腹部细长具明显的黄斑或黄环；翅透明而狭长，一些种类的雌性翅染有琥珀色；翅脉的特征包括基室无横脉，臀圈较发达，呈多边形，臀三角室2室，前翅的基臀区具4~5条横脉。

本科蜻蜓栖息于山区溪流和静水环境，包括水库、湖泊和大型池塘。它们具有极强的飞行能力，被称为“巡洋舰”。雄性可以沿着水面边缘整日巡逻。雌性在水边缘的浅滩以强有力的点水方式产卵。

The family contains over 120 species in four genera distributed throughout Eurasia, Australasia, North America and Africa. Over 20 species in two genera are recorded from China, they are widely distributed throughout the country. Members of this family are medium to large dragonflies. Eyes large with brilliant iridescent green or blue and meeting for a considerable distance above. Body black or blackish green with yellow stripes and many species have metallic reflections, the abdomen long and marked with yellow spots or rings. Wings hyaline, long and narrow becoming tinted with amber in females of some species. Venational characters include a free median space, a polygonally developed anal loop, 2-celled anal triangle, and the cubital space with 4-5 crossveins in fore wings.

Species of this family inhabit montane streams and standing waters including reservoirs, lakes and large ponds. They are strong-flying dragonflies and are called “cruisers”. Males often fly along the margin of water for a whole day. Females oviposit at the shallow margin by tapping the abdomen tip strongly onto the water surface.

黄斑丽大伪蜻　雄

Epophthalmia frontalis, male

丽大伪蜻属 Genus *Epophthalmia* Burmeister, 1839

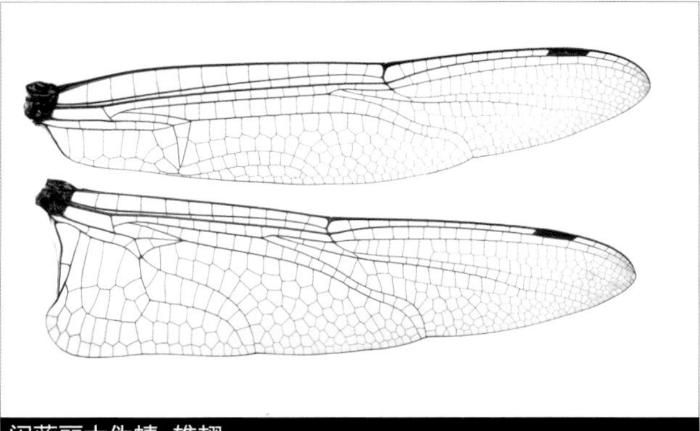

闪蓝丽大伪蜻 雄翅
Epophthalmia elegans, male wings

本属全球已知7种，东亚及东南亚广布。中国记录有4种，但描述自中国江苏的管氏丽大伪蜻存疑，而此处将描述自西双版纳的版纳丽大伪蜻作为黄斑丽大伪蜻的异名。作者认为本属中国仅2种。其中，闪蓝丽大伪蜻全国广布，在城市的湖泊、鱼塘和公园的池塘中都容易遇见。另一种黄斑丽大伪蜻仅在云南南部和西部边缘发现。

本属蜻蜓体大型，生活于大型水体。它们具有发达的绿色复眼，身体色彩较暗具黄色斑纹。雄性通常沿着池塘边缘的固定轨迹长时间巡逻。

The genus contains seven species widespread in eastern and southeastern Asia. Four species are recorded from China, but *Epophthalmia kuani* Jiang, 1998 described from Jiangsu is doubtful and *E. bannaensis* Zha & Jiang, 2010 described from Xishuangbanna of Yunnan is regarded as a synonym of *E. frontalis* Selys, 1871 here. The author considers the genus to contain only two species in China; *E. elegans* is very widespread and can be easily seen in lakes, fish ponds and city parks. *E. frontalis* can be only seen in the southern and western border of Yunnan province.

Epophthalmia species are large-sized dragonflies and frequent large bodies of water. They have large green eyes and dark body with yellow markings. Males patrol along a fixed trajectory at the margin of water for a considerable period of time.

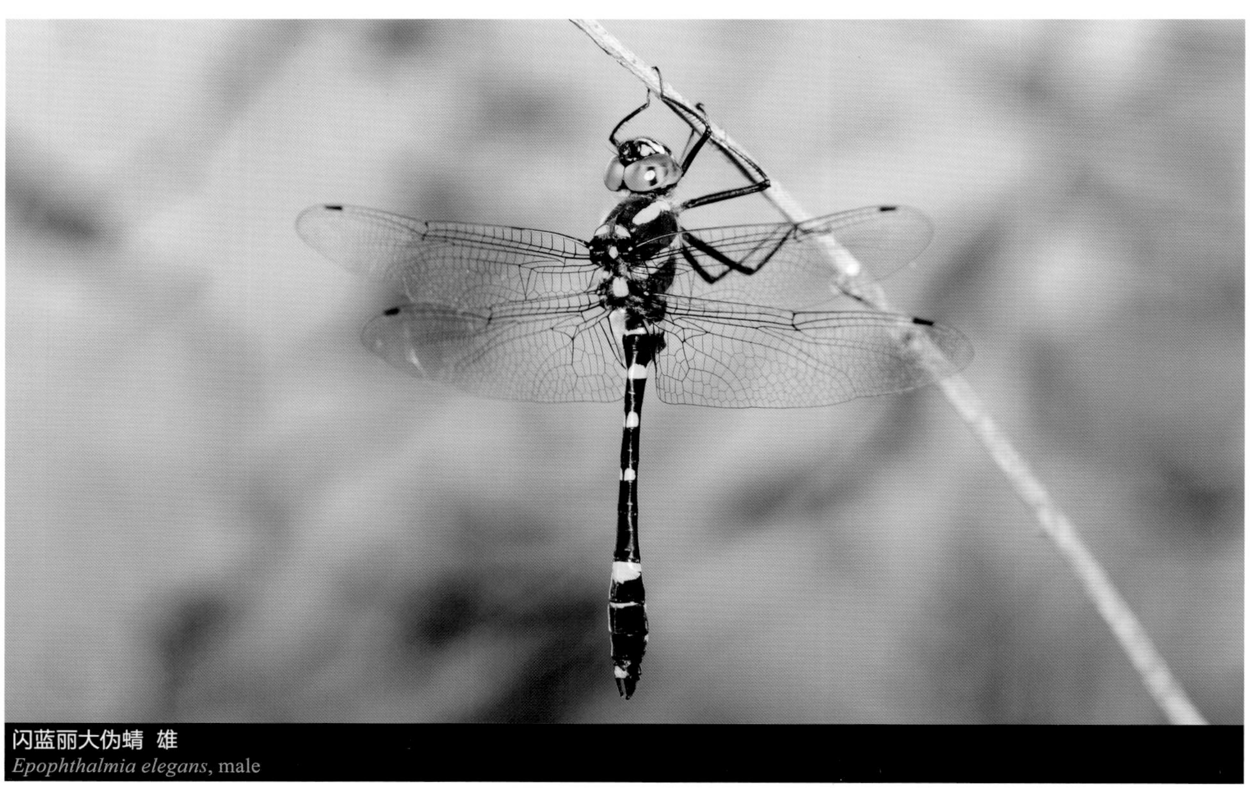

闪蓝丽大伪蜻 雄
Epophthalmia elegans, male

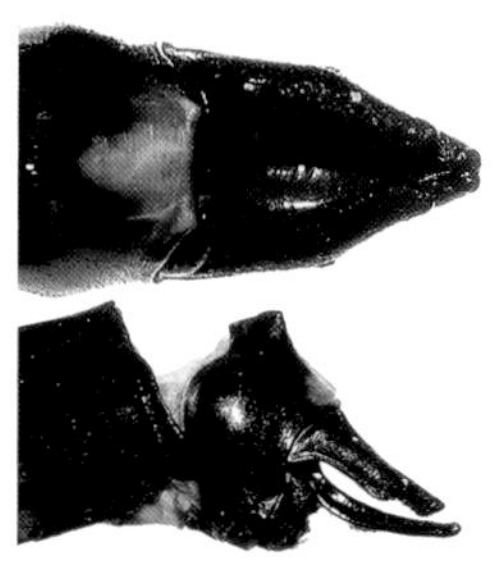

闪蓝丽大伪蜻
Epophthalmia elegans

黄斑丽大伪蜻
Epophthalmia frontalis

丽大伪蜻属 雄性肛附器
Genus *Epophthalmia*, male anal appendages

闪蓝丽大伪蜻 *Epophthalmia elegans* (Brauer, 1865)

【形态特征】雄性复眼绿色，面部黑色具黄色和白色斑纹；胸部黑绿色具金属光泽和宽阔的黄色条纹；腹部黑色具黄斑。雌性与雄性相似，翅基方具琥珀色斑。【长度】体长 76～82 mm，腹长 53～59 mm，后翅 48～51 mm。【栖息环境】海拔 2000 m以下的河流、水库、湖泊和大型池塘。【分布】全国广布；朝鲜半岛、日本、俄罗斯远东、老挝、越南、菲律宾。【飞行期】全年可见。

闪蓝丽大伪蜻 雄，贵州
Epophthalmia elegans, male from Guizhou

闪蓝丽大伪蜻 雄，贵州
Epophthalmia elegans, male from Guizhou

闪蓝丽大伪蜻 雌，贵州
Epophthalmia elegans, female from Guizhou

闪蓝丽大伪蜻 雌，贵州
Epophthalmia elegans, female from Guizhou

[Identification] Male eyes green, face black with yellow and white markings. Thorax blackish green with metallic reflections and with broad yellow stripes. Abdomen black with yellow spots. Female similar to male, wing bases with large amber tint. **[Measurements]** Total length 76-82 mm, abdomen 53-59 mm, hind wing 48-51 mm. **[Habitat]** Rivers, reservoirs, lakes and large ponds below 2000 m elevation. **[Distribution]** Widespread throughout China; Korean peninsula, Japan, Russsian Far East, Laos, Vietnam, Philippines. **[Flight Season]** Throughout the year.

黄斑丽大伪蜻 *Epophthalmia frontalis* Selys, 1871

【形态特征】雄性复眼蓝绿色，面部褐色具较小的白色斑纹；胸部黑褐色具金属光泽和黄色条纹；腹部黑色具黄斑，第10节和肛附器黄褐色。雌性与雄性相似。【长度】体长 73~80 mm，腹长 53~58 mm，后翅 45~50 mm。【栖息环境】海拔 1500 m以下的水库和大型池塘。【分布】云南（西双版纳、德宏）；印度、尼泊尔、缅甸、泰国、柬埔寨、老挝。【飞行期】全年可见。

[Identification] Male eyes bluish green, face brown with small white markings. Thorax dark brown with metallic reflections and yellow stripes. Abdomen black with yellow spots, S10 and anal appendages yellowish brown. Female similar to male. [Measurements] Total length 73-80 mm, abdomen 53-58 mm, hind wing 45-50 mm. [Habitat] Reservoirs and large ponds below 1500 m elevation. [Distribution] Yunnan (Xishuangbanna, Dehong); India, Nepal, Myanmar, Thailand, Cambodia, Laos. [Flight Season] Throughout the year.

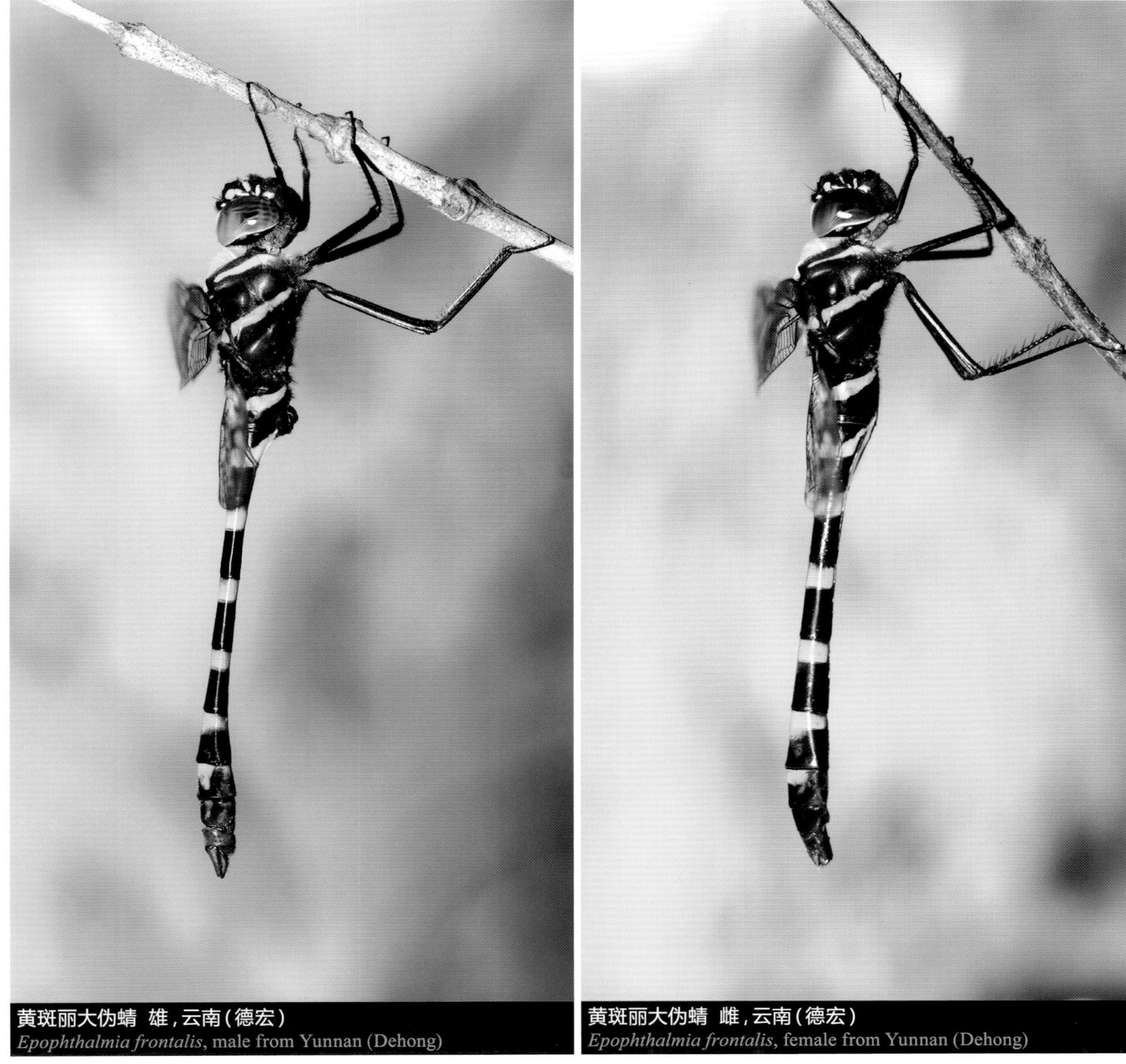

黄斑丽大伪蜻 雄，云南（德宏）
Epophthalmia frontalis, male from Yunnan (Dehong)

黄斑丽大伪蜻 雌，云南（德宏）
Epophthalmia frontalis, female from Yunnan (Dehong)

黄斑丽大伪蜻 雄，云南（德宏）
Epophthalmia frontalis, male from Yunnan (Dehong)

黄斑丽大伪蜻 雌，云南（德宏）
Epophthalmia frontalis, female from Yunnan (Dehong)

大伪蜻属 Genus *Macromia* Rambur, 1842

本属分布于欧亚大陆、北美洲和澳新界。全球已知超过80种。中国已知25种，除西北地区外全国广布。本属蜻蜓体中至大型；复眼发达，绿色或蓝色，面部通常是黑绿色、黄褐色或深褐色，额具金属光泽；胸部暗绿色具金属光泽，合胸侧面具黄色或黄白色条纹；腹部通常黑色具黄色或黄白色斑纹。本属雄性可以结合身体色彩、肛附器及后钩片的形状加以区分，但很多种类的雌性很难区分。本属后翅的三角室无横脉，可与丽大伪蜻属区分。

本属蜻蜓栖息于溪流和河流。雄性通常沿着河流和溪流边缘来回飞行，寻找配偶，有些种类的雄性则是在河流的中央巡飞。雌性在河岸边的浅滩处产卵，通常是在小范围内来回飞行并以强有力的点水方式产卵，有些种类的雌性也会将卵产在河流中央的深水区。交尾时停落到高树上。本属的大多数种类白天活动，但也有个别种类喜欢在黄昏时活动，如笛尾大伪蜻种团和弯钩大伪蜻。

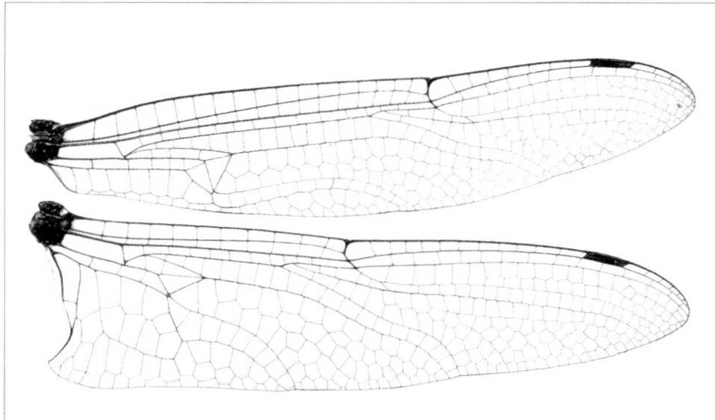

伯兰大伪蜻 雄翅
Macromia berlandi, male wings

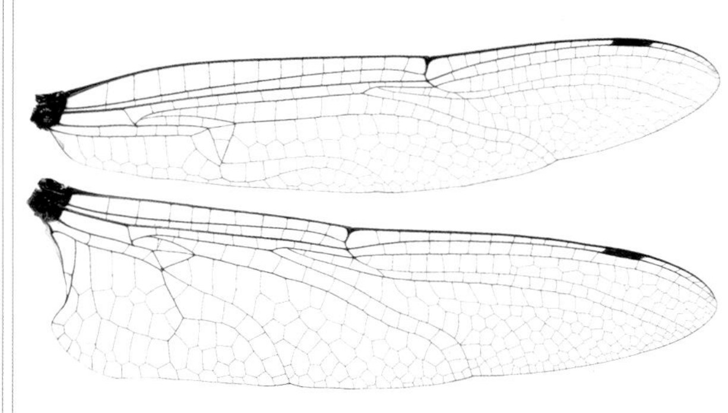

莫氏大伪蜻指名亚种 雄翅
Macromia moorei moorei, male wings

褐蓝大伪蜻 雄
Macromia cupricincta, male

弯钩大伪蜻 交尾 | 宋睿斌 摄
Macromia unca, mating pair | Photo by Ruibin Song

亮面大伪蜻 交尾 | 宋睿斌 摄
Macromia fulgidifrons, mating pair | Photo by Ruibin Song

The genus is widespread in Eurasia, North America and Australasia, over 80 species have been described. 25 species are recorded from China distributed throughout the country except the Northwest. Species of the genus are medium to large sized dragonflies. Eyes large, green or blue, face usually blackish green, yellowish brown or dark brown, frons metallic. Thorax shining blackish metallic green, synthorax laterally with yellow or yellowish white stripes. Abdomen long, black with yellow or yellowish white stripes. Males are often distinguished by the combination of body maculation, shape of anal appendages and posterior hamulus. Females of many species are more difficult to identify specifically. Species of *Macromia* lack a crossvein in hind wing triangle, differing from *Epophthalmia* species.

Macromia species inhabit streams and rivers. Males usually travel along the river and stream margin in search of females, some species patrol in the central part of water. Females lay eggs in the shallow margin by tapping the abdomen tip onto the water surface, a few species have been observed to deposit the eggs in the deep part of rivers. Most species are active during the daytime, a few are active at twilight, including species belonging to *calliope*-group and *M. unca*.

天使大伪蜻 交尾 | 宋睿斌 摄
Macromia katae, mating pair | Photo by Ruibin Song

天王大伪蜻 交尾 | 宋睿斌 摄
Macromia urania, mating pair | Photo by Ruibin Song

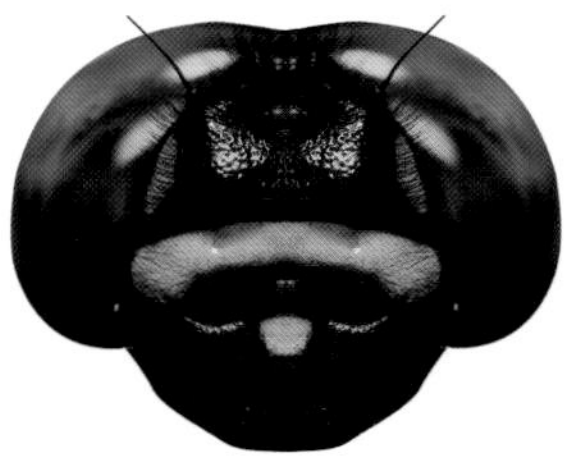
北京大伪蜻 雄
Macromia beijingensis, male

笛尾大伪蜻 雄
Macromia calliope, male

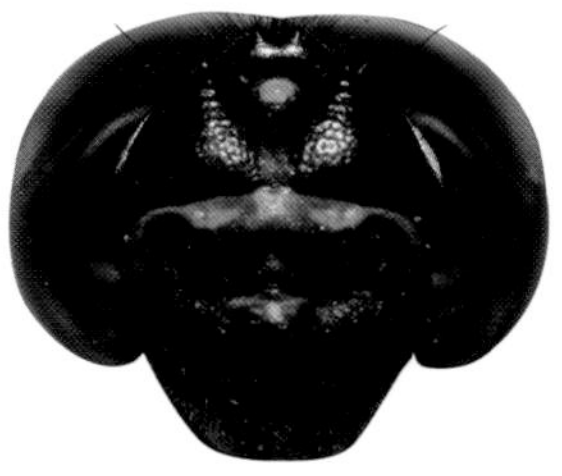
笛尾大伪蜻 雌
Macromia calliope, female

泰国大伪蜻 雄
Macromia chaiyaphumensis, male

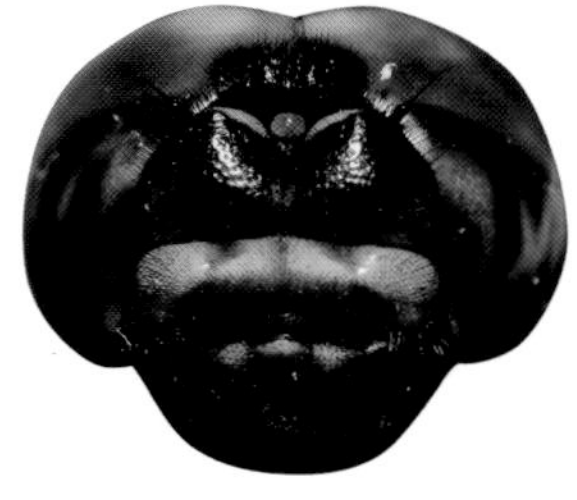
泰国大伪蜻 雌
Macromia chaiyaphumensis, female

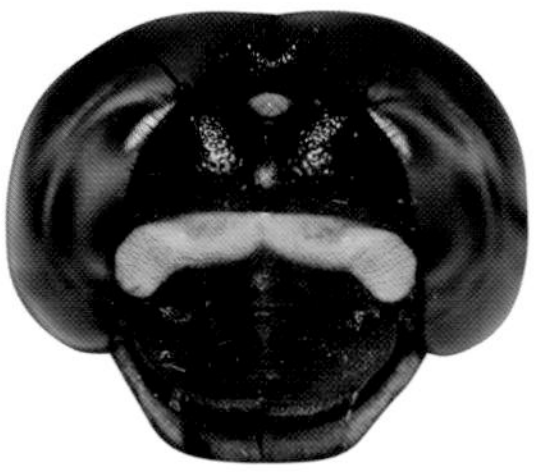
海神大伪蜻 雄
Macromia clio, male

海神大伪蜻 雌
Macromia clio, female

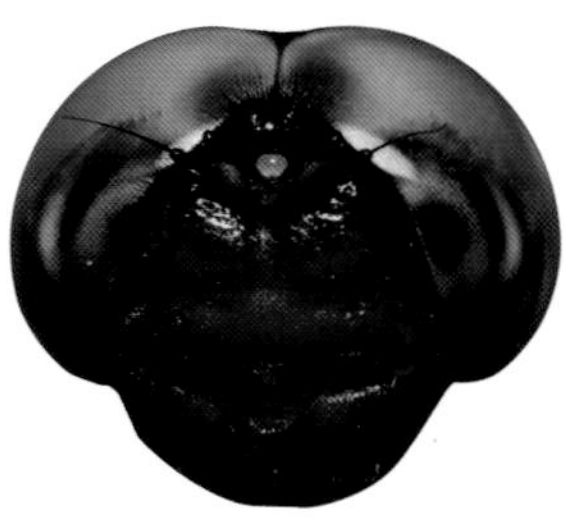
褐蓝大伪蜻 雄
Macromia cupricincta, male

大斑大伪蜻 雄
Macromia daimoji, male

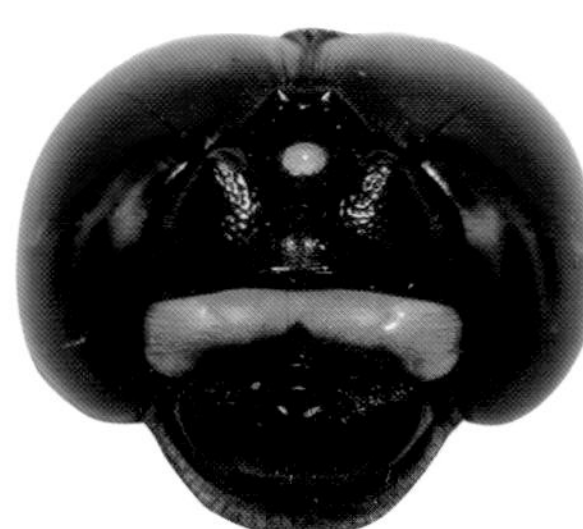
大斑大伪蜻 雌
Macromia daimoji, female

黄斑大伪蜻 雄
Macromia flavocolorata, male

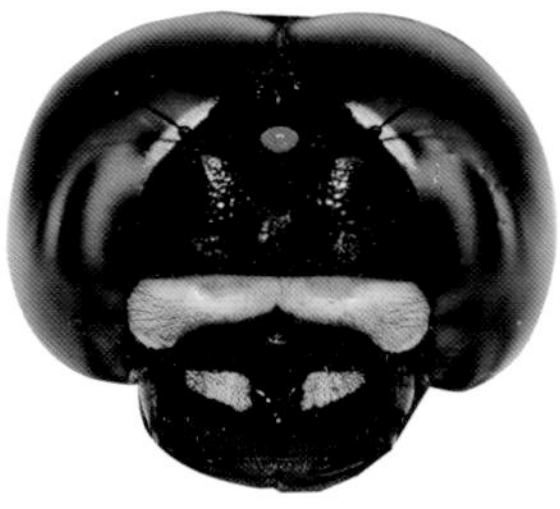
锤钩大伪蜻 雄
Macromia hamata, male

大伪蜻属 头部正面观
Genus *Macromia*, head in frontal view

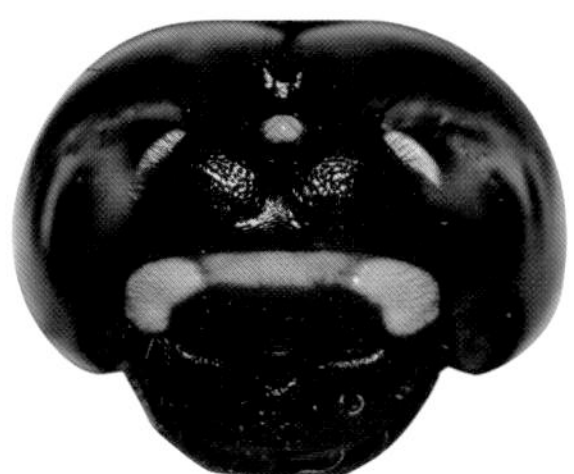
福建大伪蜻 雄
Macromia malleifera, male

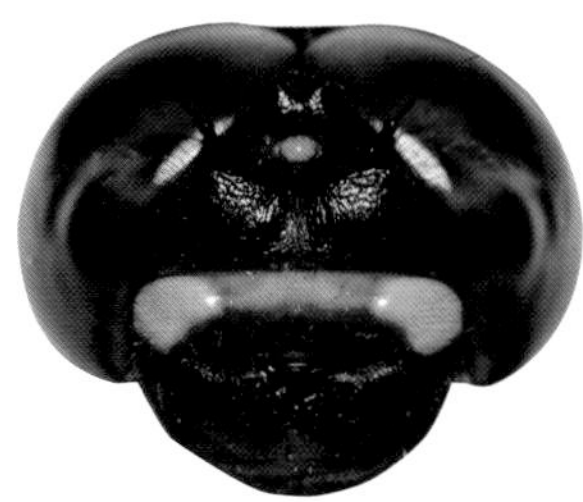
福建大伪蜻 雌
Macromia malleifera, female

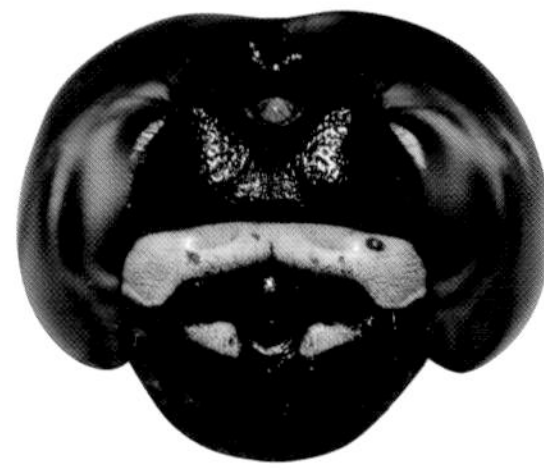
东北大伪蜻 雄
Macromia manchurica, male

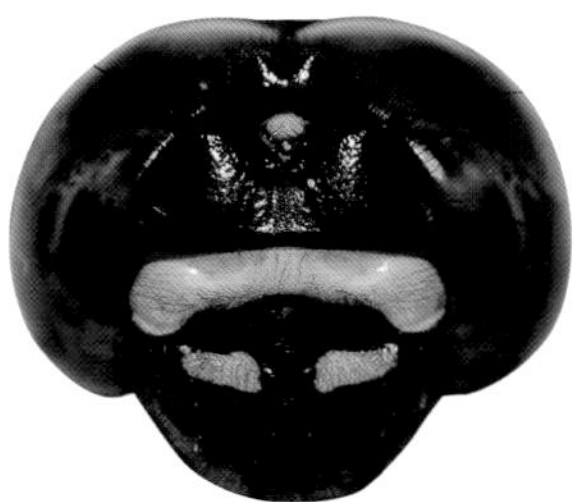
东北大伪蜻 雌
Macromia manchurica, female

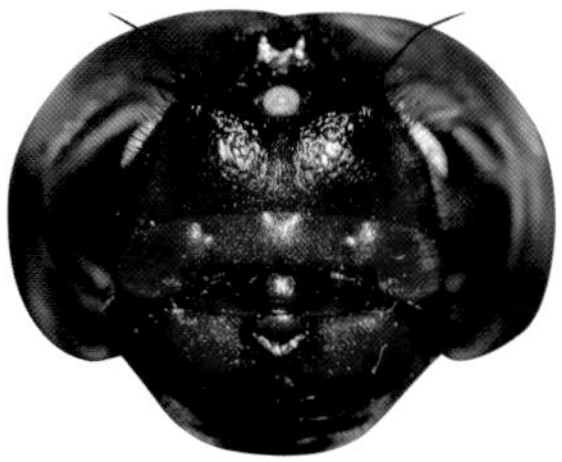
莫氏大伪蜻指名亚种 雄
Macromia moorei moorei, male

沙天马大伪蜻 雄
Macromia septima, male

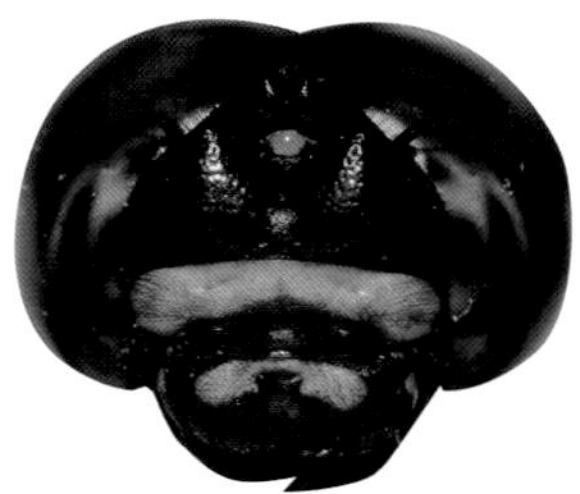
天王大伪蜻 雌
Macromia urania, female

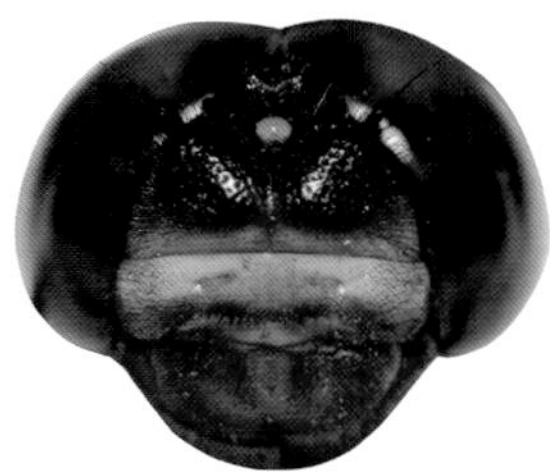
万荣大伪蜻 雄
Macromia vangviengensis, male

大伪蜻属待定种1 雄
Macromia sp. 1, male

大伪蜻属 头部正面观
Genus *Macromia*, head in frontal view

大伪蜻属待定种2 雄
Macromia sp. 2, male

大伪蜻属待定种3
Macromia sp. 3

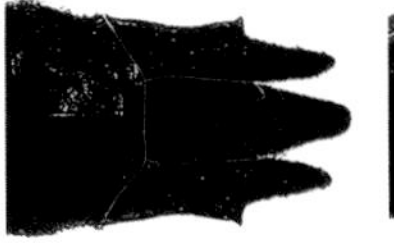

大伪蜻属待定种4
Macromia sp. 4

大伪蜻属 头部正面观
Genus *Macromia*, head in frontal view

大伪蜻属 雄性肛附器
Genus *Macromia*, male anal appendages

圆大伪蜻
Macromia amphigena

北京大伪蜻
Macromia beijingensis

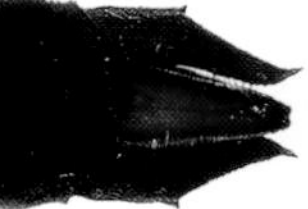

伯兰大伪蜻
Macromia berlandi

笛尾大伪蜻
Macromia calliope

泰国大伪蜻
Macromia chaiyaphumensis

海神大伪蜻
Macromia clio

褐蓝大伪蜻
Macromia cupricincta

大斑大伪蜻
Macromia daimoji

黄斑大伪蜻
Macromia flavocolorata

亮面大伪蜻
Macromia fulgidifrons

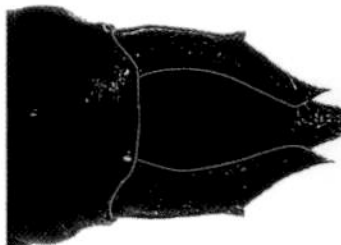

锤钩大伪蜻
Macromia hamata

天使大伪蜻
Macromia katae

福建大伪蜻
Macromia malleifera

东北大伪蜻
Macromia manchurica

莫氏大伪蜻指名亚种
Macromia moorei moorei

莫氏大伪蜻马来亚种
Macromia moorei malayana

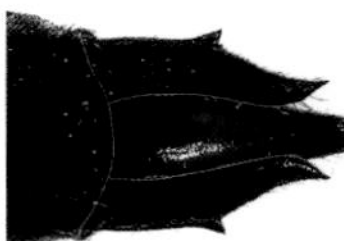

褐面大伪蜻
Macromia pinratani vietnamica

沙天马大伪蜻
Macromia septima

大伪蜻属 雄性肛附器
Genus *Macromia*, male anal appendages

弯钩大伪蜻
Macromia unca

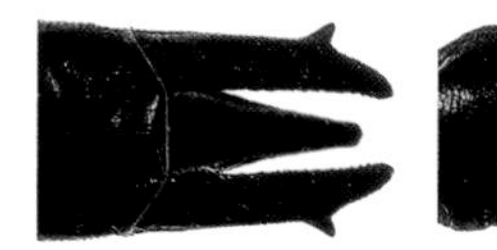

天王大伪蜻
Macromia urania

万荣大伪蜻
Macromia vangviengensis

大伪蜻属 雄性肛附器
Genus *Macromia*, male anal appendages

笛尾大伪蜻
Macromia calliope

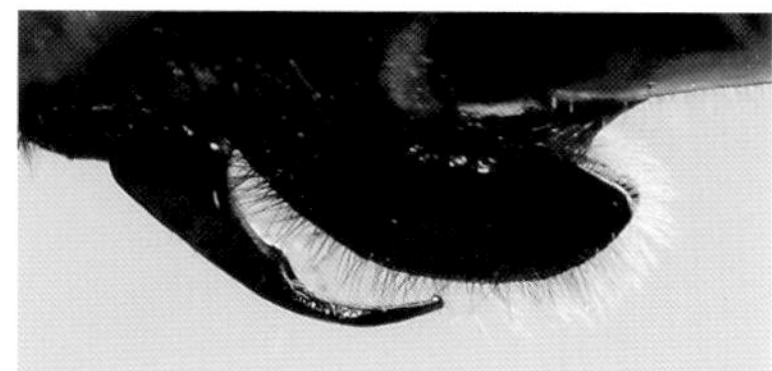

泰国大伪蜻
Macromia chaiyaphumensis

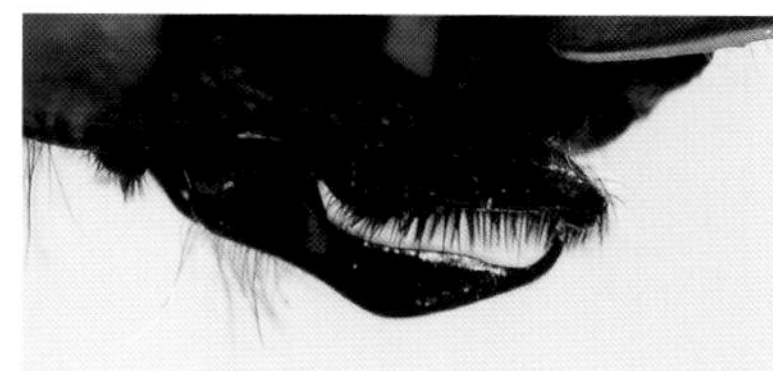

大斑大伪蜻
Macromia daimoji

黄斑大伪蜻
Macromia flavocolorata

沙天马大伪蜻
Macromia septima

天王大伪蜻
Macromia urania

笛尾大伪蜻种团 雄性后钩片
Macromia calliope-group, male posterior hamulus

圆大伪蜻 *Macromia amphigena* Selys, 1871

【形态特征】雄性上额具2对白斑；胸部肩前条纹较宽阔；腹部黑褐色，第2～3节亚基方具黄白色环纹，第4～8节亚基方或基方具黄白斑。雌性与雄性相似，身体条纹为黄色。【长度】雄性体长 63～67 mm，腹长 45～48 mm，后翅 40～41 mm。【栖息环境】海拔 500 m以下的宽阔河流。【分布】黑龙江、吉林；朝鲜半岛、日本、西伯利亚。【飞行期】5—7月。

圆大伪蜻 交尾，黑龙江 | 莫善濂 摄
Macromia amphigena, mating pair from Heilongjiang | Photo by Shanlian Mo

[Identification] Male top of frons with two pairs of white spots. Thorax with broad anterhumeral stripes. Abdomen blackish brown, S2-S3 with yellowish white rings sub-basally, S4-S8 with yellowish white spots basally or sub-basally. Female similar to male with yellow stripes. **[Measurements]**

圆大伪蜻 雄，黑龙江 | 莫善濂 摄
Macromia amphigena, male from Heilongjiang | Photo by Shanlian Mo

Male total length 63-67 mm, abdomen 45-48 mm, hind wing 40-41 mm. **[Habitat]** Broad rivers below 500 m elevation. **[Distribution]** Heilongjiang, Jilin; Korean peninsula, Japan, Siberia. **[Flight Season]** May to July.

北京大伪蜻 *Macromia beijingensis* Zhu & Chen, 2005

【形态特征】雄性上唇中央具1个圆形黄色斑点；胸部肩前条纹较宽阔；腹部黑色，第2～8节具黄白色斑，第10节背面具1对甚小的瘤状突起。雌性与雄性相似，身体条纹为黄色。【长度】雄性体长 71～75 mm，腹长 52～56 mm，后翅 45～46 mm。【栖息环境】海拔 1000 m以下的山区溪流。【分布】中国特有，分布于北京、山西、河南、四川。【飞行期】6—9月。

[Identification] Male labrum with a yellow rounded spot medially. Thorax with fairly broad antehumeral stripes. Abdomen black, S2-S8 with yellowish white markings, S10 with a pair of small tuberculiform prominences dorsally. Female similar to male with yellow markings. **[Measurements]** Male total length 71-75 mm, abdomen 52-56 mm, hind wing 45-46 mm. **[Habitat]** Montane streams below 1000 m elevation. **[Distribution]** Endemic to China, recorded from Beijing, Shanxi, Henan, Sichuan. **[Flight Season]** June to September.

北京大伪蜻 雄，四川 | 陈尽 摄
Macromia beijingensis, male from Sichuan | Photo by Jin Chen

北京大伪蜻 雌，北京 | 安起迪 摄
Macromia beijingensis, female from Beijing | Photo by Qidi An

北京大伪蜻 雄，北京 | 安起迪 摄
Macromia beijingensis, male from Beijing | Photo by Qidi An

伯兰大伪蜻 *Macromia berlandi* Lieftinck, 1941

伯兰大伪蜻 雄，广西
Macromia berlandi, male from Guangxi

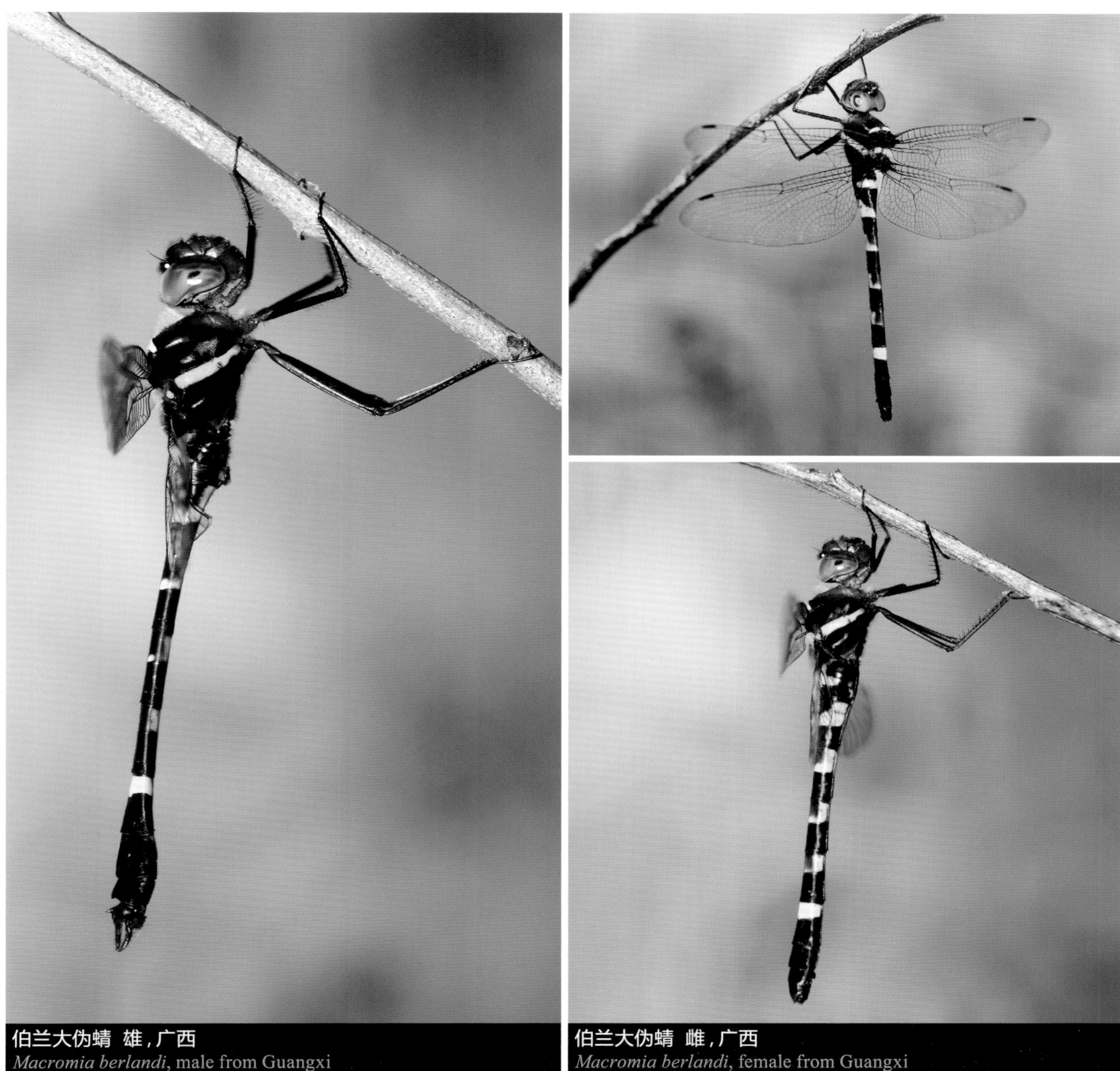

伯兰大伪蜻 雄，广西
Macromia berlandi, male from Guangxi

伯兰大伪蜻 雌，广西
Macromia berlandi, female from Guangxi

【形态特征】雄性面部褐色，额黑色；胸部无肩前条纹，后翅臀角呈尖角状；腹部黑色，第2～7节具黄斑，第10节背面具1个尖刺。雌性面部黄褐色；中胸前侧板下方和后胸后侧板下方褐色，翅基方具甚小褐斑；腹部第2～7节具发达的黄色斑纹。【长度】体长 73～77 mm，腹长 54～57 mm，后翅 44～50 mm。【栖息环境】海拔 500 m以下的溪流、沟渠和河流。【分布】福建、广东、广西、海南、香港、台湾；越南。【飞行期】4—9月。

[Identification] Male face brown, frons black. Thorax lacking antehumeral stripes, tornus angled. Abdomen black, S2-S7 with yellow spots, S10 with a sharp spine dorsally. Female face yellowish brown, lower half of mesepisternum and metepimeron brown, wing bases with small brown spots. S2-S7 with large yellow markings. **[Measurements]** Total length 73-77 mm, abdomen 54-57 mm, hind wing 44-50 mm. **[Habitat]** Streams, ditches and rivers below 500 m elevation. **[Distribution]** Fujian, Guangdong, Guangxi, Hainan, Hong Kong, Taiwan; Vietnam. **[Flight Season]** April to September.

笛尾大伪蜻 *Macromia calliope* Ris, 1916

笛尾大伪蜻 雄，广西
Macromia calliope, male from Guangxi

【形态特征】雄性面部主要黑色；胸部肩前条纹较长；腹部黑色，第2节具1条黄环，第3～5节具较小的黄斑，第7节基方具1个甚大的黄斑，第7～9节膨大。雌性与雄性相似，翅基方淡琥珀色。【长度】体长 64～66 mm，腹长 48～50 mm，后翅 38～43 mm。【栖息环境】海拔 1000 m以下的溪流和河流。【分布】贵州、广西、广东、海南；老挝、越南。【飞行期】4—10月。

[Identification] Male face mainly black. Thorax with fairly long antehumeral stripes. Abdomen black, S2 with a basal yellow ring, S3-S5 with small yellow spots, S7 with a large yellow spot basally, S7-S9 expanded. Female similar to male, wing bases with light amber tint. **[Measurements]** Total length 64-66 mm, abdomen 48-50 mm, hind wing 38-43 mm. **[Habitat]** Streams and rivers below 1000 m elevation. **[Distribution]** Guizhou, Guangxi, Guangdong, Hainan; Laos, Vietnam. **[Flight Season]** April to October.

笛尾大伪蜻 雄，广西
Macromia calliope, male from Guangxi

笛尾大伪蜻 雌，广东 | 宋睿斌 摄
Macromia calliope, female from Guangdong | Photo by Ruibin Song

笛尾大伪蜻 雌，广东 | 宋睿斌 摄
Macromia calliope, female from Guangdong | Photo by Ruibin Song

泰国大伪蜻 *Macromia chaiyaphumensis* Hämäläinen, 1986

【形态特征】雄性面部大面积黑色，上唇、后唇基和上额具黄斑；胸部肩前条纹较宽阔；腹部黑色，第2～9节具黄斑。雌性面部黄斑发达；翅基方具甚小的黑褐色斑。【长度】体长 60～62 mm，腹长 46～47 mm，后翅 36～38 mm。【栖息环境】海拔 1000 m以下森林中的溪流。【分布】云南（西双版纳）；泰国、柬埔寨、老挝。【飞行期】5—10月。

[Identification] Male face largely black, labrum, postclypeus and top of frons with yellow spots. Thorax with broad antehumeral stripes. Abdomen black, S2-S9 with yellow spots. Female face with more extensive yellow markings, wing bases with small dark brown spots. **[Measurements]** Total length 60-62 mm, abdomen 46-47 mm, hind wing 36-38 mm. **[Habitat]** Streams in forest below 1000 m elevation. **[Distribution]** Yunnan (Xishuangbanna); Thailand, Cambodia, Laos. **[Flight Season]** May to October.

泰国大伪蜻 雄，云南（西双版纳）
Macromia chaiyaphumensis, male from Yunnan (Xishuangbanna)

泰国大伪蜻 雌，云南（西双版纳）
Macromia chaiyaphumensis, female from Yunnan (Xishuangbanna)

泰国大伪蜻 雄，云南（西双版纳）
Macromia chaiyaphumensis, male from Yunnan (Xishuangbanna)

泰国大伪蜻 雌，云南（西双版纳）
Macromia chaiyaphumensis, female from Yunnan (Xishuangbanna)

海神大伪蜻 *Macromia clio* Ris, 1916

海神大伪蜻 雄，广西
Macromia clio, male from Guangxi

海神大伪蜻 雌，云南（红河）
Macromia clio, female from Yunnan (Honghe)

海神大伪蜻 雄，广西
Macromia clio, male from Guangxi

海神大伪蜻 雌，广西
Macromia clio, female from Guangxi

【形态特征】雄性后唇基黄白色；胸部肩前条纹较短；腹部黑色，第2～8节具黄斑。雌性更粗壮，腹部第2～7节具甚大的黄斑。【长度】体长 70～81 mm，腹长 50～60 mm，后翅 42～49 mm。【栖息环境】海拔 1000 m以下的溪流和河流。【分布】云南（红河）、贵州、浙江、福建、广西、广东、海南、台湾；日本、越南。【飞行期】3—9月。

[Identification] Male postclypeus yellowish white. Thorax with short antehumeral stripes. Abdomen black, S2-S8 with yellow spots. Female stouter, S2-S7 with large yellow markings. **[Measurements]** Total length 70-81 mm, abdomen 50-60 mm, hind wing 42-49 mm. **[Habitat]** Streams and rivers below 1000 m elevation. **[Distribution]** Yunnan (Honghe), Guizhou, Zhejiang, Fujian, Guangxi, Guangdong, Hainan, Taiwan; Japan, Vietnam. **[Flight Season]** March to September.

褐蓝大伪蜻 *Macromia cupricincta* Fraser, 1924

【形态特征】雄性面部完全褐色；胸部无肩前条纹，中胸前侧板下方和后胸后侧板褐色，后翅臀角呈尖角状；腹部第1～8节大面积黑色，第9～10节褐色，第2～8节具黄斑，第10节背面具1个刺突。雌性与雄性相似，腹部2～7节具宽阔的黄色环纹。【长度】体长 64～69 mm，腹长 47～50 mm，后翅 41～45 mm。【栖息环境】海拔 1000 m以下的溪流、沟渠和河流。【分布】云南（西双版纳）；印度、缅甸、泰国、柬埔寨、老挝。【飞行期】4—6月。

[Identification] Male face entirely brown. Thorax lacking antehumeral stripes, lower half of mesepisternum and metepimeron brown, tornus angled. S1-S8 largely black, S9-S10 brown, S2-S8 with yellow spots, S10 with a dorsal spine. Female similar to male, S2-S7 with broad yellow rings. **[Measurements]** Total length 64-69 mm, abdomen 47-50 mm, hind wing 41-45 mm. **[Habitat]** Streams, ditches and rivers below 1000 m elevation. **[Distribution]** Yunnan (Xishuangbanna); India, Myanmar, Thailand, Cambodia, Laos. **[Flight Season]** April to June.

褐蓝大伪蜻　雄，云南（西双版纳）
Macromia cupricincta, male from Yunnan (Xishuangbanna)

褐蓝大伪蜻 雄，云南（西双版纳）
Macromia cupricincta, male from Yunnan (Xishuangbanna)

褐蓝大伪蜻 雌，云南（西双版纳）
Macromia cupricincta, female from Yunnan (Xishuangbanna)

褐蓝大伪蜻 雌，云南（西双版纳）
Macromia cupricincta, female from Yunnan (Xishuangbanna)

大斑大伪蜻 *Macromia daimoji* Okumura, 1949

大斑大伪蜻 雄，广东
Macromia daimoji, male from Guangdong

大斑大伪蜻 雌，广东
Macromia daimoji, female from Guangdong

【形态特征】雄性后唇基白色；胸部肩前条纹较长；腹部黑色，第2～9节具黄斑，第7～9节膨大。雌性与雄性相似，腹部的黄斑更发达，翅基方稍染琥珀色。【长度】体长 67～73 mm，腹长 50～54 mm，后翅 43～49 mm。【栖息环境】海拔 1000 m以下的溪流和河流。【分布】云南（红河）、贵州、广东、广西、海南、台湾，东北地区的分布记录存疑；朝鲜半岛、日本、俄罗斯远东、越南。【飞行期】4—9月。

[Identification] Male postclypeus white. Thorax with long antehumeral stripes. Abdomen black, S2-S9 with yellow spots, S7-S9 expanded. Female similar to male, abdomen with more extensive yellow markings, wing bases slightly tinted with amber. **[Measurements]** Total length 67-73 mm, abdomen 50-54 mm, hind wing 43-49 mm. **[Habitat]** Streams and rivers below 1000 m elevation. **[Distribution]** Yunnan (Honghe), Guizhou, Guangdong, Guangxi, Hainan, Taiwan, records from Northeast China are doubtful; Korean peninsula, Japan, Russian Far East, Vietnam. **[Flight Season]** April to September.

大斑大伪蜻 雄，广东 | 宋睿斌 摄
Macromia daimoji, male from Guangdong | Photo by Ruibin Song

大斑大伪蜻 雌，广东 | 宋睿斌 摄
Macromia daimoji, female from Guangdong | Photo by Ruibin Song

黄斑大伪蜻 *Macromia flavocolorata* Fraser, 1922

黄斑大伪蜻 雄，云南（西双版纳）
Macromia flavocolorata, male from Yunnan (Xishuangbanna)

黄斑大伪蜻 雄，云南（西双版纳）
Macromia flavocolorata, male from Yunnan (Xishuangbanna)

黄斑大伪蜻 雌，云南（西双版纳）
Macromia flavocolorata, female from Yunnan (Xishuangbanna)

【形态特征】雄性后唇基黄色；胸部肩前条纹较长；腹部黑色，第2～8节具黄斑。雌性腹部的黄色条纹更发达，第2节具1条甚阔的黄色环纹。【长度】体长 59～65 mm，腹长 45～47 mm，后翅 37～43 mm。【栖息环境】海拔1000 m以下的溪流和河流。【分布】云南（西双版纳、临沧）；印度、尼泊尔、泰国、老挝、越南。【飞行期】5—9月。

[Identification] Male postclypeus yellow. Thorax with long antehumeral stripes. Abdomen black, S2-S8 with yellow spots. Female abdomen with more extensive yellow markings, S2 with a very broad yellow ring. **[Measurements]** Total length 59-65 mm, abdomen 45-47 mm, hind wing 37-43 mm. **[Habitat]** Streams and rivers below 1000 m elevation. **[Distribution]** Yunnan (Xishuangbanna, Lincang); India, Nepal, Thailand, Laos, Vietnam. **[Flight Season]** May to September.

亮面大伪蜻 *Macromia fulgidifrons* Wilson, 1998

亮面大伪蜻 雄，广西
Macromia fulgidifrons, male from Guangxi

【形态特征】雄性后唇基黄色；胸部肩前条纹较长；腹部黑色，第2节基方具1条黄色环纹，第3节具1对小黄斑，第7节基方具1个甚大黄斑，第8节侧面具小黄斑，第10节背面具1个锥形隆起。雌性与雄性相似。【长度】体长 74~77 mm，腹长 55~58 mm，后翅 50~55 mm。【栖息环境】海拔 1000 m以下森林中的溪流。【分布】中国特有，分布于广西、广东。【飞行期】4—7月。

亮面大伪蜻 雌，广东 | 宋睿斌 摄
Macromia fulgidifrons, female from Guangdong | Photo by Ruibin Song

[Identification] Male postclypeus yellow. Thorax with fairly long antehumeral stripes. Abdomen black, S2 with a basal yellow ring, S3 with a pair of small spots, S7 with a large yellow spot basally, S8 laterally with small yellow spots, S10 with a pyramidal prominence. Female similar to male. **[Measurements]** Total length 74-77 mm, abdomen 55-58 mm, hind wing 50-55 mm. **[Habitat]** Streams in forest below 1000 m elevation. **[Distribution]** Endemic to China, recorded from Guangxi, Guangdong. **[Flight Season]** April to July.

亮面大伪蜻 雌，广东 | 宋睿斌 摄
Macromia fulgidifrons, female from Guangdong | Photo by Ruibin Song

亮面大伪蜻 雄，广西
Macromia fulgidifrons, male from Guangxi

锤钩大伪蜻 *Macromia hamata* Zhou, 2003

【形态特征】雄性上唇具1对白色三角形斑点，后唇基白色；胸部肩前条纹较短；腹部黑色，第2~5节具黄白色环纹，第7节基方具1个甚大的黄白色斑，第10节背面呈锥形隆起。雌性身体条纹为黄色，腹部的黄条纹甚阔。本种与东北大伪蜻的关系尚未明确，两者仅在后钩片存在微小差异，可能是后者的异名。【长度】体长 69~78 mm，腹长 51~57 mm，后翅 43~51 mm。【栖息环境】海拔 1500 m以下的溪流和河流。【分布】中国特有，分布于四川、贵州、湖北、福建、广西、广东。【飞行期】5—10月。

锤钩大伪蜻 雄，贵州
Macromia hamata, male from Guizhou

锤钩大伪蜻 雌，贵州
Macromia hamata, female from Guizhou

[Identification] Male labrum with a pair of white triangular spots, postclypeus white. Thorax with short antehumeral stripes. Abdomen black, S2-S5 with yellowish white rings, S7 with a large yellowish white spot basally, S10 with a pyramidal prominence. Female with yellow markings, abdomen with broad yellow markings. The species is similar to *M. manchurica*, differing only by a slight difference in the shape of poster hamulus. It maybe a synonym of *M. manchurica*. **[Measurements]** Total length 69-78 mm, abdomen 51-57 mm, hind wing 43-51 mm. **[Habitat]** Streams and rivers below 1500 m elevation. **[Distribution]** Endemic to China, recorded from Sichuan, Guizhou, Hubei, Fujian, Guangxi, Guangdong. **[Flight Season]** May to October.

锤钩大伪蜻 雄，贵州
Macromia hamata, male from Guizhou

锤钩大伪蜻 雌，贵州
Macromia hamata, female from Guizhou

天使大伪蜻 *Macromia katae* Wilson, 1993

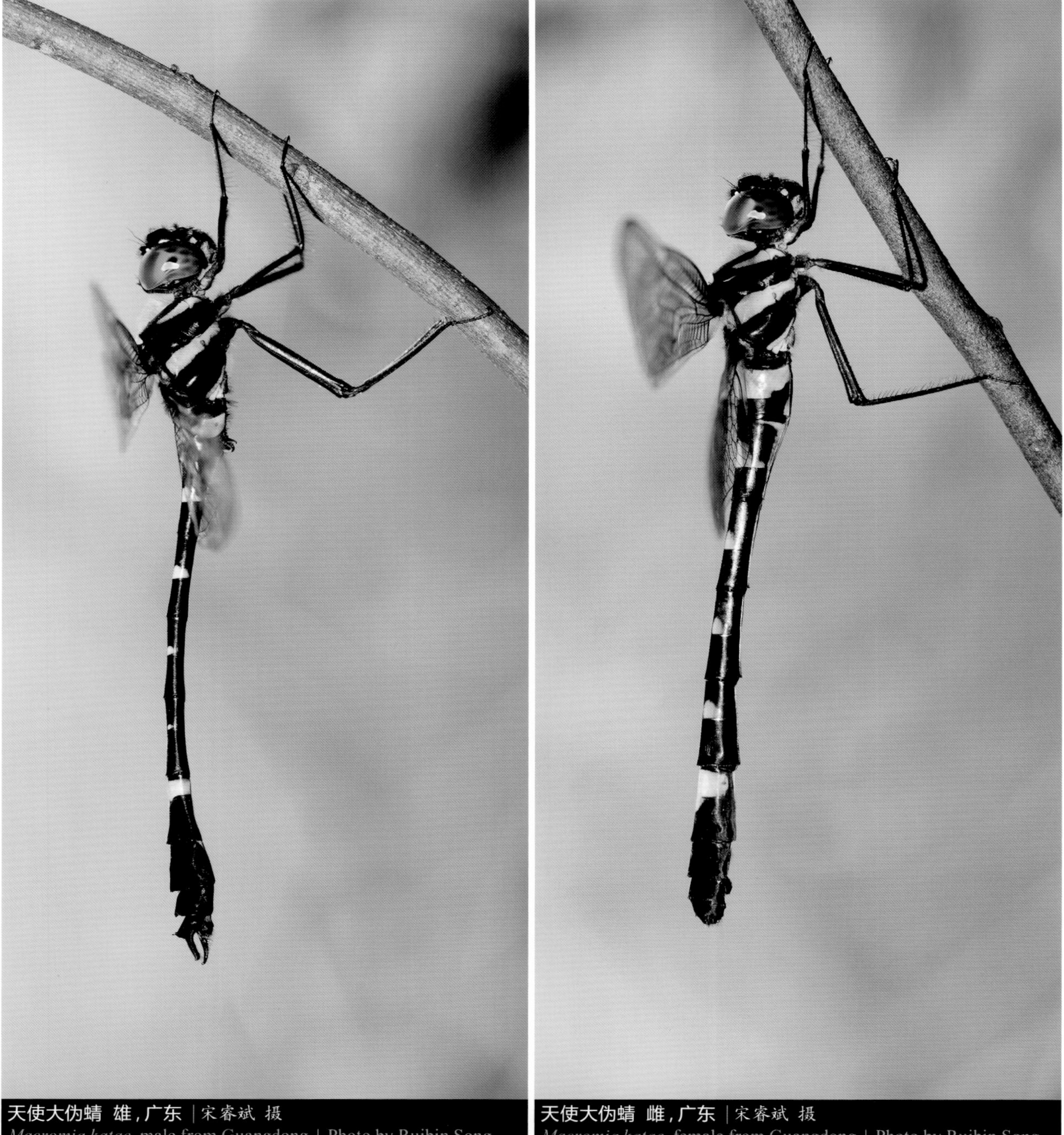

天使大伪蜻 雄，广东 | 宋睿斌 摄
Macromia katae, male from Guangdong | Photo by Ruibin Song

天使大伪蜻 雌，广东 | 宋睿斌 摄
Macromia katae, female from Guangdong | Photo by Ruibin Song

【形态特征】雄性后唇基具白色斑点，前额具黄白色条纹；胸部肩前条纹较长；腹部黑色，第2~7节具黄斑，第10节背面具1个刺突。雌性翅基方和端部具大面积的琥珀色斑；腹部甚细，黄条纹甚阔。【长度】体长 70~71 mm，腹长 54~56 mm，后翅 43~47 mm。【栖息环境】海拔 500 m以下的林荫溪流和沟渠。【分布】广东、广西、海南、香港；老挝、越南。【飞行期】5—8月。

天使大伪蜻 雄，广西
Macromia katae, male from Guangxi

天使大伪蜻 雌，广西
Macromia katae, female from Guangxi

[Identification] Male postclypeus with white spots, antefrons with yellowish white stripes. Thorax with fairly long antehumeral stripes. Abdomen black, S2-S7 with yellow spots, S10 with a dorsal spine. Female wings infused with large amber markings basally and apically. Abdomen narrow with broad yellow markings. **[Measurements]** Total length 70-71 mm, abdomen 54-56 mm, hind wing 43-47 mm. **[Habitat]** Shady streams and ditches below 500 m elevation. **[Distribution]** Guangdong, Guangxi, Hainan, Hong Kong; Laos, Vietnam. **[Flight Season]** May to August.

福建大伪蜻 *Macromia malleifera* Lieftinck, 1955

【形态特征】雄性后唇基黄色；胸部肩前条纹甚短；腹部黑色，第2～8节具黄斑，第10节背面稍微隆起。雌性腹部粗壮，第2～7节具黄色环纹，翅浅褐色。【长度】体长 77～82 mm，腹长 57～61 mm，后翅 50～55 mm。【栖息环境】海拔 1500 m以下的山区溪流。【分布】中国特有，分布于浙江、湖南、福建、广东。【飞行期】4—10月。

[Identification] Male postclypeus yellow. Thorax with short antehumeral stripes. Abdomen black, S2-S8 with yellow spots, dorsum of S10 slightly protruded. Female abdomen stout with yellow rings on S2-S7, wings tinted with light brown. **[Measurements]** Total length 77-82 mm, abdomen 57-61 mm, hind wing 50-55 mm. **[Habitat]** Montane streams below 1500 m elevation. **[Distribution]** Endemic to China, recorded from Zhejiang, Hunan, Fujian, Guangdong. **[Flight Season]** April to October.

福建大伪蜻 雄，广东 | 宋睿斌 摄
Macromia malleifera, male from Guangdong | Photo by Ruibin Song

福建大伪蜻 雄，广东 | 宋睿斌 摄
Macromia malleifera, male from Guangdong | Photo by Ruibin Song

福建大伪蜻 雌，广东
Macromia malleifera, female from Guangdong

福建大伪蜻 雌，广东
Macromia malleifera, female from Guangdong

东北大伪蜻 *Macromia manchurica* Asahina, 1964

【形态特征】雄性上唇具1对白色三角形斑点，后唇基白色；胸部肩前条纹较短；腹部黑色，第2～8节具黄斑，第10节背面具1个锥形突起。雌性身体条纹为黄色，腹部第2～7节具甚阔的黄条纹。【长度】体长 70～73 mm，腹长 50～54 mm，后翅 43～46 mm。【栖息环境】海拔 1000 m以下的溪流和河流。【分布】黑龙江、吉林、辽宁、北京；朝鲜半岛、俄罗斯远东。【飞行期】6—9月。

[Identification] Male labrum with a pair of white triangular spots, postclypeus white. Thorax with short antehumeral stripes. Abdomen black, S2-S8 with yellow spots, S10 with a pyramidal prominence. Female with yellow markings, S2-S7 with broad yellow stripes. [Measurements] Total length 70-73 mm, abdomen 50-54 mm, hind wing 43-46 mm. [Habitat] Streams and rivers below 1000 m elevation. [Distribution] Heilongjiang, Jilin, Liaoning, Beijing; Korean peninsula, Russian Far East. [Flight Season] June to September.

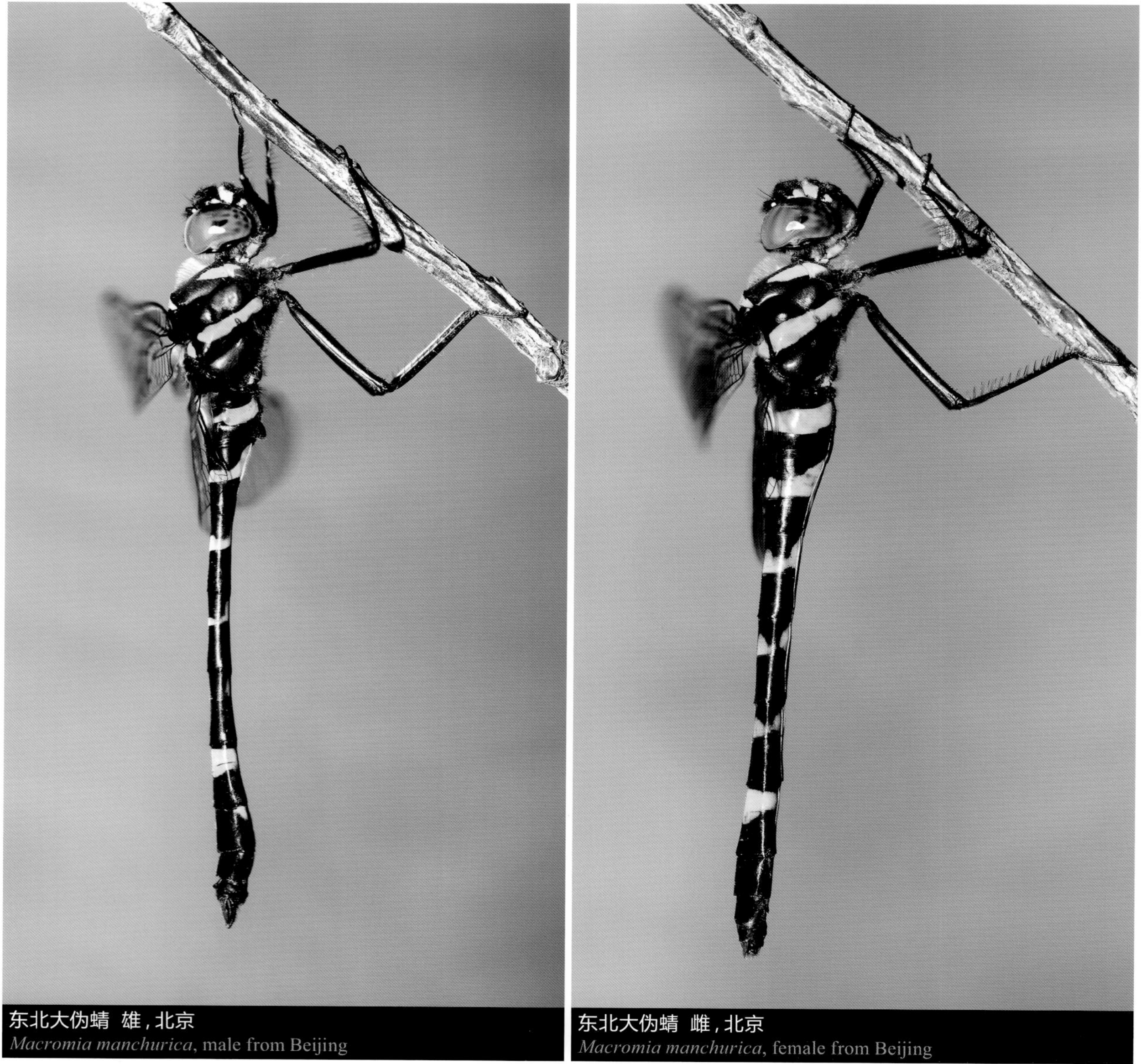

东北大伪蜻 雄，北京
Macromia manchurica, male from Beijing

东北大伪蜻 雌，北京
Macromia manchurica, female from Beijing

东北大伪蜻 雄，北京
Macromia manchurica, male from Beijing

东北大伪蜻 雌，北京
Macromia manchurica, female from Beijing

莫氏大伪蜻指名亚种 *Macromia moorei moorei* Selys, 1874

莫氏大伪蜻指名亚种 雌，贵州
Macromia moorei moorei, female from Guizhou

莫氏大伪蜻指名亚种 雄，贵州
Macromia moorei moorei, male from Guizhou

莫氏大伪蜻指名亚种 雄，广西
Macromia moorei moorei, male from Guangxi

莫氏大伪蜻指名亚种 雌，贵州
Macromia moorei moorei, female from Guizhou

【形态特征】雄性面部褐色；胸部无肩前条纹，中胸前侧板下方和后胸后侧板下方褐色；腹部黑色，第2~8节具黄白斑。雌性较粗壮，腹部较短，第2~7节具宽阔的黄色环纹。本种与伯兰大伪蜻相似，但本种雄性后翅臀角圆弧形，而伯兰大伪蜻雄性臀角呈角状。【长度】体长 71~78 mm，腹长 52~57 mm，后翅 47~53 mm。【栖息环境】海拔 500~2500 m的溪流和河流。【分布】云南（昆明、大理、保山）、四川、贵州、湖北、广西；南亚。【飞行期】5—9月。

[Identification] Male face brown. Thorax without antehumeral stripes, lower part of mesepisternum and metepimeron brown. Abdomen black, S2-S8 with yellowish white spots. Female more robust, abdomen shorter, S2-S7 with broad yellow rings. Similar to *M. berlandi*, but male with tornus rounded, tornus of male *M. berlandi* angled. **[Measurements]** Total length 71-78 mm, abdomen 52-57 mm, hind wing 47-53 mm. **[Habitat]** Streams and rivers at 500-2500 m elevation. **[Distribution]** Yunnan (Kunming, Dali, Baoshan), Sichuan, Guizhou, Hubei, Guangxi; South Asia. **[Flight Season]** May to September.

莫氏大伪蜻马来亚种 *Macromia moorei malayana* Laidlaw, 1928

【形态特征】本亚种与指名亚种相似，但体型稍小，腹部的黄色斑纹不如指名亚种丰富。莫氏大伪蜻的身体条纹变异较大，亚种的界限也不清晰，或许仅是由于地理分布造成的变异。【长度】体长 66~77 mm，腹长 48~56 mm，后翅 43~51 mm。【栖息环境】海拔 1500 m以下的山区溪流。【分布】云南（德宏、临沧、普洱、西双版纳）、广西、广东、海南；南亚、东南亚。【飞行期】3—12月。

莫氏大伪蜻马来亚种 雄，云南（普洱）
Macromia moorei malayana, male from Yunnan (Pu'er)

[Identification] The subspecies is similar to nominate subspecies but size smaller, abdominal yellow markings less extensive. The body maculation of *M. moorei* is variable, and subspecific diagnose are not clear, they may just be the result of geographic variation. **[Measurements]** Total length 66-77 mm, abdomen 48-56 mm, hind wing 43-51 mm. **[Habitat]** Montane streams below 1500 m elevation. **[Distribution]** Yunnan (Dehong, Lincang, Pu'er, Xishuangbanna), Guangxi, Guangdong, Hainan; South and Southeast Asia. **[Flight Season]** March to December.

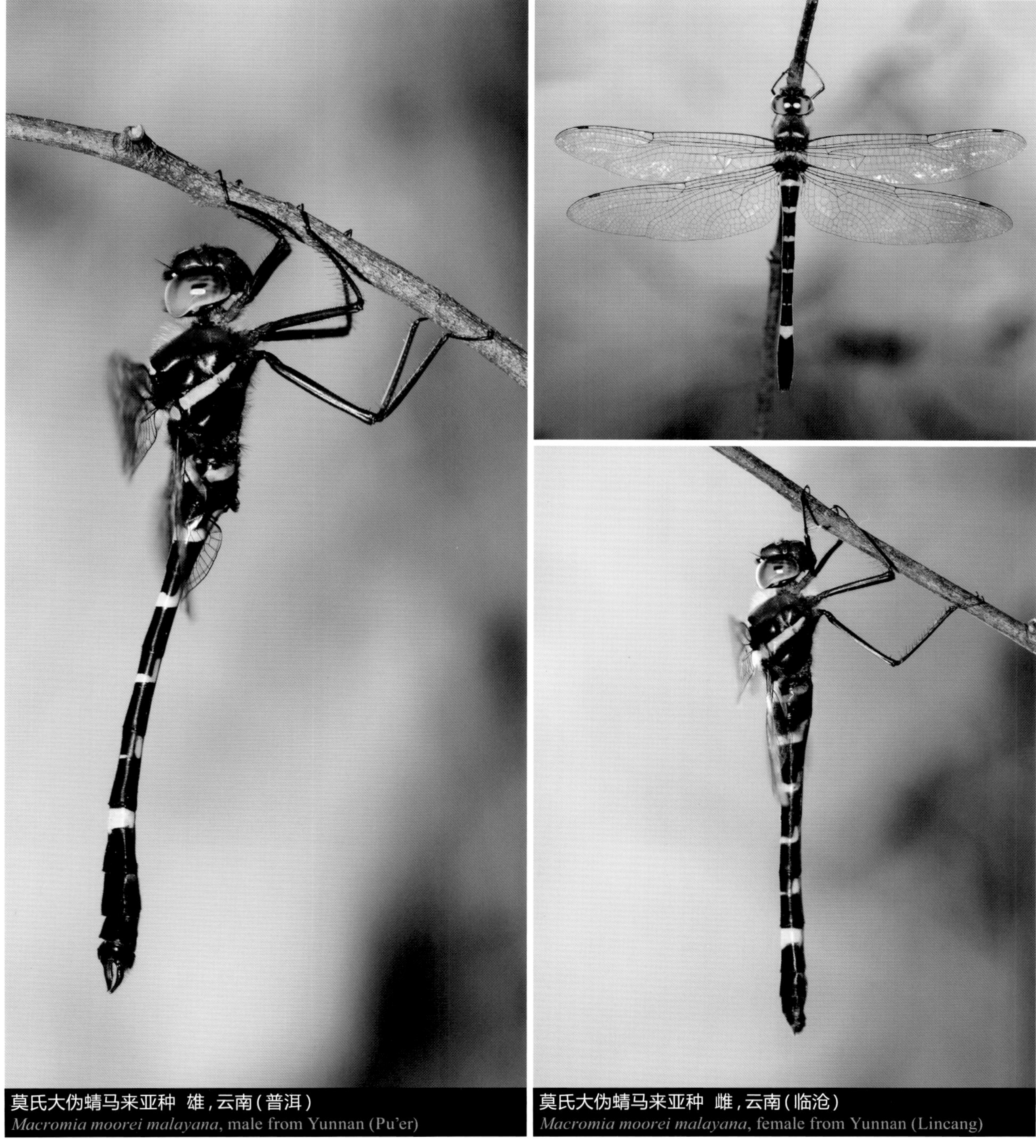

莫氏大伪蜻马来亚种 雄，云南（普洱）
Macromia moorei malayana, male from Yunnan (Pu'er)

莫氏大伪蜻马来亚种 雌，云南（临沧）
Macromia moorei malayana, female from Yunnan (Lincang)

褐面大伪蜻 *Macromia pinratani vietnamica* Asahina, 1996

褐面大伪蜻 雄，云南（普洱）
Macromia pinratani vietnamica, male from Yunnan (Pu'er)

褐面大伪蜻 雌，云南（普洱）
Macromia pinratani vietnamica, female from Yunnan (Pu'er)

【形态特征】雄性面部褐色；胸部无肩前条纹，中胸前侧板和后胸后侧板褐色；腹部黑色，第2～8节具黄斑，第10节背面具1个锥形突起，雌性与雄性相似。【长度】体长 67～76 mm，腹长 50～56 mm，后翅 47～48 mm。【栖息环境】海拔 1500 m以下的山区溪流。【分布】云南（西双版纳、普洱、红河）；老挝、越南。【飞行期】4—6月。

[Identification] Male face brown. Thorax without antehumeral stripes, mesepisternum and metepimeron brown. Abdomen black, S2-S8 with yellow spots, S10 with a pyramidal prominence. Female similar to male. **[Measurements]** Total length 67-76 mm, abdomen 50-56 mm, hind wing 47-48 mm. **[Habitat]** Montane streams below 1500 m elevation. **[Distribution]** Yunnan (Xishuangbanna, Pu'er, Honghe); Laos, Vietnam. **[Flight Season]** April to June.

褐面大伪蜻 雄，云南（普洱）
Macromia pinratani vietnamica, male from Yunnan (Pu'er)

褐面大伪蜻 雌，云南（普洱）
Macromia pinratani vietnamica, female from Yunnan (Pu'er)

沙天马大伪蜻 *Macromia septima* Martin, 1904

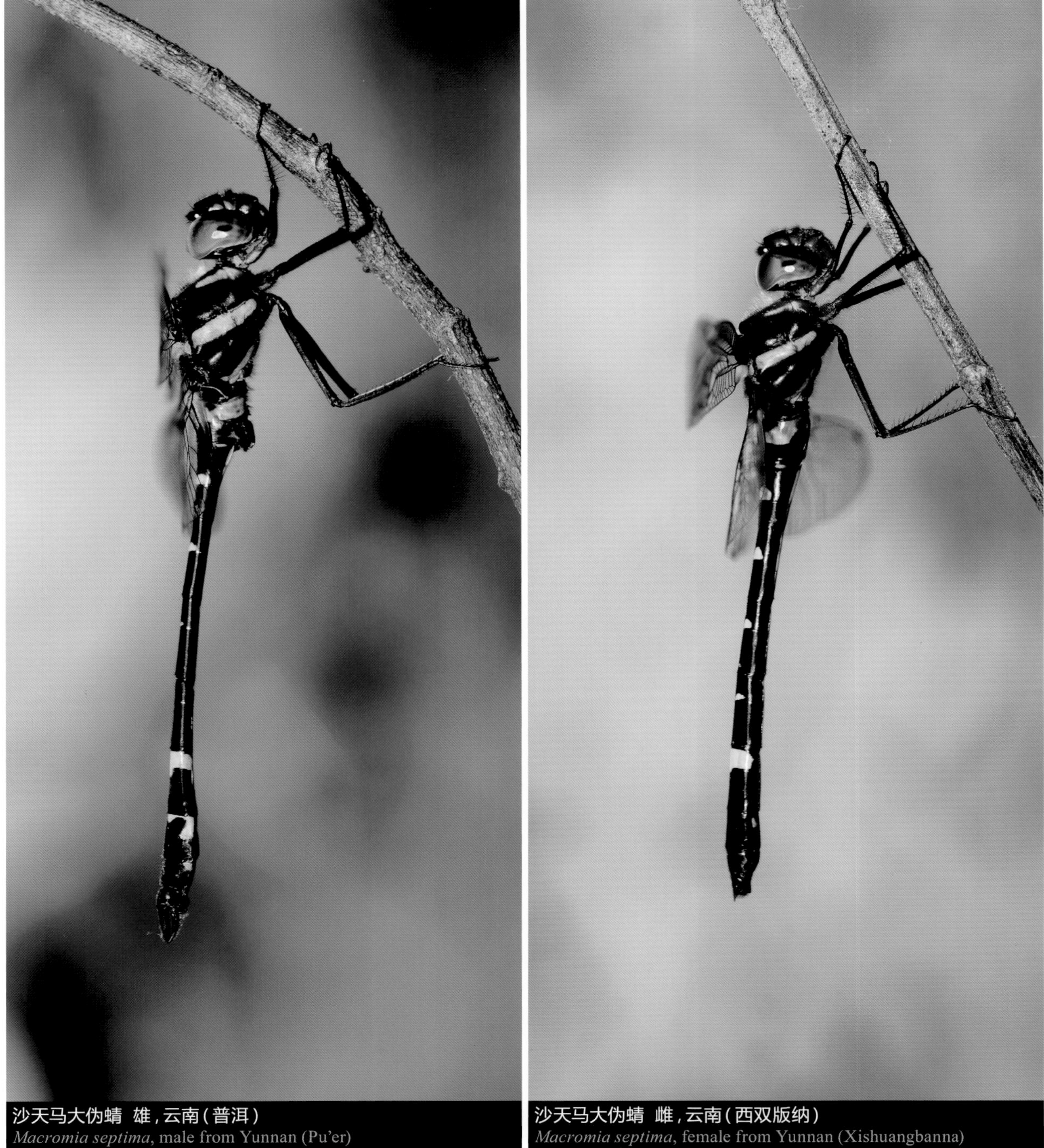

沙天马大伪蜻 雄，云南（普洱）
Macromia septima, male from Yunnan (Pu'er)

沙天马大伪蜻 雌，云南（西双版纳）
Macromia septima, female from Yunnan (Xishuangbanna)

【形态特征】雄性面部褐色；胸部肩前条纹较宽阔；腹部黑色，第2～8节具黄斑。雌性面部色彩稍淡，翅基方琥珀色。【长度】体长 58～64 mm，腹长 43～47 mm，后翅 39～44 mm。【栖息环境】海拔 1500 m以下的山区溪流。【分布】云南（西双版纳、普洱）、福建、广东、海南；泰国、老挝、越南、印度尼西亚（爪哇）。【飞行期】5—12月。

沙天马大伪蜻 雄，云南（普洱）
Macromia septima, male from Yunnan (Pu'er)

沙天马大伪蜻 雌，云南（西双版纳）
Macromia septima, female from Yunnan (Xishuangbanna)

[Identification] Male face brown. Thorax with fairly broad antehumeral stripes. Abdomen black, S2-S8 with yellow spots. Female face paler, wing bases amber. **[Measurements]** Total length 58-64 mm, abdomen 43-47 mm, hind wing 39-44 mm. **[Habitat]** Montane streams below 1500 m elevation. **[Distribution]** Yunnan (Xishuangbanna, Pu'er), Fujian, Guangdong, Hainan; Thailand, Laos, Vietnam, Indonesia (Java). **[Flight Season]** May to December.

弯钩大伪蜻 *Macromia unca* Wilson, 2004

【形态特征】雄性上唇黑色，后唇基黄色，前额黄色具1个甚大的黑色斑；胸部肩前条纹较宽阔；腹部黑色，第2~8节具黄斑，第7~10节稍微膨大，第10节背面具1个锥形突起。雌性与雄性相似。【长度】体长 60~66 mm，腹长 44~49 mm，后翅 41~48 mm。【栖息环境】海拔 1500 m以下的山区溪流。【分布】贵州、湖北、浙江、福建、广西、广东；越南。【飞行期】4—8月。

[Identification] Male labrum black, postclypeus yellow, antefrons yellow with a large black spot. Thorax with fairly broad antehumeral stripes. Abdomen black, S2-S8 with yellow spots, S7-S10 slightly expanded, S10 with a pyramidal prominence. Female similar to male. **[Measurements]** Total length 60-66 mm, abdomen 44-49 mm, hind wing 41-48 mm. **[Habitat]** Montane streams below 1500 m elevation. **[Distribution]** Guizhou, Hubei, Zhejiang, Fujian, Guangxi, Guangdong; Vietnam. **[Flight Season]** April to August.

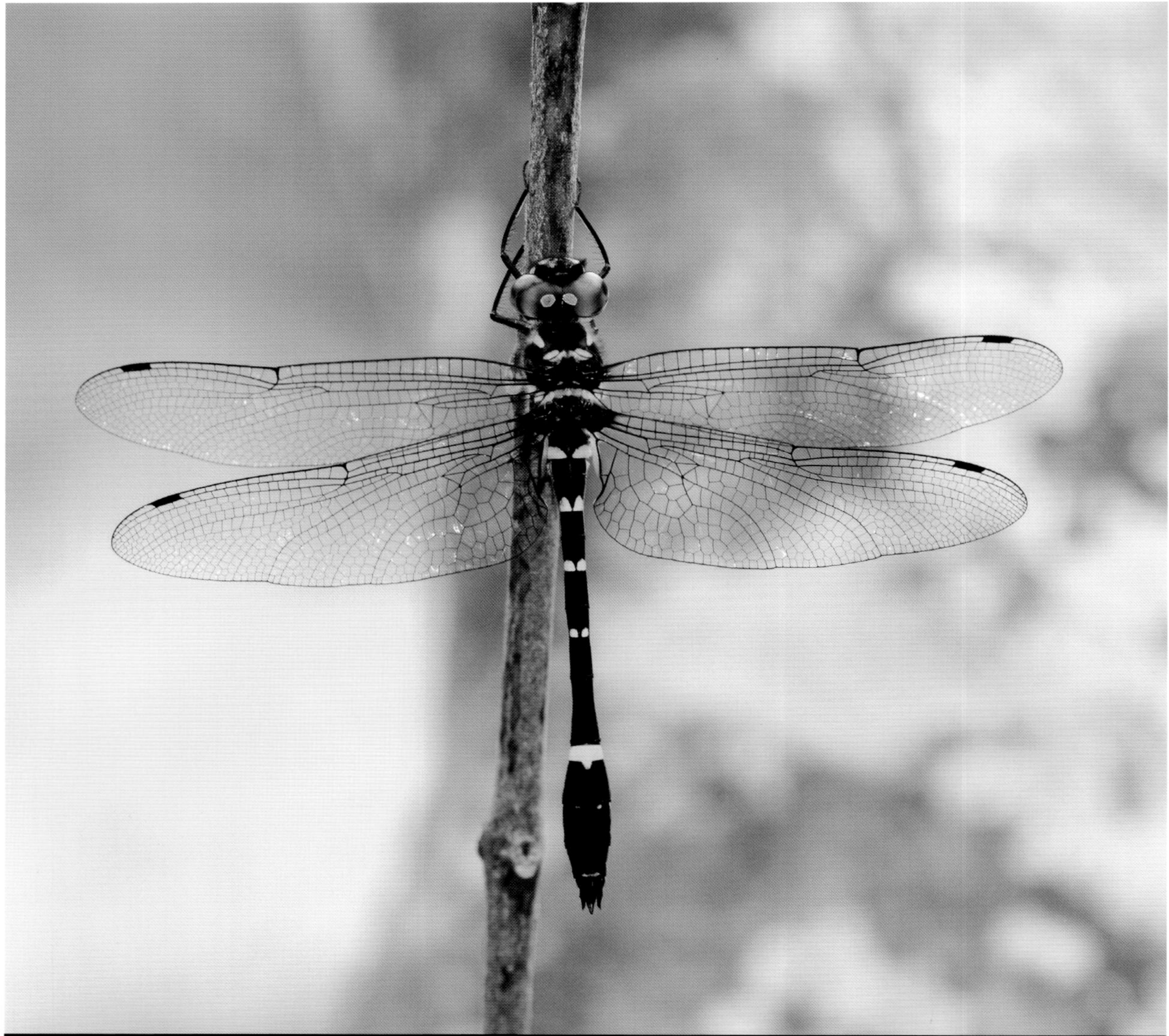

弯钩大伪蜻 雄，广东 | 宋睿斌 摄
Macromia unca, male from Guangdong | Photo by Ruibin Song

弯钩大伪蜻 雄，广东 | 宋睿斌 摄
Macromia unca, male from Guangdong | Photo by Ruibin Song

弯钩大伪蜻 雌，广东 | 宋睿斌 摄
Macromia unca, female from Guangdong | Photo by Ruibin Song

弯钩大伪蜻 雌，浙江
Macromia unca, female from Zhejiang

天王大伪蜻 *Macromia urania* Ris, 1916

【形态特征】雄性后唇基黄色；胸部肩前条纹较长；腹部黑色，第2～8节具黄斑，第7～9节膨大。雌性翅淡琥珀色，基方具褐斑，腹部第7～9节膨大更显著。【长度】体长 66～69 mm，腹长 50～53 mm，后翅 39～44 mm。【栖息环境】海拔 1000 m以下的溪流和河流。【分布】云南（红河）、贵州、福建、广西、广东、海南、香港、台湾；日本、越南。【飞行期】3—10月。

天王大伪蜻 雄，广西
Macromia urania, male from Guangxi

天王大伪蜻 雌，广东 | 宋睿斌 摄
Macromia urania, female from Guangdong | Photo by Ruibin Song

天王大伪蜻 雄，广西
Macromia urania, male from Guangxi

天王大伪蜻 雌，广西
Macromia urania, female from Guangxi

[Identification] Male postclypeus yellow. Thorax with long antehumeral stripes. Abdomen black, S2-S8 with yellow spots, S7-S9 expanded laterally. Female wings with light amber tint, bases with brown spots, S7-S9 more strongly expanded. **[Measurements]** Total length 66-69 mm, abdomen 50-53 mm, hind wing 39-44 mm. **[Habitat]** Streams and rivers below 1000 m elevation. **[Distribution]** Yunnan (Honghe), Guizhou, Fujian, Guangxi, Guangdong, Hainan, Hong Kong, Taiwan; Japan, Vietnam. **[Flight Season]** March to October.

万荣大伪蜻 *Macromia vangviengensis* Yokoi & Mitamura, 2002

万荣大伪蜻 雄，云南（普洱）
Macromia vangviengensis, male from Yunnan (Pu'er)

【形态特征】雄性面部主要黄色，前额具1个甚大的黑褐色斑；胸部主要黑褐色，中胸前侧板褐色；腹部黑褐色，第2～8节具黄斑，第10节背面具1个锥形突起。【长度】雄性体长 59 mm，腹长 43 mm，后翅 38 mm。【栖息环境】海拔 1000 m的山区溪流。【分布】云南（普洱）；老挝。【飞行期】4—6月。

[Identification] Male face mainly yellow, antefrons with a large blackish brown spot. Thorax mainly blackish brown, mesepisternum brown. Abdomen blackish brown, S2-S8 with yellow spots, S10 with a pyramidal prominence. [Measurements] Male total length 59 mm, abdomen 43 mm, hind wing 38 mm. [Habitat] Montane streams at 1000 m elevation. [Distribution] Yunnan (Pu'er); Laos. [Flight Season] April to June.

大伪蜻属待定种1 *Macromia* sp. 1

大伪蜻属待定种1 雄，云南（西双版纳）
Macromia sp. 1, male from Yunnan (Xishuangbanna)

【形态特征】雄性面部黄褐色，前额具1个甚大的黑色斑；胸部肩前条纹较长；腹部黑色，第2～8节具黄斑。【长度】雄性体长 53 mm，腹长 40 mm，后翅 32 mm。【栖息环境】海拔 1000 m 以下的溪流。【分布】云南（西双版纳）。【飞行期】5—11月。

[Identification] Male face yellowish brown, antefrons with a large black spot. Thorax with fairly long antehumeral stripes. Abdomen black, S2-S8 with yellow spots. **[Measurements]** Male total length 53 mm, abdomen 40 mm, hind wing 32 mm. **[Habitat]** Streams below 1000 m elevation. **[Distribution]** Yunnan (Xishuangbanna). **[Flight Season]** May to November.

大伪蜻属待定种2 *Macromia* sp. 2

大伪蜻属待定种2 雄，北京
Macromia sp. 2, male from Beijing

【形态特征】雄性后唇基黄白色，上额具2对黄斑；胸部肩前条纹宽阔；腹部黑色，第2～9节具黄斑。【长度】雄性体长 74～77 mm，腹长 54～56 mm，后翅 44～45 mm。【栖息环境】海拔 1000 m以下的溪流和河流。【分布】北京。【飞行期】7—9月。

[Identification] Male postclypeus yellowish white, top of frons with two pairs of yellow spots. Thorax with broad antehumeral stripes. Abdomen black, S2-S9 with yellow spots. **[Measurements]** Male total length 74-77 mm, abdomen 54-56 mm, hind wing 44-45 mm. **[Habitat]** Streams and rivers below 1000 m elevation. **[Distribution]** Beijing. **[Flight Season]** July to September.

大伪蜻属待定种2 雄，北京
Macromia sp. 2, male from Beijing

大伪蜻属待定种3 *Macromia* sp. 3

大伪蜻属待定种3 雄，云南（德宏）
Macromia sp. 3, male from Yunnan (Dehong)

【形态特征】雄性面部褐色；胸部无肩前条纹，中胸前侧板下方和后胸后侧板下方褐色；腹部黑色，第2~8节具黄斑，第10节背面具1个锥形突起。雌性腹部较短，第2~7节具黄环，翅基方具甚小的褐斑。【长度】体长 69~70 mm，腹长 50~52 mm，后翅 43~45 mm。【栖息环境】海拔 1000 m以下的河流。【分布】云南（德宏、西双版纳）。【飞行期】10月—次年4月。

[Identification] Male face brown. Thorax lacking antehumeral stripes, lower part of mesepisternum and metepimeron brown. Abdomen black, S2-S8 with yellow spots, S10 with a pyramidal prominence. Female abdomen short, S2-S7 with yellow rings, wing bases with small brown spots. **[Measurements]** Total length 69-70 mm, abdomen 50-52 mm, hind wing 43-45 mm. **[Habitat]** Rivers below 1000 m elevation. **[Distribution]** Yunnan (Dehong, Xishuangbanna). **[Flight Season]** October to the following April.

大伪蜻属待定种3 雄，云南（德宏）
Macromia sp. 3, male from Yunnan (Dehong)

大伪蜻属待定种3 雌，云南（德宏）
Macromia sp. 3, female from Yunnan (Dehong)

大伪蜻属待定种3 雌，云南（德宏）
Macromia sp. 3, female from Yunnan (Dehong)

大伪蜻属待定种4 *Macromia* sp. 4

大伪蜻属待定种4 雄，云南（西双版纳）
Macromia sp. 4, male from Yunnan (Xishuangbanna)

大伪蜻属待定种4 雌，云南（西双版纳）
Macromia sp. 4, female from Yunnan (Xishuangbanna)

【形态特征】雄性面部主要黄褐色，额黑褐色；胸部肩前条纹甚小；腹部黑色，第2～8节具黄斑，第7～9节稍微膨大。雌性面部色彩较淡，肩前条纹稍长，腹部具宽阔的黄色条纹。【长度】体长 55～56 mm，腹长 40～41 mm，后翅 36～37 mm。【栖息环境】海拔 1000 m以下的河流。【分布】云南（西双版纳）。【飞行期】5—7月。

[Identification] Male face mainly yellowish brown, frons blackish brown. Thorax with short antehumeral stripes. Abdomen black, S2-S8 with yellow spots, S7-S9 slightly expanded laterally. Female face paler, antehumeral stripes longer, abdomen with broad yellow stripes. **[Measurements]** Total length 55-56 mm, abdomen 40-41 mm, hind wing 36-37 mm. **[Habitat]** Rivers below 1000 m elevation. **[Distribution]** Yunnan (Xishuangbanna). **[Flight Season]** May to July.

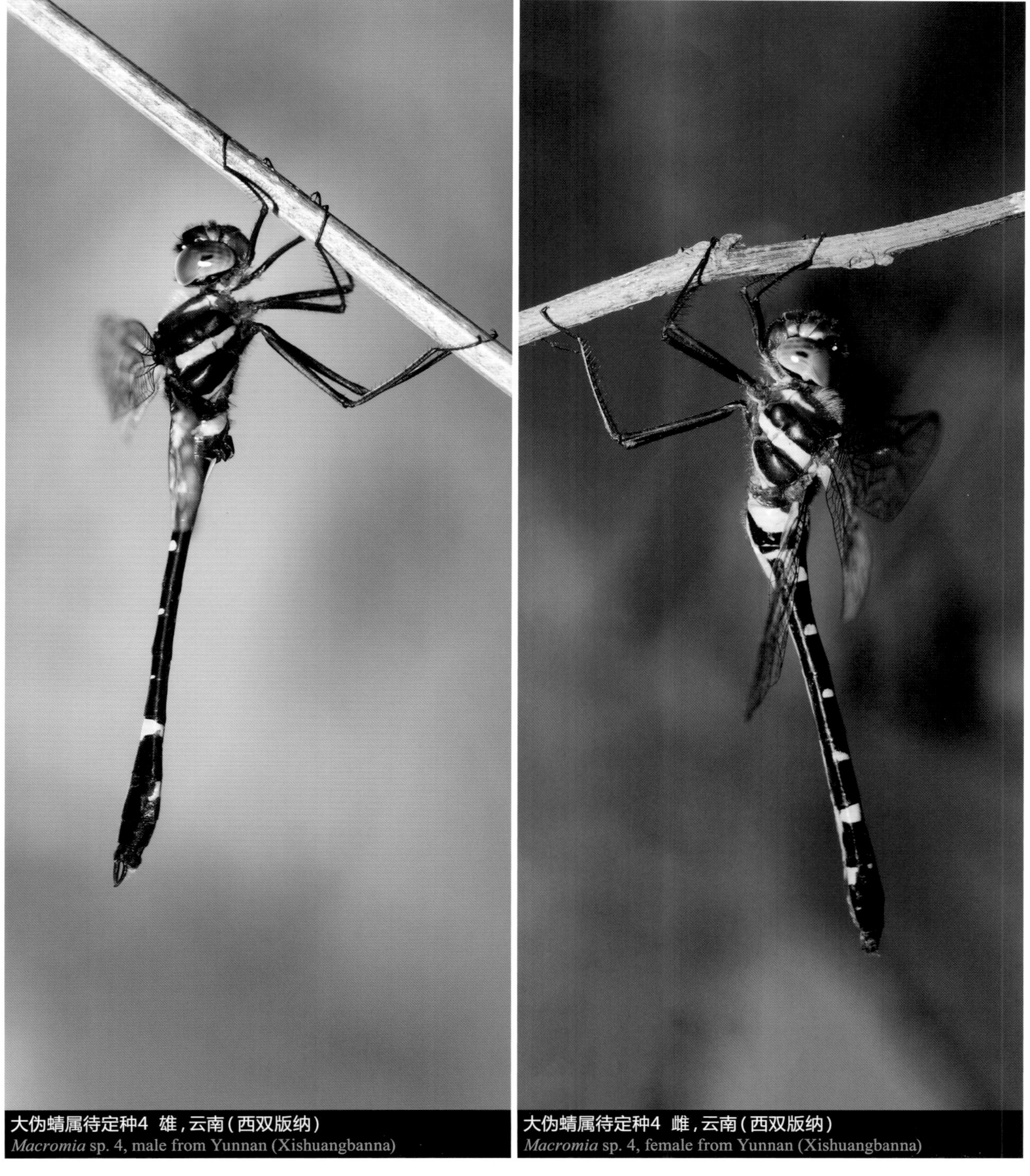

大伪蜻属待定种4 雄，云南（西双版纳）
Macromia sp. 4, male from Yunnan (Xishuangbanna)

大伪蜻属待定种4 雌，云南（西双版纳）
Macromia sp. 4, female from Yunnan (Xishuangbanna)

7 综蜻科 Family Synthemistidae

本科世界性分布，全球已知28属150种。中国已知2属16种，分布于华南和西南地区。中国分布的综蜻体中型；复眼发达，亮绿色，身体墨绿色具金属光泽和黄色条纹；翅透明，翅脉较稀疏，基室无横脉，三角室仅1室。

本科在中国分布的蜻蜓主要栖息于茂盛森林中的溪流。多数种类喜欢阴暗的环境，白天停落在茂盛的森林中躲避阳光，黄昏时较活跃。

The family is widespread all over the world, including over 150 species in 28 genera. 16 species in two genera are recorded from China distributed in the South and Southwest. The Chinese species are medium-sized. Eyes large and brilliant green, body primarily blackish green with metallic reflections and yellow markings. Wings hyaline, venation sparse, median space free and triangle 1-celled.

The Chinese species inhabit streams in dense forests. Most prefer dark habitats and perch on branches within the dense forest during the daytime but become active at twilight.

赛丽异伪蜻 雄

Idionyx selysi, male

飓中伪蜻 雄

Macromidia rapida, male

异伪蜻属 Genus *Idionyx* Hagen, 1867

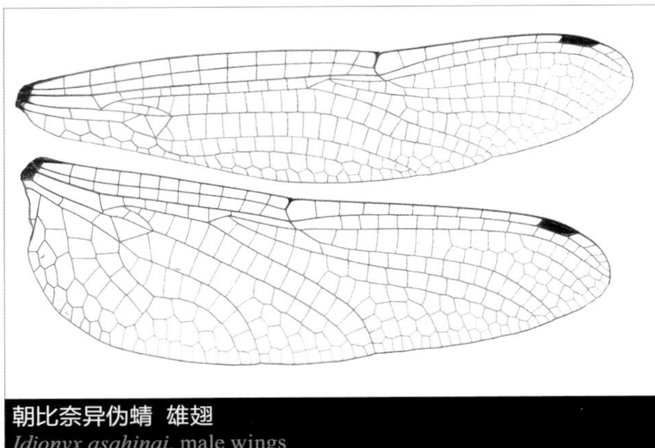
朝比奈异伪蜻 雄翅
Idionyx asahinai, male wings

本属全球已知约30种，主要分布在东洋界。中国已知约10种，分布于华南和西南地区。本属蜻蜓翅透明，翅脉较稀疏，翅痣较短，臀三角室通常2室，前翅基臀区具1条横脉，后翅2条，后翅臀圈袋状，雄性后翅基方圆弧形。本属蜻蜓雄性可以通过面部色彩及肛附器的形状来区分。雌性头部通常具有各种奇形怪状的角状结构，也成为了重要的辨识特征。

本属蜻蜓主要栖息于茂盛森林中的林荫溪流。雄性没有领域行为，属于游窜型，经常沿着森林的林荫小路或者溪流的边缘飞行。很多种类的雄性很难遇见，但雌性经常被发现在森林中集群捕食。

The genus contains about 30 species mainly distributed in the Oriental region. Ten species are recorded from China, found in the South and Southwest. Wings hyaline, venation relatively sparse, pterostigma fairly short, anal triangle usually 2-celled, in fore wings the cubital space with only one crossvein and in hind wings two crossveins, anal loop sack-shaped, male tornus rounded. Males of the genus can be distinguished by the color of face and shape of anal appendages. Female vertex usually armed with various horns or prominences providing a good means for separating species.

Idionyx species inhabit shady streams in dense forests. Male territory behavior has not yet been observed, they are travellers, usually flying along paths at stream margins. In many species males are seldom seen, but females are often seen foraging in small group.

威异伪蜻 交尾 | 宋睿斌 摄
Idionyx victor, mating pair | Photo by Ruibin Song

朝比奈异伪蜻
Idionyx asahinai

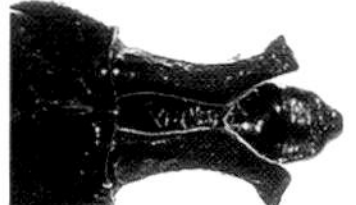
长角异伪蜻
Idionyx carinata

郁异伪蜻
Idionyx claudia

三角异伪蜻
Idionyx optata

赛丽异伪蜻
Idionyx selysi

黄面异伪蜻
Idionyx stevensi

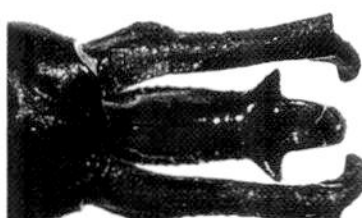
威异伪蜻
Idionyx victor

异伪蜻属待定种1
Idionyx sp.1

异伪蜻属 雄性肛附器
Genus *Idionyx*, male anal appendages

朝比奈异伪蜻 雌
Idionyx asahinai, female

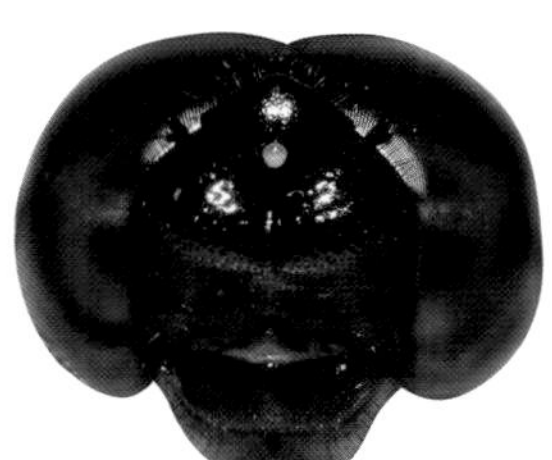
长角异伪蜻 雄
Idionyx carinata, male

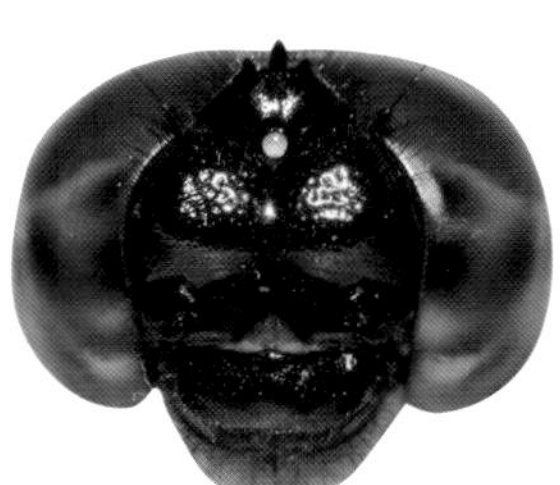
长角异伪蜻 雌
Idionyx carinata, female

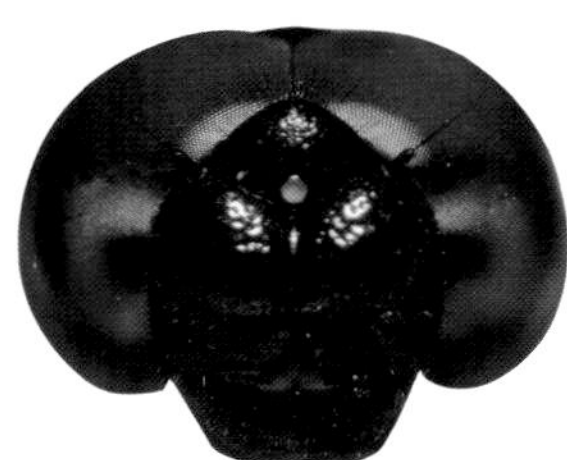
郁异伪蜻 雄
Idionyx claudia, male

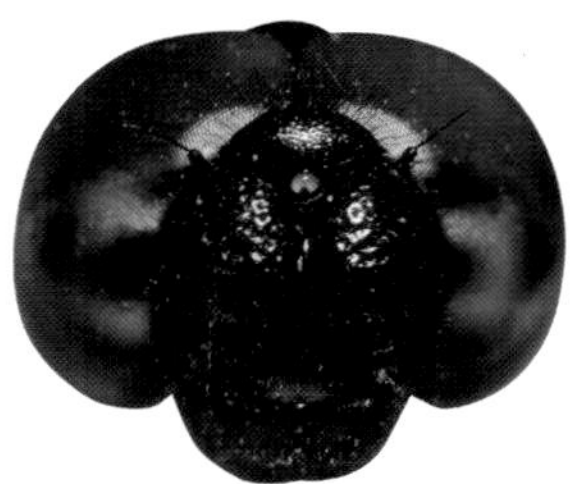
郁异伪蜻 雌
Idionyx claudia, female

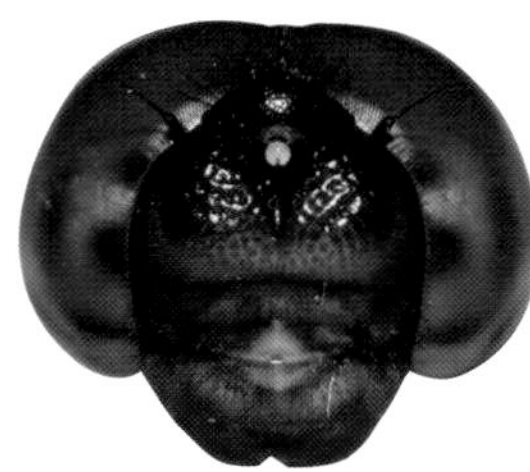
三角异伪蜻 雄
Idionyx optata, male

异伪蜻属 头部正面观
Genus *Idionyx*, head in frontal view

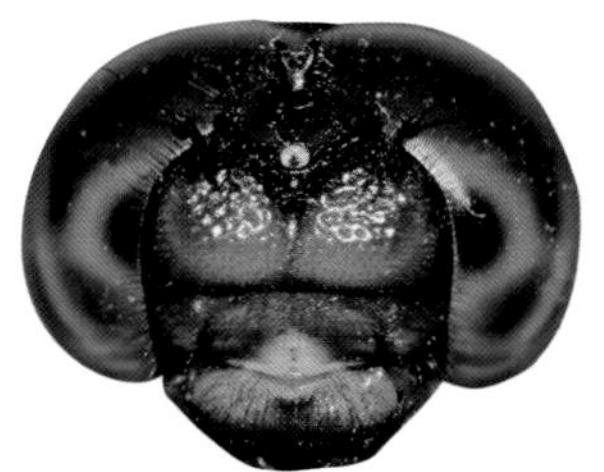
三角异伪蜻 雌
Idionyx optata, female

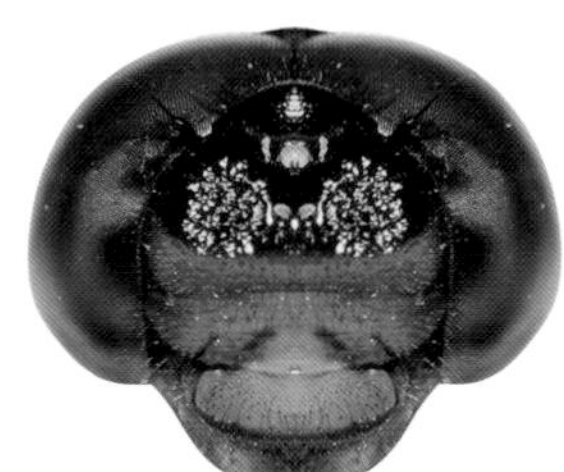
黄面异伪蜻 雄
Idionyx stevensi, male

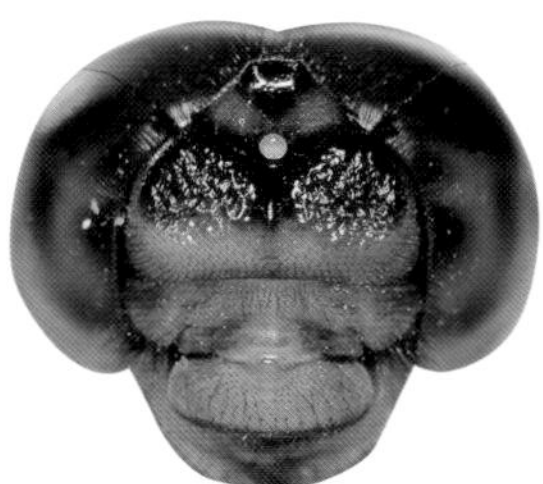
黄面异伪蜻 雌
Idionyx stevensi, female

云南异伪蜻 雌
Idionyx yunnanensis, female

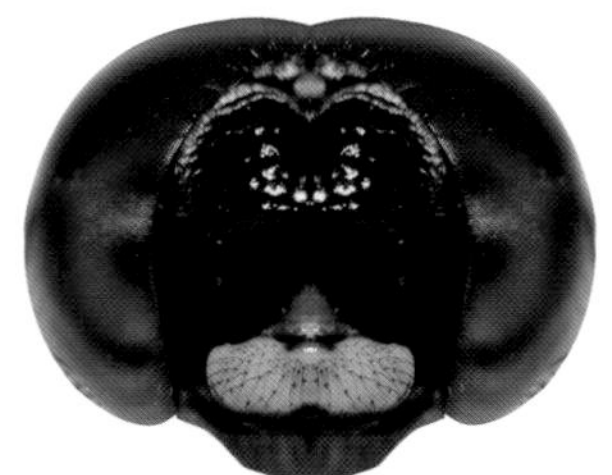
赛丽异伪蜻 雄
Idionyx selysi, male

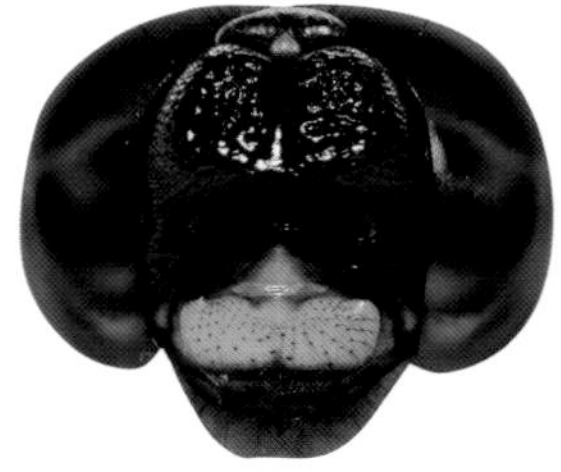
赛丽异伪蜻 雌
Idionyx selysi, female

异伪蜻属 头部正面观
Genus *Idionyx*, head in frontal view

朝比奈异伪蜻 *Idionyx asahinai* Karube, 2011

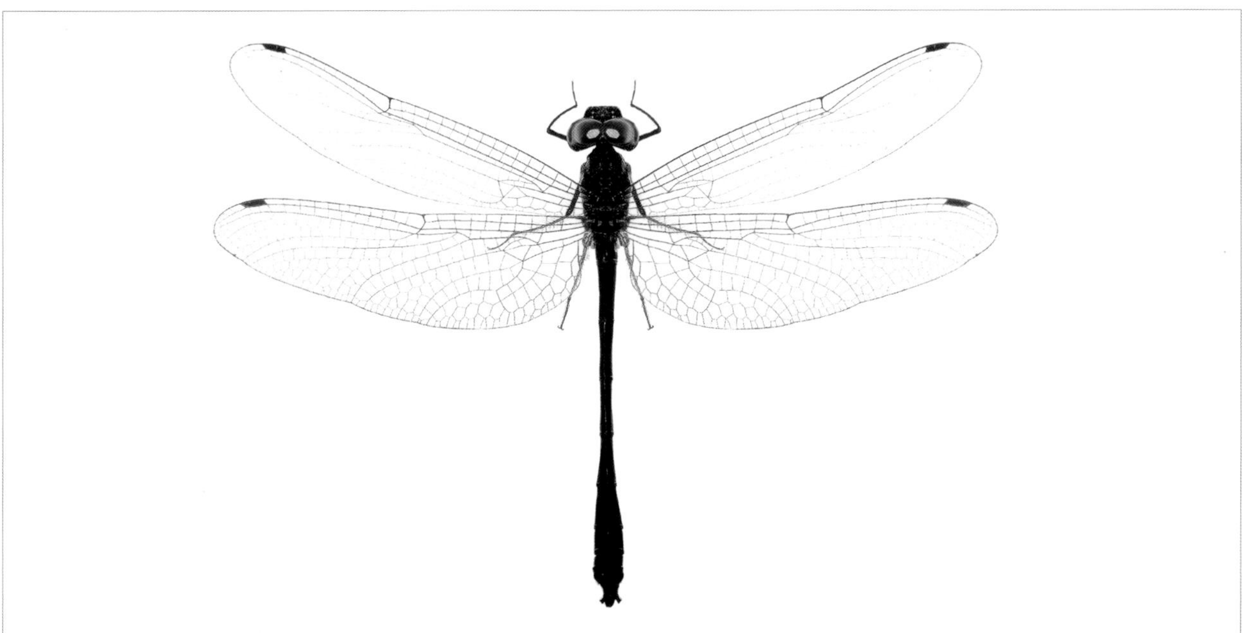
朝比奈异伪蜻 雄，越南
Idionyx asahinai, male from Vietnam

【形态特征】雄性面部金属黑绿色；胸部金属黑绿色，侧面具2条黄色条纹；腹部黑色，第1~3节侧面具黄色条纹，腹部第10节背面具1个角状突起。雌性头顶具1个长角。【长度】体长 40~48 mm，腹长 30~37 mm，后翅 33~35 mm。【栖息环境】海拔 1000 m以下森林中的溪流。【分布】云南（红河）；越南。【飞行期】5—7月。

[Identification] Male face blackish green with metallic luster. Thorax blackish green with metallic luster, laterally with two yellow stripes. Abdomen black, S1-S3 with lateral yellow stripes, S10 with a long horn dorsally. Female vertex with a very long horn. [Measurements] Total length 40-48 mm, abdomen 30-37 mm, hind wing 33-35 mm. [Habitat] Streams in forest below 1000 m elevation. [Distribution] Yunnan (Honghe); Vietnam. [Flight Season] May to July.

朝比奈异伪蜻 雌，云南（红河）
Idionyx asahinai, female from Yunnan (Honghe)

长角异伪蜻 *Idionyx carinata* Fraser, 1926

【形态特征】雄性面部黑色具金属光泽；胸部金属黑绿色，侧面具2条黄色条纹；腹部黑色，第1～4节背面后方具甚细的黄色条纹，第10节背面具1个刺突。雌性头顶具3个角状突起，翅浅褐色，基方具琥珀色斑。【长度】体长46～50 mm，腹长 35～37 mm，后翅 35～38 mm。【栖息环境】海拔 1500 m以下森林中的溪流。【分布】贵州、浙江、福建、湖南、广西、广东；老挝、越南。【飞行期】5—8月。

长角异伪蜻 雄，广东 | 莫善濂 摄
Idionyx carinata, male from Guangdong | Photo by Shanlian Mo

[Identification] Male face black with metallic luster. Thorax blackish green with metallic luster, laterally with two yellow stripes. Abdomen black, S1-S4 with very narrow yellow stripes apically, S10 with a dorsal spine. Female vertex with three horns, wings tinted with light brown, bases with amber spots. **[Measurements]** Total length 46-50 mm, abdomen 35-37 mm, hind wing 35-38 mm. **[Habitat]** Streams in forest below 1500 m elevation. **[Distribution]** Guizhou, Zhejiang, Fujian, Hunan, Guangxi, Guangdong; Laos, Vietnam. **[Flight Season]** May to August.

长角异伪蜻 雌，广东
Idionyx carinata, female from Guangdong

郁异伪蜻 *Idionyx claudia* Ris, 1912

【形态特征】雄性面部黑色具金属光泽；胸部黑绿色具金属光泽，侧面具2条黄色条纹；腹部黑色，第1~3节背面后方具甚细的黄色条纹，第10节背面具1个刺突。雌性与雄性相似，翅浅褐色，基方具琥珀色斑。【长度】体长42~45 mm，腹长 31~34 mm，后翅 33~35 mm。【栖息环境】海拔 1500 m以下森林中的溪流。【分布】中国特有，分布于云南（红河）、贵州、浙江、福建、广西、广东、香港。【飞行期】5—9月。

[Identification] Male face black with metallic luster. Thorax blackish green with metallic luster, laterally with two yellow stripes. Abdomen black, S1-S3 with very narrow yellow stripes apically, S10 with a dorsal spine. Female similar to male, wings tinted with light brown, bases with amber spots. **[Measurements]** Total length 42-45 mm, abdomen 31-34 mm, hind wing 33-35 mm. **[Habitat]** Streams in forest below 1500 m elevation. **[Distribution]** Endemic to China, recorded from Yunnan (Honghe), Guizhou, Zhejiang, Fujiang, Guangxi, Guangdong, Hong Kong. **[Flight Season]** May to September.

郁异伪蜻 雄，广东
Idionyx claudia, male from Guangdong

郁异伪蜻 雄，浙江
Idionyx claudia, male from Zhejiang

郁异伪蜻 雌，广东
Idionyx claudia, female from Guangdong

郁异伪蜻 雌，广东
Idionyx claudia, female from Guangdong

三角异伪蜻 *Idionyx optata* Selys, 1878

【形态特征】雄性面部黄褐色；胸部肩前条纹较短，侧面具2条黄色条纹；腹部黑色，第1~5节具甚细的黄色条纹，第10节背面具1个刺突。雌性头顶具3个角状突起，腹部第1~6节具黄色条纹。【长度】体长 42~44 mm，腹长 30~32 mm，后翅 30~32 mm。【栖息环境】海拔 1500 m以下森林中的溪流。【分布】云南（西双版纳、普洱）；印度、泰国、老挝。【飞行期】5—10月。

[Identification] Male face yellowish brown. Thorax with short antehumeral stripes, laterally with two yellow stripes. Abdomen black, S1-S5 with narrow yellow stripes, S10 with a dorsal spine. Female vertex with three horns, S1-S6 with yellow markings. **[Measurements]** Total length 42-44 mm, abdomen 30-32 mm, hind wing 30-32 mm. **[Habitat]** Streams in forest below 1500 m elevation. **[Distribution]** Yunnan (Xishuangbanna, Pu'er); India, Thailand, Laos. **[Flight Season]** May to October.

三角异伪蜻 雄，云南（西双版纳）
Idionyx optata, male from Yunnan (Xishuangbanna)

三角异伪蜻 雌，云南（西双版纳）
Idionyx optata, female from Yunnan (Xishuangbanna)

三角异伪蜻 雄，云南（西双版纳）
Idionyx optata, male from Yunnan (Xishuangbanna)

三角异伪蜻 雌，云南（西双版纳）
Idionyx optata, female from Yunnan (Xishuangbanna)

赛丽异伪蜻 *Idionyx selysi* Fraser, 1926

【形态特征】雄性面部黑色具金属光泽，上唇和前唇基白色；胸部肩前条纹甚短，侧面具2条黄色条纹；腹部黑色，第1～6节黄色条纹，第10节背面具1个刺突。雌性翅基方具琥珀色斑，腹部第1～9节具黄色条纹。【长度】体长42～46 mm，腹长 30～33 mm，后翅 28～32 mm。【栖息环境】海拔 1500 m以下森林中的溪流。【分布】云南（西双版纳、临沧、普洱）；缅甸、泰国、老挝。【飞行期】4—10月。

[Identification] Male face black with metallic luster, labrum and anteclypeus white. Thorax with very short antehumeral stripes, laterally with two yellow stripes. Abdomen black, S1-S6 with yellow stripes, S10 with a dorsal spine. Female wing bases with amber spots, S1-S9 with yellow stripes. **[Measurements]** Total length 42-46 mm, abdomen 30-33 mm, hind wing 28-32 mm. **[Habitat]** Streams in forest below 1500 m elevation. **[Distribution]** Yunnan (Xishuangbanna, Lincang, Pu'er); Myanmar, Thailand, Laos. **[Flight Season]** April to October.

赛丽异伪蜻　雄，云南（临沧）
Idionyx selysi, male from Yunnan (Lincang)

赛丽异伪蜻　雌，云南（西双版纳）
Idionyx selysi, female from Yunnan (Xishuangbanna)

黄面异伪蜻 *Idionyx stevensi* Fraser, 1924

【形态特征】雄性面部大面积黄色，额金属墨绿色；胸部肩前条纹较短，侧面具2条黄色条纹；腹部黑色，第1~5节具黄色条纹或斑点，第10节背面具1个刺突。雌性额具1对较矮的锥状突起。【长度】体长 46~48 mm，腹长35~36 mm，后翅 35~37 mm。【栖息环境】海拔 1000~1500 m茂盛森林中的溪流。【分布】云南（德宏）；印度、尼泊尔。【飞行期】4—6月。

黄面异伪蜻 雄，云南（德宏）
Idionyx stevensi, male from Yunnan (Dehong)

黄面异伪蜻 雌，云南（德宏）
Idionyx stevensi, female from Yunnan (Dehong)

黄面异伪蜻 雄，云南（德宏）
Idionyx stevensi, male from Yunnan (Dehong)

黄面异伪蜻 雌，云南（德宏）
Idionyx stevensi, female from Yunnan (Dehong)

[Identification] Male face largely yellow, frons dark green with metallic luster. Thorax with short antehumeral stripes, laterally with two yellow stripes. Abdomen black, S1-S5 with yellow stripes or spots, S10 with a dorsal spine. Female frons with a pair of short pyramidal prominences. **[Measurements]** Total length 46-48 mm, abdomen 35-36 mm, hind wing 35-37 mm. **[Habitat]** Streams in forest at 1000-1500 m elevation. **[Distribution]** Yunnan (Dehong); India, Nepal. **[Flight Season]** April to June.

威异伪蜻 *Idionyx victor* Hämäläinen, 1991

威异伪蜻 雄，广东
Idionyx victor, male from Guangdong

【形态特征】雄性面部黑色具金属光泽，上唇白色，前唇基中央具白斑；胸部黑绿色具金属光泽，合胸侧面具2条黄色条纹；腹部黑色，第1～4节背面后方具甚细的黄色条纹或斑点。雌性与雄性色彩相似，翅基方具琥珀色斑。【长度】体长 42～43 mm，腹长 31～32 mm，后翅 29～33 mm。【栖息环境】海拔 500 m以下森林中的溪流。【分布】云南（红河）、福建、广西、广东、海南、香港；越南。【飞行期】4—8月。

[Identification] Male face black with metallic luster, labrum white, anteclypeus with white spot. Thorax blackish green with metallic luster, laterally with two yellow stripes. Abdomen black, S1-S4 with very narrow yellow stripes or spots apically. Female similar to male but wing bases with amber spots. **[Measurements]** Total length 42-43 mm, abdomen 31-32 mm, hind wing 29-33 mm. **[Habitat]** Streams in forest below 500 m elevation. **[Distribution]** Yunnan (Honghe), Fujian, Guangxi, Guangdong, Hainan, Hong Kong; Vietnam. **[Flight Season]** April to August.

威异伪蜻 雌，广东
Idionyx victor, female from Guangdong

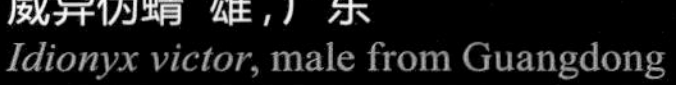

威异伪蜻 雄，广东
Idionyx victor, male from Guangdong

威异伪蜻 雌，广东 | 宋黎明 摄
Idionyx victor, female from Guangdong | Photo by Liming Song

云南异伪蜻 *Idionyx yunnanensis* Zhou, Wang, Shuai & Liu, 1994

【形态特征】雌性面部大面积黄色，头顶具1个角锥形突起；胸部黑色，肩前条纹甚短，合胸侧面具2条黄色条纹，翅基方具琥珀色斑；腹部黑色，第1～9节具黄色条纹。【长度】雌性体长 44～47 mm，腹长 32～34 mm，后翅34～35 mm。【栖息环境】海拔 1000～1500 m茂盛森林中的溪流。【分布】中国云南（德宏、临沧）特有。【飞行期】5—7月。

[Identification] Female face largely yellow, vertex with a pyramidal horn. Thorax black with short antehumeral stripes, laterally with two yellow stripes, wing bases with amber spots. Abdomen black, S1-S9 with yellow stripes. **[Measurements]** Female total length 44-47 mm, abdomen 32-34 mm, hind wing 34-35 mm. **[Habitat]** Streams in forest at 1000-1500 m elevation. **[Distribution]** Endemic to Yunnan (Dehong, Lincang) of China. **[Flight Season]** May to July.

云南异伪蜻 雌，云南（临沧）
Idionyx yunnanensis, female from Yunnan (Lincang)

异伪蜻属待定种1 *Idionyx* sp.1

【形态特征】雄性面部黑色具金属光泽；胸部肩前条纹甚短，合胸侧面具2条黄白色条纹；腹部黑色，第2～4节背面后方具甚细的黄色斑纹，第10节背面具1个刺突。雌性头顶具1对角状突起，翅基方具琥珀色斑。【长度】体长47～50 mm，腹长 35～37 mm，后翅 35～36 mm。【栖息环境】海拔 500 m以下森林中的溪流。【分布】云南（红河）。【飞行期】4—6月。

[Identification] Male face black with metallic luster. Thorax with very short antehumeral stripes, sides with two yellowish white stripes. Abdomen black, S2-S4 with narrow yellow markings apically, S10 with a dorsal spine. Female vertex with a pair of horns, wing bases with amber spots. **[Measurements]** Total length 47-50 mm, abdomen 35-37 mm, hind wing 35-36 mm. **[Habitat]** Streams in forest below 500 m elevation. **[Distribution]** Yunnan (Honghe). **[Flight Season]** April to June.

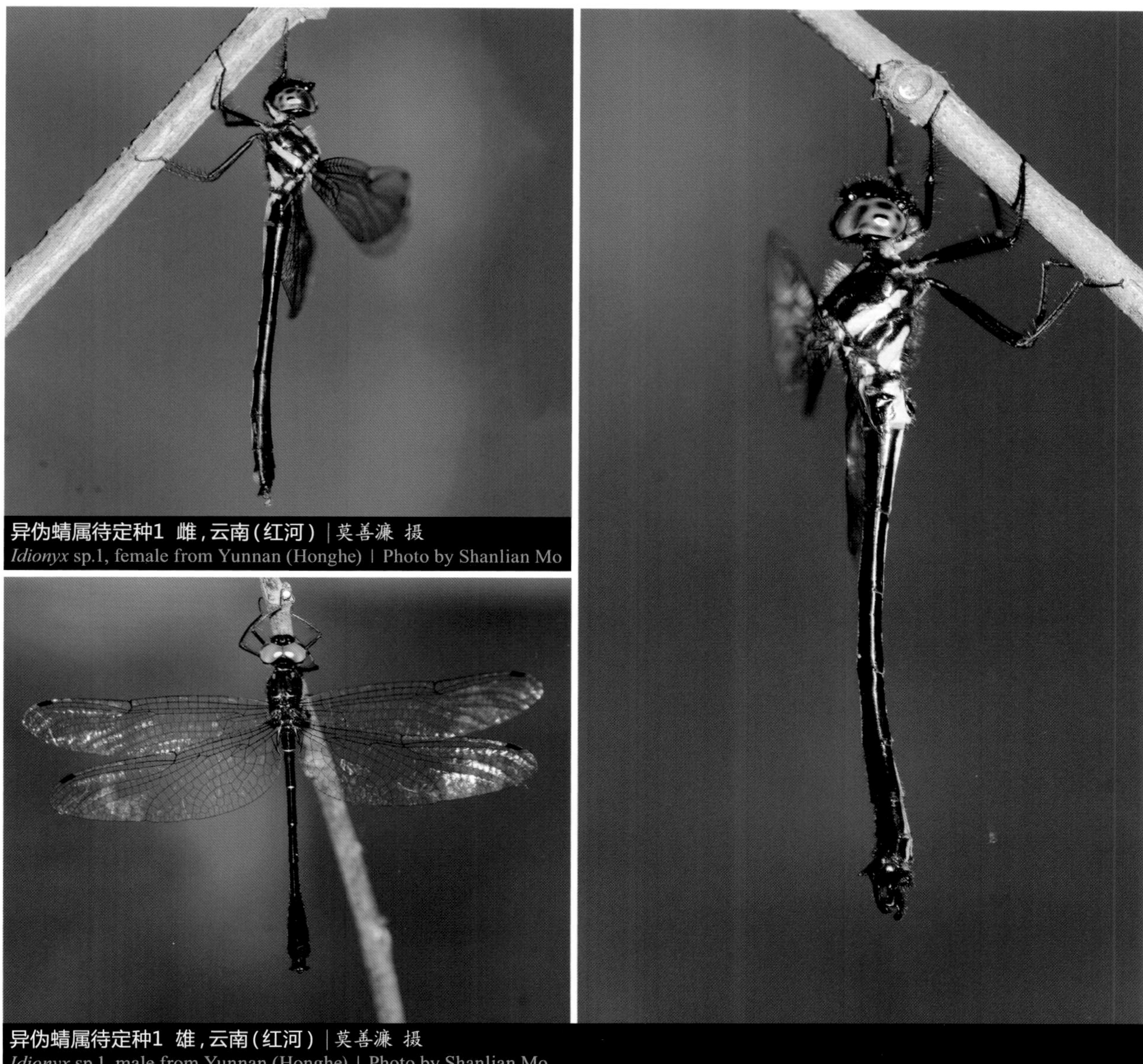

异伪蜻属待定种1 雌，云南（红河）| 莫善濂 摄
Idionyx sp.1, female from Yunnan (Honghe) | Photo by Shanlian Mo

异伪蜻属待定种1 雄，云南（红河）| 莫善濂 摄
Idionyx sp.1, male from Yunnan (Honghe) | Photo by Shanlian Mo

异伪蜻属待定种2 *Idionyx* sp.2

【形态特征】本种与长角异伪蜻相似，但体型稍小，具较短的肩前条纹，雄性肛附器的构造不同。【长度】体长43～46 mm，腹长 31～35 mm，后翅 33～34 mm。【栖息环境】海拔 1500 m以下森林中的溪流。【分布】云南（普洱）。【飞行期】5—7月。

[Identification] Similar to *I. carinata*, but size slightly smaller, thorax with short antehumeral stripes, male anal appendages of different shape. **[Measurements]** Total length 43-46 mm, abdomen 31-35 mm, hind wing 33-34 mm. **[Habitat]** Streams in forest below 1500 m elevation. **[Distribution]** Yunnan (Pu'er). **[Flight Season]** May to July.

异伪蜻属待定种2 雌，云南（普洱）
Idionyx sp.2, female from Yunnan (Pu'er)

中伪蜻属 Genus *Macromidia* Martin, 1907

本属全球已知11种，分布于亚洲的热带和亚热带地区。中国已知5种，主要分布于华南和西南地区。本属蜻蜓体中型，和异伪蜻属体型相当；复眼发达，亮绿色，身体墨绿色或黑色具黄色斑纹；翅透明，翅痣较长，臀三角室通常2室，前翅基臀区具2条横脉，后翅3条，臀圈近椭圆形，雄性后翅基方略呈角状。雄性的肛附器构造相对简单，雌性头顶无角状突起。中伪蜻属蜻蜓可以通过身体色彩结合肛附器的构造区分。

本属蜻蜓主要栖息于较低海拔茂盛森林中的溪流。它们白天藏匿在茂盛的森林中躲避阳光，有时悬挂在近水面的树丛中，黄昏时较活跃。本属的雄性具领域行为，黄昏时在溪流上来回巡飞。

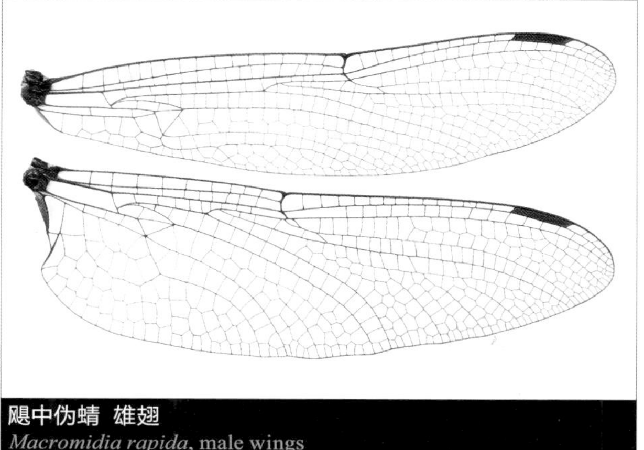

飓中伪蜻 雄翅
Macromidia rapida, male wings

The genus contains 11 species distributed in tropical and subtropical Asia. Five species are recorded from China, mainly found in the South and Southwest. Species of the genus are medium-sized, similar in size to *Idionyx* species. Eyes large and brilliant green, body metallic dark green or black with yellow markings. Wings hyaline, pterostigma fairly long, anal triangle usually 2-celled, in fore wings the cubital space with two crossveins and in hind wings three, anal loop almost oblong in shape and male tornus slightly angled. Male anal appendages is simple in structure, no horns present on vertex of female. Species of the genus can be distinguished largely by arrangement of body markings and male anal appendage morphology.

伊中伪蜻 雄 | 宋睿斌 摄
Macromidia ellenae, male | Photo by Ruibin Song

Macromidia species inhabit lowland streams in forest. They perch in shady forest during the daytime, often hanging on the low tree branches above water surface, they are active at twlight. Territorial males fly back and forth over streams at twilight.

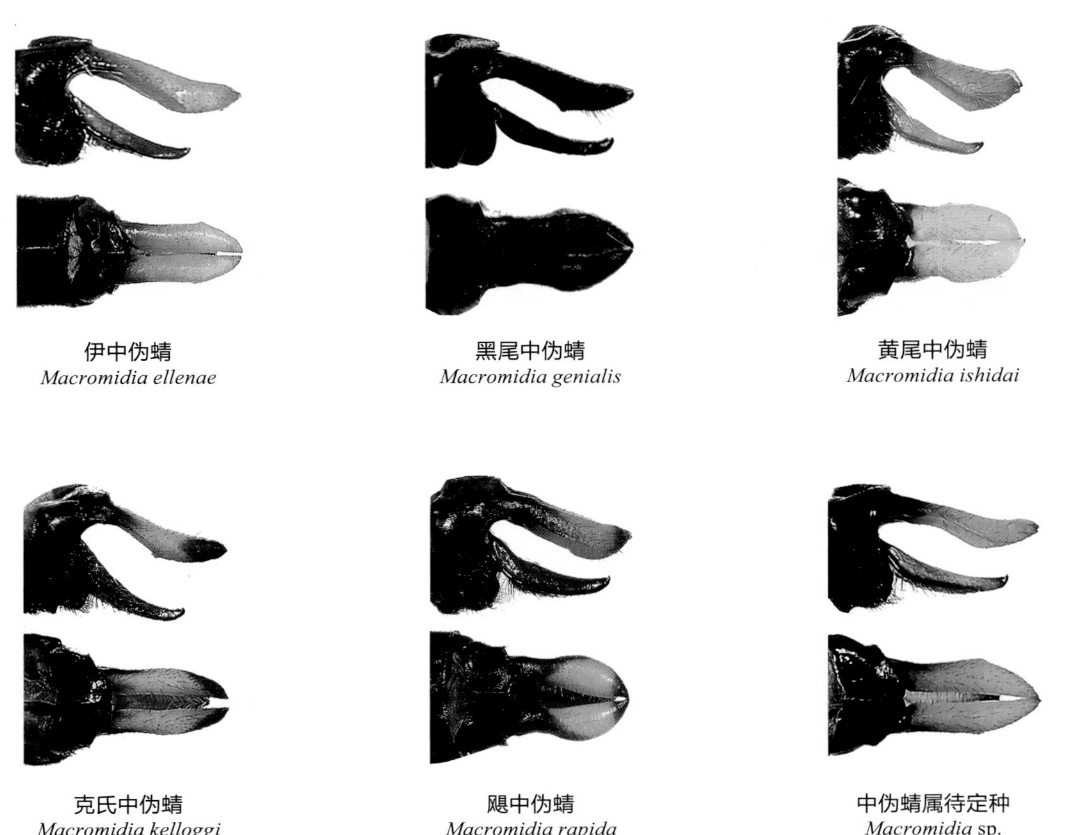

伊中伪蜻 *Macromidia ellenae*
黑尾中伪蜻 *Macromidia genialis*
黄尾中伪蜻 *Macromidia ishidai*
克氏中伪蜻 *Macromidia kelloggi*
飓中伪蜻 *Macromidia rapida*
中伪蜻属待定种 *Macromidia* sp.

中伪蜻属 雄性肛附器
Genus *Macromidia*, male anal appendages

伊中伪蜻 *Macromidia ellenae* Wilson, 1996

【形态特征】雄性面部黑色，下唇、上唇下缘和后唇基中央黄色；胸部肩前条纹较短，合胸侧面具5个形状各异的黄斑；腹部黑色，第1～6节具黄色条纹，肛附器黄色。雌性与雄性相似，翅基方具琥珀色斑。【长度】体长 42～45 mm，腹长 31～34 mm，后翅 28～30 mm。【栖息环境】海拔 500 m以下的林荫溪流。【分布】中国特有，分布于广东、香港。【飞行期】3—6月。

[Identification] Male face black, labium, lower margin of labrum and postclypeus with yellow markings. Thorax with short antehumeral stripes, laterally with five yellow spots of different shape. Abdomen black, S1-S6 with yellow markings, anal appendages yellow. Female similar to male, wing bases with amber tint. **[Measurements]** Total length 42-45 mm, abdomen 31-34 mm, hind wing 28-30 mm. **[Habitat]** Shady streams below 500 m elevation. **[Distribution]** Endemic to China, recorded from Guangdong, Hong Kong. **[Flight Season]** April to June.

伊中伪蜻 雄，广东 | 吴宏道 摄
Macromidia ellenae, male from Guangdong | Photo by Hongdao Wu

伊中伪蜻 雌，广东 | 吴宏道 摄
Macromidia ellenae, female from Guangdong | Photo by Hongdao Wu

伊中伪蜻 雄，广东 | 宋睿斌 摄
Macromidia ellenae, male from Guangdong | Photo by Ruibin Song

伊中伪蜻 雌，广东 | 宋睿斌 摄
Macromidia ellenae, female from Guangdong | Photo by Ruibin Song

黑尾中伪蜻 *Macromidia genialis* Laidlaw, 1923

黑尾中伪蜻 雄，云南（西双版纳）
Macromidia genialis, male from Yunnan (Xishuangbanna)

黑尾中伪蜻 雄，云南（西双版纳）
Macromidia genialis, male from Yunnan (Xishuangbanna)

黑尾中伪蜻 雌，云南（西双版纳）
Macromidia genialis, female from Yunnan (Xishuangbanna)

黑尾中伪蜻 雌，云南（西双版纳）
Macromidia genialis, female from Yunnan (Xishuangbanna)

【形态特征】雄性面部金属黑绿色；胸部具较短的肩前条纹，侧面具2条黄色条纹；腹部黑色，第1~7节背面具黄条纹，肛附器黑色。雌性与雄性相似。【长度】体长 46~48 mm，腹长 35~37 mm，后翅 30~35 mm。【栖息环境】海拔 1000 m以下森林中的溪流和沟渠。【分布】云南（西双版纳）；缅甸、泰国、老挝、马来半岛、印度尼西亚。【飞行期】5—7月。

[Identification] Male face blackish green with metallic luster. Thorax with short antehumeral stripes, laterally with two yellow stripes. Abdomen black, S1-S7 with yellow stripes dorsally, anal appendages black. Female similar to male. **[Measurements]** Total length 46-48 mm, abdomen 35-37 mm, hind wing 30-35 mm. **[Habitat]** Streams and ditches in forest below 1000 m elevation. **[Distribution]** Yunnan (Xishuangbanna); Myanmar, Thailand, Laos, Peninsular Malaysia, Indonesia. **[Flight Season]** May to July.

黄尾中伪蜻 *Macromidia ishidai* Asahina, 1964

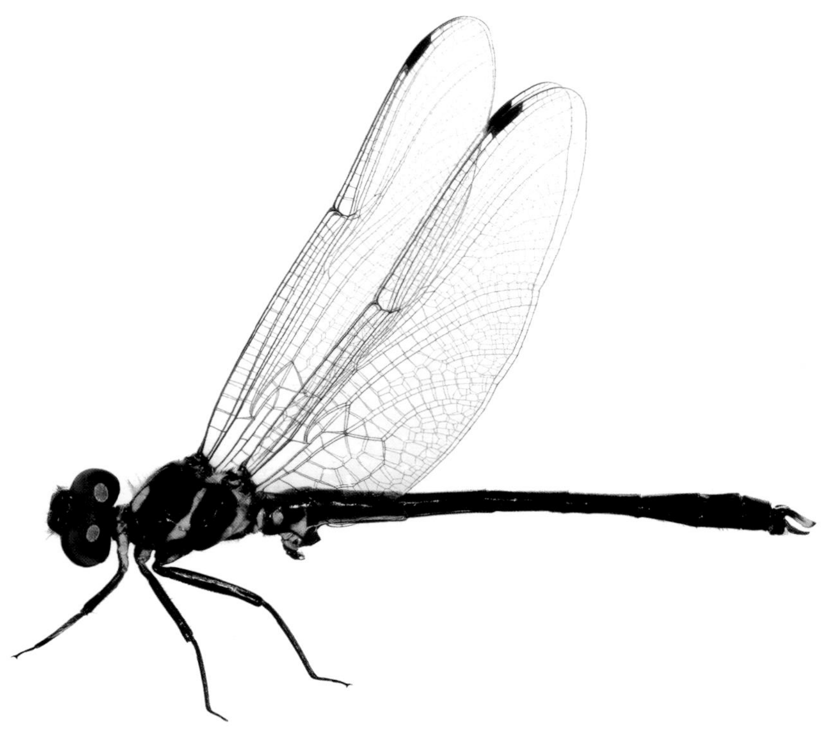

黄尾中伪蜻 雄，台湾
Macromidia ishidai, male from Taiwan

【形态特征】雄性面部黑色，下唇黄色；胸部肩前条纹较短，侧面具2条黄色条纹；腹部黑色，第1~7节具黄色条纹，上肛附器黄色，基方和末端黑色。【长度】体长 54 mm，腹长 40 mm，后翅 37 mm。【栖息环境】海拔 500 m以下森林中的溪流。【分布】中国台湾；日本。【飞行期】4—8月。

[Identification] Male face black, labium yellow. Thorax with short antehumeral stripes, laterally with two yellow stripes. Abdomen black, S1-S7 with yellow stripes, superior appendages largely yellow with bases and tips black. **[Measurements]** Total length 54 mm, abdomen 40 mm, hind wing 37 mm. **[Habitat]** Streams in forest below 500 m elevation. **[Distribution]** Taiwan of China; Japan. **[Flight Season]** April to August.

克氏中伪蜻 *Macromidia kelloggi* Asahina, 1978

【形态特征】雄性面部黑色，下唇黄色；胸部肩前条纹较短，侧面具2条黄色条纹，翅稍染褐色；腹部黑色，第1~7节具黄色斑纹，上肛附器黄色。雌性与雄性相似但较粗壮。【长度】体长 49~54 mm，腹长 36~39 mm，后翅 35~40 mm。【栖息环境】海拔 1000 m以下森林中的溪流和沟渠。【分布】中国特有，分布于浙江、福建、广东。【飞行期】4—7月。

[Identification] Male face black, labium yellow. Thorax with short antehumeral stripes, laterally with two yellow stripes, wings slightly tinted with brown. Abdomen black, S1-S7 with yellow markings, superior appendages yellow.

Female similar to male but stouter. **[Measurements]** Total length 49-54 mm, abdomen 36-39 mm, hind wing 35-40 mm. **[Habitat]** Streams and ditches in forest below 1000 m elevation. **[Distribution]** Endemic to China, recorded from Zhejiang, Fujian, Guangdong. **[Flight Season]** April to July.

克氏中伪蜻 雄，浙江
Macromidia kelloggi, male from Zhejiang

克氏中伪蜻 雌，浙江
Macromidia kelloggi, female from Zhejiang

克氏中伪蜻 雄，广东 | 宋睿斌 摄
Macromidia kelloggi, male from Guangdong | Photo by Ruibin Song

飓中伪蜻 *Macromidia rapida* Martin, 1907

【形态特征】雄性面部黄褐色，额和头顶黑绿色具金属光泽；胸部无肩前条纹，侧面具2条黄色条纹；腹部黑色，第1~7节具黄色条纹，上肛附器基方黄色，端方黑色。雌性与雄性相似，翅基方具黑褐色条纹。【长度】体长50~53 mm，腹长38~42 mm，后翅32~38 mm。【栖息环境】海拔1000 m以下森林中的溪流和沟渠。【分布】云南（红河）、广东、广西、海南、香港；泰国、老挝、越南。【飞行期】4—8月。

[Identification] Male face yellowish brown, frons and vertex metallic dark green. Thorax without antehumeral stripes, laterally with two yellow stripes. Abdomen black, S1-S7 with yellow stripes, superior appendages yellow basally, black apically. Female similar to male, wing bases with blackish brown stripes. **[Measurements]** Total length 50-53 mm, abdomen 38-42 mm, hind wing 32-38 mm. **[Habitat]** Streams and ditches in forest below 1000 m elevation. **[Distribution]** Yunnan (Honghe), Guangdong, Guangxi, Hainan, Hong Kong; Thailand, Laos, Vietnam. **[Flight Season]** April to August.

飓中伪蜻 雄，广东
Macromidia rapida, male from Guangdong

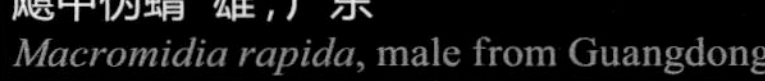

飓中伪蜻 雄，广东
Macromidia rapida, male from Guangdong

飓中伪蜻 雌，广东
Macromidia rapida, female from Guangdong

中伪蜻属待定种 *Macromidia* sp.

中伪蜻属待定种 雄，广东
Macromidia sp., male from Guangdong

中伪蜻属待定种 雄，广东
Macromidia sp., male from Guangdong

【形态特征】雄性面部黑绿色具金属光泽，下唇黄色；胸部肩前条纹较短，侧面具2条黄色条纹；腹部黑色，第1～6节具黄斑，肛附器黄白色。雌性与雄性相似，翅染褐色。【长度】体长 45～47 mm，腹长 34～35 mm，后翅 29～33 mm。【栖息环境】海拔 500 m以下的河流和溪流。【分布】贵州、广东。【飞行期】5—8月。

[Identification] Male face metallic blackish green, labium yellow. Thorax with short antehumeral stripes, laterally with two yellow stripes. Abdomen black, S1-S6 with yellow markings, anal appendages yellowish white. Female similar to male, wings tinted with brown. [Measurements] Total length 45-47 mm, abdomen 34-35 mm, hind wing 29-33 mm. [Habitat] Rivers and streams below 500 m elevation. [Distribution] Guizhou, Guangdong. [Flight Season] May to August.

中伪蜻属待定种 雄，广东
Macromidia sp., male from Guangdong

中伪蜻属待定种 雌，广东
Macromidia sp., female from Guangdong